Parallel Computers 2

T0187602

Parallel Computers 2

ARCHITECTURE, PROGRAMMING AND ALGORITHMS

R W Hockney

Emeritus Professor, University of Reading

C R Jesshope

Reader in Computer Architecture
Department of Electronics and Computer Science,
University of Southampton

CRC Press
Taylor & Francis Group
Boca Raton London New York

CRC Press is an imprint of the
Taylor & Francis Group, an **informa** business

CRC Press
Taylor & Francis Group
6000 Broken Sound Parkway NW, Suite 300
Boca Raton, FL 33487-2742

First issued in paperback 2020

© 1988 by Taylor & Francis Group, LLC
CRC Press is an imprint of Taylor & Francis Group, an Informa business

No claim to original U.S. Government works

ISBN-13: 978-0-367-45604-7 (pbk)
ISBN-13: 978-0-85274-811-4 (hbk)

Visit the Taylor & Francis Web site at
http://www.taylorandfrancis.com

and the CRC Press Web site at
http://www.crcpress.com

British Library Cataloguing in Publication Data

Hockney, R. W.
    Parallel computers 2: architecture,
    programming and algorithms.——2nd ed.
    1. Parallel processing (Electronic computers)
    I. Title   II. Jesshope, C. R.   III. Hockney,
    R. W. Parallel computers
    004'.35   QA76.6

    ISBN 0-85274-811-6
    ISBN 0-85274-812-4 (pbk.)

Library of Congress Cataloging in Publication Data

Hockney, Roger W.
    Parallel computers.
    Bibliography: p.
    Includes index.
    1. Parallel processing (Electronic computers)
2. Electronic digital computers.   I. Jesshope, C. R.
II. Title
QA76.5.H57   1988   004'.35   87-28849

ISBN 0-85274-811-6
ISBN 0-85274-812-4 (pbk.)

Typeset by P & R Typesetters Ltd

To our wives, Judith and Susie

Contents

Preface to the First Edition

The 1980s are likely to be the decade of the parallel computer, and it is the purpose of this book to provide an introduction to the topic. Although many computers have displayed examples of parallel or concurrent operation since the 1950s, it was not until 1974–5 that the first computers appeared that were designed specifically to use parallelism in order to operate efficiently on vectors or arrays of numbers. These computers were based on either executing in parallel the various subfunctions of an arithmetic operation in the same manner as a factory assembly line (pipelined computers such as the CDC STAR and the TIASC), or replicating complete arithmetic units (processor arrays such as the ILLIAC IV). There were many problems in these early designs but by 1980 several major manufacturers were offering parallel computers on a commercial basis, as opposed to the previous research projects. The main examples are the CRAY-1 (actually first installed in 1976) and the CDC CYBER 205 pipelined computers, and the ICL DAP and Burroughs BSP processor arrays. Unfortunately, since we wrote this material Burroughs have experienced problems in the production of the BSP and, although a prototype machine was built, Burroughs have withdrawn from this project. However, this still remains a very interesting design and with its demise perhaps gives more insight into this field. Pipelined designs are also becoming popular as processors to attach to minicomputers for signal processing or the analysis of seismic data, and also as attachments to micro based systems. Examples are the FPS AP-120B, FPS-164, Data General AP/130 and IBM 3838.

Parallelism has been introduced in the above design because improvements in circuit speeds alone cannot produce the required performance. This is evident also in the design studies produced for the proposed National Aeronautics Simulation Facility at NASA Ames. This is to be based on a computer capable of 10^9 floating-point operations per second. The CDC proposal for this machine comprises four high-performance pipelined units, whereas the Burroughs design is an array of 512 replicated arithmetic units. The advent of very large-scale integration (VLSI) as a reliable chip manufacturing process, provides a technology that can produce very large arrays of simple processing elements (PES). The ICL DAP (4096 PES) and the Goodyear Aerospace MPP are early examples of processor arrays that are likely to be developed within the decade to take advantage of VLSI technology. It may be claimed that the above designs are of interest only to large scientific laboratories and are not likely to make an impact on the mass of computer users. Most experience shows, however, that advances in computer architecture first made for the scientific market do later become part of the general computing scene.

It seems therefore that the need for greater parallelism in design (i.e. the demand for greater performance than circuit speed alone can give), coupled with the technology for implementing highly parallel designs (VLSI), is likely to make parallelism in computer architecture a growth area in the 1980s. It is the purpose of this book to explain the principles of and to classify both pipelined and array-like designs, to show how these principles have been actually implemented in a number of successful current designs (the CRAY-1, CYBER 205, FPS AP-120B, ICL DAP and Burroughs BSP), and to compare the performance of the different designs on a number of substantial applications (matrix operations, FFT, Poisson-solving). The advent of highly parallel architectures also introduces the problem of designing numerical algorithms that execute efficiently on them, and computer languages in which these algorithms can be expressed. We regard the algorithmic and language aspects of parallelism as being equally important as the architectural, and devote chapters specifically to them.

A feature of our treatment that may be novel is the concept of a two-parameter description of the performance of a computer: in addition to the usual maximum performance in floating-point operations per second r_∞, we introduce the half-performance length $n_{1/2}$, which is the length of vector that is required to achieve half the maximum performance. This variable may also be regarded as measuring the amount of parallelism in a design and varies from $n_{1/2} = 0$ for a serial computer to $n_{1/2} = \infty$ for an infinite array of processors. Thus the second parameter nicely classifies parallel designs into a spectrum characterised by a quantitative measure. It therefore answers the question: how parallel is my computer? Of equal importance is the fact that $n_{1/2}$ also determines quantitatively the choice of the most efficient algorithm. It therefore also answers the question: what is the best algorithm to use on my parallel computer? We have also introduced an algebraic-style notation, that enables the overall architecture of a computer to be expressed in one line, thus circumventing the need for verbose descriptions and aiding the classification of designs by allowing generic descriptions by a formula.

Most material in this book has been collected for a lecture course given in the Computer Science Department at Reading University entitled 'Advanced Computer Architecture'. This course has evolved into a 40-lecture unit over the last five years, and it was the lack of a suitable text that provided the motivation for writing this book. The lecture course is given as an option to third year undergraduates but would also be suitable for a specialised course at MSc level, or as taught material to PhD students preparing for a thesis in the general area of parallel computation.

Many people have helped, by discussion and criticism, with the preparation of the manuscript. Amongst these, we would like to mention our colleagues at Reading University: particularly Jim Craigie, John Graham, Roger Loader, John Ogden, John Roberts and Shirley Williams; and Henry Kemhadjian of Southampton University. Dr Ewan Page, Vice-Chancellor of Reading University, and the series editor Professor Mike Rogers of Bristol University have also suggested improvements to the manuscript. We have also received very generous assistance with information and photographs from representatives of the computer manufacturers, amongst whom we wish to thank Pete Flanders, David Hunt and Stewart Reddaway of ICL Research and Advanced Development Centre, Stevenage, and John Smallbone of ICL, Euston; Professor Dennis Parkinson of ICL and Queen Mary College, London University; Stuart Drayton, Mick Dungworth and Jeff Taylor of CRAY research, Bracknell; David Barkai and Nigel Payne at Control Data Corporation (UK), and Patricia Conway, Neil Lincoln and Chuck Purcell of CDC Minneapolis; J H Austin of Burroughs, Paoli, and G Tillot of Burroughs, London; C T Mickelson of Goodyear Aerospace, Akron; and John Harte, David Head and Steve Markham of Floating Point Systems, Bracknell. Many errors were also corrected by Ms Jill Dickinson (now Mrs Contla) who has typed our manuscript with her customary consummate skill.

We have dedicated this book to computer designers, because without their inspiration and dedication we would not have had such an interesting variety of designs to study, classify and use. The design of any computer is inevitably a team effort but we would like to express our particular appreciation to Seymour Cray, Neil Lincoln, George O'Leary and Stewart Reddaway, who were the principal designers for the computers that we have selected for detailed study.

R W Hockney, C R Jesshope
1981

Preface to the Second Edition

In the seven years that have passed since the publication of *Parallel Computers* sufficient changes have occurred to warrant the preparation of a second edition. Apart from the evolution of architectures described in the first edition (e.g. CYBER 205 to ETA[10], and CRAY-1 to CRAY X-MP and CRAY-2), many novel multi-instruction stream (MIMD) computers have appeared experimentally, and some are now commercially available (e.g. Intel iPSC, Sequent Balance, Alliant FX/8). In addition, microprocessor chips (e.g. the INMOS transputer) are now available that are specifically designed to be connected into large networks, and many new systems are planned to be based upon them.

Whilst keeping the overall framework of the first edition, we have included these developments by expanding the Introduction (including necessary extensions to the algebraic architecture notation, and the classification of designs), and selecting some architectures for more detailed description in the following chapters. In the chapter on vector pipelined computers (Chapter 2), we have included a description of the highly successful Japanese vector computers (Fujitsu VP, Hitachi S810, and NEC SX2), as well as the new generation of multiple vector computers from the USA (the CRAY-2 and ETA[10]). In the chapter on multiprocessors and processor arrays, we have included the Denelcor HEP, the first commercially available MIMD computer, and the Connection Machine.

Although the HEP is no longer available, we feel that its architecture, based on multiple instruction streams time-sharing a single instruction pipeline, is sufficiently novel and interesting to warrant inclusion; and the Connection Machine represents the opposite approach of connecting a very large number (approximately 65 000) of small processors in a network.

The INMOS transputer is the first commercial chip which has been conceived of as a building block for parallel computers; its architecture and application are therefore covered in some detail in Chapter 3. There are currently around one dozen commercial computers on the market making use of multiple transputers, and probably many thousands of companies worldwide investigating the possibilities of this chip, either as a potential product, or as a component for a parallel embedded system. This interest is not surprising, when it is considered that just a few dozen T800 transputer chips are capable of delivering the performance of the CRAY-1, described in the first edition of this book. This chip product is symptomatic of the developments that have occurred over the last five to ten years in semiconductor technology; gate density has doubled every two years or so and clock periods improved by 50% in the same period. Currently there is no sign that this progression will flatten out significantly. These trends and their implication are considered in Chapter 6.

Developments in parallel languages since the first edition are covered by the inclusion of the proposed FORTRAN 8X standard for the expression of vector computation which will be available on all vector computers, CMLISP, the programming language for the Connection Machine, and OCCAM, the programming language for the INMOS transputer.

The advent of MIMD computers has necessitated the definition of performance parameters to take into account the cost of the synchronisation of the multiple instruction streams, and the cost of communicating data between processors. In defining the parameters ($s_{1/2}$ for synchronisation, and $f_{1/2}$ for communication), we have followed the philosophy of the (r_∞, $n_{1/2}$) parameters which we introduced in the first edition to characterise the behaviour of a vector pipeline, and which are now quite widely known and adopted. That is to say we have chosen parameters that can be related to some property of the user's program (e.g. the amount of arithmetic in a parallelised block of code). This leads to the development of methods of algorithm analysis based on these parameters which are given in Chapter 5.

It would take too much space to thank all who have helped with information or illustrations for the new edition, but we would like to thank particularly: David Dent (ECMWF); Paul Elstone, John Larson and William White (Cray Research); Shaun Powell, P J Elms, Alain Hochedez, and Jean-Claude Lignac (CDC); Meg Saline and Cliff Arnold (ETA Systems); David Snelling (ex Denelcor) and Ian Curington (ex FPS); David May (INMOS); Geoff Manning (AMT Ltd); Leon Bentley, Andrew Rushton, Jimmy Stewart, Adriano Cruz, Russel O'Gorman, Gadge Panesar, Ernest Ng and Charles Askew (Southampton University).

It is easy for a book on computers to become a rather tedious description of endless different architectures, indeed some of our readers may feel we have also fallen into that trap. However, as with the first edition, we have attempted to put the presentation in a historic perspective, and within a theoretical framework of a notation, a classification and a theory of performance which we hope may still be applicable when the particular computers described are past history. In this way we hope the new edition will remain as popular as a teaching text as was the first edition, presenting principles and methodology as well as rapidly outdated facts. Nevertheless we have described many of the computers, languages and algorithms that will be used by computer practitioners in the next five years, both professional computer scientists and computational scientists of various disciplines (e.g. physics, chemistry, biology and engineering). We therefore expect this new edition to be equally useful to those various groups as they try to obtain the best performance from the many and varied parallel computer architectures now available to them.

R W Hockney
C R Jesshope
(Reading and Southampton Universities, August 1987)

1 Introduction

Large-scale parallelism is the principal innovation to appear in the design of large commercially available computers in the 1980s. In this introductory chapter we trace the history of this development (§1.1), then lay out the principles that can be applied to classify these parallel computers (§1.2) and characterise their relative performance (§1.3). In order to make efficient use of such parallel computers the programmer must be aware of the overall organisation of the computer system, which we aptly describe as its architecture: that is to say, the number and type of processors, memory modules and input/output channels, and how these are controlled and interconnected. Chapters 2 and 3 are devoted respectively to describing the architecture of the two main classes of parallel computers, namely the *pipelined computer* and the *multiprocessor array*. It would be impractical to give a comprehensive description of all designs in these categories; instead we have selected principally commercially available computers that have or are likely to have a significant number of sales, and which differ sufficiently from each other to illustrate alternative approaches to the same problem. The pipelined computers described in Chapter 2 are the CRAY X-MP, CRAY-2, CYBER 205, ETA[10] and the FPS 164/MAX. The processor arrays and multiprocessors described in Chapter 3 are the ICL DAP, Burroughs BSP, Denelcor HEP and the Connection Machine. In our discussion of future developments in Chapter 6 we compare the characteristics of Bipolar and MOS technologies, discuss the technological issues arising in the design of parallel computers, and consider the potential of wafer-scale integration.

The advent of very large-scale chip integration (VLSI), culminating in the single-chip microprocessor, has led to a large number of computer designs in which more than one instruction stream cooperate towards the solution of a problem. Such multi-instruction stream or MIMD computers are reviewed (§1.1.8) and classified (§1.2.5) in Chapter 1 and one of them, the Denelcor HEP (§3.4.4), is described in detail in Chapter 3.

1

New architectural features require new computer languages and new numerical algorithms, if the user is to take advantage of the development. In the case of parallel computers that are designed to work most efficiently on one- or sometimes two-dimensional lists of data, this is most naturally achieved by introducing the mathematical concept of vector or matrix into the computer language and by analysing algorithms in terms of their suitability for execution on vectors of data. A good computer language also greatly influences the ease of programming and Chapter 4 is devoted to *parallel languages* from the point of view of implicit parallelism (§4.2), structure parallelism (§4.3) and process parallelism (§4.4). The usefulness of a parallel computer depends on the invention or selection of suitable *parallel algorithms* and in Chapter 5 we establish principles for measuring the performance of an algorithm on a parallel computer. These are then applied to the selection of algorithms for recurrences, matrix multiplication, tridiagonal linear equations, transforms and some partial differential equations.

Although our presentation is necessarily based on the situation *circa* 1987, we try to establish principles that the reader can apply as computer architecture develops further. Our objective is to enable the reader to assess new designs by classifying their architecture and characterising their performance; and thereby be able to select suitable algorithms and languages for solving particular problems.

1.1   HISTORY OF PARALLELISM AND SUPERCOMPUTING

Our history will be primarily a study of the influence of parallelism on the architecture of top-of-the-range high-performance computers designed to solve difficult problems in science and engineering (e.g. the solution of three-dimensional time-dependent partial differential equations). The requirement in such work is for the maximum arithmetic performance in floating-point (i.e. real as opposed to integer) arithmetic, which has been achieved by a combination of technological advances and the introduction of parallelism into the architecture of the computers. Computers satisfying the above requirement are now generally known as *supercomputers*, so that this section might equally well be regarded as a history of supercomputing.

We start our discussion of the history of parallelism in computer architecture by considering the reduction in the time required for a simple arithmetic operation, for example a floating-point multiplication, in the period since the first commercially produced computer, the UNIVAC 1, appeared in 1951. This is shown in figure 1.1 and demonstrates roughly a ten-fold increase in

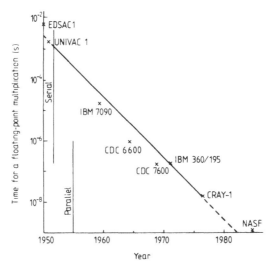

FIGURE 1.1 The history of computer arithmetic speed since 1950, showing an increase of a factor of 10 in 5 years.

arithmetic computing speed every five years. This sensational increase in computer speed has been made possible by combining the technological improvements in the performance of the hardware components with the introduction of ever greater parallelism at all levels of the computer architecture.

The *first generation* of electronic digital computers in the 1950s used electronic valves as their switching components with gate delay times† of approximately 1 μs. These were replaced about 1960 by the discrete germanium transistor (gate delay time of approximately 0.3 μs) in such *second-generation* machines as the IBM 7090. Bipolar planar integrated circuits (ICS) on silicon, at the level of small-scale integration (SSI), with a few gates per chip and gate delay time of about 10 ns, were introduced in about 1965 and gradually improved until around 1975 gate delay times of slightly less than 1 ns were reliably possible. Although about five to ten times slower, much greater packing densities are possible with the alternative metal oxide silicon (MOS) semiconductor technology. By the beginning of the 1980s microprocessors, with a speed and capacity about equal to the first-generation valve computers, were available on a single chip of silicon a few millimetres

---

† The gate delay time is the time taken for a signal to travel from the input of one logic gate to the input of the next logic gate (see e.g. Turn 1974 p 147). The figures are only intended to show the order of magnitude of the delay time.

square. This engineering development obviously makes possible the implementation of various highly parallel architectures that had previously remained only theoretical studies.

Looking at the period from 1950 to 1975 one can see that the basic speed of the components, as measured by the inverse of the gate delay time, has increased by a factor of about 10^3, whereas the performance of the computers, as measured by the inverse of the multiplication time, has increased by a factor of about 10^5. This additional speed has been made possible by architectural improvements, principally the introduction of parallelism, which is the main subject of this volume. For economic reasons, different technologies favour the introduction of parallelism in different ways. For example, in Chapter 6 we discuss the influence of VLSI and wafer-scale integration on the way increased parallelism is likely to be introduuced into the architecture of future scientific supercomputers. Technological development, including some interesting predictions for the 1980s, is also discussed by Turn (1974) in his book entitled *Computers in the 1980s* and by Sumner (1982) in the Infotech *State of the Art Report: Supercomputer Systems Technology,* Our discussion that follows centres on the organisational aspect of introducing parallelism rather than the details of its implementation in hardware.

The above comparison does not take into account the possible overlapping of different arithmetic and logical operations, and we may alternatively compare the clock period, as a measure of the speed of the technology, with the number of useful arithmetic operations performed per second on an actual problem as a measure of performance. Wilkes *et al* (1951) quote an execution time of $18n$ ms for evaluating a series of n terms ($2n$ arithmetic operations) on the EDSAC1 (Wilkes and Renwick 1949) which had a clock period of $2\ \mu s$. This is an average rate of 100 arithmetic operations per second, which we may compare with a performance of 130 million arithmetic operations per second on the CRAY-1 (Russell 1978) for matrix multiplication. Since the CRAY-1 has a clock period of 12.5 ns, we have an improvement in total performance of 10^6 over the period of about three decades, of which only about a factor of 160 can be attributed to improvements in technology. One should also note a great improvement in the quality of an arithmetic operation, from 36-bit fixed-point arithmetic on the EDSAC1 to 64-bit floating-point arithmetic on the CRAY-1.

The overall architecture of the first generation of computers is described as serial and follows the fundamental ideas of the stored program computer laid down by Burks *et al* (1946), and usually referred to simply as the von Neumann organisation. Such a computer comprises an input and output (I/O) device, a single memory for storing both data and instructions, a single control unit for interpreting the instructions, and a single arithmetic and logical unit for processing the data. The latter two units are referred to as the central processing unit or CPU. The important feature in the present context is that each operation of the computer (e.g. memory fetch or store, arithmetic or logical operation, input or output operation) had to be performed sequentially, i.e. one at a time. Parallelism refers to the ability to overlap or perform simultaneously many of these tasks.

The principal ways of introducing parallelism into the architecture of computers are described fully in §1.2. They may be summarised as:

(a) *Pipelining*—the application of assembly-line techniques to improve the performance of an arithmetic or control unit;

(b) *Functional*—providing several independent units for performing different functions, such as logic, addition or multiplication, and allowing these to operate simultaneously on different data;

(c) *Array*—providing an array of identical processing elements (PES) under common control, all performing the same operation simultaneously but on different data stored in their private memoric^c—i.e. lockstep operation;

(d) *Multiprocessing* or *MIMD*—the provision of several processors, each obeying its own instructions, and usually communicating via a common memory, or connected in a network.

Of course, individual designs may combine some or all of these parallel features. For example, a processor array may have pipelined arithmetic units as its PES, and one functional unit in a multi-unit computer might be a processor array.

Up to 1980, multiprocessor designs have been concerned primarily with methods of connecting several independent computers so as to maximise the throughput of a computer installation. This is certainly important but is outside the scope of this book. Consequently we will point out examples of multiprocessors when they arise as part of the parallelism of a computer, but refer the reader to the reviews by Enslow (1974, 1977), Infotech (1976) and Jones and Schwarz (1980) for the discussion of multiprocessors in the wider context of system performance. The advent of the cheap microprocessor around 1975 has made realistic the concept of many linked microprocessors cooperating on the solution of one problem and such systems are likely to become increasingly important in the 1980s. A number of experimental systems have been proposed and implemented (Satyanarayanan (1980), Hockney (1985b, d) and §1.1.8). An excellent discussion of parallelism as it affects all aspects of computer design (arithmetic units, memory organisation, instruction scheduling, input/output, multiprocessing and multiprogramming) is given by Lorin (1972) in his book *Parallelism in Hardware and Software*:

Real and Apparent Concurrency. A most comprehensive survey of all types of parallel architectures appears in the book by Hwang and Briggs (1984) entitled *Computer Architecture and Parallel Processing.* Supercomputers are specifically considered in the book by Lazou (1986) entitled *Supercomputers and their Use.*

In our presentation of the history of parallelism we do not attempt a complete survey but focus attention only on developments that represented significant steps in the introduction of parallelism. Nor do we attempt to mention all machines or papers. Instead we have selected only those machines that either achieved significant sales or were particularly influential within the computer science community, and similarly only a few of the most influential scientific papers are mentioned. The main evolutionary trends and connections, evident in the introduction of parallelism into computer architecture, are illustrated in figure 1.2. We now describe separately the main lines of development.

Readers interested in additional information on the early history of computers, mostly prior to 1950, are directed to the collection of original papers edited by Randell (1975) entitled *The Origins of Digital Computers: Selected Papers*, and to Hartree's (1950) book *Calculating Instruments and Machines*. A collection of essays written by many of the early pioneers of computing has been compiled by Metropolis, Howlett and Rota (1980) under the title *A History of Computing in the Twentieth Century*, and makes absorbing reading. Wartime work in the USA and UK is described, as well as developments in the USSR, Japan, Germany and Czechoslovakia. A highly readable account of the history of computer design from Babbage to the early 1950s is given by Kuck (1978) in his book entitled *The Structure of Computers and Computations*. This includes a very interesting account of the interplay between the personalities involved in the early development of the digital computer, and the engineering problems that they encountered. A particularly valuable historical survey is given by Rosen (1969) who describes the development of computers from the ENIAC in 1946 to the CDC 7600 in 1968. Detailed information on the architecture of many of the computers that we mention is given by Bell and Newell (1971) in their book *Computer Structures: Readings and Examples*, which includes papers on most of the important families of computers. It covers the period approximately 1950–70, and gives design and historical details not readily available elsewhere. This book has been revised and brought up to date by Siewiorek, Bell and Newell (1982) under the title *Computer Structures: Principles and Examples*. These two books are primarily concerned with the architecture of the most successful serial computers (e.g. PDP 8 and 11, IBM 360), but the latter does include a description of some parallel systems (ILLIAC IV, STARAN, C.mmp,

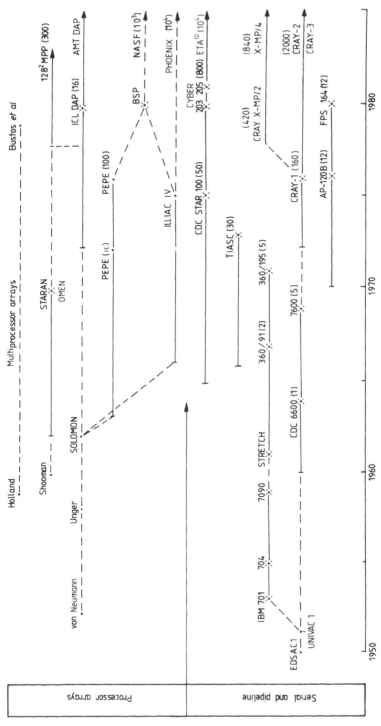

FIGURE 1.2 Evolutionary tree showing the architectural connections and influences during the development of parallel computers from the early 1950s to the mid-1980s. Numbers in parentheses are a guide only to the performance in Mflop/s of the computer when used sensibly. Full lines indicate design and manufacture, the star the delivery of the first operational system. Broken lines indicate strong architectural relationships.

CRAY-1, TIASC). Other machines (MU5, ATLAS, IBM 370, UNIVAC 1100, DEC 10, CRAY-1) are described in the special issue of the *Communications of the Association of Computing Machinery* devoted to computer architecture (ACM 1978) and in the book by Ibbett (1982) entitled *The Architecture of High Performance Computers*, which also discusses the CDC 6600 and 7600, the IBM 360 models 91 and 195, the TIASC and CDC Cyber 205. Useful summaries of the architecture of most commercially available computers are given in a series of reports on computer technology published by Auerbach (1976a). A recent review of parallelism and array processing including programming examples is given by Zakharov (1984).

1.1.1 Parallelism prior to 1960
The earliest reference to parallelism in computer design is thought to be in General L F Menabrea's publication in the Bibliothèque Universelle de Genève, October 1842, entitled *Sketch of the Analytical Engine Invented by Charles Babbage* (see Morrison and Morrison 1961 p 244, Kuck 1977). In listing the utility of the analytic engine he writes:

> Secondly, the economy of time: to convince ourselves of this, we need only recollect that the multiplication of two numbers, consisting each of twenty figures, requires at the very utmost three minutes. Likewise, when a long series of identical computations is to be performed, such as those required for the formation of numerical tables, the machine can be brought into play so as to give several results at the same time, which will greatly abridge the whole amount of the processes.

It does not appear that this ability to perform parallel operation was included in the final design of Babbage's calculating engine or implemented; however it is clear that the idea of using parallelism to improve the performance of a machine had occurred to Babbage over 100 years before technology had advanced to the state that made its implementation possible. Charles Babbage undoubtedly pioneered many of the fundamental ideas of computing in his work on the mechanical difference and analytic engines which was motivated by the need to produce reliable astronomical tables (Babbage 1822, 1864, Babbage 1910, Randell 1975).

The first general-purpose electronic digital computer, the ENIAC, was a highly parallel and highly decentralised machine. Although conceived and designed principally by J W Mauchly and J P Eckert Jr, ENIAC was described mainly by others (Goldstine and Goldstine 1946, Hartree 1946, 1950, Burks and Burks 1981). Since it had 25 independent computing units (20 accumulators, 1 multiplier, 1 divider/square rooter, and 3 table look-up units), each following their own sequence of operations and cooperating towards the solution of a single problem, ENIAC may also be considered as the first example of MIMD computing. Furthermore, it used decimal arithmetic internally, and computed on all ten decimal digits of a number in parallel. ENIAC was designed and built in the Moore School of Engineering at the University of Pennsylvania between 1943 and 1946, under contract to the US Army. It was subsequently moved to the Aberdeen Ballistics Research Laboratory where it was used successfully for the calculation of projectile trajectories and firing tables until 1955.

The overall architecture of ENIAC was conceived as an electronic version of the mechanical differential analysers which were, at that time, the most advanced machines for the solution of differential equations (Bush 1931, 1936, Hartree 1950). These machines used a vertical wheel in contact with a horizontal disc to form a mechanical integrator. If y is the distance of the wheel from the centre of the disc, and the disc rotates by an angle dx, then the wheel rotates by an angle y dx. Hence if y is varied proportionally to the integrand, the total accumulated rotation of the wheel is $\int y\, dx$. In order to obtain a solution, several such integrators were linked together by shafts in a pattern corresponding to the differential equation that was to be solved (see Hartree 1950). The mechanical differential analyser was thus an analogue device in which the angular rotation of the shafts was proportional to the values of the variables in the problem. As was also the case with later electronic analogue computers, all the integrators (then called amplifiers) worked simultaneously and in parallel on the problem. This pattern of parallel working was maintained in the ENIAC. However, since finite differences had to be used to represent differential coefficients, and integrals became finite sums, the integrators of the differential analyser became accumulator units (i.e. adders) in the ENIAC.

Using present-day terminology one could describe the programming of ENIAC as follows: the variables in the finite difference formulation of the problem were assigned to different accumulators, and the outputs and inputs of the accumulators were connected together as demanded by the equations, using large plugboards; switches were then set on each accumulator to specify the particular sequence of operations it was to perform (i.e. a different microprogram could be given to each of the 20 accumulators); finally a master program was set on switches in the master control unit, which initiated the microprograms in the accumulators at the appropriate times, and synchronised their actions. Note that there was no concept of a single stored program that described the whole algorithm, and which was subsequently executed by a fixed computer architecture—this important idea of the stored program computer was implemented first in the next generation of computers.

In contrast, the architecture of ENIAC was rearranged for each problem, by using the plugboard to rewire the connections between the units. One could say that the algorithm was literally wired into the computer. It is interesting that such ideas are beginning to sound very 'modern' again in the 1980s in the context of MIMD computing, reconfigurable VLSI arrays, and special-purpose computers executing very rapidly a limited set of built-in algorithms. However, the time was not ripe for this type of parallel architecture in the 1940s, as can be seen in the following quotation from Burks (1981) who was a member of the ENIAC design team.

> The ENIAC's parallelism was relatively short-lived. The machine was completed in 1946, at which time the first stored program computers were already being designed. It was later realized that the ENIAC could be reorganized in the centralized fashion of these new computers, and that when this was done it would be *much easier to put problems on the machine*. This change was accomplished in 1948.... Thereafter the plugboard of the ENIAC was never modified, and the machine was programmed by setting switches at a central location. Thus the first general-purpose electronic computer, built with a parallel decentralized architecture, *operated for most of its life as a serial centralized computer*! [Author's italics.]

The difficulty of programming parallel computers is a recurring theme that is still with us today, and it remains to be seen whether the second coming of the parallel computer in the 1980s will be more successful than the first!

The first stored program computers—EDSAC at the University of Cambridge (Wilkes and Renwick 1949) and EDVAC at the University of Pennsylvania (completed in 1952)—and their commercial derivatives (e.g. UNIVAC 1: see Eckert *et al* 1951) used mercury delay line memories of a few 100 words. Since, in such storage, the successive bits of a number are presented to the reading head in a time sequence starting with the least significant bit, it was natural and economic to perform the arithmetic on two numbers bit by bit. The addition, for example, of two 32-bit numbers was performed in 32 machine cycles by very simple circuitry capable only of adding one bit to another. Arithmetic performed in this way is termed *bit-serial* and an early use of parallelism in computers was to speed up the arithmetic by performing the operations on all bits of a number in parallel.

It is interesting to note that the serial organisation of arithmetic was considered as one of the advantages gained by introducing electronic components into computers, which had traditionally performed their arithmetic in parallel. For example, serial arithmetic had been rejected by Babbage for his difference engine because of the long execution time involved (Babbage 1864, Morrison and Morrison 1961 p 34), and his analytic engine was to have performed arithmetic on 50 decimal digits in parallel. A typical desk calculator of the 1940s worked with about 12 decimal digits in parallel. The extra speed of electronics over the mechanical and electromechanical components of previous machines allowed an improvement of performance by a factor of between 10^3 and 10^4 together with the economy of equipment that went with processing only a single digit at a time. The same was also true if one compared the parallel arithmetic on ENIAC, which was a previous electronic machine, with the serial arithmetic on machines like EDVAC (von Neumann 1945). Circuit techniques had improved to a situation where the time between pulses on EDVAC was 1 μs compared with 10 μs on ENIAC, and this enabled a 32-bit serial addition to be performed in 32 μs compared with an equivalent ten decimal digit addition of 200 μs on ENIAC. Multiplication by successive addition on EDVAC was also faster than multiplication on ENIAC, although ENIAC could multiply ten decimal digits by one decimal digit in parallel.

A first-generation valve computer containing several parallel features was the pilot ACE and its commercial derivative, the English Electric DEUCE. The design of the pilot model ACE (Wilkinson 1953) was undertaken during a visit by Dr H D Huskey to the UK National Physical Laboratory in 1947, following the ideas of Dr A M Turing (see his biography, Turing 1959), and first operated in 1951. This machine had 11 mercury delay lines, each with a capacity of 32 32-bit words and a circulation time of 1 ms. The arithmetic was serial with one bit occurring every clock period of 1 μs. Features of interest included a card reader, card punch and multiplier that could operate in parallel with the rest of the machine, and instructions (which we would now class as vector instructions) that could perform a limited number of operations on all the 32 numbers in a delay line. For example, decimal to binary or binary to decimal conversion was undertaken by the CPU whilst the rows of a card were passing through the reader or punch. Also the sign adjustment required to complete a signed multiplication was computed using the adder during the 2 ms that were required by the multiplier to perform an unsigned multiplication. These are examples of functional parallelism. An instruction could be held active (in effect repeated) for up to 32 clock periods thus, for example, allowing the sum of all numbers in a delay line to be computed by one instruction. This facility was frequently used for computing check sums, and is one of the earliest examples of a vector instruction. In the production version, the English Electric DEUCE (Haley 1956) which appeared in 1954, the fast store was increased to 12 delay lines and backed up by an 8000-word magnetic drum. Parallelism was also evident during drum transfers, because calculations in the rest of the machine could proceed at the same time. An autonomous hardware fixed-point divider (2 ms to complete) was also provided on DEUCE which, like the multiplier, could work in parallel with the rest of the machine. However it used the same registers as the multiplier. Three single-, three double- and two quadruple-word delay lines were provided as fast-access registers on DEUCE. One single- and one double-word delay line were associated with fixed-point adders and could also act as accumulators.

Bit-parallel arithmetic became a practical part of computer design with the availability of static random-access memories from which all the bits of a word could be read conveniently in parallel. The first experimental machine to use parallel arithmetic was finished at the Institute of Advanced Studies (IAS) in 1952, and this was followed in 1953 by the first commercial computer with parallel arithmetic, the IBM 701. Both these machines used electrostatic cathode ray tube storage devised by Williams and Kilburn (1949), and were followed in a few years by the first machines to use magnetic-core memory. The most successful of these was undoubtedly the IBM 704 of which about 150 were sold. This machine had not only parallel arithmetic but also the first hardware floating-point arithmetic unit, thus providing a significant speed-up over previous machines that provided floating-point arithmetic, if at all, by software. The first IBM 704 was commissioned in 1955 and the last machine switched off in 1975. The remarkable history of this first-generation valve machine which performed useful work for 20 years is described by McLaughlin (1975).

In the IBM 704, along with other machines of its time, all data read by the input equipment or written to the output equipment had to pass through a register in the arithmetic unit, thus preventing useful arithmetic from being performed at the same time as input or output. Initially the equipment was an on-line card reader (150 to 250 cards per minute), card punch (100 cards per minute) and line printer (150 lines per minute). Soon these were replaced by the first magnetic tape drives as the primary on-line input and output equipment (at 15 000 characters per second, at least 100 times faster than the card reader or line printer). Off-line card-to-tape and tape-to-printer facilities were provided on a separate input/output (I/O) computer, the IBM 1401. However these tape speeds were still approximately 1000 times slower than the processor could manipulate the data, and input/output could be a major bottleneck in the overall performance of the IBM 704 and of the computer installation as a whole.

The I/O problem was at least partially solved by allowing the arithmetic and logic unit of the computer to operate in parallel with the reading and printing of data. A separate computer, called an *I/O channel*, was therefore added whose sole job was to transfer data to or from the slow peripheral equipment, such as card readers, magnetic tapes or line printers, and the main memory of the computer. Once initiated by the main control unit, the transfer of large blocks of data could proceed under the control of the I/O channel whilst useful work was continued in the arithmetic unit. The

I/O channel had its own instruction set, especially suited for I/O operations, and its own instruction processing unit and registers. Six such channels were added to the IBM 704 in 1958 and the machine was renamed the IBM 709. This is therefore an early case of multiprocessing. The machine still used electronic valves for its switching logic and had a short life because, by this time, the solid state transistor had become a reliable component. The IBM 709 was re-engineered in transistor technology and marketed in 1959, as the IBM 7090. This machine, together with the upgraded versions (IBM 7094 and 7094 II), was extremely successful and some 400 were built and sold.

During the early development of any new device it is usual to find a wide range of innovative thought and design, followed by a period dominated by a heavy investment in one particular type of design. After this period, further innovation tends to be very difficult due to the extent of this investment. The pattern can be seen in the development of the motor car, with a wide range of engine principles used around 1900 including petrol, steam and rotary engines, and the subsequent huge investment in the petrol driven internal combustion engine that one can now scarcely imagine changing. Similarly many novel architectural principles for computer design were discussed in the 1950s although, up to 1980, only systems based on a single stream of instructions and data had met with any commercial success. It appears likely, however, that very large-scale integration (VLSI) technology and the advent of cheap microprocessors may enable the realisation of some of these architectures in the 1980s.

In 1952 Leondes and Rubinoff (1952) described a multi-operation computer based around a rotating drum memory. The machine DINA was a digital analyser for the solution of the Laplace, diffusion and wave equations. A somewhat similar concept was put forward later by Zuse (1958) in the design for a 'field calculating machine'. The principle of spatially connected arrays of processors was established by von Neumann (1952) who showed that a two-dimensional array of computing elements with 29 states could perform all operations, because it could simulate the behaviour of a Turing machine. This theoretical development was followed by proposals for a practical design by Unger (1958) that can be considered as the progenitor of the SOLOMON, ILLIAC IV and ICL DAP computers that were to appear in the 1970s.

Similarly the paper by Holland (1959) describing an assembly of processors each obeying their own instruction stream, can be considered the first large-scale multiprocessor design and the progenitor of later linked microprocessor designs such as those proposed about 20 years later by Pease (1977), and by Bustos et al (1979) for the solution of diffusion problems. The indirect binary n-cube, proposed by Pease, was a paper design for 2^n microprocessors connected topologically as either a 1-, 2-, up to n-dimensional cube suitable for a wide range of common numerical algorithms, such as the fast Fourier transform. It was envisaged that up to 16 384 microprocessors could be used ($n = 14$). The Hypercube announced by IMS Associates (Millard 1975) was based on Pease's ideas: it comprised a 4-dimensional cube with two microprocessors at each node, one for internode communication and the other for data manipulation. This particular hypercube came to nothing at this time; however, the idea reappeared in the early 1980s as the Cosmic Cube and the Intel Personal Supercomputer (see §1.1.8).

1.1.2   Fast scalar computers

A scalar computer is one that provides instructions only for manipulating data items comprising single numbers, in contrast to the vector computer that also has instructions for manipulating data items comprising an ordered set of numbers (that is to say a vector). The history of the development of fast scalar computers in the 1960s and 1970s is largely the history of the introduction of more and more parallelism into the overall serial design of the single instruction stream/single data stream computers, such as the IBM 7090, that had become a fairly standard architecture by the end of the 1950s.

A computer that had a profound influence on both architecture and software was the ATLAS (Kilburn *et al* 1962, Sumner *et al* 1962, Howarth *et al* 1961, 1962, Lavington 1978). This machine was conceived at the University of Manchester in about 1956 and designed by a combined industrial and university group under the direction of Professor Kilburn. The prototype was working at the University in 1961, and the first production model was completed by Ferranti Ltd (later to become part of International Computers and Tabulators, and afterwards International Computers Limited) in 1963. The ATLAS was known principally for pioneering the use of a complex multiprogramming operating system based on a large virtual one-level store and an interrupt system. The operating system organised the allocation of resources to the programmes currently in various stages of execution. The virtual address space appeared to the user as a single-level store with approximately 10^6 words; however this was automatically translated to references to the multilevel physical store comprising magnetic cores (16 K words, where K = 1024), magnetic drum (96 Kwords) and finally magnetic tape as backing store. This was made possible by a paging system, which transferred information between the different levels of the store hierarchy in units of 512 words called pages. The interrupt system enabled slow I/O devices to work autonomously and only to interrupt calculations in the cpu when it was absolutely necessary.

In addition to the above organisational improvements, the ATLAS also made early use of parallelism to improve its calculational performance. The 2 μs main core memory was divided into four independent banks (called stacks) which made possible, in favourable circumstances, the retrieval in one store cycle of two instructions and their two operands in parallel. The arithmetic itself was performed in a bit-parallel fashion (Kilburn *et al* 1960). Functional parallelism was present in the provision of a separate autonomous 24-bit adder for index calculations, called the B-arithmetic unit, in addition to the main fixed- and floating-point arithmetic unit which worked on 48 bits (40-bit mantissa and 8-bit power-of-eight exponent) using a single main accumulator. The B-arithmetic unit worked in association with 128 24-bit index registers (called B lines), built from fast magnetic cores with a cycle time of 0.7 μs. The instructions were single-address and specified an operation between the contents of that address and the contents of the main accumulator. The 48 bits of an instruction contained a 10-bit function code, a 24-bit main core address and two 7-bit index-register addresses. The operand address was obtained by adding the contents of one or both of the specified index registers to the main address. The principle of pipelining was used in order to overlap the following phases of instruction execution: instruction fetch, 24-bit operand address calculation in the B unit, operand fetch and 48-bit arithmetic operation in the main accumulator. The effectiveness of the pipelining can be judged from the fact that, in an example of a series of floating-point additions, the 6.0 μs per operation that would be required for the sequential execution of all the phases of the instruction is reduced to an average of 1.6 μs per operation with pipelining. Between two and four instructions are in various stages of execution at any time. The corresponding time for floating-point multiplication was 5 μs, and the average time per order for FORTRAN programs was measured to be approximately 3 μs. The logic of the Ferranti ATLAS was implemented in discrete germanium transistors and diodes with a typical gate delay of 12 ns. About 80 000 transistors were used in the computer.

In order to make effective use of such parallel features as multiple arithmetic units, registers and memories, it is necessary to look ahead in the instruction stream and determine which instructions can be executed concurrently without altering the logic of the program. Having detected this parallelism in the program, it is then necessary to schedule the issue of instructions to the arithmetic units for execution in the optimal way. Both these aspects of 'look-ahead' are considered by Keller (1976) and Kuck (1978), and are included in most fast scalar computers such as the CDC 6600 and IBM 360/91.

The gradual introduction of functional parallelism and pipelining can be seen in the evolution of the serial computer concept at Control Data Corporation under the influence of Seymour R Cray (founder member of the Company in 1957, Vice President and leading designer). The CDC 6600, first delivered in 1964, was the first computer to employ functional parallelism as a major feature of its design (Thornton 1964). The development of the machine is well described by Thornton (1970), who was responsible for most of the detailed design, in his book *Design of a Computer—The Control Data 6600*. The CDC 6600 comprised a 1 μs magnetic-core memory, divided into 32 independent banks that could work in parallel; 10 separate functional units for multiplication (duplicated), division, addition, long addition, shift, boolean, branch and increment (duplicated); and 10 peripheral processors forming a very flexible link with slow input/output devices. The 10 peripheral processors, although they timeshared the use of a single arithmetic and control unit, executed independent programs on data in separate memories. They therefore acted logically as independent processors and were an example of multiprocessing. To suit the scientific market *all* the arithmetic was floating-point (60-bit words), except in the increment unit (18-bit addition and subtraction only) which was for integer loop counting and address calculation. The CDC 6600 was such a success that it took much of the scientific market that had previously been dominated by IBM with their 7090 and 7094. In 1969, Control Data followed the 6600 with an upgraded version, the CDC 7600. This was about four times faster because the clock period was reduced from 100 ns in the 6600 to 27.5 ns. The ten serially organised functional units of the 6600 were replaced by eight pipelined functional units and one serial unit for division that could not be pipelined. Because of the extra speed of the pipelined units, it was no longer necessary to duplicate the multiplication and increment units. An extra functional unit for counting the number of ones in a word was added. The architectures of the CDC 6600 (renamed CYBER 70 model 74) and the CDC 7600 (renamed CYBER 70 model 76) must be two of the most successful produced for the scientific market. Over 50 of the latter machines have been installed.

The history of IBM production computers during the 1960s and 1970s also showed a gradual introduction of further parallelism. This development started in 1956 with much enthusiasm and a contract from the Los Alamos Scientific Laboratory to build a computer 100 times as fast as the IBM 704. This computer was called STRETCH (Dunwell 1956, Bloch 1959), and was later marketed for a brief period as the IBM 7030. Its principal novel features were a look-ahead facility to pick up, decode, calculate addresses and fetch operands several instructions in advance, and the division of memory into two independent banks that could send data to the arithmetic units in parallel.

The maximum transfer rate of data to and from memory was thereby increased by a factor equal to the number of memory banks. This was the first use of parallelism in memory, and enabled a relatively slow magnetic-core memory to be matched more satisfactorily to the faster processor. Almost all subsequent large computers have used banked (sometimes called interleaved) memory of this kind. The first STRETCH was delivered to Los Alamos in 1961, but did not achieve its design goals. Its manufacture also proved financially unsatisfactory to the company, and after seven systems were built (one being installed at AWRE Aldermaston UK) the computer was withdrawn from the product range. Potential customers were then sold the slower but very popular IBM 7090 series.

After the experience of STRETCH, it seemed that IBM had lost interest in high-speed computing. The IBM 360 series of computers was announced in 1964, but it contained no machine with a performance comparable to that of the CDC 6600 which was first installed the same year. The dramatic success of the CDC 6600 in replacing IBM 7090s and converting most large scientific centres to a rival company, made IBM respond. It was not until 1967, however, that the IBM 360/91 (Anderson *et al* 1967) arrived with a performance of about twice that of the CDC 6600. This machine had the look-ahead facility of STRETCH, and like the CDC 6600 had separate execution units for floating-point and integer address calculation each of which was pipelined and could operate in parallel. The principle of pipelining was also introduced to speed up the processing of instructions, the successive operations of instruction fetch, decode, address calculation and operand fetching being overlapped on successive instructions. In this way several instructions were simultaneously in different phases of their execution as they flowed through the pipeline. However the CDC 7600 appeared in 1969 and outperformed the 360/91 by about a factor two. IBM's reply in 1971 was the 360/195 which had a comparable performance to the CDC 7600. The IBM 360/195 (Murphy and Wade 1970) combined the architecture of the 360/91 with the cache memory that was introduced in the 360/85. The idea of introducing a high-speed buffer memory (or cache) between the slow main memory and the arithmetic registers goes back to the Ferranti ATLAS computer (Fotheringham 1961). The cache, 32 768 words of 162 ns semiconductor memory in the 360/85, held the most recently used data in blocks of 64 bytes. If the data required by an instruction were not in the cache, the block containing it was obtained from the slower main mempry (4 Mbytes of 756 ns core storage, divided into 16 independent banks) and replaced the least frequently used block in the cache. It is found in many large-scale calculations that memory references tend to concentrate around limited regions of the address space. In this case most references will be to data in the fast cache memory and the performance of the 4 Mbyte slow memory will be effectively that of the faster cache memory.

Gene Amdahl, who was chief architect of the IBM 360 series (Amdahl *et al* 1964), formed a separate company in 1970 (the Amdahl Corporation) to manufacture a range of computers that were compatible with the IBM 360 instruction code, and could therefore use IBM software. These machines were an important step in the evolution of computer technology and parallelism. The first machine in the range, the AMDAHL 470V/6, was the first to use large-scale integration (LSI) technology for the logic circuits of the CPU (bipolar emitter coupled logic, ECL, with 100 circuits per chip), and for this reason is sometimes called a fourth-generation technology computer. Six such machines were delivered in 1975 of which the first two were to NASA and the University of Michigan. The use of LSI reduced the machine to about one-third of the size of the comparable IBM 360/168, which used a much smaller level of integration. Although the arithmetic units in this machine were not pipelined, a high throughput of instructions was obtained by pipelining the processing of instructions. The execution of instructions was divided into 12 suboperations that used 10 separate circuits. When flowing smoothly, a new instruction could be taken every two clock periods (64 ns) and therefore up to six instructions were simultaneously in different phases of execution, and could be said to be in parallel execution. A high-speed buffer (or cache) bipolar memory of 16 Kbytes (65 ns access) improved the effective access time to the main memory of up to 8 Mbytes of MOS store (650 ns access).

1.1.3 Pipelined vector computers

In 1972 Seymour Cray left Control Data Corp. to start his own company, Cray Research Inc., with the aim of producing the fastest computer in the world. In the extraordinarily short time of four years the CRAY-1 computer was designed and built, the first model being delivered to the Los Alamos Scientific Laboratory in 1976. The CRAY-1 (Russell 1978, Dungworth 1979) followed the evolutionary trend from the 6600 and 7600. It provided 12 functional units, now all pipelined, a faster clock with period 12.5 ns, and a 16-bank one-million-word bipolar memory with a 50 ns cycle time. The principal novel feature was the provision of eight vector registers, each capable of holding 64 floating-point numbers (64 bits long), and a set of about 32 machine instructions for manipulating and performing arithmetic on these vectors. Three functional units were reserved for vector operations (shift, logical and addition) and three shared with scalar instructions (floating-point addition, multiplication and reciprocal approximation). The CRAY-1 was referred to as a vector computer, because it provided instructions for manipulating vectors. It was the first pipelined vector computer to be commercially successful, and remains by far the most successful; by the end of 1984 88 CRAY-1s and its sucessor, the CRAY X-MP, had been sold. The CRAY-1 regularly achieved measured rates of 130 Mflop/s (millions of floating-point operations per second) on suitable problems (e.g. matrix multiplication).

The main weaknesses of the original CRAY-1 design were (a) inadequate I/O capability, (b) low transfer rate from main memory to vector registers (i.e. bandwidth), and (c) the absence of a hardware scatter/gather instruction. These resulted in the actual performance on real problems often being significantly less than the arithmetic pipelines themselves could deliver. The CRAY-1S, announced in 1979, solved the first problem by introducing an I/O Subsystem to relieve the main CRAY-1 CPU of I/O interrupt handling and buffer management tasks, associated with front-end computers, discs and other peripheral devices. The Subsystem comprises two to four I/O processors and a buffer memory of 1/2 or 1 Mbyte. The I/O Subsystem has a direct channel to main memory with a transfer rate of 850 million bits per second. The CRAY-1S also increased the main memory capacity from 1 Mwords to 4 Mwords in the same physical space by using 4 Kbit memory chips instead of 1 Kbit chips.

The next major step in the evolution of the CRAY-1 series of computers was the announcement of the CRAY X-MP in 1982. This is a multiprocessor design that can be described as two CRAY-1 computers working from a common memory of 2 or 4 Mwords. Each CPU has four ports in memory, one for I/O and three for use by the vector registers, thus increasing the maximum memory bandwidth for data transfer by a factor of eight, and solving the limited bandwidth problem (b). In the absence of memory bank conflicts, two vector input streams and one output stream may now access memory simultaneously in both CPUs, and each may support an arithmetic pipeline at close to its maximum speed with data coming from and returning to main memory. The two CPUs may work on different parts of one problem, or on independent problems (jobs). Synchronisation is either via main memory locations, or a set of special synchronisation registers. An increased level of integration (16 gates per chip as compared with 4 or 5 gates per chip on the CRAY-1) allows the two CPUs to fit in the same physical space as the original CRAY-1. High-speed secondary storage is provided in the form of a solid state device (SSD), comprising up to 32 Mwords in 64 banks of volatile MOS memory. The data transfer rate between SSD and main memory is 1250 Mbyte per second. The initial version of the CRAY X-MP thus removes the first two weaknesses of the CRAY-1. In 1984 a four-CPU model was announced which removes the third weakness by including a hardware scatter/gather instruction. The top of this range, the CRAY X-MP/48, has a peak performance of 840 Mflop/s and an ECL bipolar central memory of 8 Mwords (64-bit) arranged in 64 interleaved banks. The CRAY X-MP is chosen for detailed description in Chapter 2 (§2.2).

In 1981 Seymour Cray stepped down as chairman of Cray Research in order to devote his time to the development of the CRAY-2 and other future machines. The new liquid immersion technology of the CRAY-2 was demonstrated in 1981 (Cray 1981). This involves submerging the whole machine in a bath of a clear inert liquid fluorocarbon, which acts as the cooling fluid. When encased in a transparent container—creating the so-called 'goldfish bowl' computer—quite beautiful effects are observed due to turbulence in the fluid. The improved cooling efficiency allows a greater density of component packaging, and a smaller CPU. The whole machine is contained in a cylinder 4 feet high and $4\frac{1}{2}$ feet in diameter. Circuit boards are packed closely together in three dimensions, giving a maximum wire length of 16 inches—compared with the two-dimensional packing and wire length of four feet on the CRAY-1. The CRAY-2 is designed to have up to four CPUs and a clock period of 4 ns. If the two floating-point pipelines ($+$, $\times$) in each of the four CPUs all work simultaneously, then two arithmetic operations are completed every nanosecond, corresponding to 2 Gflop/s. As with the CRAY X-MP, more CPUs may be added in later machines. Each CPU has a similar architecture to the CRAY-1, but the addition of 16 Kwords of local memory for intermediate results makes the machine incompatible with the CRAY-1 series at the machine code level. The CRAY-2 will use the UNIX operating system, and 32-bit addressing provides a total address space of 256 Mwords. Prototype CRAY-2 machines with one CPU and 16 Mword memories were delivered to NMFECC Livermore and the University of Minnesota in 1984/5, and three four-CPU machines were delivered in 1985, including one to NASA Ames Research Laboratory, California, with 256 Mwords of memory. Four installations were delivered in 1986 with four CPUs and either 64 or 256 Mwords of memory. The first CRAY-2 in Europe was ordered in 1985 by the State of Baden-Würtemberg in Germany, for installation at the University of Stuttgart. By 1987 CRAY-2s had also been installed at the Ecole Polytechnique, Paris, and at the UKAEA Harwell Laboratory in the UK. Initially the CRAY-2 will use the same silicon 16-gate logic chips that are used on the CRAY X-MP. The next major development is likely to be the use of gallium arsenide chips which should allow the clock period to be reduced to about one nanosecond. The gallium arsenide machine will probably be called the CRAY-3, and is scheduled for 1990. It is likely to have 16 processors, a shared memory of 1 Gword, and a performance in excess of 15 Gflop/s. The CRAY-2 is described in detail in §2.2.7.

Two other pipelined vector computers have earlier origins than the CRAY-1, namely the CDC STAR 100 and the Texas Instruments Advanced Scientific Computer TIASC (Theis 1974, Hockney 1977). The STAR 100 (Hintz and Tate 1972) was conceived about 1964 as a processor for vectors with an instruction set based on Iverson's (1962) APL language, that would with pipelining be able to sustain rates of 100 Mflop/s on the long vectors that were common to many scientific problems at the Lawrence Livermore Laboratory (LLL). A letter of intent was received in 1967 and design started. After a long gestation period of six years the first machine was operational in 1973 and two machines were delivered to LLL in 1974–5. There were two major hiatuses, one technological and one in the design itself, which led to the slow development, and when available the STAR 100 suffered from its very early start in the mid-1960s. Its 1.2 μs magnetic-core memory had been surpassed by semiconductor memory, and its clock with an 80 ns period was slow compared with its competitors by the time the machine was available in the mid-1970s. Furthermore, the best serially organised computers such as the CDC 7600 and IBM 360/195 had much faster arithmetic units for scalar operations, and for operations on any but the longest vectors. The STAR 100 was only able to outperform these more general-purpose competitors on very carefully prepared code, with vectors of several hundred or thousand elements. Consequently only about four machines were built and these went to government laboratories (two at LLL and one at NASA Langley) or were kept by the company as a facility available over the CYBERNET communications network. No STAR 100 was sold to a genuine commercial customer. The STAR 100, however, has been completely re-engineered in LSI with semiconductor memory, and was introduced to the market in 1979 as the CYBER 203E (Kascic 1979). This machine has been renamed the CYBER 205 and is now highly competitive with the CRAY-1. We take it as a second example of pipelined architecture in §2.3. The CYBER 205 differs from the CRAY-1 in processing all vector instructions to and from main memory (there are no vector registers); however it has multiple general-purpose pipelines, as opposed to the specialised pipelines of the CRAY-1. The first delivery was to the UK Meteorological Office in 1981.

In September 1983 CDC founded a separate company, ETA Systems Inc., in order to speed up the further development of their supercomputers. Control Data retained 40% ownership in ETA Systems, who announced the target of 10 Gflop/s for their ETA[10] machine. This is described as having eight processors with the architecture of the CYBER 205 but three to five times faster. Thus, program compatibility will be maintained with all the software developed for the CDC CYBER 205. The extra speed and greater level of integration are achieved by using CMOS technology cooled to liquid nitrogen temperatures ($-200\,°C$). A CMOS gate array chip made by Honeywell with up to 20 000 gates is used which dissipates 2 watts, as compared with 4 watts dissipation for the 250-gate ECL chips used on the CYBER 205. This will enable the CPU to be contained in about five square feet of floor space. The ETA[10] is described in more detail in §2.3.7.

The TIASC (Watson 1972, Ibbett 1982) was started about 1966 as a computer suitable for the high-speed processing of seismic data (involving very heavy use of Fourier transforms). It was an interesting design based on one, two, or four identical general-purpose pipelines, each capable of performing all the elementary instructions on vector operands. Instructions could be taken from one or two instruction processing units. We therefore have here another example of multiprocessing. The semiconductor memory had eight banks and a cycle time of 320 ns. With four pipes operating optimally a design rate of 50 Mflop/s was theoretically achieved. The number of arithmetic pipes and instruction units was varied in order to meet the particular demands of the customer. The first TIASC was operational in 1973 and about seven were built, most of which were used within the company and its associates. The most important sales were a four-pipe configuration (the only one) to the Geophysical Fluid Dynamics Laboratory (GFDL) at Princeton, and a two-pipe configuration to the Naval Research Laboratory (NRL) in Washington. However, after installing about seven systems, Texas Instruments discontinued manufacture of the TIASC. It also suffered, like the STAR 100, from a scalar unit that was significantly slower than other competitive computers like the CDC 7600.

The first venture of the Japanese manufacturers into the supercomputer field was by Fujitsu with their FACOM VP-100 and VP-200 pipelined vector computers, which were announced in 1982 (Miura and Uchida 1984). These machines are clearly designed to combine the best features of the CRAY-1 and CYBER 205. They have separate scalar and vector units that may work simultaneously, as in the CYBER 205; however, the vector unit is register-oriented, as in the CRAY-1. There are separate pipelines for add/logical, multiply and divide, working with data held in 64 Kbytes of vector register memory. A feature unique to these machines is the ability to reconfigure this vector storage dynamically under program control. The storage may be configured either as eight vector registers each of length 1024 elements, or as 256 vector registers of length 32 elements, or by factors of two between these limits. There are two load/store pipelines between the vector registers and the main memory (compared with one on the CRAY-1 and three on the CYBER 205 and CRAY X-MP). The maximum main memory size is 256 Mbytes (eight times that of the CYBER 205 or CRAY-1). The FACOM VPs provide masking instructions similar to those of the CYBER 205 and they have a hardware indirect vector addressing instruction (i.e. a random scatter/gather instruction), present on the CYBER 205 but missing on the CRAY-1. The clock period of the vector unit is 7.5 ns which gives a maximum processing rate of 533 Mflop/s for the VP-200. This rate and the memory sizes given above are halved for the VP-100. The first delivery was a VP-100 to the University of Tokyo at the end of 1983. Two other Japanese pipelined vector computers were announced in 1983, the HITACHI S-810 model 10/model 20 (630 Mflop/s maximum) and the NEC SX-1/SX-2 (1300 Mflop/s maximum). All these Japanese computers are described in more detail in Chapter 2, §2.4.

IBM's first venture into vector processing was announced in 1985 (nine years after the first CRAY-1 was delivered) as a part of the System/370 range of computers. This comprises a multiprocessor IBM 3090 scalar processor with a maximum performance of about 5 Mflop/s, to which can be optionally attached vector facilities which have about three to four times the scalar performance. This is a much lower ratio of vector to scalar speed than is provided by competitive machines, but is thought by IBM to be the best cost-effective choice. Each processor of the 3090 can support only one vector facility, and the current (1987) maximum number of processors is six, although this number will presumably increase. The IBM 3090-VF therefore provides both vector (SIMD) processing in the vector facility, and multi-tasking (MIMD) programming using the multiple processors. Each 3090 processor has a 64 Kbyte cache buffer memory that is used for instruction and data by both the scalar and vector units. Behind the cache, the four-processor model 400 has a central memory of 128 Mbyte and one extended memory of 256 Mbyte. Each vector facility has 16 vector registers, each holding 128 32-bit numbers, which may be combined in pairs for storing 128 64-bit numbers. The cycle time is 18.5 ns corresponding to a peak performance of 54 Mflop/s per pipeline. The vector facility has independent multiply and add pipelines giving a peak theoretical performance of 108 Mflop/s per vector facility. However, these rates can only be approached in highly optimised codes which keep most of the required data in registers or cache memory most of the time, thereby making minimal access to central or extended memory which is a severe bottleneck in the system. For example, the rate observed for a single dyadic vector operation in FORTRAN (see §1.3.3) with all data in cache memory is 13 Mflop/s. If the data are in central memory, this is reduced to

7.3 Mflop/s for a stride between successive elements of the vectors of unity (compare 70 Mflop/s on the CRAY X-MP, see §2.2.6) and to 1.7 Mflop/s for a stride of eight.

1.1.4 SOLOMON and after

A major milestone in the history of parallelism is certainly the paper of Slotnick *et al* (1962) entitled 'The SOLOMON computer'. This computer concept has also been described by Gregory and McReynolds (1963), and can be traced back to the ideas of Unger (1958). The acronym stands for Simultaneous Operation Linked Ordinal MOdular Network, and describes a two-dimensional array of 32 × 32 processing elements, each with a memory for 128 32-bit numbers and an arithmetic unit working in a bit-serial fashion, under control of a single instruction stream in a central control unit. Contrary to the evolutionary development of the serial computer to the vector pipeline computer, the SOLOMON concept was a radical change in thinking on computer architecture, and had a substantial influence on computer science research as well as on computer design. The SOLOMON computer was never built exactly as described in the 1962 paper, but gave birth not only to the ILLIAC IV and Burroughs PEPE floating-point processor arrays, but also to the Goodyear Aerospace STARAN and ICL DAP arrays of one-bit processors which are often called associative processors. We will discuss these developments separately.

The US Department of Defense's Advanced Research Projects Agency (ARPA) awarded a contract to the University of Illinois for the design of a SOLOMON-type computer in 1966, and this computer then became known as the ILLIAC IV. The original design is described by Slotnick (1967) in another influential paper entitled 'Unconventional systems', and by Barnes *et al* (1968) and McIntyre (1970). It was to comprise four quadrants, each with a control unit interpreting a single stream of instructions for 64 floating-point processing elements (PES). Each PE was to have 2000 64-bit words of thin-film memory and the PES in each quadrant were connected as an 8 × 8 array. The four quadrants were to be connected by a highly parallel I/O bus, and backed up by a large disc as secondary memory from which jobs would be read, and to which results would be written. Each of the 256 PES was envisaged to produce a floating-point operation in 240 ns so that a maximum rate of 1 Gflop/s was planned. Although only one quadrant was built and a maximum processing rate of approximately 50 Mflop/s was measured, the manufacture of the ILLIAC IV had a considerable influence on the development of computer technology and architecture. The design of the ILLIAC IV as actually built and its use in realistic applications is described by Feierbach and Stevenson (1979a) and by Slotnick (1971).

The tortuous story of the ILLIAC IV is described by Falk (1976) in his article 'Reaching for the gigaflop'. The machine was the first to use semiconductor memory chips (256-bit bipolar logic gates from Fairchild) for all its main memory, after it was discovered that there was insufficient space for the thin-film memories that were originally proposed. The bipolar logic gates were originally intended to be packed about 20 to the chip (MSI) but this had to be reduced to about 7 gates per chip (SSI). This was due largely to the decision to pioneer the new and faster emitter-coupled logic (ECL) rather than the established transistor–transistor logic (TTL). ILLIAC IV also pioneered the use of 15-layer circuit boards and computer-aided layout methods that proved necessary to wire them. The manufacture of ILLIAC IV (now reduced to one quadrant) was entrusted to Burroughs and the machine was delivered to NASA Ames Research Center, California in 1972. It was not until 1975, however, that a useable service could be offered and this was with the original clock period lengthened from 40 ns to 80 ns, with the consequential slowing down of all the operations.

The ILLIAC IV may be regarded as a failure in that it cost four times the original contract figure and did not come even within a factor of 10 of its originally proposed performance. However its influence was profound and we will probably see in the 1980s computer architectures rather similar to that of ILLIAC IV now that technology has advanced sufficiently to make such architectures practicable. ILLIAC IV was, like many major changes, too ambitious for the technology of its time. The computer software developed in association with ILLIAC IV was also considerable, including four computer languages that could express the parallelism of the computer, and much work on the development of suitable algorithms for standard mathematical problems which contain parallelism, such as matrix manipulations and the solution of partial differential equations (see, for example, Kuck 1968). The computer languages were the ALGOL-like TRANQUIL (Abel *et al* 1969) and GLYPNIR (Lawrie *et al* 1975), the PASCAL-like ACTUS (Perrott 1978), and CFD FORTRAN (Stevens 1975).

The Burroughs Corporation has played a major role in the development of array-like parallel processors. This began with the ILLIAC IV, for which Burroughs was the system contractor in the period 1969–73, continued with the PEPE parallel processor for the US Army (delivery began in 1976), and culminated with the announcement in 1977 of the Burroughs Scientific Processor (BSP) as a commercial venture for the general scientific market, in competition with the pipelined designs from Cray Research Inc. (CRAY-1) and Control Data Corporation (CYBER 205). We will now describe the relation of the PEPE and BSP designs to the ILLIAC IV.

The ILLIAC IV was designed for the solution of partial differential equations and can be described as an 8×8 array of 64-bit floating-point processing elements each with 2 Kwords of memory, working in lockstep fashion with nearest-neighbour connections, and controlled by a single instruction stream process in a central control unit. PEPE, the Parallel Element Processor Ensemble, was designed, on the other hand, to control a ballistic missile defense system of radar detectors and missile launchers (Berg *et al* 1972, Cornell 1972, Vick and Cornell 1978). PEPE had its origins in work at Bell Laboratories, Whippany, on content-addressable distributed logic memories (Lee and Paull 1963) combined with floating-point processing (Crane and Githens 1965), which led in about 1972 to the building of an experimental machine (PEPE IC) with 16 32-bit floating-point processing elements (Crane *et al* 1972). The full-size PEPE was manufactured by Burroughs and comprised a loosely coupled system of 288 PEs, each containing three processors (one each for input of radar signals, processing of data, and output of control signals) controlled in lockstep fashion by three control units, one for each type of processor in the PES. The three control units were connected to three standard I/O channels of a CDC 7600 which acted as host to the complete system. When operating, each target that was identified became the responsibility of one PE and, because there were no ordered connections between the targets, no direct connections were provided at all between the PES. When necessary, communication between the PES took place via the memories of the control units. The array of processors was then said to be unstructured and the word *ensemble* was coined for this arrangement. Since each PE had a floating-point processing rate of 1 Mflop/s, the complete PEPE could be rated as having a potential maximum computing rate of 288 Mflop/s. More realistic estimates for actual problems lead to estimates of about 100 Mflop/s (Vick and Cornell 1978). A 36-element version of the PEPE hardware was delivered to the BMDATC Advanced Research Center in Huntsville, Alabama in 1976.

One of the problems with the ILLIAC IV is the delay in routing data long distances across the array, caused by the limited nearest-neighbour connections between the 64 processors and the 64 banks of PE memory. In their commercial design, the BSP (Jensen 1978, Austin 1979), Burroughs have reduced the number of processors to 16 and the number of memory banks to 17. This smaller number makes it possible to provide connections via an 'alignment network' between any processor and any of the memory banks. The choice of a prime number of memory banks different from the number of processors allows the use of mapping algorithms that reduce the number of memory conflicts that may arise in common matrix manipulations. The PES themselves are serially organised floating-point processors with an addition or multiplication time of 320 ns for the production of 16 results (one in each PE). The careful overlapping of the read, write and arithmetic, together with the provision of all memory bank to PE connections, is expected to eliminate most bottlenecks; and the BSP is designed thereby to sustain a large fraction of its maximum processing rate of 50 Mflop/s on the majority of problems, The BSP is chosen for detailed study in §3.4.3, but was withdrawn from the market in 1980 before any had been sold.

ARPA and NASA jointly established an Institute of Advanced Computation (IAC) to support the ILLIAC IV, and in 1977 this institute published a design proposal for a machine called PHOENIX (Feierbach and Stevenson 1976b) to replace the ILLIAC IV in the mid-1980s. IAC foresaw the need for a 10 Gflop/s machine in order to solve three-dimensional problems in aerodynamic flow with sufficient resolution. The PHOENIX design can be described as 16 ILLIAC IVs each executing their own instruction stream under the control of a central control unit. If each PE can produce a result in 100 ns (a reasonable assumption with 1980 technology) the total of 1024 PES could produce the required 10^{10} operations per second.

NASA also commissioned two other design studies, from Control Data Corporation and Burroughs, for machines to replace the ILLIAC IV and form a Numerical Aerodynamic Simulation Facility (NASF) for the mid-1980s (Stevens 1979). The CDC design was based on an uprated four-pipe CYBER 205, operating in lockstep fashion, plus a fifth pipe as an on-line spare that can be electronically switched in if an error is detected. Each pipeline can produce one 64-bit result or two 32-bit results every 8 ns. However each result may be formed from up to three operations, leading to a maximum computing rate of 3 Gflop/s. There was also a fast scalar processor clocked at 16 ns. In contrast, the Burroughs design may be regarded as an upgrade to the BSP architecture, being based on 512 PES connected to 521 memory banks (the nearest larger prime number). Unlike the ILLIAC IV, each PE has its own instruction processor. The same instructions are assigned to each processor, but the arrangement does permit them to be executed in different sequences depending on the result of data-dependent conditions that may differ from processor to processor. With a planned floating-point addition time of 240 ns, multiplication time of 360 ns and 512 processors, a maximum processing rate of about 1–2 Gflop/s could be envisaged. ECL technology was planned with a 40 ns clock period.

The initial SOLOMON computer design (Slotnick et al 1962) was a 32 × 32 array of one-bit processors each with 4096 bits of memory, conducting its arithmetic on 1024 numbers in parallel in a bit-serial fashion. This describes quite closely the pilot model of the ICL Distributed Array Processor (DAP), that was started in 1972 and commissioned in 1976 (Flanders et al 1977, Reddaway 1979, 1984). The first production model of the machine was installed at Queen Mary College London in 1980. It comprises a 64×64 array of PES and forms a memory module of a host ICL 2980. As with the SOLOMON and the ILLIAC IV, the PES are connected in a two-dimensional network with nearest-neighbour connections. The original DAP used small-scale integration with 16 PES and their memory per circuit board. We take the ICL DAP as an architecture for detailed study in §3.4. In October 1986, Active Memory Technology (AMT) was founded to develop further the DAP concept. It markets the VLSI DAP 500 as processors attached to work stations (§3.5).

An important feature of the engineering design of the ICL DAP is that the PE logic is mounted on the same printed circuit board as the memory to which it belongs. An attractive possible development using VLSI technology, is to include the PE and its memory on the same chip. In the von Neumann concept the logic and memory are both conceptually and materially in different units (or even separate cabinets), which can lead to severe bottlenecks if the transfer rate between the two units is inadequate. In the ICL DAP, as its name implies, the logic is distributed into the memory where it is adjacent to the data it is to manipulate. The advent of VLSI which allows approximately 10^5 logic gates to be included on one chip, has made the distribution of logic into memory a practicable proposition. Indeed this has become almost a necessity because of the problems associated with interconnections between chips. The AMT DAP 510, for example, is a 32×32 array with 64 PES per VLSI chip.

Stone (1970) proposed a logic-in-memory computer in which each of the 16 Kbyte segments of the cache memory of an IBM 360/85 was to have special-purpose logic that would process the 16 segments in parallel. This could be used as an associative memory or for simple arithmetic operations on 16 elements in parallel. In a limited sense the proposal was for a 16-PE DAP without the processor interconnections. Kautz (1971) proposed a more extensive logic-in-memory computer called an Augmented Content-Addressed Memory (ACAM) with special circuitry for sorting, matrix inversion, fast Fourier transform, correlation etc. The engineering of the LSI chips was to be by cellular arrays with a universal logic capability, implementing as far as possible in hardware special algorithms that had been devised for parallel computation: for example, fast Fourier transform (Pease 1968) and matrix inversion (Pease 1967).

1.1.5   Orthogonal and associative computers

Another influential paper in the history of parallelism is that of Shooman (1960) entitled 'Parallel computing with vertical data', in which he describes an 'orthogonal computer' organisation (see also Shooman 1970). The arithmetic units of a conventional serial computer take their data serially from memory in the form of words (e.g. 32-bit floating-point numbers) and by 1960 most scientific computers would process the bits of the word in parallel. Such a procedure may be described as word-serial and bit-parallel processing. Shooman recognised that many problems involving information retrieval required searches on only a few bits of each word and that conventional word-serial processing was inefficient. He proposed, therefore, that the memory should also be referenced in the orthogonal direction, i.e. across the words by bit slice. If the bits in memory are thought of as a two-dimensional array, with the bits of the kth word forming the kth horizontal row, then the lth bit slice is the bit sequence formed from the lth bit of each number—that is to say the lth vertical column of the array. In the orthogonal computer, one PE is provided for each word of memory and all the bits of a bit slice can be processed in parallel. This is called bit-serial and word-parallel processing. The orthogonal computer provided a 'horizontal unit' for performing word-serial/bit-parallel operation and a separate 'vertical unit' for bit-serial/word-parallel operation.

The idea of performing tests in parallel on all words has led to the idea of the *associative* or *content-addressable* memory, in which an item is referenced by the fact that part of its contents match a given bit pattern (or mask), rather than by the address of its location in store. In a purely associative memory, there is no facility to address a data item by its position in store. However many systems provide both forms of addressing. The multitude of different processors based on associative memories have been reviewed by Thurber and Wald (1975) and Yau and Fung (1977), and the reader is also referred to the books by Foster (1976) and Thurber (1976) for a more complete treatment. We will describe here only two of these, the OMEN and STARAN, that have been marketed commercially.

The OMEN (Orthogonal Mini EmbedmeNt) series of computers were a commercial implementation of the orthogonal computer concept, manufactured by Sanders Associates for signal processing applications. They are described by Higbie (1972). The OMEN-60 series used a PDP-11 for the conventional horizontal arithmetic unit and an array of 64 PEs for the associative vertical arithmetic unit which operated on byte slices, rather than bit slices. Depending on the model, either bit-serial arithmetic with eight bits of storage were provided with each PE, or alternatively hardware floating-point with eight 16-bit registers and five mask registers. Logic was provided between the PES to reverse the order of the bytes within a slice, or perform a perfect shuffle or barrel shift.

Another computer derived from the orthogonal computer concept was the Goodyear STARAN (Batcher 1979) which was conceived in 1962, completed in 1972 and by 1976 about four had been sold. The STARAN typically comprised four array modules, each 256 one-bit PES and between 64 Kbits and 64 Mbits of total storage, controlled by a sequential PDP-11. Unlike the SOLOMON, however, the storage was not assigned to specific PES; instead, a flexible 'FLIP' network was interposed between the PES and the memory. A slice of 256 bits was selected from memory in a pattern specified, under program control, by a 256-bit code. The pattern selected may, for example, have treated the store as a multidimensional array with a varying number of dimensions, or shuffled the data in the manner required by the fast Fourier transform and other important numerical algorithms. Connections between the PES were achieved by passing the 256-bit slices of data through the FLIP network, thus achieving in minimum time a highly flexible effective interconnection pattern that could be varied from problem to problem by the programmer. The STARAN, like other bit-oriented computers, was most effective when performing logic and short word-length integer arithmetic. A particularly suitable application is the digital processing of pictures, in which the image is divided into millions of pixels (picture elements) each of which is represented by 6–12 bits. The first STARAN was delivered to the Rome Air Force Base for such an application and it has also been proposed for air traffic control.

The STARAN and DAP concepts have been brought together in proposals from Goodyear Aerospace Corporation for a Massively Parallel Processor (MPP), which comprises a 132×128 array of one-bit PES connected two-dimensionally. The machine is designed principally for picture processing of satellite photographs at rates of 10^6 pixels per second. Eight- and twelve-bit integer arithmetic is designed for execution at the rate of a few Gflop/s and 32-bit floating-point arithmetic between 200 and 400 Mflop/s.

1.1.6   Array and attached processors

Another line of computer evolution involving parallelism has been the development of relatively cheap, special-purpose computers for processing arrays of data, which are therefore often referred to as 'array processors'. Note that, in this context, this name does *not* imply that the computers are architecturally arrays of processors; in fact most designs use the pipeline principle. Such computers normally, but not always, require a host, and are therefore also called *attached processors*. The main application is to signal processing and the analysis of seismological data, and the most ubiquitous algorithm, round which the hardware is designed, is the fast Fourier transform (FFT). The first such processors were augmented 16-bit minicomputers, which permitted parallel access to both data and instructions. Bipolar technology was used to enhance the performance to rates of about 5×10^4 operations per second. The next phase of development was the design of special-purpose 'function boxes' for the FFT and similar algorithms. Parallelism was extensively used in these function boxes, which typically comprised multiple program, data and coefficient memories that could be simultaneously accessed, and multiple adders and multipliers interconnected in the way required for the particular algorithm. The function boxes were controlled by microprograms that were installed at the factory. The machines were thus almost totally inflexible, but they were able to provide a factor of 10 improvement in performance to about 0.5 Mop/s.

In 1973 a new generation of array processor emerged in which multiple processing units were interconnected by a limited number of data buses and ran asynchronously under the overall supervision of a central control unit. Processing speeds of around 5 Mop/s were achieved; however the asynchronous operation led to major difficulties of timing and non-reproducibility of conditions in program check-out.

By far the most successful of the array processors is the Floating Point Systems AP-120B which has returned to the principle of synchronous operation (Harte 1979). The company was founded in 1970, the machine launched in 1976 and by 1985 about 5000 systems had been installed. The FPS AP-120B, which may be attached to either minicomputers or main-frame computers as hosts, performs 38-bit floating-point arithmetic in separate pipelined multiplication and addition units, and 16-bit counting and address calculation in an independent integer arithmetic unit. Three memories (for data, tables and program) and two 'scratch pads' of registers are provided with multiple paths between each memory and each arithmetic unit. All units of the machine are controlled in each clock cycle by a 64-bit instruction that gives precisely reproducible conditions, and unlike many earlier machines, allows the computer to be programmed for a variety of uses. Thus the special-purpose line of development has evolved in this machine to a cheap general-purpose processor of arrays. Typically, processing rates of 5–10 Mflop/s may be achieved, at costs of approximately £50 000 (1980 prices). This is highly cost-effective computing and may enable many calculations presently prohibitively expensive for industrial use to become commonplace. In many respects the FPS AP-120B may be considered as the 'poor-man's' CRAY-1, being about one-tenth as fast and one-fiftieth of the price. The overall architecture of both machines is quite similar, both being based on multiple independent pipelined functional units. In 1980 Floating Point Systems announced the FPS-164 Attached Processor, a 64-bit version of the AP-120B, with a main memory expandable to 1.5 million 64-bit words. The peak processing rate is marginally reduced to 11 Mflop/s in this machine (clock period lengthened to 182 ns) but the larger main memory eliminates the need for a host computer and the time lost in the AP-120B waiting for slow transfers to and from the host. Computation rates between 1 and 8 Mflop/s are typical, depending on the problem and the care with which it is programmed. There are no vector instructions, and such operations must be coded as a tight scalar loop. In 1985 an implementation of the FPS 164 in ECL technology was announced. This machine, called the FPS 264, is four to five times faster than the FPS 164, but otherwise the same.

A novel enhancement to the FPS 164 was announced in 1984, the FPS 164/MAX which stands for Matrix Algebra Accelerator. This machine has a standard FPS 164 as a master, and may add up to 15 MAX boards, each of which is equivalent to two additional FPS 164 CPUs. In total there is therefore the equivalent of 31 FPS 164 CPUs or 341 Mflop/s. The FPS 164 and 164/MAX are considered in detail in §2.5.7 and §2.5.8, respectively.

1.1.7   Enter the Japanese with the Fifth Generation

The Japanese shocked the computer world in October 1981 with a vision of computing in the 1990s that they called the Fifth Generation (Feigenbaum and McCorduck 1983). Their ideas were announced to the world at an international conference in Tokyo entitled 'Fifth Generation Computer Systems' (FGCS) (Moto-oka 1982, 1984). The Japanese identified expert and knowledge-based systems (KBS) as the main application area for computing in the 1990s. These systems use a database of rules to represent the knowledge of an expert, and use this to draw conclusions and inferences from facts supplied interactively by a user. The system is intended to behave very much like a consultation between a human expert and someone with a problem requiring the expert's knowledge. The most successful applications to date have been in limited areas of well-defined knowledge such as medical diagnosis in specific areas. Other examples include the machine translation of foreign languages, machine maintenance and economic analysis. Some large companies are concerned with the loss of expertise that arises when a skilled long-service member of staff retires. The capturing of his knowledge within a computer expert system is proposed as a way of partially alleviating this problem.

In order to provide the above capability it is envisaged that an FGCS would comprise three main parts: a knowledge database machine with 100 to 1000 Gbyte of storage to contain and manage the accumulated knowledge; a problem-solving and inference machine to manipulate the knowledge database and respond to questions; and an intelligent interface machine to communicate with the human user in speech and images. The Japanese Ministry of International Trade and Industry (MITI) together with the eight leading electronic manufacturers (including Fujitsu, Hitachi, NEC, Mitsubishi, OKI and Toshiba) are collaborating in a ten-year national project to build such an FGCS by 1992. The project is led by Professor Moto-oka of Tokyo University and has a budget of about 400 million US dollars.

The basic computer requirement of the inference machine can be expressed as the number of logical inferences made per second (LIPS), and it is estimated that a fifth-generation computer would need to perform between 100 MLIPS and 1 GLIPS. Present computers have the capability of between 0.01 and 0.1 MLIPS, so that one is looking for about four orders of magnitude improvement in speed of logical operations over current computers. Such factors are only likely to be achieved by a combination of improved technology, architecture and methods. It is concluded that an increase in device speeds will not alone be sufficient, and that computers will have to become more parallel and use architectures such as *data-flow* which automatically make use of the inherent parallelism in a problem. In such computers an operation takes place whenever the input data for it are available and there is a functional unit free (see §3.2.2 and Treleaven (1979), Treleaven, Brownbridge and Hopkins (1982) and Dennis (1980)). In contrast to computers controlled by instruction streams, which are described as *control-flow* computers, the only limits to the order of execution are those imposed by data dependencies. Subject to these constraints, that can be identified by the hardware, a data-flow computer may perform as many operations in parallel as it has functional units. Of much greater importance is the fact that the programmer is not required to identify the parallelism in his problem nor to program it explicitly into his instruction stream. The use of high levels of parallelism by replication is also felt to be necessary in order to take advantage of the opportunities offered by VLSI mass production.

Another important part of the fifth-generation concept is a vast improvement in the interface between the human user and the computer, which is known as the man–machine interface (MMI). Current computers are weak in speech recognition, textual manipulation, graphical communication and image processing; all of which are important capabilities for a good interface to the human user. Great importance is attached to improvements in these areas, which lend themselves well to parallel computation; indeed, already highly parallel SIMD computers like the ICL DAP and STARAN have proved to be most effective in image processing. As well as being able to communicate via images and sound, it is envisaged that the MMI should also make use of the inference machine to appear to be as 'intelligent' as possible to the user.

The fifth-generation computer project does not address itself to the provision of improved facilities for numerical processing, because this is the concern of a separate project. In 1982 MITI started another national project to develop 'High Speed Computing Systems for Scientific and Technological Use', with the target of producing a 10 Gflop/s supercomputer by 1989 (Kashiwagi 1984). The project is lead by Dr H Kashigawi of the Electro-Technical Laboratory and has a budget of 100 million US dollars. It is in collaboration with the six electronic manufacturers mentioned above in connection with the fifth-generation project. The usual applications to weather prediction, aerodynamics and nuclear energy research were cited to justify the programme. The project involves research into parallel computer architecture and software, new switching devices with high speed and level of integration, and finally the construction and evaluation of the supercomputer. New logical devices with a 10 ps gate delay time (using a Josephson junction operating at liquid helium temperatures, and the high-electron-mobility transistor, HEMT, operating at liquid nitrogen temperatures) and a 30 ps gate delay (using gallium arsenide) are the target, together with new memory devices with 10 ns access time. The level of integration envisaged for logic is 3000 gates per chip and for memory 16 Kbits per chip. In addition to the national project, several company supercomputers appeared in 1984/5. These are all multiple pipelined designs with similarities to the CRAY-1 and CDC CYBER 205. They are the Fujitsu FACOM VP-100/VP-200, the HITACHI S-810 model 10/model 20 and the NEC SX-1/SX-2, with peak performances of 533 Mflop/s, 630 Mflop/s and 1300 Mflop/s respectively. Several experimental machines have been built with other architectures in the universities and national research institutes (processor arrays, MIMD and data-flow computers), and it will be interesting to see what architecture is finally chosen for the national supercomputer.

1.1.8   MIMD computers

It is sometimes said that, spurred on by the advance of VLSI technology and the cheap microprocessor, the 1980s will be the decade of the MIMD computer. Such multiple instruction stream/multiple data stream computers are those controlled by more than one stream of instructions. Here we also limit the term to tightly coupled systems in which the instruction streams can be programmed to cooperate together on the solution of a single problem. Thus we exclude loosely coupled computers connected by slow message passing networks (e.g. ARPANET), and multiprocessor configurations that are hidden from the user such as the four processors in the IBM 3081. We have already met small-scale examples of MIMD computing in the 4-CPU CRAY X-MP and CRAY-2, and the 8-CPU ETA[10]. However, MIMD computers are usually thought of as larger collections of computers (most frequently minis or micros) and memory modules, connected together by a switch, or arranged in a network. Logically, the same effect can be obtained by passing multiple instruction streams through a single instruction processing unit, as is done in a single-PEM Denelcor HEP computer. Both of these are examples of

MIMD computing, because both have multiple instruction streams. There are now so many proposed MIMD architectures that the inclusion of a complete list here, even if possible, would be inappropriate as well as boring. Rather few have actually been built and successfully operated, and this has guided our choice of those to describe below. A more complete survey is given in Hockney (1985b, d, 1986).

(i) Carnegie–Mellon C.mmp and Cm*

One of the most ambitious early examples of MIMD computing was the C.mmp computer at Carnegie–Mellon University (Wulf and Bell 1972, Mashburn 1982). This comprised 16 DEC PDP-11 minicomputers connected to 16 memory modules by a 16×16 crossbar switch. The switch provided a direct electrical connection between every computer and every memory module (see figure 3.8). The computers communicated by sharing a 32 Mbyte logical address space, of which 2.7 Mbytes was implemented by the memory modules. All computers are also connected to a common interrupt bus. The design started in 1971 and the machine was completed in 1975. It remained operational until 1980. Its successor, the Cm*, was an entirely different concept, using the microprocessors that had now become available (Swan, Fuller and Siewiorek 1977, Fuller et al 1978). Unlike the C.mmp, the memory in Cm* is subdivided amongst the microprocessors, and made local to them; and a hierarchical packet-switching network provides communication between the microprocessors. The primary unit is the 'computer module' which comprises a DEC LSI-11 microprocessor with 64 Kbytes of dynamic MOS memory, and perhaps other peripherals, attached to its LSI-11 bus. This module may act as an independent computer; however, up to 14 may be linked to a common intracluster bus to form a tightly coupled cluster within which data transfer is by direct memory access. The system is built up by linking clusters together via two intercluster buses to form a loosely coupled network exchanging data by packet-switching techniques. Both the intra- and intercluster buses are controlled by microprogrammed computers. The tightness of the coupling between the computer modules is in inverse proportion to the time to transfer data between them, and this varies by a factor of ten depending on the relative positions of the modules. Transfers within the local memory of the same module take 3 μs, those between different modules of the same cluster take 9 μs, and those between different clusters take 26 μs. All component computers of Cm* share a common virtual address space of 256 Mbytes, and the design is expandable to an arbitrary extent by adding more clusters. It is hoped to take advantage of the latter feature in a future VLSI implementation of the architecture containing many hundreds or even thousands of computers. The detailed hardware design of Cm* began in 1975, and a single cluster of ten computers was operational in 1977. A five-cluster Cm* containing 50 computer modules is now operational, and this has been proposed as a flexible test-bed for simulating other proposed MIMD computer designs. Both C.mmp and Cm* are described by Satyanarayanan (1980), and operational experience is reported by Jones and Schwarz (1980).

(ii) UK experiments

The many small experimental university MIMD systems developed in the late 1970s and early 1980s are too numerous to enumerate in full. However, the following systems are typical. Shared-memory MIMD systems have been built at the University of Loughborough under Professor Evans, and used extensively for the development of parallel MIMD algorithms (Barlow, Evans and Shanehchi 1982). The first system used two Interdata 70 minicomputers sharing 32 Kbytes of their address space. Later, in 1982, four TI 990/10 microcomputers sharing memory were commissioned as an asynchronous parallel processor (Barlow et al 1981). A larger MIMD system, called CYBA-M, has been built at the University of Manchester Institute of Science and Technology (UMIST) under Professor Aspinall, and consists of 16 Intel 8080 micro-processors sharing a multiport memory through part of their address space (Aspinall 1977, 1984, Dagless 1977). Other interesting UK projects are the Manchester Data-Flow computer, the Imperial College ALICE reduction machine and the University of Southampton RPA and Supernode projects. The last two are discussed in Chapter 3.

(iii) New York University Ultracomputer and IBM RP3

Most of the above UK computers have not been well publicised but were built and operated successfully for several years. A much more ambitious university project was the NYU Ultracomputer which was extensively publicised but remained for many years a paper machine. However, the publications have been influential in the theory and design of large-scale MIMD systems. In response to the potential of VLSI replication, the original concept of Schwartz (1980) defined a family of computer architectures, called ultra-computers, which linked together thousands of processing elements in a 'perfect-shuffle nearest-neighbour (PSNN)' connection network (see figure 3.21). This network requires each PE to be connected to four others at the most, and it was shown that it was suitable for implementing the many parallel algorithms that are based on 'divide and conquer' strategies. A 16K PE ultracomputer with these connections can be used as a 128×128 2D array, or as a $32 \times 32 \times 16$ 3D array, or as a $16 \times 16 \times 8 \times 8$ 4D hyper-array. The memory was to be distributed throughout the network, with 132K 16-bit words being associated as local memory with each PE. Separate data-routing chips make the PSNN connections between PES in groups of eight. In a later design for the ultracomputer (Gottlieb *et al* 1983), the memory was removed from the processing elements and concentrated in memory modules (MMS). The PES are then connected to the MMS by a multistage switch in the form of an Omega network (Lawrie 1975). If there are N PES and N MMS then there are log2N stages in the switch. A design for a 4096-PE machine was considered for implementation in the 1990s. Initially it is intended to build a 64-PE prototype from commercially available microprocessors and memories, and custom-made switching elements. IBM is building a 512-PE version of the Ultracomputer design at their Yorktown Heights Research Laboratory, which is called the Research Parallel Processor Project (RP3). This comprises four quadrants of 64 PES each. The first quadrant is scheduled for service in 1988.

A unique feature of the Ultracomputer design is the provision of a 'fetch-and-add' operation (FA), which is partly implemented by the nodes of the switch. If many PES wish to add data into the same main memory location, they may all independently issue FA instructions to that location, perhaps even at the same time. The switch hardware combines these requests as they meet at nodes in the switch, and finally the memory is incremented as if all the requests had been issued in an arbitrary order. Thus the necessary sequence of add operations is maintained even if the order, which is not important, is not known. Fetch-and-add operations are of fundamental importance in many MIMD algorithms: e.g. for counting down a synchronisation variable common to many processes, and for charge accumulation in a particle-pushing plasma simulation. The provision of special hardware for this operation in the Ultracomputer is an important innovation which will significantly speed up the operation of many codes. If every PE issues an FA on the same variable, all results are obtained in the time required for just one FA.

(iv) Illinois Cedar project

The Cedar project that is being built at the University of Illinois by David Kuck and his team is aimed towards achieving a multi-gigaflop/s machine by 1990 (Gajski *et al* 1983). The overall architecture is broadly similar to that of the Ultracomputer, but Cedar does not have the fetch-and-add instruction. Sixteen clusters of eight PES are connected via an extended Omega global switching network to 256 global memory modules of 4 to 16 Mword each. The time between a PE data request to global memory and the data being available at the PE is about 2 μs. Each cluster has eight PES, each with 16 Kwords of local memory. These PES are pipelined, and are interconnected via a local switching network. All memory locations have a full–empty flag (like the Denelcor HEP) to allow synchronisation of the PES. The prototype Cedar 32 was planned to have four clusters of eight PES and use a 400 ns clock period. This provides 2.5 Mflop/s per PE in a low-speed technology, giving a total maximum performance of 80 Mflop/s (comparable to the CRAY-1) for the desk-top-sized prototype. This is planned to be expanded to a 16-cluster Cedar 128 (320 Mflop/s, in 1988) and to a 64-cluster Cedar 512 (1.2 Gflop/s, in 1990). An alternative engineering is planned using a 40 ns clock period and 25 Mflop/s per PE, giving a four-cluster Cedar 32H (800 Mflop/s, in 1989) and a 16-cluster Cedar 128H (3.2 Gflop/s, in 1991). Funding began in 1983 and industrial participation is expected and required in order to produce the larger machines. In fact, the above plans have not been followed exactly, because in 1985 it was decided to use the commercially available Alliant FX/8 as the Cedar cluster. This machine is described in §1.1.9.

(v) Erlangen EGPA

Perhaps the most original and imaginative MIMD architecture was developed under Professor Händler at the University of Erlangen, West Germany, and is called the Erlangen General Purpose Array or EGPA (Händler, Hofmann and Schneider 1975, Händler 1984). The connections between the computers in EGPA are topologically similar to a pyramid, with computers at the corners and connections along the edges. The control C-computer, at the top of the pyramid, controls four B-computers at the corners of its base. The four B-computers also have direct connections along the edges of the base. This five-computer system has been working since 1981, and uses AEG 80-60 minicomputers each with 512 Kbytes of memory. The idea is expandable to further levels by making each B-computer itself the top of a further pyramid with four A-computers at its base. There are then 16 A-computers in total, which are connected amongst themselves at the lowest level as a 4×4 array with nearest-neighbour connections, as in the ICL DAP. The advantage of the EGPA pyramidal topology is that short-range communication can take place most effectively along the nearest-neighbour connections, whereas long-range communication can take place most effectively by sending the data higher up the pyramid. For example, if the bottom computers form an $n \times n$ array, then the furthest computers can communicate in $2\log_2 n$ steps via the top of the whole pyramid, compared with $2n$ steps if the nearest-neighbour connections alone were provided. The five-computer system has been used for contour and picture processing, and has proved to be about three times faster than a single computer of the same type (Herzog 1984). If the hierarchical development is taken one stage further by making each A-computer the top of a further pyramid, there will be 1, 4, 16 and 64 computers respectively at the four levels, making a total of 85 computers. This is the basis of the Erlangen Multiprocessor 85 project (Händler *et al* 1985).

(vi) Livermore S-1

By far the largest MIMD project is the Livermore S-1 computer (Widdoes and Correll 1979, Farmwald 1984). Sponsored by the US Navy and Department of Energy, and supported by Edward Teller, the complete design for the S-1 computer comprises 16 CRAY-1-class pipelined vector computers connected to 16 memory banks by a full crossbar switch which provides a direct logical connection between each computer and each memory bank. The S-1 can therefore be regarded as a 'grown-up' version of the C.mmp. An overall performance equivalent to about ten CRAY-1 computers is expected, that is to say close to 1 Gflop/s. Each of the component computers (called uniprocessors) is provided with a data cache (64 Kbytes) and instruction cache (16 Kbytes) in order to limit traffic through the switch. All the computers are also directly connected to a common bus (the synchronisation box) in order to facilitate the transfer of small amounts of data and the synchronisation of messages rapidly between them. Each memory module may contain up to 1 Gbyte (2^{30}) of storage, giving a total physical storage of 16 Gbyte (9-bit bytes). The programmer, however, sees a uniform virtual memory of up to 2 Gbyte. The hardware paging mechanism translates virtual addresses to physical addresses without the need for programmer intervention. The normal arithmetic mode is 36-bit floating-point, but 18-bit and 72-bit numbers are also included.

Single instructions are provided for some common mathematical functions (e.g. sine, exponential, log) which take about two multiply times. In addition to the usual element-by-element vector instructions, matrix transpose, matrix multiply, and fast Fourier transform may be performed by single instructions. There are no vector registers, and all vector operations are to and from main memory assuming contiguous storage of successive elements, as in the CYBER 205. Also, there is no hardware scatter/gather instruction to assist the assembly of contiguous vectors from disordered data, although the matrix transpose instruction can be used if the data are ordered. The S-1 project provides for a continual upgrading of the technology within a constant architectural design. A Mark I uniprocessor using ECL 10K MSI was commissioned in 1977, followed by the Mark IIA using ECL 100K in 1983. The Mark V version calls for a supercomputer on a wafer.

(vii) MIDAS

The Modular Interactive Data Analysis System (MIDAS) at the University of California, Berkeley (Maples *et al* 1981), is a hierarchical system consisting of a primary computer controlling several secondary computers, each of which controls a multiple processor array (MPA). The primary computer handles system control and user communication, and allocates resources to each job. The secondary computers handle individual problems, compute the sequential parts and allocate the parallel parts for parallel execution within the MPA. At the lowest level the MPAs are clusters of eight PES (Modcomp 7860s) which form the computational heart of the system. Each MPA also has an input and an output processor and a crossbar switch connecting the processors to 16 switchable memory modules, each of 256 Kbytes. The allocation of a memory module to a PE can be changed in 50 ns, after which the memory module may be directly addressed by the PE. The MPA is therefore a crossbar-switched shared-memory MIMD computer. A prototype with four PES and eight memory modules has been operational since January 1982, and a complete subsystem with a primary computer, one secondary computer and one MPA has been operational since February 1983. The Modcomp 7860 is rated at approximately 85% of a VAX 11/780, which is equivalent to 0.3 Mflop/s, giving a maximum performance of approximately 3 Mflop/s (equivalent to about a CDC 7600) for the one-MPA system. This has been used to solve a variety of problems in computational physics, particularly nuclear science.

(viii) NASA Finite Element Machine

The NASA Finite Element Machine (FEM) (Jordan 1978) is a two-dimensional MIMD array of PES controlled by a TI990 minicomputer. The processing elements are TI9900 microcomputers with an AM9512 floating-point arithmetic coprocessor and 32 Kbytes RAM plus 16 Kbytes ROM of memory. One-bit data paths connect each PE to its four nearest neighbours and its four second nearest neighbours (i.e. both the orthogonal and the diagonal connections are present in the two-dimensional lattice). All PES are also connected to a 16-bit global bus that can sense the state of eight flag registers in each PE. 'All', 'any' and other logical functions can be derived from the state of these PE flags, and used to synchronise the operation of the array. The design of the machine describes a 6 × 6 array of 36 PES. Initially a 4 × 2 array was built and became operational in 1983. A second 4 × 2 array was added to this in 1984 to make a 4 × 4 array. The initial eight-PE array was extensively used and much early MIMD experience was obtained on it.

(ix) Hypercubes (Cosmic Cube, Intel iPSC, FPS T-Series)

Binary hypercubes have been extensively studied as possible interconnection networks for MIMD computers since the important paper by Pease (1977) entitled 'The indirect binary n-cube microprocessor array'. However, they were not realised in a practical way until the Cosmic Cube was built at Cal Tech by Geoffrey Fox and Charles Seitz in 1984. Following the publicity received by this machine, several commercial versions have appeared, and this field of development is growing rapidly.

The point, line, square and cube with nodes at their corners and connections along their edges, are respectively the zeroth-, first-, second-, and third-order hypercube networks. Each one is made by taking two copies of the next lower-order hypercube and joining corresponding corners (i.e. nodes). For example, the fourth-order hypercube is made from two cubes by connecting corresponding corners. In this sequence the dth-order hypercube has $n = 2^d$ nodes, and $d = \log_2 n$ connections to each node. Thus one attraction of this connection scheme is that the number of connections grows relatively slowly with the size of the hypercube. Another attraction is that the dth-order hypercube has, obviously, as subsets the connections of all lower-order hypercubes, as well as all the connections needed for a fast Fourier transform (FFT) of n data. That is to say, if the n data are assigned one value to each node of a dth-order hypercube, then the data combined at any stage of the FFT algorithm are always in adjacent nodes. We now describe the architecture of some hypercube computers in more detail.

(x) Cosmic Cube

The Cosmic Cube at the California Institute of Technology (Seitz 1985) and its commercial derivative, the Intel iPSC (personal supercomputer), are both hypercube networks of microprocessors. The Cosmic Cube is a 2^6 hypercube hosted by a VAX 11/780. Each node of the cube comprises an Intel 8086 with an Intel 8087 floating-point coprocessor, and 128 Kbytes RAM plus 8 Kbytes ROM of local memory. Communication between PEs is by queued message passing along the edges of the hypercube at a rate of 2 Mbit/s, each PE being connected to six neighbours. There are independent asynchronous communication channels in the send and receive directions (full duplex), each with queues for a 64-bit message packet. The communication latency between nodes is admitted to be large compared to the instruction time, but is said to be comparable to the overhead time taken by a node to deal with a 64-bit message packet. This makes the machine a more accurate simulator of future machines with single-chip nodes. The machine can be loosely described as the equivalent of 64 IBM PCs collaborating on a problem.

The 8087 has a maximum performance of 50 Kflop/s, hence the 64-processor Cosmic Cube has a maximum of about 3 Mflop/s. This is described as approximately ten times a VAX 11/780 for about the same cost. However, it is only 1/10 to 1/100 the performance of the large supercomputers. The machine has been working since October 1983 and a number of computational physics problems have been reformulated and successfully computed on it. An extended 2^{10} node hypercube, called the Homogeneous Machine, is planned using at each node the faster Intel 80286 plus 80287 plus 80186 chips and 256 Kbytes of local memory (using 64 Kbit chips), expandable to 1 Mbyte with 256 Kbit chips. The maximum performance of the 2^{10} hypercube is estimated to be about 100 Mflop/s, that is to say, about the same as the large supercomputers (CRAY X-MP and CYBER 205).

(xi) Intel iPSC and iPSC-VX, and N-Cube

In 1985 Intel announced the iPSC hypercubes in sizes 2^5, 2^6 and 2^7 rated at 2, 4 and 8 Mflop/s. Each processing node contains an Intel 80286 micro-processor with an 80287 arithmetic coprocessor rated at $1/16$ Mflop/s. This gives 512 Kbytes of RAM and 64 Kbytes EPROM. The 32-node hypercube occupies about 2 ft × 2 ft × 4 ft and is controlled by an Intel 310 micro-computer. Another company, N-Cube, is marketing hypercubes based on specially designed 32-bit VLSI nodal chips with a peak performance of 0.3 Mflop/s and 128 Kbytes of local memory. Their first produce N-Cube/10 can be expanded to a 10-cube of 1024 nodes.

In 1986 Intel announced the iPSC-VX in which a pipelined vector processor board is attached to each node, increasing the peak theoretical vector performance to 20 Mflop/s per node in 32-bit mode and 6.7 Mflop/s in 64-bit mode. However, since two boards are now required per node, the maximum number of nodes is reduced to 64. As there has been no accompanying increase in the rate of internode communication, the performance of this machine is likely to be determined primarily by the time spent in communication rather than the arithmetic (i.e. the value of $f_{1/2}$ will be very high, see §1.3.6). Because of this, very careful problem formulation and programming will be required to realise problem performances anywhere near the peak rates.

(xii) FPS T-Series

Also in 1986, Floating Point Systems announced the T-Series, or Tesseract, which is a hypercube theoretically expandable to $2^{14} = 16\,384$ nodes. However, the first machines delivered in 1986 to Michigan Tech University, the University of Grenoble and the UK SERC Daresbury Laboratory were 16- or 32-node machines. Each node has an INMOS transputer (see §3.5.5) to handle communication and 1 Mbyte of dual port video DRAM as memory. The DRAM can feed a 256-element contiguous vector of 32-bit numbers in parallel to one of four vector registers which themselves feed two WEITEK 64-bit pipelined floating-point arithmetic chips for multiplication and addition respectively. With a 125 ns clock, the theoretical peak performance is

16 Mflop/s per node. However, this rate assumes that the vectors are contiguously stored in the DRAM and that both the multiplier and adder are being used simultaneously. The INMOS transputer has the job of rearranging data into contiguous form, which it may do in parallel with the operation .of the vector pipelines. However, the ratio of the time to perform an arithmetic operation to the time to rearrange data in the DRAM to the time to obtain data from a neighbouring node is 1:26:256. Again we find that communication between nodes is at least two orders of magnitude slower than the arithmetic, and the caveats applied above to the performance of the Intel iPSC, clearly also apply to the T-series.

(xiii) Denelcor HEP—pipelined MIMD
It is interesting, and perhaps significant, that the first MIMD computer to be marketed commercially, the Denelcor HEP (Smith 1978), is quite dissimilar to any of the computers that we have described above. Rather than having a number of distinct instruction processing units, each with its own instruction stream, in HEP we find many instruction streams sharing a single pipelined instruction processing unit. The overall architecture of a full-size heterogeneous element processor (HEP) can be described as 16 process execution modules (PEMS) connected to 128 data memory modules (DMMS) via a packet-switching network. Each PEM contains 1 Mwords of program memory, 2 Kwords of register memory and 4 Kwords of constant memory. The creation of instruction streams is under program control: the machine instruction (or FORTRAN command) CREATE initiates a new instruction stream, and up to 50 user instruction streams can be created. Instructions are taken in turn from each stream into the top of an eight-stage instruction pipeline. When the pipeline is full an instruction is completed every clock period of 100 ns, giving a maximum performance of 10 Mips per PEM, or 160 Mips for a complete HEP. If we assume that about five instructions are issued on average for every floating-point operation, these rates correspond approximately to 2 Mflop/s per PEM, and 32 Mflop/s per HEP. Higher processing rates of 5 to 6 Mflop/s per PEM can be obtained if the average number of instructions per floating-point operation can be reduced. This is possible for certain matrix computations if the PEM registers are programmed to act as vector registers for intermediate results. This limits the number of references to data memory through the switch (Sorensen 1984, Dongarra and Sorensen 1985).

The method of implementing MIMD computing within the PEM of a HEP is much more flexible than that used in designs based on a fixed number of microcomputers. The number of instruction streams can be changed from one problem to the next by appropriate programming, thus the number of streams can be chosen to suit the problem being solved—it is no longer a number fixed by the hardware that may not suit the problem or algorithm being implemented. Each DMM may contain up to 1M 64-bit words, giving a maximum capacity of 1 Gbyte for a complete system. The switch is a multilevel packet-switching network with a 50 ns propagation time between nodes. A typical time to get data to a PEM from memory via the switch is 2 μs. Because of its interesting architecture we have chosen the Denelcor HEP for detailed study in §3.4

The development of the HEP was sponsored by the US Army, and culminated in the delivery of a 4 PEM × 4 DMM system to the Ballistics Research Laboratories, Aberdeen, Maryland in 1982. This was especially fitting because BRL had also sponsored and taken delivery of ENIAC in 1946, which can be considered as the first MIMD computer. Single PEM/DMM systems have been delivered to the University of Georgia Research Foundation (1982), Messerschmidt Research Munich (1983), and Los Alamos Research Laboratories (1983). The original HEP as described above, or HEP-1, was built with conservative ECL 10K technology, because it was not desired to pioneer a revolutionary architecture at the same time as a new technology. However, the HEP-2, which was announced in 1983 for delivery in 1986, was to have used upgraded ECL VLSI technology with a switching time of 300 ps and 2500 gates per chip. The clock period would have been 20 ns and there would have been a combined multiply–add pipeline, with a peak rate of 100 Mflop/s per PEM. It was to be based on the HEP-1 MIMD architecture and designed to have a performance ranging from 250 Mips to 12 000 Mips, corresponding roughly to 50 Mflop/s and 2.4 Gflop/s. Regrettably, these plans did not come to fruition because the company went out of business in 1985 due to financial problems. However, the design of the HEP is so novel and interesting that we describe it in more detail in §3.4.4.

(xiv) CDC CyberPlus—a ring system
The CyberPlus, from Control Data, is derived from the Advanced Flexible Processor that was built for military applications such as the rapid analysis of photographs taken from aircraft. The architecture of CyberPlus is unusual and interesting because it is based on communication via a multiple-ring topology.

The CDC CyberPlus multi-parallel processor comprises from 1 to 16 CyberPlus processors connected in a ring to a channel of a host CDC Cyber 170/800. It is therefore a ring network MIMD computer in our classification (see §1.2.6). Up to four such rings can be attached to the host. The CyberPlus processor itself has 256 or 512 Kwords of 64-bit memory (80 ns access) for floating-point data, 16 Kwords of 16-bit memory (20 ns access) for integers, and a program memory of 4 Kwords of 240-bit instructions. There are fifteen independent functional units for integer, boolean and read/write operations and an option of three functional units for floating-point add, multiply and divide/square root may be added. The floating-point capability is rated at 65 Mflop/s in 64-bit mode and 103 Mflop/s in 32-bit mode. The clock period is 20 ns and the long 240-bit instruction specifies the destination of the output from each functional unit at every cycle of the machine.

The CyberPlus processors form stations on a ring, and communication between processors is achieved by sending information packets to the ring. The packets move round the ring at the rate of one station per clock period, until their destination is reached. A packet may be read from and written to the ring by every processor every clock period. The time to communicate between processors therefore depends on their relative separation on the ring. A second ring connects the CyberPlus processors to the Cyber 170 host and its memory.

(xv) Bus-connected systems (ELX SI 6400, FPS 5000, Sequent Balance)

By far the most common commercial MIMD computer systems are based on attaching multiple computing elements (CEs) and multiple memory modules to a common bus. We describe some of these below. Other computers with a similar architecture are the FLEX/32 (20/cabinet), the Encore Multimax (20) and the Culler PSC (2), where we have given the maximum number of CEs in parentheses.

(xvi) ELXSI 6400

The ELXSI 6400 is an example of a MIMD system in which the processors and memory are connected by a fast shared bus. In ELXSI (Taylor 1983) there may be from one to ten CPUs and one to four I/O processors accessing from one to six memory systems (4 to 192 Mbytes) via a common high-speed databus, known as the Gigabus. The 64-bit CPUs deliver 4 Mips, so that the maximum system could potentially achieve 40 Mips. The Gigabus is 64-bits wide and has a cycle time of 25 ns, giving a maximum transfer rate of 320 Mbyte/s of addresses and data. Usable data rates of 160 to 220 Mbyte/s are claimed. The memory cycle time is 400 ns (for two 64-bit reads and one write) and the floating-point multiply time is 300 ns (64-bit). Also taking into account the bus cycles required, the total time to complete a floating-point multiply on one CPU is about 800 ns, giving a performance of 1.2 Mflop/s. Actually the Livermore kernels are executed at between 0.3 and 1.4 Mflop/s on a one-CPU system. By overlapping memory access with arithmetic, it appears that one memory system can support two CPUs without interference, and therefore six memory systems are more than adequate to support the full ten-CPU system, giving a peak performance of 12 Mflop/s. An ELXSI 6400 has been ordered by the NASA Ames Dryden Data Analysis Facility.

(xvii) FPS-5000

The Floating Point Systems 5000 series of computers is, like ELXSI, a bus-connected shared-memory MIMD system (Cannon 1983). An 8 or 12 Mflop/s control processor and up to three 18 Mflop/s XP32 arithmetic coprocessors are connected via a 6 Mword/s bus to a system-common memory of up to 1 Mword. The control processor is either an FPS AP-120B or FPS-100 'array processor', but the XP32 is a new design, using the WEITEK 32-bit floating-point multiplier and adder VLSI chips. The latter are eight-stage pipelines operating on a 167 ns clock period, giving a peak performance of 6 Mflop/s per chip. One multiplier chip feeds its result into two adder chips in a manner suited to the computation of the fast Fourier transform, giving a peak performance of 18 Mflop/s per coprocessor, and 62 Mflop/s for the maximum system. Independent programs can be executed in the control processor and each of the coprocessors. The FPS-5000 is described in more detail in §2.5.10.

(xviii) Sequent Balance 8000, 21000

Designed as a superminicomputer with about six times the performance of a DEC VAX 11/780, the Sequent Balance is a 32-bit bus-connected shared-memory architecture. Up to 12 PES (model 8000) or 24 PES (model 21 000) and up to 28 Mbyte of shared common memory are attached to a 52-bit wide pipelined packet bus with a sustained bandwidth of 26.7 Mbyte/s. There is a virtual memory address space of 16 Mbyte for each user process. Each PE is an NS 32032 processor operated at 10 MHz with a floating-point unit, memory management unit, and 8 Kbyte of cache memory. Two PES are mounted per board and connected to the bus by a custom IC chip which manages interprocessor communication, synchronisation and interrupts. First shipments of the Sequent Balance began in 1984 and by the end of 1985 some 80 systems had been manufactured. The computer is popular as an economic vehicle for investigating parallel algorithms for use on shared-memory systems, and is used for this purpose, for example, in the Department of Computer Studies at Loughborough University of Technology.

(xix) IBM /CAP and Cornell Supercomputer Center

IBM's love–hate relationship with large-scale scientific supercomputing, which has already been commented upon, took a positive turn in the early 1980s under the leadership of Dr Enrico Clementi who had been appointed an IBM Fellow and allowed to pursue his own interests in computational chemistry with IBM's support. Finding disappointing computer facilities for such work in existing IBM products, Clementi conceived the idea of attaching as many as ten Floating Point Systems 164 computers to channels on an IBM host machine. The computers form a star network centred on the host, and there is no direct connection between the ten FPS 164 computers. Although the channel transfer rate is slow compared with computation, Clementi and his colleagues found that many scientific problems could be broken down into subtasks that did not need to communicate too frequently over the slow channels. Computing rates comparable to the CRAY-1 are claimed for this *loosely coupled array of processors* (called /CAP) at very economic cost. The /CAP is described in more detail in §2.5.9.

The first /CAP system installed at IBM Kingston, and a second was installed in 1985/6 at the IBM European Center for Scientific and Engineering Computing (ECSEC), Rome, for use by European scientists. A rather similar facility is also being set up at Cornell University by the US National Science Foundation as one of four advanced scientific computing centres. This is in association with the Cornell Center for Theory and Simulation in Science and Engineering, headed by Kenneth G Wilson, the 1982 Nobel Laureate in Physics.

The initial configuration which uses ten FPS 164 computers to achieve CRAY-1-like speeds is, perhaps, not very exciting, but experience using it has demonstrated how to break down large problems onto this type of loosely coupled MIMD system. There are, however, three impending developments that could transform this picture. First, a high-speed bus is under development that will allow the FPS 164s to communicate directly with each other, without having to use the slow channels to the host; secondly, the FPS 164s can be replaced by FPS 264s with a four- to five-fold increase in speed and no program changes; thirdly, up to 15 FPS MAX boards can be added to each FPS, giving a theoretical peak computation rate of 341 Mflop/s for each FPS, and 3.4 Gflop/s for the complete /CAP configuration.

(xx) BBN Butterfly

This computer differs from the other MIMD systems described in this section in that it is a distributed-memory switched system (see §1.2.6 and figure 1.9). That is to say, all the memory of the system is distributed as local memory to the multiple CEs which are then connected to each other by a multistage packet switch. However, the distributed nature of the memory is hidden from the programmer who may write his programs as though all the memory was shared by all the CEs. The only difference between accessing data in the CE's local memory or the local memory of another CE is the time for the messages to pass through the switch. This is stated to be 4 μs which is a very long time in computer terms. The topology of the switch is a Banyan network

(Feng 1981) which is similar to that required to bring data together in the $\log_2 n$ stages of a fast Fourier transform on n data (the so-called butterfly operation, see also §3.2.2 and §5.5.1). Hence the name butterfly switch.

The first model of the BBN Butterfly (1985) contained 128 CEs out of a design maximum of 256, each based on the Motorola 68000 with 1 Mbyte of local memory. Later enhancements include hardware floating-point using the Motorola 68020 with 68881 coprocessor, and a local memory of 4 Mbyte. On CE with its memory is contained on a board. The original 128-CE Butterfly computer has achieved 26 Mflop/s on the multiplication of two 400×400 matrices, and 3 Mflop/s on the solution of 1200 linear equations by Gaussian elimination.

(xxi) The INMOS Transputer

An important stage in the development of MIMD networks of computers was the announcement of the INMOS Transputer in 1985 (INMOS 1985). This is a family of VLSI components specifically designed for supporting concurrent operation and the associated OCCAM parallel programming language. For example, the T800 transputer provides, on a single 1×1 cm^2 chip, a 32-bit microprocessor with a floating-point coprocessor, 4 Kbits of on-chip memory, and four 1-bit serial links for connection to other transputers in a network. A number of MIMD computer systems based on the transputer are now under construction, and clearly the transputer will have a significant impact on the development of MIMD networks. For this reason the transputer is described in detail in Chapter 3 (§3.5.5), and the OCCAM language in Chapter 4 (§4.4.2).

(xxii) The WEITEK floating-point VLSI chips

Another VLSI development which has had a significant influence on the implementation, and particularly the performance of MIMD computer systems has been the availability of off-the-shelf VLSI chips for pipelined floating-point arithmetic. Prior to 1984 a fast 64-bit floating-point adder requiring about 13 000 gates was implemented in medium-scale integration (MSI, with 10 to 100 gates per chip). It needed about 2000 chips and occupied seven large 16 inch $\times$ 22 inch boards (Charlesworth and Gustafson 1986). In mid-1984, however, WEITEK Inc. of Sunnyvale, California, successfully made the transition to very large-scale integration (VLSI, with more than 1000 gates per chip), and offered the capability of fast pipelined 64-bit floating-point arithmetic in a set of nine VLSI chips (Ware et al 1984). By mid-1985 other manufacturers (e.g. Analog Devices) were offering similar products. The WEITEK 64-bit chips include a seven-stage floating-point multiplier and a six-stage floating-point adder, clocked at 125 ns. The theoretical peak performance is therefore 16 Mflop/s from just two VLSI chips. It is therefore possible for any designer to obtain a theoretical peak performance of 16 Mflop/s per CE (or 1 Gflop/s for a 64-CE system) by adding these off-the-shelf chips to every CE of a MIMD system. Of course, the fraction of this theoretical peak that is actually achieved may be very small indeed unless adequate communication bandwidth is also provided to feed data to the floating-point chips, which is rarely the case. Nevertheless, the arithmetic potential of MIMD systems has been revolutionised by these components which are found at the heart of many systems (e.g. Alliant FX/8, FPS T-Series, FPS 5000, FPS-164/MAX).

1.1.9 Minisupercomputers (Convex C-1, SCS-40)

Sometimes called 'affordable supercomputers', the minisupercomputer might be defined as a 64-bit floating-point vector computer with 1/8 to 1/4 the performance of a supercomputer but at about 1/10 the price, i.e. at the price of a minicomputer. For example, the Convex C-1, announced in 1984, has a peak performance of 20 Mflop/s (64-bit) or 40 flop/s (32-bit) for a price of $495 000 (Linback 1984). (The performance of this and other mini-supercomputers is compared with mini- and supercomputers in table 1.1.) The appearance of this computer is another aspect of the computer revolution introduced by VLSI chips with 10 000 or more gates per chip. This technology not only allows all the logic of a simple serial computer to be contained on a single chip, producing the microprocessor, it also allows all the logic of a sophisticated pipelined vector computer to be placed on a few hundred chips which in turn can be mounted on a few boards. As an extreme example, the ETA Corporation have shown that the logic of a two-pipe CYBER 205 can be placed on one 19 inch × 24 inch board using CMOS VLSI chips containing 20 000 logic gates per chip (see §2.3.7).

(i) Convex C-1

The Convex C-1 uses 8000 gates per chip and packages the machine with 128 Mbyte of memory in a single 5-foot high 19 inch rack. A second rack contains a tape drive and disc storage. With a total power dissipation of 3.2 kW, fan cooling is adequate, and an office environment, as used for the typical minicomputer, is satisfactory. In contrast, the CRAY X-MP super-computer uses 16-gate per chip integration, generates about 200 kW, and requires a separate freon refrigeration plant for cooling; hence one reason for the difference in costs. The relative computation speeds should be roughly inversely proportional to the clock periods which are 12.5 ns for the CRAY-1 and 100 ns for the C-1, giving a ratio of eight, much as is observed.

The performance on a real problem can be judged and set in context by the LINPACK benchmark (Dongarra 1986) in which a set of 300 linear equations is solved in FORTRAN using matrix–vector techniques (see table 1.1). The relative performances are 66 Mflop/s for the CRAY-1S, 8.7 Mflop/s for the Convex C-1 and 0.1 Mflop/s for the DEC VAX 11/780 with floating-point accelerator (a much used and typical 1984 minicomputer). Hand optimisation of critical coding improves the performance of the C-1 to 14 Mflop/s. On the other hand a 1984 supercomputer, the CRAY X-MP/4

Table 1.1 Some benchmark comparisons between minisupercomputers. The average performance is given in Mflop/s for the stated problem in 64-bit floating-point arithmetic (except where stated). Comparable figures for top-of-the-range super-computers are given in table 2.7. (Data from Dongarra 1985, Dongarra and Sorensen 1985, 1987.)

	MINI	MINISUPER			SUPER
Problem	VAX[b] 11/780 FPA	Alliant FX/(p)[d]	Convex C-1	SCS 40	CRAY[d] X-MP/(p)
Theoretical peak performance		5.9 (1) 47 (8)	20 40 (32-bit)	44	210 (1) 420 (2) 840 (4)
$(r_\infty, n_{1/2})$ FORTRAN §1.3.3		(0.9, 151) (1) (1.1, 23) (8)			$(70, 53)(1)^f$ $(130, 5700)(2)^f$
Livermore 3 inner product					96 (1)
Livermore 6 tridiagonal					8 (1)
Livermore 14 particle pusher					7 (1)
LINPACK[a] FORTRAN $n = 100$		1.3 (1) 2.5 (8) 6.2 (8)[e]	2.9	7.3	24 (1)
Assembler inner-loop $n = 100$		1.7 (1) 2.6 (8) 8.5 (8)[e]	3.2		44 (1)
Matrix–vector best assembler 0.11[c] $n = 300$		7.3 (8)[c] 14 (8)	8.7[c] 14	26	171 (1) 257 (2) 480 (4)

Notes

[a] Solution of n linear equations (Dongarra *et al* 1979).

[b] Typical 1984 32-bit minicomputer for comparison.

[c] All FORTRAN.

[d] Number of CES or CPUS used in parentheses.

[e] FORTRAN with compiler directives.   [f] Hockney (1985a)

using all the four CPUs, can achieve 480 Mflop/s on this problem. The significance of the development of minisupercomputers is thus that engineering firms currently doing technical calculations on a VAX 11/780 or similar minicomputer can enhance their calculational capability probably by about two orders of magnitude by installing a minisupercomputer, without significantly increasing their costs. This means that complex engineering simulations, previously confined to costly specialist supercomputer centres, can now for the first time be performed in-house. Furthermore, the quality of the arithmetic is improved from 32-bit for a VAX to 64-bit on a minisupercomputer. However, although the 1984 Convex C-1 can justifiably claim to have between 1/8 and 1/4 of the performance of a 1976 supercomputer (the CRAY-1) the benchmark results also show that it has only 1/50 of the performance of a 1984 supercomputer (the CRAY X-MP/4) on the LINPACK benchmark.

The architecture of the Convex C-1 is broadly similar to the CRAY-1 in that it is based on a set of functional units working from vector registers. However, the detailed architecture is quite different; the machine does not use the CRAY instruction set. Unlike the CRAY-1, the C-1 instruction set allows addressing to an individual byte, and the 32-bit address (compare 24-bit on the CRAY-1) can directly address a virtual memory space of 4 Gbyte, or 500 Mword (64-bit). There are three functional units (for load/store/vector edit, add/logical and multiply/divide), and eight vector registers holding 128 64-bit elements each. Each functional unit comprises two identical pipes, one for odd and the other for even elements. Each pipe performs a 64-bit operation every 200 ns or a 32-bit operation every clock period of 100 ns. This leads to an effective processing rate of one 64-bit result every 100 ns, or one 32-bit result every 50 ns. Since only two of the pipes perform floating-point arithmetic, this corresponds to a peak performance of 20 Mflop/s in 64-bit mode or 40 Mflop/s in 32-bit mode. The vector registers are supplied with data from either a 64 Kbyte, 50 ns cache memory, or directly from a 16 Mword, 16-bank dynamic RAM main memory. The bandwidth for transfers of 64-bit data words from main memory to the cache is 10 Mword/s, compared to 80 Mword/s on the CRAY-1 and 315 Mword/s on the CRAY X-MP.

(ii) SCS-40

The second minisupercomputer to be announced was the Scientific Computer Systems SCS-40 which appeared in 1986. Like the Convex C-1, the manufacturers claim that the SCS-40 delivers 25% of the performance of the CRAY X-MP/1 at 15% of the cost. However, unlike the C-1, this machine uses the CRAY instruction set, and CRAY programs should run without alteration. The physical architecture comprises 16 main memory banks

(4 Mwords maximum configuration) connected via multiple buses and a vector crossbar switch to eight 64-element vector registers. These in turn are connected by multiple buses to a set of pipelined functional units, corresponding to those of the CRAY series. Each functional unit has a maximum performance of one result per clock period of 45 ns, giving a peak performance of 44 Mflop/s if the floating-point add and multiply pipelines are working simultaneously.

The bus cycle time is 22.5 ns, allowing a single physical bus to move two 64-bit words in the machine clock period, and thereby act as two logical buses. Four such logical buses are provided between main memory and the vector registers, giving a bandwidth of 4 words per clock period of 89 Mword/s (compare 10 Mword/s on the C-1). The six logical buses provided between the vector registers and the functional units allow two vector dyadic operations to be performed simultaneously. In order to provide these speeds of operation the machine logic is engineered from MSI and LSI integrated circuits using ECL logic, in contrast to the slower VLSI CMOS logic used in the Convex C-1 (clock period 100 ns). The speed difference is evident in the LINPACK benchmark for matrices of order 100, when the SCS-40 reaches 7.3 Mflop/s and the Convex C-1 2.9 Mflop/s; a ratio almost exactly that between the clock periods of the machines. Recoding of the LINPACK benchmark in terms of matrix–vector, rather than vector–vector operations, improves the performance to 26 Mflop/s for a matrix of order 300.

(iii) Alliant FX/8

The Alliant FX/8 is the only minisupercomputer described here which is based on a MIMD architecture. That is to say that there are eight computational elements (CEs) sharing a common memory. Each CE is a vector computer with eight vector registers holding 32 64-bit words each, and separate pipelined functional units for floating-point addition, multiplication and division. With a cycle time of 170 ns each CE has a peak vector performance of 11.8 Mflop/s (32-bit) or 5.9 Mflop/s (64-bit), and an eight-CE machine has therefore a theoretical peak performance of 94 Mflop/s (32-bit) or 47 Mflop/s (64-bit). The CEs are connected via a crossbar switch to two 64 Kbyte caches with a bandwidth of 376 Mbyte/s. The caches in turn access the shared memory via a 188 Mbyte/s memory bus. The shared memory is expandable to 64 Mbyte in 8 Mbyte modules, each divided into four banks. The benchmark performance of the Alliant FX/8 is compared with the other minisupercomputers in table 1.1. The performance is seen to be very similar to that of the Convex C-1, and less than that of the SCS-40.

The machine also contains a 'concurrency bus' that is directly connected to all the CEs. This is used to synchronise the CEs with the minimum amount of overhead. Each CE has a concurrency control unit (CCU) which is its interface to the concurrency bus. The CCUs distribute the work among the CES at run-time, and synchronise the calculation by hardware. In this way, for example, data dependencies between different instantiations of a DO loop are maintained correctly by hardware without any intervention from the programmer, even though different loop indices are assigned different CES. The Alliant FX/8 is to be used as the eight-PE cluster in the Illinois Cedar project that has already been described (§1.1.8). It is also sold separately, for example, as an Apollo DOMAIN workstation. First deliveries were made in 1985. A one-CE model is marketed as the Alliant FX/1. This has a 32 Kbyte cache and one or two 8 Mbyte memory modules.

1.2   CLASSIFICATION OF DESIGNS

We have seen during our discussion of the history of parallelism that a wide variety of different parallel architectures have been proposed, and a fair number have been realised, at least in experimental form. Attempts to bring some order into this picture by classifying the designs have not however met with any general success, and there is (c1987) no useful and accepted classification scheme or accompanying notation. We will however present the taxonomy of Flynn (1972) in §1.2.2 and that of Shore (1973) in §1.2.3 since both of these have been discussed quite widely and some of the associated terminology has become part of the language of computer science. The problem with these classifications is that several well established architectures, particularly the highly successful *pipelined computer*, do not fit into them at all clearly, and others such as the ICL DAP may fit equally well into several different groups. An alternative approach is to focus attention on the principal ways in which parallelism appears in the architecture of actual computers, namely: pipelining, processor replication and functional parallelism. These divisions themselves, springing as they do from the engineering reality, form the basis of a taxonomy that is easier to apply than the more theoretical concepts of Flynn and Shore.

    The first stage in the development of any successful classification is, however, the ability to describe reasonably accurately and concisely the essential features of a particular architecture, and this requires a suitable notation. We give therefore in §1.2.4 such a notation that enables an architecture to be described within one line of text. It plays much the same role in discussing computer architecture as the chemical formulae of large molecules do in chemistry. This notation is then used in §1.2.5 to give a structural classification of the serial and parallel computers that are discussed in this book.

1.2.1   Levels of parallelism

History shows that parallelism has been used to improve the effectiveness of computers since the earliest designs, and that it has been applied at several distinct levels which might be classified as:

(1) *Job level*
    (i)   between jobs;
    (ii) between phases of a job;
(2) *Program level*
    (i)   between parts of a program;
    (ii) within DO loops;
(3) *Instruction level*
    (i)   between phases of instruction execution;
(4) *Arithmetic and bit level*
    (i)   between elements of a vector operation;
    (ii) within arithmetic logic circuits

At the highest level the objective of a computer installation is to maximise the rate of processing of jobs. In the simplest analysis each job may be considered as being divided into several sequential phases, each of which requires a different system program and system resources. The typical phases might be: input FORTRAN source code from a disc or tape; compile FORTRAN into object code; link object code with various library sub-routines; execute the resulting module; print output files. Any I/O operation is very slow compared with program execution, hence any large computer installation provides several I/O channels or peripheral processors which can perform I/O in parallel with program execution, and provides a battery of disc and tape drives. Most installations have a single processor for program execution, but some contain two or more processors. It is the purpose of the *operating system*, which is the name given to the computer program that controls the flow of work through the computer, to organise the sharing of the system resources between the different jobs. Usually several programs (e.g. 5–10) reside in the fast memory of the computer, and in the case of a single processor only one will be in execution. As soon as this program requires a slow I/O facility, for example a read from disc, this operation is initiated in the channel and another program put into execution. The first program waits until the data are available and control passes back to it when the other programs are similarly obliged to wait. In this way the I/O of one job is overlapped with the execution of another, in a dynamic manner according to the needs of the jobs that happen to be sharing the computer at a particular time.

Such operating systems, because they do not know the internal logical inter-relations of the programs being executed, must assume that the different phases of one job have to be executed in sequence and that any I/O statement in the program must be completed before the next statement is executed. In some circumstances a programmer can maintain control over his I/O operations and arrange to read his data in blocks from backing store. In this case he may apply buffering in order to overlap the I/O in his program with the execution of his program. In three-stage buffering, for example, two channels are used and three blocks of data are present in the store. Simultaneously a new block is read into buffer 1 via the input channel, new values are calculated by the processor from the block of data in buffer 2, and the last block of values calculated are being written from buffer 3 via the output channel. The calculation proceeds by cyclically changing the roles of the buffers.

We can see from the above that the main requirement of computer architecture in allowing parallelism at the job level is to provide a correctly balanced set of replicated resources, which comes under the general classification of functional parallelism applied overall to the computer installation. In this respect it is important for the level of activity to be monitored well in all parts of the installation, so that bottlenecks can be identified, and resources added (or removed) as circumstances demand.

We next consider the types of parallelism that arise during the execution of a program that constitutes one of the job phases considered above. Within such a program there may be sections of code that are quite independent of each other and could be executed in parallel on different processors in a multiprocessor environment (for example a set of linked microprocessors). Some sections of independent code can be recognised from a logical analysis of the source code, but others will be data-dependent and therefore not known until the program is executed. In another case, different executions of a loop may be independent of each other, even though different routes are taken through the conditional statements contained in the loop. In this case, each microprocessor can be given the full code, and as many passes through the loop can be performed in parallel as there are microprocessors. This situation arises in Monte-Carlo scattering calculations for non-interacting particles, and has important applications in nuclear engineering. The programming problems associated with such linked microprocessors are an active area of current research, and many such systems have become operational in the 1980s (see §§1.1.8 and 1.2.6).

All manufacturers of computers designed for efficient operations on vectors of numbers (Burroughs, Texas Instruments and CRAY) have produced FORTRAN compilers that recognise when a DO loop can be replaced by one or several vector instructions. This is the recognition by software that a particular DO loop is just the scalar representation of a set of vector operations, and can therefore be executed much more effectively by machine instructions that are especially engineered to perform such operations efficiently. The architectural features used are then either the pipelined arithmetic units in the case of pipelined vector computers (e.g. CYBER 205 and CRAY-1) or the replicated processing elements in the case of the processor array (e.g. ICL DAP and BSP).

At a lower level still, we have already noted that the processing of any instruction may be divided into several suboperations, and that pipelining may be used to overlap the different suboperations on different instructions. Such pipelining is widely used on all fast scalar processors (e.g. ICL 2900, AMDAHL 470V/6) and on some processor arrays (e.g. BSP). Part of the processing of an instruction is the performance of the arithmetic and this itself may be divided into suboperations and pipelined (see §1.3.1), and therefore represents a lower-level example of parallelism.

At the lowest level, we have the choice in the arithmetic logic itself, whether to perform this arithmetic in a bit-serial fashion, or on all the bits of a number in parallel. Intermediate possibilities exist: for example, performing the logic on all bits of a byte† in parallel, but taking the bytes of the number sequentially. Such issues of parallelism, at the arithmetic and bit level, were an active area of development in the first-generation computers of the 1950s. After bit-parallel arithmetic in floating-point became the standard in about 1955, the existence of parallelism at this lowest level was largely forgotten. However, the advent in the late 1970s of the microprocessor, which may be regarded as a first-generation computer on one or a few chips, has again made these issues important. Furthermore the desire to produce cost-effective large arrays of processing elements, has seen a return to bit-serial arithmetic in some large computers such as the STARAN and ICL DAP.

1.2.2   Flynn's taxonomy

Flynn does not base his macroscopic classification of parallel architecture on the structure of the machines, but rather on how the machine relates its instructions to the data being processed. A *stream* is defined as a sequence of items (instructions or data) as executed or operated on by a processor. Four broad classifications emerge, according to whether the instruction or data streams are single or multiple:

(1) SISD—single instruction stream/single data stream. This is the conventional serial von Neumann computer in which there is one stream of

†1 byte is a sequence of 8 binary digits (bits).

instructions (and therefore, in practice, only one instruction processing unit) and each arithmetic instruction initiates one arithmetic operation, leading to a single data stream of logically related arguments and results. It is irrelevant whether pipelining is used to speed up the processing of instructions or the arithmetic. It is what we have previously called a serial scalar computer. Examples are: CDC 6600 (unpipelined); CDC 7600 (pipelined arithmetic); AMDAHL 470V/6 (pipelined instruction processing).

(2) SIMD—single instruction stream/multiple data stream. This is a computer that retains a single stream of instructions but has vector instructions that initiate many operations. Each element of the vector is regarded as a member of a separate data stream hence, excepting the degenerate case of vectors of length one, there are multiple data streams. This classification therefore includes all machines with vector instructions. Again it is irrelevant whether the capability of vector processing is realised by pipelining or by building arrays of processors. Examples: CRAY-1 (pipelined vector computer); ILLIAC IV (processor array); ICL DAP (processor array); OMEN-64 (processor array).

(3) MISD—multiple instruction stream/single data stream. This class seems to be void because it implies that several instructions are operating on a data item simultaneously. However Flynn (1972) states that it includes specialised streaming organisations using multiple instruction streams on a single sequence of data. However no examples are given.

(4) MIMD—multiple instruction stream/multiple data stream. Multiple instruction streams imply the existence of several instruction processing units and therefore necessarily several data streams. This class therefore includes all forms of multiprocessor configurations, from linked main-frame computers to large arrays of microprocessors.

From our point of view, the problem with the above classification scheme is that it is too broad: since it lumps *all* parallel computers except the multiprocessor into the SIMD class and draws no distinction between the pipelined computer and the processor array which have entirely different computer architectures. This is because it is, in effect, a classification by broad function (whether or not explicit vector instructions are provided) rather than a classification of the design (i.e. architecture). It is rather like classifying all churches in one group because they are places of worship. This is certainly a valid broad grouping, and distinguishes churches from houses, but is not very useful to the architect who wishes to distinguish between the different styles of church architecture by the shape and decoration of their arches and windows. In this book we are, like the architect, interested in studying the details of the organisation, and therefore need a finer classification.

1.2.3   Shore's taxonomy

Unlike Flynn, Shore (1973) based his classification on how the computer is organised from its constituent parts. Six different types of machine were recognised and distinguished by a numerical designator (see figure 1.3).

Machine I. The conventional von Neuman architecture with a single control unit (CU), processing unit (PU), instruction memory (IM) and data memory (DM). A single DM read produces all bits of any word for processing in parallel by the PU. The PU may contain multiple functional units which may or may not be pipelined; hence this group includes both the pipelined scalar computer (e.g. CDC 7600) and the pipelined vector computer (e.g. CRAY-1), whose similarity in architecture has already been noted.

Machine II. This is the same as machine I except that a DM read fetches a bit slice from all words in memory instead of all bits of one word; and the PU is organised to perform its operations in a bit-serial fashion. If the memory

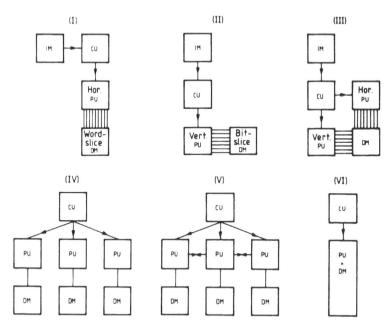

FIGURE 1.3   Schematic representation of the six machine classes defined by Shore (1973). Machine I, word-serial, bit-parallel; II, word-parallel, bit-serial; III (= I + II), orthogonal computer; IV, unconnected array; V, connected array; VI, logic-in-memory array.

is regarded as a two-dimensional array of bits with one word stored per row, machine II reads a vertical slice of bits, whereas machine I reads a horizontal slice. Examples are the ICL DAP and STARAN.

Machine III. This is a combination of machines I and II. It comprises a two-dimensional memory from which may be read either words or bit slices, a horizontal PU to process words and a vertical PU to process bit slices; in short it is the orthogonal computer of Shooman (1970). Both the ICL DAP and STARAN may be programmed to provide the facilities of machine III, but since they do *not* have separate PUs for words and bit slices they are not in this class. The Sanders Associates OMEN-60 series of computers is an implementation of machine III exactly as defined (Higbie 1972).

Machine IV. This machine is obtained by replicating the PU and DM of machine I (defined as a processing element, PE) and issuing instructions to this ensemble of PEs from a single control unit. There is no communication between the PEs except through the control unit. A well known example is the PEPE machine. The absence of connections between the PEs limits the applicability of the machine, but makes the addition of further PEs relatively straightforward.

Machine V. This is machine IV with the added facility that the PEs are arranged in a line and nearest-neighbour connections are provided. This means that any PE can address words in its own memory and that of its immediate neighbours. Examples are the ILLIAC IV, which also provides short-cut communication between every eight PEs.

Machine VI. Machines I to V all maintain the concept of separate data memory and processing units, with some databus or switching element between them, although some implementations of one-bit machine II processors (e.g. ICL DAP) include the PU and DM one the same IC board. Machine VI, called a logic-in-memory array (LIMA), is the alternative approach of distributing the processor logic throughout the memory. Examples range from simple associative memories to complex associative processors.

Forgetting for the moment the awkward case of the pipelined vector computer, we can see that Shore's machines II to V are useful subdivisions of Flynn's SIMD class, and that machine I corresponds to the SISD class. Again the pipelined vector computer, which clearly needs a category of its own, is not satisfactorily covered by the classification, since we find it in the same grouping as unpipelined scalar computers with no internal parallelism above the requirement to perform arithmetic in a bit-parallel fashion. We also find it unsatisfactory that the machine classes are designated by numbers that have no mnemonic or ordinal significance, in much the same way that one would not find it helpful for a zoologist to describe the reptiles as animal class VI.

1.2.4   An algebraic-style structural notation (ASN)

The first step in the creation of a classification scheme is the definition of a suitable notation. In order to differentiate between the computers described in this book, it should be more detailed than that of Flynn or Shore, and take explicit account of the successful pipelined designs. The notation is structural and based on a shorthand indicating the number of instruction, execution and memory units, and the manner of their interconnection and control. In this respect it is rather similar to a chemical formula, but the description of a computer is more complex: for example, in chemistry there is only one type of carbon atom,† whereas in a computer we must distinguish between many types of execution unit (integer, floating-point, pipelined, bit-serial, etc). An exact mathematical definition of the syntax of the notation is given in Backus Naur form (BNF) in Appendix 1. We give below a tutorial exposition with examples, and describe the semantics of the notation.

The notation regards a computer as a number of functional units that manipulate data, connected by data paths and operating under the control of instruction units. The simplest example of a von Neumann serial computer in the structural notation is

$$C = I[E—M]$$

which defines the computer C to be a single instruction processing unit I controlling the units in the brackets. These are a single execution unit E for performing arithmetic, connected by a single data path (the dash or so-called von Neumann bottleneck) to a single or unbanked memory unit M.

One can see from the above example that our notation is similar in intent to the PMS (processor–memory–switch) level notation of Bell and Newell (1971); however it differs in some important ways that make it more suitable for our purpose. Our primary objective has been to allow concise one-line or few-line descriptions of the overall architecture in an algebraic style that lends itself to printing, typing and input to computer programs. In particular we desire a neat notation for arrays of processors of variable dimension and size, and are most interested in the number and type of data-manipulating units and how they are controlled.

---

† Note that the three naturally occurring isotopes of carbon all have the same electronic structure and therefore the same chemistry.

The detailed description of our structural notation is treated under the headings:

A The units (rules 1–10)—define the symbols used to represent units and to combine them into groups.

B Connections between units (rules 11–16)—define the notation for data paths and the manner of describing complex networks.

C Comments (rule 17)—allows additional information to be included in a flexible and open-ended way.

D Control of units (rules 18–20)—define the different types of control that make a collection of units into a computer.

E Examples—show how different computers may be described.

A The units

(1) Symbols—the following is an alphabetical list of the symbols which designate the different types of units comprising a computer or processor:

B	An integer, fixed-point or boolean execution unit.
C	A computer which is any combination of units including at least one I unit.
Ch	An I/O channel which may send data to or from an I/O device interface and memory, independently of the other units.
D	An I/O device, e.g. card reader, disc, VDU. The nature of the device is given as a comment in parentheses.
E	An execution unit which manipulates data. That is to say it performs the arithmetic, logical and bit-manipulation functions on the data streams. It is subdivided into F and B units and often called an ALU or arithmetic and logical unit.
F	A floating-point execution unit.
H	A data highway or switching unit. Transfers data without change, other than possibly reordering the data items (e.g. the FLIP network of STARAN).
I	An instruction unit which decodes instructions and sends (or issues) commands to execution units where the instructions are carried out. Often called an IPU or instruction processing unit.
IO	An I/O device interface which collects data from a device and loads it to a local register or vice versa.
M	A one-dimensional memory unit where data and instructions

are stored, e.g. registers, buffer memory, main memory, disc.

O   A two-dimensional or orthogonal memory.

P   A processor which is defined as any collection of units including an E unit but not including an I unit.

S   A switching unit which interconnects other units, such as a multistage omega switch. In most cases a switch would perform no computation on the user's data.

U   An unspecified unit. A symbol to be used when none of the above categories applies. The nature of the unit is specified by a comment in parentheses following the symbol. Examples might be various intermediate control units within a complex system.

(2) Pipelining is indicated by a lower case p following the symbol for the unit, or a structure enclosed in braces. This means that some or all of the operations or suboperations performed by the unit or structure are overlapped in time, e.g.

Ip   a pipelined instruction unit,

Ep   a pipelined execution unit,

$\{E < --,-- > M\}p$ a structure in which the operations of memory read and memory write are overlapped.

(3) Vector instructions, if provided in the instruction set, may be indicated by a final lower case v in the symbol for an I unit. In the case of processor arrays, in which all instructions are operations on vectors, the symbol is understood and therefore omitted, e.g.

Ipv   a pipelined instruction unit with vector instructions,

Iv   an instruction unit with vector instructions that is not pipelined.

(4) Different units of the same kind can be distinguished by a final integer, for example:

Ep1   pipelined execution unit number 1,

Ep3   pipelined execution unit number 3.

(5) Substitution is encouraged, in order to promote clarity through a hierarchical notation. For example, the different types defined under rule (4) are to be defined in subsequent equations, separated by semicolons, e.g.

I[E1, E2];   E1 =   ;   E2 =   .

(6) Multiple units—the number of units of the same kind that can operate simultaneously is indicated by an initial integer. Note particularly that a unit, however multipurpose and complex, is not counted more than once unless it can perform more than one operation at a time, e.g.

E a single multifunction execution unit for multiplication, addition, logical etc, performing only one operation at a time, as in the IBM 7090,

10E 10 independent function units for multiplication, addition, logical etc, that may operate simultaneously, as in the CDC 6600.

(7) Replication—a bar over a symbol, or over a structure delimited by braces { }, is used to indicate that all the units in a group are identical, e.g.

$$64\overline{P} = 64\{\overline{E-M}\}$$ 64 identical processing elements, as in the ILLIAC IV.

(8) Groups of units may be defined by enclosing the units in braces { }. If the units are separated by commas (the concurrent separator) they may operate simultaneously and therefore in parallel. If the units are separated by a slash (the sequential separator) the units may operate only one at a time, that is to say sequentially or serially, e.g.

{4Fp, 2B} four floating-point arithmetic pipelines and two unpipelined integer units that may operate simultaneously,

{E1/E2/E3} three execution units that may operate only one at a time.

Only one type of separator applies within any set of braces; however a multiple unit still implies simultaneous operation, e.g.

{3F1/B1} three floating-point units that may operate simultaneously amongst themselves but sequentially with a fixed-point unit,

$3F1 = \{F(+), F(*), F(\div)\}$ the floating-point units are an adder, multiplier and divider that operate concurrently,

$B1 = \{B(+)/B(\text{shift})\}$ the fixed-point unit has an adder and shifter that operate one at a time.

In the above the symbols in parentheses are comments (see rule (17)).

(9) The number of bits that are operated on in parallel by the unit is indicated by a subscript, e.g.

I_{16} a 16-bit instruction unit,

E_{64} a 64-bit execution unit.

Multiple instruction streams processed by a single I-unit, as in the Denelcor HEP, can be indicated by a multiplier showing the possible number of streams, e.g.

Ip_{50*64} the instruction processing unit within a PEM of the Denelcor HEP, which processes 50 64-bit user instruction streams in a pipelined fashion.

If the multiplier is omitted, one stream is implied. The following notation is used for memory units. Note that each section of memory that operates independently (i.e. each memory bank) is treated as a separate memory unit. An asterisk is used for 'times' or multiply, e.g.

nM_{w*b} a one-dimensional memory divided into n banks; each bank contains w words of b bits and when accessed delivers b bits in parallel,

M_{1K*32} a one-dimensional memory circuit with 1024 32-bit words,

$8M_{64*64}$ the eight vector registers of the CRAY-1, each holding 64 64-bit numbers,

O_{w*b} a two-dimensional orthogonal memory of w words of b bits, that delivers either a word slice of b bits or a bit slice of w bits.

(10) The characteristic time associated with the operation of a unit is indicated by a superscript. The unit of measurement is nanoseconds unless otherwise stated in supporting text or as a comment in parentheses; e.g.

I^{40} instruction unit with clock period of 40 ns,

E^{200} execution unit with average operation time of 200 ns,

M^{650} memory unit with access time of 650 ns.

B Connections between units

(11) Databus connections between two units are indicated by the following connectors:

	a connection of unspecified type,
$<-$	a simplex connection to the left,
$->$	a simplex connection to the right,
$<->$	a full duplex connection,
$<-/->$	a half duplex connection.

A simplex connection can transfer data only in the direction shown. A full duplex connection may transfer data in both directions at the same time. A half duplex connection may transfer data in either direction, but not both at the same time. Using the concurrent and sequential separators we have:

$$< - > \qquad \text{an abbreviation for } \{ <-, -> \},$$
$$< -/- > \qquad \text{an abbreviation for } \{ <-/-> \}.$$

In the above the dash may be replicated (or printed as a line of arbitrary length) in order to improve the appearance of a structural description, e.g.

$$E ----- > M; \qquad E \text{———} > M; \qquad E < --/-- > M.$$

Thus a multiple dash in the notation is semantically and logically equivalent to a single dash. Arrows may be typed using 'less than' $<$ and 'greater than' $>$ symbols. Note the use of the semicolon (as in ALGOL60) as a separator between structural expressions. The width of the databus in bits may be written below the connector in the format 'number of data bits' + (optionally) 'number of address bits'. Thus:

$\frac{100}{64+16}$ a databus transferring 64 data bits and 16 address bits every 100 ns

$\overline{8} >$ a simplex connection to the right which is 8 data bits wide.

Multiple data buses can be described with a multiplier and an asterisk ($*$) as in FORTRAN. Braces are used for algebraic grouping because parentheses are reserved for comments, e.g.

$$\underset{4*\{64+16\}}{< ------- >} \text{ four identical full duplex databuses, each with 64 data bits and 16 address bits.}$$

A complex data path or switch can be identified by a highway identifier and explained in a subsequent definition, e.g.

$$E - H3 - M; \qquad H3 = \{\{\overline{16} >, \ <\overline{8}\}/<\overline{24}\}.$$

The right-hand side of such a definition should contain only data paths.

(12) Series connections—a chain of units linked by databuses such as $E - M1 - M2 - M3$ describes a set of units that are connected logically in series, in the sense of an electrical circuit.

(13) Parallel connections, in the logical sense of a parallel connection in an electrical circuit, can be described using the concurrent (,) or sequential (/) separators between units (or descriptions of the parallel paths) and enclosing the list in braces { }. Greater flexibility is obtained by introducing the

no-connection symbol (|). A unit may be connected outside the braces as follows:

U	connection outside brace unspecified,
—U|	connection outside brace to left, no connection to right,
|U—	connection outside brace to right, no connection to left,
—U—	connection outside brace to left and right.

The no-connection symbol may be omitted if no ambiguity arises: for example if there is nowhere the connection could be made in any case, e.g.

—{U1, U2, U3}   a group of units that may operate concurrently, connected in an unspecified way to the databus on the left,

—{U1/U2/U3}—   a group of units working one at a time with unspecified connections to the databuses on the left and right,

—{—U1—, —U2—, —U3—}—   three concurrently operating parallel paths which would be drawn:

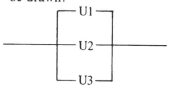

—{—U1—, |U2—, —U3—}—   as above but U2 is not connected to the left, thus:

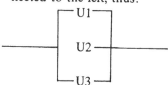

—{—U1—/—U2—/U3—}—   three parallel *alternative* paths, working one at a time. Since this is a distinction in time it cannot be drawn differently from the third example above.

In order to illustrate the use of descriptions of paths instead of simple units, and the nesting of such parallel connections, consider the complex of connections

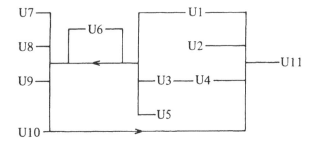

which can be expressed as

$$\{U7—, U8—, U9—, U10—\}—\{\{—U6—/<—\}$$

$$—\{—U1—, |U2—, —U3—U4—, —U5|\}/—>\}—U11$$

where the arrowed paths are unidirectional and alternative paths.

(14) *Connection points* are specified by lower case letters within or at the end of connectors, e.g.

$$E—a—M, \qquad U<—/—>q.$$

They may be needed to specify long connections or structures that cannot be represented in the manner of the last paragraph. For example, the structure

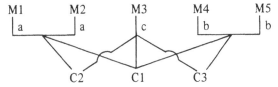

showing three computers C1, C2, C3 connected to five memory banks M1, M2, M3, M4, M5 is most clearly expressed by

$$C = \{C1—\{a,b,c\}, C2—\{a,c\}, C3—\{b,c\},$$

$$\{a—M1, a—M2, c—M3, b—M4, b—M5\}\}$$

where connection points a, b and c have been defined as shown on the memory banks. We note the use of a list of connection points in braces in the case of multiple connections. Since any connection can be included in this manner, connection points provide the facility of describing an arbitrary network of units.

(15) *Arrays of processors* are for the most part arranged in multi-dimensional rectangular or square arrays. It is natural therefore to express such an arrangement in the multiplier preceding the description of the

processor. We use the asterisk for multiplication (as in FORTRAN) and powers as appropriate, e.g.

$128*64\bar{P}$ an 8192 array of identical processors arranged as a 128 times 64 array,

$64^2\bar{P}$ a 64×64 array of processors,

$2^4\{2\bar{C}\}$ a four-dimensional binary hypercube with 16 nodes. Each node has two computers.

The extent of connectivity between the processors may be given as a superscript to braces { } or brackets [ ] in the form '*c-nn*'. This means that the processors may transfer data directly to or from processors out to and including their *c*th nearest neighbour. If *c* is zero there is no direct connection. For example:

$288\{\overline{3E-M}\}^{0\text{-nn}}$ PEPE with 288 unconnected processors, each with three execution units and memory,

$C[64^2\bar{P}]^{1\text{-nn}}$ the ICL DAP with 1st nearest-neighbour connections to four processors,

$\{32^2\bar{P}\}^{2\text{-nn}}$ the CLIP image processing computer (Duff 1978) that has connections to 2nd nearest neighbours. That is to say to the eight neighbours:

Other more complex connection patterns may be specified as comments between parentheses.

(16) Cross connections (such as crossbar and other switching networks) between multiple execution and memory units are indicated by a cross. The details of the connection network must be explained in separate text. Transfer times and bit widths may be added as above. For example:

$Ip[16\bar{F} \times 17\bar{M}]$ the BSP with 16 floating-point execution units cross-connected to 17 memory banks.

The cross symbol may be replicated to improve legibility, and thus multiple crosses are logically and semantically identical to a single cross. The exact action of a switch can be defined by including the switching function as a comment between parentheses (see rule (17) below). For example:

$$n\text{E} \times \left(\left\lfloor \frac{x}{2^k} \right\rfloor 2^k + ||x|_{2^k} + 2^{k-1}|_{2^k}, \, k = 1, \ldots, \log_2 n \right) \times m\text{M}$$

represents the exchange permutation considered in §3.3.2 and shown in figure 3.9. In the above expression $\lfloor f \rfloor$ is the integer floor function of f, and $|x|_{2^k}$ means that the argument x is taken modulo 2^k (see §3.3). Alternatively, substitution may be used to achieve greater clarity, e.g.

$$n\text{E} \times \text{H3} \times m\text{M}; \qquad \text{H3} = \text{H}(|x|_{2^{k+1}}).$$

A switch may also be specified by using the unit symbol S. This is useful when the switch connection is one-sided: for example when a number of identical units are connected by a switch in a MIMD system, or when the switch is a major unit of the system which should be emphasised. For example

$$\text{C(Cedar cluster)} = 8\bar{\text{C}} \times \text{S (local omega switch)}.$$

C Comments
The notation is open-ended to the extent that comments may be added throughout to clarify the meaning of a unit or connection.

(17) Comments are enclosed in parentheses ( ) and contain additional optional information about the immediately preceding symbol. The information may have any syntax, e.g.

$\text{M1}^{10}(\text{bipolar}) — \text{M2}^{400}(\text{MOS}) — \text{M3}^{(1\text{ms})}(\text{disc})$ a memory hierarchy,
$\text{Fp}(*, \text{ECL})$ a pipelined floating-point multiplier in ECL technology,
$\text{Ip}(4 \text{ segs})[\ ]$ a four-segment instruction pipeline.

Note the use of a comment as a superscript in order to allow nonstandard time units. The unspecified unit symbol (U) allows any unit to be defined, e.g.

$\text{U}(2803)$ an IBM 2803 control unit.

In the case of data connections the parentheses are inserted between a pair of dashes or crosses, e.g.

$\text{E}—(\text{half mile coaxial line})—\text{M}; \qquad \text{E} \times (\text{Banyan network}) \times \text{M}.$

The following symbols are used: bit is denoted by (b) and byte (B), with the usual SI unit prefixes and the convention $K = 1024$, $M = K^2$, $G = K^3$, $T = K^4$.

D Control of units
A set of units under the control of an instruction stream defines a computer.

(18) Computers and processors. Within the notation we maintain the convention that a *computer* is a group of units which can process instructions and therefore contains at least one I unit. In the context of overall architecture, an I unit is programmable and processes a user's instruction stream. With this definition a microprocessor is more aptly described as a microcomputer. The simplest computer is described as:

$$C = I[E—M].$$

A *processor*, on the other hand, is any set of units that can process data but does not itself process the user instruction stream. It therefore contains an E unit but no I unit. The simplest case would be:

$$P = E—M.$$

This definition of a processor fits with the common usage in the term processor array for an array of E—M units under common control of an external I unit, as in the ICL DAP.

The above distinction is not absolutely clear cut, because many units which we would regard as execution units, are in fact controlled by microprograms, the instructions of which are processed by the E unit. From the point of view of overall architecture—which is our main interest here—the important point is whether a unit is programmable by the user. If it is, then we would regard it as having an I unit. If it is not user programmable then there is no I unit, even though internally a fixed micro-instruction stream may be involved. Similarly, from the overall architectural view we would only describe programmable registers as part of the description of a computer, even though there are many other internal registers in the computer (for example between the stages of a pipelined execution unit).

There is no reason why the notation should not be used to describe the internal structure of a microprogrammed general-purpose arithmetic pipeline, or a microprocessor in detail. In this case all the internal registers, programmable or not, would be described and the unit processing the microprogram would be usefully classified as an I unit. It is clear, therefore, that the meaning of an I unit depends on the use to which the notation is being put, and should be made clear in the supporting text.

(19) The extent of control exercised by an I or a C unit is shown by the control brackets [   ]. The units controlled are listed inside the brackets, separated by commas if they may operate simultaneously, or by slashes if the units operate only one at a time, e.g.

I[4I2[64$\bar{\text{P}}$]]   a master instruction unit (I1) controlling 4
                quadrants, each comprising an instruction

	unit (I2) controlling 64 identical processors: as in the original plans for ILLIAC IV,
Ip[C1,C2,C3]	a pipelined instruction unit controlling three computers that may operate simultaneously,
C1[Ep1/Ep2]; C1 = I[B—M]	a control computer C1 which performs boolean operations and contains memory, controlling two pipelined execution units that operate one at a time.

(20) The type of control exercised by an I unit or a C unit may optionally be indicated by a lower case subscript to the control bracket. The following modes are adequate to describe the computers in this volume. Other modes can be described in comments:

asynchronous (a)	The units controlled have more than one clock. The clocks are *not* synchronised and communication between units must be coordinated by flags and agreed 'hand-shaking' protocols.
horizontal (h)	A single composite instruction controls the action of a set of different units at each clock period, e.g. the FPS AP-120B.
lockstep (*l*)	A set of identical processors are controlled synchronously to perform the same operation at the same time, e.g. the ICL DAP.
issue-when-ready (r)	Instructions are issued to execution units as soon as the required unit and registers are free, e.g. the CRAY-1.

For example:

I[10F, 10C̄]$_r$	The CDC 6600 with 10 different independent floating-point functional units and 10 identical I/O computers. Instructions are issued when units are ready.
C[64P̄]$_l$	The ILLIAC IV with 64 identical processors controlled in lockstep mode.
I[4C]$_{(description of control)}$	Four computers controlled by an I unit in the manner described in the comment within parentheses.

E Examples
In order to illustrate the use of the above notation we give below one-equation or one-line descriptions of a variety of computers. The detail of the description

can be varied according to need, and for clarity a hierarchical description in several equations is to be preferred—with the overall architecture described in the first equation, and the detail filled in by substitution in subsequent equations, if necessary down to the level of individual registers. More complex structures may require the extension of the formula into the second dimension, in the manner of a structural chemical formula.

$$C(Z80 + memory) = C(Z80)_{\bar{8}} M_{64K \cdot 8}; \; C(Z80) = I_8^{250}[B_{8\bar{8}} M_{18 \cdot 8}]$$

$$C(INTEL\ 8086) = I_8[B_{16\,\overline{16}} M_{13 \cdot 8}]_{\overline{16}} M_{1M \cdot 16}$$

$$C(EDSAC1) = I[B_1 - M]; \; M = M_{512 \cdot 35}$$

$$C(IBM\ 7090) = I[F_{36} - M]$$

$$C(CDC\ 6600) = I[10E - 32M - 10\bar{P}]_r; \; 10E = \{4F_{60}, 6B\}$$

$$C(CDC\ 7600) = I[9Ep - M - 16\bar{P}]_r; \; 9Ep = \{3Fp, 6B\};$$
$$M = 32M1 - 8M2$$

$$C(IBM\ 360/195) = Ip^{54}[2C - 32M1^{160}(1KB) - 16M2^{800}];$$
$$2C = \{C1.C2\}; C1 = I[3Fp - M]; C2 = I[B - M]$$

$$C(CRAY\text{-}1) = Iv^{12}[12Ep^{12} - 16M^{50}]_r; \; 12Ep = \{3Fp_{64}, 9B\}$$

$$C(CYBER\ 205) = Iv^{20}[4\bar{F}p_{64} - 512M_{16K \cdot 32}^{80}]$$

$$C(TIASC(2IPU,\ 4\ pipe)) = 2Ivp[2\bar{F}p] - 8M - 8\bar{P}$$

$$C(ILLIAC\ IV(4\ quadrant)) = C1[4C2^{80}[64\bar{P}]_i^{1\text{-}nn}];$$
$$P = F_{64}^{600} - M_{2K \cdot 64}^{400}$$

$$C(PEPE) = C1[3I[288\{\overline{3E - M}\}]_i^{0\text{-}nn}]; C1 = C(CDC7600)$$

$$C(BSP) = Ip[16\bar{F} \times 17\bar{M}]_i$$
$$C(64 \times 64\ ICL\ DAP) = C[64^2\bar{P}]_i^{1\text{-}nn}; \; P = B_1 - M_{4K \cdot 1}$$

$$C(STARAN) = I[32[256\bar{B}_1 \times O_{256 \cdot 256}]_i] - M$$

$$C(OMEN\text{-}64) = I_{16}[64\bar{B}_1 - O_{64 \cdot 16} - E_{16}]_i$$

$$C(HYPERCUBE) = I[2^4\bar{C}]; \; C = C(2*INTEL\ 8080)$$

$$C(HEP) = 16\ \bar{C}p_{50 \cdot 64}^{100}(PEM) \times (packet\ switch) \times 128\ \bar{M}_{1M \cdot 64}^{100}(DMM)$$

$$C(EGPA) = C(control)[2^2\bar{C}(boundary)[2^2\bar{C}(array)]]^{1\text{-}nn}$$

$$C(/CAP) = C(IBM\ host)[10\{-(channel) - C(FPS\ 164/MAX)\}]$$

$$C(Cedar) = 16\bar{C}1(cluster) \times S1(global\ omega\ switch) \times 256\ \bar{M}_{16M \cdot 64}$$

$$C1(cluster) = 8\bar{C}2 \times S2(local\ omega\ switch)$$

or
$$C(Cedar) = 16\{8\bar{C}2 \times S2\} \times S1 \times 256\bar{M}$$

The SE England regional star computer network:

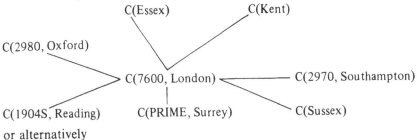

C(Essex)   C(Kent)

C(2980, Oxford)

C(7600, London)   C(2970, Southampton)

C(1904S, Reading)   C(PRIME, Surrey)   C(Sussex)

or alternatively

$$C(7600, London)-\{-C(2980, Oxford), -C(Essex), -C(Kent), -C(2970,$$
$$Southampton), -C(Sussex), -C(PRIME, Surrey), -C(1904S, Reading)\}$$

A 'Benzene' ring of 6 computers.

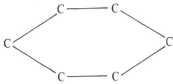

or, alternatively
$$\{-C-C-C-C-C-C-,-\}\ or\ 2\{-C-C-C-\}.$$

1.2.5 A structural taxonomy

We will now formulate, with the aid of the above notation, a structural taxonomy of both serial and parallel computers. The overall subdivisions of this taxonomy are shown in figure 1.4. These are further subdivided for serial computers in figure 1.5 and parallel computers in figures 1.6 to 1.10. These figures are tree structures and therefore there is only one route from the top of the diagram to any of the classes of computer that are defined on the bottom line. For a computer to belong to a particular class it must possess all the properties stated in the boxes that are traversed to reach the definition. A well known computer in each class is given, together with the canonical definition of the class in the structural notation and a descriptive name for computers in that class. It is unavoidable that certain large computers will have properties belonging to more than one class. One will then have to decide which is the more dominant property. We hope that this circumstance

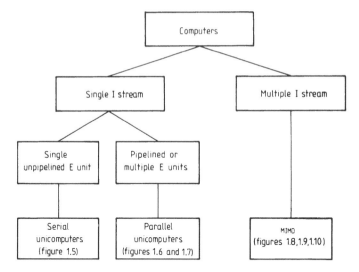

FIGURE 1.4   The broad subdivisions in computer architecture.

is relatively rare and that the subdivisions are fine enough to differentiate those computers that one feels should be treated separately.

At the highest level we follow the functional classification of Flynn and divide computers into those with single instruction streams (SI) and those with multiple instruction streams (MIMD). A taxonomy for MIMD computers

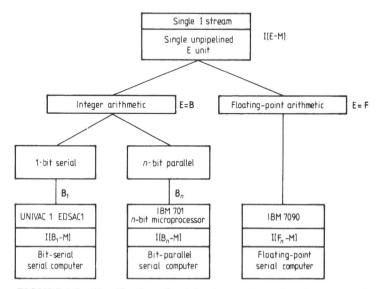

FIGURE 1.5   Classification of serial unicomputers using the structural notation of §1.2.4.

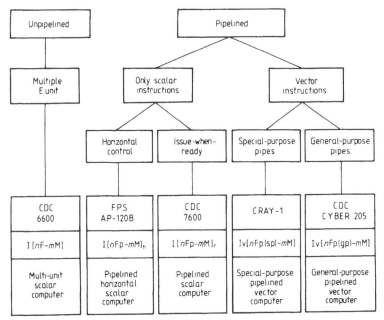

FIGURE 1.6   Parallel unicomputers based on functional parallelism and pipelining. A classification of computers discussed in this book.

is given in the next section, §1.2.6. Here we further subdivide sɪ machines into those with a single unpipelined E unit and those with multiple and/or pipelined E units. Remembering that an E unit may only execute one function at a time (even though it may be able to compute many functions), the single unpipelined E unit subdivision (figure 1.5) leads to sequential operation and all the serial computers, whereas the multiple E units or pipelined subdivision allows different types of overlapping and the family of parallel unicomputers (figures 1.6 and 1.7).

The next level of subdivision is based on the type of arithmetic performed by the E units. The difference in complexity between a one-bit arithmetic unit (less than 10 logic gates) and a floating-point arithmetic unit (many thousands of logic gates) is sufficiently large to constitute a qualitative difference which should be recognised. The extra space required for floating-point circuitry also places very different constraints on assembling large arrays of processors, from those associated with one-bit processors. Consequently arrays of floating-point processors and arrays of one-bit processors are different in their engineering, appearance and properties. In order to include the historic evolution of serial computers, this class has been divided into integer arithmetic and floating-point arithmetic, and the former class into serial and parallel arithmetic. This side of the classification is of more than

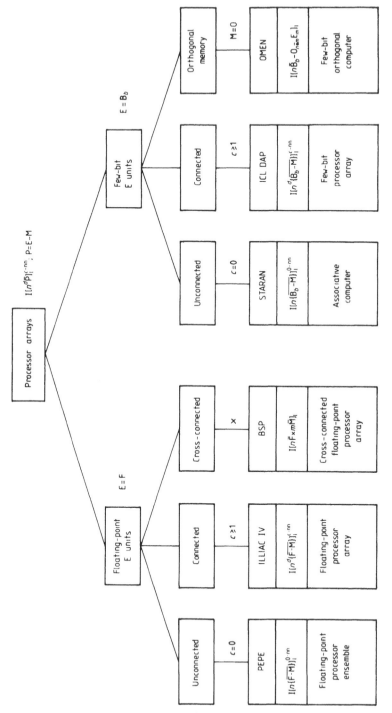

FIGURE 1.7 A classification of the processor arrays discussed in this book. (STARAN is classified for the situation when there is no data permutation in the FLIP network.)

historical interest because it includes the early generations of microprocessor (e.g. an 8-bit microprocessor).

Figure 1.6 covers the introduction of functional parallelism and pipelining into the traditional serial computer concept, and we first differentiate into separate classes computers with and without pipelined E units. On the left is the unpipelined multi-unit scalar computer, such as the CDC 6600, which obtains its performance entirely by functional parallelism. On the right the pipelined computers are first divided into those with or without explicit vector instructions. We find that this division is necessary to separate high-performance scalar computers such as the CDC 7600 from the pipelined vector computers such as the CRAY-1. In other respects these machines are remarkably similar. The pipelined computers with vector instructions are further subdivided into those with separate special-purpose pipelines for each type of arithmetic operation (e.g. the CRAY-1), and those with one or more general-purpose pipelines each capable of performing more than one type of operation (e.g. the CYBER 205). The pipelined computers with only scalar instructions are subdivided into those in which one instruction controls all units at each cycle (horizontal control as in the FPS AP-120B), and those in which instructions are issued to units individually when they are ready to carry out an operation (as in the CDC 7600).

The alternative of obtaining parallelism by the replication of processors under lockstep control is considered in figure 1.7. Again this is divided into the floating-point class and the few-bit class (thus including the lockstep array of 8-bit microprocessors). The next subdivision concerns the connections between the processors, whether these be unconnected (i.e. connectivity $c = 0$) or connected to neighbours ($c \geqslant 1$) in a d-dimensional array. Other forms of connections are lumped loosely under the class of cross-connected processors and memory. Under the few-bit category it seems appropriate to identify the unconnected processors as the classical associative computer and make explicit mention of computers based on an orthogonal memory. Many computers can be programmed to reference both word slices and bit slices, but note that to fit our definition an orthogonal computer must have separate word-slice and bit-slice processors. That is to say it must be specifically engineered to be an orthogonal computer, in order to be classed as one.

1.2.6 A taxonomy of MIMD computers

The vast number of different proposals for multi-instruction stream computers that have appeared in the last few years, only some of which were described in §1.1.8, form a confusing menagerie of computer designs. In order to put these into some kind of order, we show in figures 1.8, 1.9 and 1.10 a possible taxonomy for such MIMD computers (Hockney 1985b, d). In this taxonomy

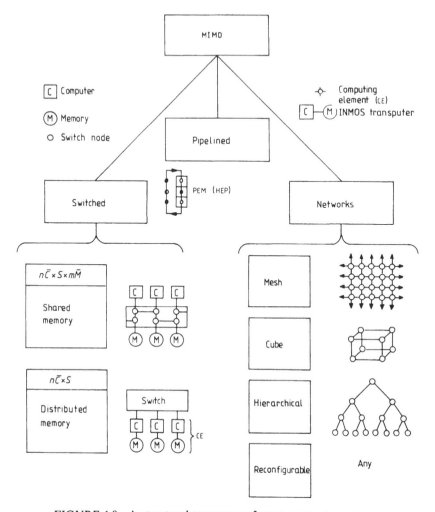

FIGURE 1.8   A structural taxonomy of MIMD computer systems.

we have only included computers controlled by multiple streams of conventional instructions—the so-called *control-flow* computers. Other more novel forms of control for exploiting parallelism, such as *data-flow* and *reduction* computers are considered in §3.2.2.

Figure 1.8 shows the broad division into *pipelined*, *switched* and *network* systems. Multiple instruction streams may be processed either by time-sharing a single sophisticated pipelined instruction processing unit, or by providing separate (and necessarily much simpler) instruction processing hardware for each stream. The first alternative is described as *pipelined MIMD* and is found within a single PEM of the Denelcor HEP (see §3.4.4). MIMD systems using

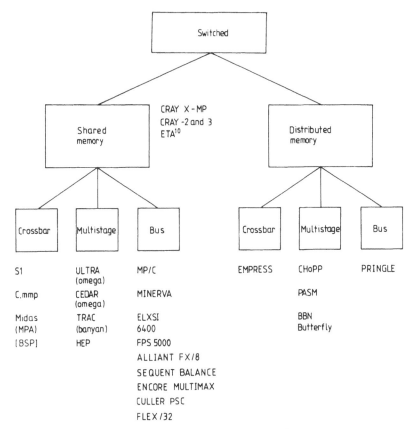

FIGURE 1.9   Subdivisions within the *switched* class of MIMD computer systems. Examples of computers within each class are given below each class name. Those in square brackets show the structure of the class but have SIMD rather than MIMD control. The type of multistage switch is given in parentheses.

the second alternative naturally divide into those with a separate and identifiable switch (*switched MIMD*) and those in which computing elements are connected in a recognisable and often extensive network (*MIMD networks*). In the former all connections between the computers are made via the switch, which is usually quite complex and a major part of the design. In the latter, individual computing elements (CES) may only communicate directly with their neighbours in the network, and long-range communication across the network requires the routing of information via, possibly, a large number of intermediate CES. The CE must therefore provide a small computer (e.g. a microprocessor), a portion of the total system memory, and a number of links for connection to neighbouring elements in the network. Early *network* systems provided these facilities on a board. However, the INMOS Transputer

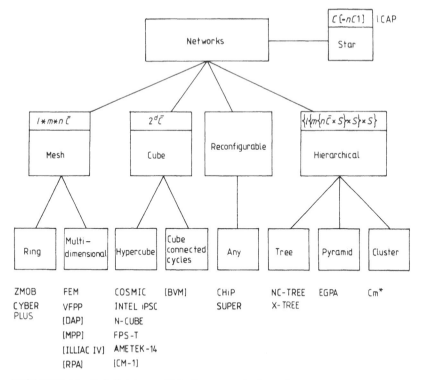

FIGURE 1.10   Subdivisions within the *network* class of MIMD computer systems.

(see §3.5.5) now provides a CE on a chip, and is an ideal building block for *MIMD networks*.

In *network* systems the CES are the nodes of the network and may also be called nodal computers or processors, or processing elements. Within our notation and classification, however, they should be called computing elements in order to indicate that they are complete computers with an instruction processing unit. The term PE is reserved for the combination of arithmetic unit and memory without an instruction processing unit, as is found in SIMD computers such as the ICL DAP (see §3.4.2).

Switched systems are further subdivided in figure 1.9 into those in which all the memory is distributed amongst the computers as local memory and the computers communicate via the switch (*distributed-memory MIMD*); and those in which the memory is a shared resource that is accessed by all computers through the switch (*shared-memory MIMD*). A further subdivision is then possible according to the nature of the switch, and examples are given in figure 1.9 of *crossbar*, *multistage* and *bus* connections in both *shared-* and *distributed*-memory systems. Many larger systems have both shared common memory and distributed local memory. Such systems could be considered as hybrids or simply as *switched MIMD*. However, we prefer to classify them as variations with the shared-memory section, and reserve the *distributed-memory* section for systems with *no* separate shared memory. The classification is based on the location of the memory that is intended for permanent storage of the main data of a problem. Local (or cache) memory that is present in almost all systems and used for temporary storage during calculation is not relevant to the classification.

All *MIMD networks* appear to be distributed-memory systems, but they may be further subdivided according to the topology of the network, as is shown in figure 1.10. The simplest network is the *star*, in which several computers are connected to a common host as in the /CAP configuration of IBM Kingston and Rome. *Single and multidimensional meshes* are exemplified by the CyberPlus 1D ring design and the 2D meshes of the NASA FEM and Columbia VFPP. The ICL DAP, Goodyear MPP and ILLIAC IV are examples of 2D meshes but they are controlled by a single instruction stream, and are therefore SIMD rather than MIMD computers. Binary *hypercube networks* in which there are only two computers along each dimension form an interesting class which is receiving a lot of attention in the Cosmic Cube of CIT and its commercial derivative, the Intel iPSC. Examples also exist of *hierarchical networks* based on trees, pyramids (the EGPA) and bus-connected clusters of computers (Cm*). The most suitable computer network will certainly depend on the nature of the problem to be solved, hence it is attractive to have a MIMD network which may be *reconfigured* under program control. The CHiP computer and the Southampton ESPRIT Supernode computer (see §3.5.5(v)) are designed to satisfy this requirement.

1.3 CHARACTERISATION OF PERFORMANCE

In the last section we have discussed the classification of many varied computer architectures from the simple serial scalar microprocessor, through the multi-unit pipelined vector computers, to the highly replicated processor array. These designs appear to have little in common, yet it is the purpose of this section to define two parameters that characterise the performance of all these computers; and to look upon the serial, pipelined and array-like processor as different members of a continuous spectrum of designs, rather than representing fundamentally different computers. Of course, such a simplistic view can only be regarded as a first-order description, but we believe that it is sufficient to characterise quantitatively the nature of a computer and thereby to determine the best algorithms to use on it. In order to derive this simple generic description of *all* serial and parallel computers we must first examine in more detail the principal ways of increasing the speed of an arithmetic unit.

1.3.1 Serial, pipelined and array architectures

Figure 1.11 illustrates the different ways of performing an arithmetic operation on serial, pipelined and array architectures. As an example, we take the problem of adding two floating-point vectors x_i and y_i $(i = 1, 2,\dots, n)$ to obtain the sum vector $z_i = x_i + y_i$ $(i = 1, 2,\dots, n)$. The operation of adding any pair of the above elements $(x = e \times 2^p$ and $y = f \times 2^q)$ may be divided into four suboperations which, for simplicity, we will assume take the same time to complete. These are: (1) compare exponents, i.e. form $(p - q)$; (2) shift x with respect to y, $(p - q)$ places in order to line up the binary points; (3) add the mantissa of x to the mantissa of y; and (4) normalise by shifting the result z to the left until the leading non-zero digit is next to the binary point. In the serial computer the four suboperations must be completed on the first element pair x_1, y_1 to produce the first result z_1, before the second

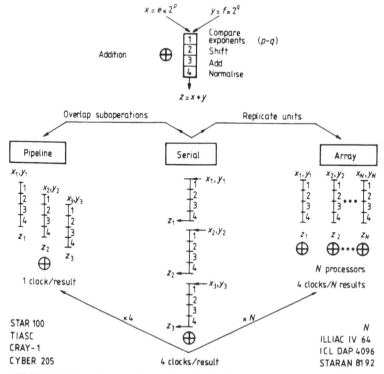

FIGURE 1.11   Comparison of serial, pipelined and array architectures

element pair enters the arithmetic unit. This sequential calculation of the elements of the result vector is illustrated in the centre of figure 1.11, where a time axis running from top to bottom is understood. If l is the number of suboperations (in this case, $l = 4$) and τ is the time required to complete each (usually the clock period) then the time to perform an operation on a vector of length n is

$$t_{\text{serial}} = l\tau n, \tag{1.1a}$$

and the maximum rate of producing results is

$$r_{\infty\,\text{serial}} = (l\tau)^{-1}. \tag{1.1b}$$

In serial operation we notice that the circuitry responsible for each of the l suboperations is only active for $1/l$ of the total time. This in itself represents an inefficiency, which is more obvious if we draw the instructive analogy with a car assembly line. The suboperations that are required to manufacture the sum of two numbers are analogous to the suboperations that are required to manufacture a car: for example, (1) bolt the body to the chassis, (2) attach the engine, (3) attach the wheels and (4) attach the doors. Serial operation corresponds to only one group of men working at a time, and only one car being present on the assembly line. Clearly, in our example, three-quarters of the assembly-line workers are always idle.

The car assembly line obtains its efficiency by allowing a new car to begin assembly in suboperation (1) as soon as the first car has gone on to suboperation (2). In this way a new car is started every τ time units and, when the line is full, a car is completed every τ time units. We often speak of the suboperations as forming a pipeline, and in our example there are four cars at various stages of assembly within the pipeline at any time, and none of the assembly-line workers is ever idle. This is the principle that is used to speed up the production of results in a pipelined arithmetic unit, and is illustrated to the left of figure 1.11. The timing diagram shows that the speed-up is obtained by overlapping (i.e. performing at the same time, or in parallel) different suboperations on different pairs of arguments. The time to perform the operation on a vector of length n is therefore

$$t_{\text{pipe}} = [s + l + (n - 1)]\tau \tag{1.2a}$$

where $s\tau$ is a fixed set-up time that is required to set up the pipeline for the vectors in question, i.e. to compute the first and last addresses for each vector and other overheads. It also includes the fixed time for numbers to be transferred between memory and the arithmetic pipeline. The number of sub-operations (stages or segments) in the pipeline is l and therefore differs for different arithmetic operations. When full, and therefore operating smoothly, a pipeline delivers one result every clock period τ, hence

$$r_{\infty\,\text{pipe}} = \tau^{-1}. \tag{1.2b}$$

Comparing this with equation (1.1b) one sees that pipelining of an operation increases the speed at most by a factor of l—the number of suboperations that are overlapped.

It is obvious from the above that any operations that can be subdivided into roughly equal suboperations can be pipelined. A very common example is the pipelining of instruction processing, in which the overlapped sub-operations might be: (1) instruction decode; (2) calculate operand addresses; (3) initiate operand fetch; (4) send command to functional unit; and (5) fetch next instruction. Computers with pipelined instruction processing units are the IBM 360/91 (one of the earliest), ICL 2980, AMDAHL 470V/6. Other computers, such as the BSP, pipeline on a more macroscopic scale and overlap the operations of (1) memory fetch, (2) unpipelined arithmetic operation and (3) storage of results. Machines that do have an arithmetic pipeline, do not necessarily have vector instructions in their repertoire (e.g. CDC 7600, IBM 360/195). The most notable machines with arithmetic pipelines and vector instructions are the CDC STAR 100 (the first), its derivative the CYBER 205, the TIASC and the CRAY-1.

An alternative way of increasing the speed of arithmetic is to replicate the execution units and form an array of processing elements (PES) under the common control of a single instruction stream. The PES all perform the same arithmetic operation at the same time, but on different data in their own memories. If there are N such processors ($N < n$), the first N argument pairs (x_i, y_i) can be sent to the PES and the first N results found simultaneously in a time of one parallel operation on all elements of the array, say $t_\parallel$ (4τ in our example). The next N elements can then be loaded and also computed in parallel in a further $t_\parallel$ time units. This is repeated until all the elements are computed. The timing diagram is shown on the right of figure 1.11 and we conclude that the time to compute a vector of length n on such a processor array is

$$t_{\text{array}} = t_\parallel \lceil n/N \rceil \tag{1.3a}$$

and

$$r_{\infty\,\text{array}} = N/t_\parallel \tag{1.3b}$$

where $\lceil x \rceil$—the ceiling function of x—is the smallest integer that is either equal to or greater than x. The function gives the number of repetitions of the array operation that are required if there are more elements in the vector than there are processors in the array. The maximum rate of producing results occurs when the vector length n is an integer multiple of the number of processors; and represents a speed-up of N over the operation of a serial processor with the same type of arithmetic unit.

Computers have appeared both in the form of relatively few unpipelined floating-point processors: the ILLIAC IV (64 PEs), the BSP (16 PEs); and in the form of much larger arrays of very simple one-bit processors: the STARAN (256 PEs) and the ICL DAP (4096 PEs). The latter alternative becomes particularly attractive when implemented in VLSI technology, which is ideally suited to the large-scale replication of simple logic. Intermediate possibilities exist for arrays based on commercially available n-bit microprocessors (typically, $n = 4$, 8, 16). Also, for a given performance, there is always the choice between a few highly optimised arithmetic pipelines, and a larger array of simpler processors. This contrast is evident in the following sequence of competing commercial designs: the CRAY-1 (two floating-point pipelines), the BSP (an array of 16 unpipelined floating-point processors) and the ICL DAP (an array of 4096 one-bit processors). The ideas of pipelining and replication can conceptually be combined into an array of pipelined processors. *Circa* 1985, although there are examples of multiple pipelines (e.g. the CDC CYBER 205 and CRAY X-MP), there are no examples of pipelines interconnected as multidimensional arrays.

A good introduction to the principles of pipelining and replication is given by Graham (1970). A more detailed survey of pipeline architecture, including the discussion of several actual designs, can be found in Ramamoorthy and Li (1977) and Kuck (1978). The engineering aspects of effective pipelining of digital systems are discussed in Jump and Ahuja (1978) and in the comprehensive book by Kogge (1981) entitled *The Architecture of Pipelined Computers*. Replicated architectures are surveyed by Thurber and Wald (1975) and Thurber (1976).

1.3.2 The performance parameters $(r_\infty, n_{1/2})$

We shall use as our unifying criterion for the varied architectures described in the last section, the performance of the computer during a single arithmetic operation on a vector of length n. This will be fitted as closely as possible to the following generic formula for the time of the operation, t, as a function of the vector length:

$$t = r_\infty^{-1}(n + n_{1/2}). \tag{1.4a}$$

The two parameters r_∞ and $n_{1/2}$ completely describe the hardware performance of the idealised generic computer and give a first-order description of any real computer. These characteristic parameters are called:

(a) *the maximum or asymptotic performance r_∞*—the maximum rate of computation in floating-point operations performed per second. For the generic computer this occurs asymptotically for vectors of infinite length, hence the subscript. The common unit for floating-point execution is millions of floating-point operations per second (megaflop/s or Mflop/s).

(b) *the half-performance length $n_{1/2}$*—the vector length required to achieve half the maximum performance;

Alternatively if $n < n_{1/2}$ it is often more convenient to express the time by the equivalent formula

$$t = \pi_0^{-1}[1 + (n/n_{1/2})] \tag{1.4b}$$

where $\pi_0 = r_\infty/n_{1/2}$ is called the *specific performance*.

When deriving average values for r_∞ and $n_{1/2}$ from a timing expression for a sequence of vector operations, it is important to remember that t is defined as the time *per vector operation of length n*. Thus if a vector arithmetic operations take a time

$$T = b + cn, \tag{1.4c}$$

then

$$t = \frac{T}{a} = \frac{c}{a}(n + b/c).$$

Hence by comparison with equation (1.4a),

$$r_\infty = a/c, \qquad n_{1/2} = b/c, \tag{1.4d}$$

from which it is clear that speeding up all the circuits of a computer by the same factor f (by dividing b and c by f), for example by decreasing the clock period, increases the asymptotic performance by this factor but does *not* alter $n_{1/2}$.

The significance of the above parameters is quite different. The maximum performance (r_∞) is primarily a characteristic of the computer technology used. It is a scale factor applied to the performance of a particular computer architecture reflecting the technology in which a particular implementation of that architecture is built. Furthermore if we are taking the ratios of the performances (defined as inversely proportional to the time for execution) of different algorithms on the same computer, r_∞ cancels and plays *no* role in the choice of the best algorithm. Therefore this parameter does not appear in our discussion of algorithms in Chapter 5.

The half-performance length ($n_{1/2}$), on the other hand, is a measure of the amount of parallelism in the computer architecture and is not affected by changes in technology. We shall see that it varies from $n_{1/2} = 0$ for serial computers with no parallel operation to $n_{1/2} = \infty$ for an infinite array of processors. It therefore provides a quantitative one-parameter measure of the amount of parallelism in a computer architecture. Because $n_{1/2}$ does not appear as a factor in equation (1.4a), the relative performance of different algorithms on a computer is determined by the value of $n_{1/2}$. The vector length (or average length), n, may be said to measure the parallelism in the problem, and the ratio $v = n_{1/2}/n$ measures how parallel a computer appears to a particular problem. If $v = 0$ or small then an algorithm designed for a sequential or serial environment will be the best; however if v is large an algorithm designed for a highly parallel environment will prove the most suitable. Chapter 5 is therefore mainly a discussion of the influence of $n_{1/2}$ or v on the performance of an algorithm.

It is evident from the timing equation (1.2a) for a pipelined computer that any overhead, such as the set-up time $s\tau$, contributes to the value of $n_{1/2}$, even though it may not represent any parallel features in the architecture. The number of pipeline stages, l, in the same expression does, however, measure hardware parallelism because it is the number of suboperations that are being performed in parallel. It is not therefore strictly true that $n_{1/2}$ always measures hardware parallelism, but we may describe it as measuring the *apparent parallelism* of the hardware. From the user's point of view the behaviour of the computer is determined by the timing expression (1.4a) and the value of $n_{1/2}$, however it arises. A pipelined computer with a large value of $n_{1/2}$ appears and behaves as though it has a high level of real hardware parallelism, even though it may be due to a long set-up time. And it simply does not matter to the user how much of the apparent parallelism is real. For this reason we will not draw a distinction between real and apparent parallelism in the rest of this book; we simply refer to $n_{1/2}$ as measuring the parallelism of the computer.

Expressing the timing equation (1.4a) in terms of a start-up time, t_0, and a time per result, τ, we have

$$t = t_0 + n\tau = \tau(n + t_0/\tau) \qquad (1.4e)$$

which, comparing with equation (1.4a), corresponds to

$$n_{1/2} = t_0/\tau = t_0 r_\infty \qquad (1.4f)$$

where

$$r_\infty = \tau^{-1}.$$

Comparing (1.4a) and (1.4e) leads to two other interpretations of the meaning of $n_{1/2}$. First, $n_{1/2}$ measures the number of floating-point operations which could have been done in the time of the vector start-up, t_0. It therefore measures the importance, in terms of lost floating-point operations, of the start-up time to the user. Secondly, when the vector length equals $n_{1/2}$, the first and second terms of equation (1.4e) are equal, and half the time is being lost in vector start-ups (first term), and only half the time is being used to perform useful arithmetic (second term).

Previous analyses of vector timings by Calahan (1977), Calahan and Ames (1979), Heller (1978) and Kogge (1981) all use linear timing relations like (1.4e), rather than our expression (1.4a). All the above authors recognise the importance of the ratio (t_0/τ), but they do not single it out as a primary parameter. In our view, however, it is not the absolute value of the start-up time that is of primary importance in the comparison of computers and algorithms, but its ratio to the time per result. It is for this reason that we signify this ratio with a descriptive symbol, $n_{1/2}$, and make it central to the analysis.

It will be clear from the above that an $n_{1/2}$ timing analysis can be applied to any process that obeys a linear timing relation such as (1.4a) or (1.4e). Although we do not pursue it in this book, an obvious case is that of input and output operations (I/O). In many large problems I/O dominates the time of calculation. Obtaining data from a disc is characterised by a long start-up time for the movement of arms and track searching, before the first data element is transferred, followed by the rest of the elements in quick succession. That is to say, the time to access n elements obeys equation (1.4e) which is best interpreted using the equivalent equation (1.4a) and the two parameters r_∞ and $n_{1/2}$. The maximum rate, r_∞, is usually measured in Mbytes per second, and $n_{1/2}$ would be the length of the block transferred in bytes required to produce an average transfer rate of $\frac{1}{2}r_\infty$. Values of $n_{1/2}$ obtained for I/O systems are usually very large, indicating that one should transfer large blocks of data $(n > n_{1/2})$ as few times as possible.

1.3.3 The $(r_\infty, n_{1/2})$ benchmark

The maximum performance and half-performance length of a computer are best regarded as experimentally determined quantities obtained by timing the performance of a computer on a test problem which we call the $(r_\infty, n_{1/2})$ benchmark:

```
CALL SECOND(T1)
CALL SECOND(T2)
T0 = T2 - T1

DO 20 N = 1,NMAX
CALL SECOND(T1)

DO 10 I = 1,N
10    A(I) = B(I)*C(I)

CALL SECOND(T2)
20    T = T2 - T1 - T0
```

$$(1.5)$$

or its equivalent assembler code. The DO 10 loop in the above FORTRAN code would be replaced by a vector instruction by any vectorising compiler. The measurement and subtraction of the subroutine-call overhead, T0, is necessary in order to obtain an accurate measurement. NMAX is the maximum vector length and SECOND is a subroutine giving the CPU time in seconds. If the time t is then plotted against the vector length n a straight-line graph such as figure 1.12 arises. The negative of the intercept of the line with the n-axis gives the value of $n_{1/2}$ and the reciprocal of the slope of the line gives the value of r_∞. The results shown are for the CDC CYBER 205 and are typical of a pipeline machine.

We may also derive the value of $n_{1/2}$ from the timing data given by manufacturers. Typically this will be in the form of equation (1.6a)

$$t = \tau[s + l + (n - 1)] \tag{1.6a}$$

whence, by comparison with equation (1.4a), one obtains for the pipeline computer

$$n_{1/2} = s + l - 1 \tag{1.6b}$$

and

$$r_\infty = \tau^{-1}. \tag{1.6c}$$

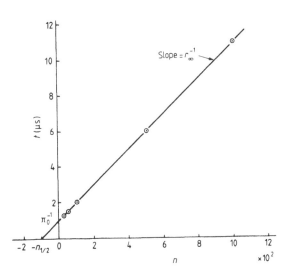

FIGURE 1.12 Measurement of $n_{1/2}$ and r_∞ for a computer from its performance as a function of vector length. Data from Kascic (1979) for a two-pipe CDC CYBER 203E, $n_{1/2} = 100$, $r_\infty = 100$ Mflop/s.

Since the length of the pipeline, l, depends on the operation being carried out, we do not expect $n_{1/2}$ to be absolutely constant for a particular computer. It will depend to some extent on the operations being performed, and how the computer is used. In particular we note that any unnecessary overheads in loop control that are introduced by a compiler, will appear as a software addition to the value of s, and therefore as an increased value of $n_{1/2}$. Notwithstanding these reservations, we regard $n_{1/2}$ as a useful characterisation of vector performance.

For serial computers we compare the timing equation (1.1a) with the generic form (1.4a) and obtain

$$n_{1/2} = 0 \qquad \text{and} \qquad r_\infty = (l\tau)^{-1}. \tag{1.6d}$$

The characterisation of a processor array by the parameter $n_{1/2}$ is less obvious, because the timing formula (1.3a) is discontinuous, as shown in figure 1.13. It is best to distinguish two cases, depending on whether the vector length, n, is less than or greater than the number of processors in the array, N. If $n \leqslant N$, the array is filled or partially filled only once. Thus the time for a parallel operation is independent of n and equal to $t_\parallel$. In this circumstance, from the point of view of the problem, the array acts as though it has an infinite number of processors. Appropriately, the correct limit is obtained in the generic formula (1.4b) if we take

$$n_{1/2} = \infty \qquad \text{and} \qquad \pi_0 = t_\parallel^{-1} \qquad \text{for } n \leqslant N. \tag{1.7a}$$

On the other hand if $n > N$, the processor will have to be filled several times and the best characterisation will be obtained if we take as the generic approximation a line that represents the average behaviour of the array. This is the broken line in figure 1.13, that passes through the centres of the steps which represent the actual performance. We obtain from the intercept and slope of this line

$$n_{1/2} = N/2 \qquad \text{and} \qquad r_\infty = N/t_\parallel \qquad \text{for } n > N. \tag{1.7b}$$

A more complicated situation arises in the case of an array of pipelined processors—for example the CDC NASF design. In this case the actual execution time is given by

$$t = \tau[(s + l - 1) + \lceil n/N \rceil], \tag{1.8a}$$

where N is the number of identical arithmetic pipelines, and $s\tau$ and l are the set-up time and number of segments of each pipe. This result can be understood by thinking of the system as one pipeline that performs its suboperations on superwords, each of which comprises N numbers. Then $\lceil n/N \rceil$ is the number of such superwords that must be processed, and

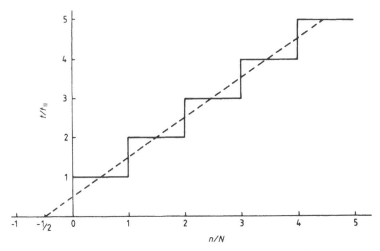

FIGURE 1.13   The time t for a vector calculation of length n on an array of N processors (full line). The nearest linear approximation (broken line) shows that $n_{1/2} = N/2$ and $r_\infty = N/t_\parallel$, where $t_\parallel$ is the time for one parallel operation on the array.

replaces n in the pipeline equation (1.6a). This timing equation is plotted in figure 1.14 (full line) together with the generic approximation (broken line), from which we conclude

$$n_{1/2} = N(s + l - \tfrac{1}{2}) \qquad \text{and} \qquad r_\infty = N/\tau. \qquad (1.8b)$$

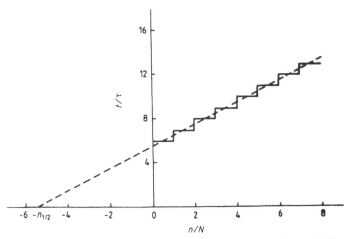

FIGURE 1.14   The measurement of $n_{1/2}$ for an array of N pipelines. The full line is the actual timing data and the broken line is the approximate generic description. In this example $n_{1/2} = 5.5N$, $r_\infty = N/\tau$ and $s + l = 6$.

Neglecting the unimportant difference between 1 and $1/2$ in the formulae (1.6b) and (1.8b) for $n_{1/2}$, we see that replicating the pipelines N times, multiplies not only the maximum rate by N as expected, but also the half-performance length by N. The latter result is also to be expected if one interprets $n_{1/2}$ as the total amount of parallelism in the system, i.e. the total number of things that are done in parallel. The total parallelism in N pipelines is then expected to be N times the parallelism in a single pipeline.

1.3.4   The spectrum of computers

The two parameters $n_{1/2}$ and r_∞ provide us with a quantitative means of comparing the parallelism and maximum performance of *all* computers. It is most instructive to examine the relative position of different computers on the $(n_{1/2}, r_\infty)$ plane. Such a diagram, which may be regarded as separating the computers into a spectrum, is shown in figure 1.15 for most of the computers in this book. The vertical axis (r_∞) is the maximum performance in megaflop/s of the computer and the horizontal axis $(n_{1/2})$ is the amount of parallelism in the architecture. A computer is then represented by a dot in the diagram, except that in most cases a computer may be used in different modes with different levels of parallelism. We thus find that one computer is often represented by a solid line connecting the points representing its different modes of operation. We also show by arrowed lines computer designs that are evolutionary developments of each other. Since we use a logarithmic scale, we arbitrarily plot serial designs $(n_{1/2} = 0)$ on the line $n_{1/2} = 1$. This is quite reasonable because serial computers, when programmed to process a vector, do display a small non-zero value of $n_{1/2}$.

We have noted that replicating a processor N times multiplies both $n_{1/2}$ and r_∞ by N; consequently computer designs that are related by replication without change in technology form diagonal lines at approximately 45°. Examples are the scalar, vector and matrix modes of operation for the ICL DAP, at the bottom right; the 1-, 2- and 4-pipe versions of the TIASC and the 2- and 4-pipe versions of the CYBER 205, in the centre of the diagram; and the evolution of the 16-PE BSP to the 512-PE NASF design of Burroughs. Along such lines the ratio $r_\infty/n_{1/2}$, i.e. the performance per unit parallelism, is constant. This ratio, which we denote by π_0, is more characteristic of a particular computer family than either the maximum performance or half-performance length considered separately. A high value of this ratio is desirable because less parallelism is required in order to achieve a desired performance. However, more advanced technology and cooling methods are required to increase π_0, and the cost rises quickly. The maximum specific performances π_0 for a range of computer architectures are given in table 1.2, in order of decreasing value. The specific performance is also the parameter

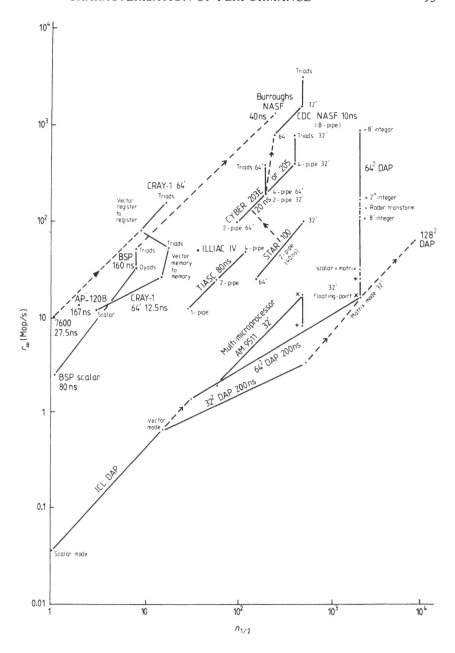

FIGURE 1.15   The two-dimensional spectrum of computers plotted on the $(n_{1/2}, r_\infty)$ plane. Diagonal lines at roughly 45° correspond to development by increased replication. Operations are floating-point except where stated. A prime denotes the number of bits in the representation of a number.

TABLE 1.2   The specific performance, π_0, for a range of parallel computer architectures. Maximum values are quoted.

Computer	Maximum specific performance $\pi_0 = r_\infty/n_{1/2}$ (M/s)
CDC 7600, CRAY-1	10
BSP, Burroughs NASF	7
CYBER 205, CDC NASF	1–4
TIASC	0.4
STAR 100	0.2
ICL DAP	0.008–0.04
Multi-microprocessors (c 1980) e.g. AMD AM9511	~ 0.008–0.02

that determines the performance of a computer on short vectors (see §1.3.5). Hence short-vector performance is unchanged as we pass along these diagonal lines, and is only improved as one passes orthogonally to them, towards the top left of the diagram.

Improvements in technology, in particular a shorter clock period, increase r_∞ without changing $n_{1/2}$, and cause vertical movement in the diagram. This is apparent in the evolution of the CYBER 205 (20 ns clock) to the CDC NASF design (10 ns clock). A reduction in start-up time causes a reduction in $n_{1/2}$ and a movement to the left in the diagram. This also contributes to the change from the CYBER 205 to the CDC NASF design.

The range of computers shown varies from very highly replicated slow arithmetic units of the ICL DAP (high $n_{1/2}$), to computers with a few high-performance pipelined arithmetic units such as the CRAY-1 (low $n_{1/2}$). Roughly the same maximum megaflop performance rate might be expected from a 128 × 128 DAP and a CRAY-1; however the half-performance length of these machines differs by two orders of magnitude and they would have quite different programming characteristics—the CRAY-1 almost having the characteristics of the serial machine (low $n_{1/2}$) and the DAP almost those of an infinite parallel machine (high $n_{1/2}$). The two machines are therefore separated by a large amount horizontally in our spectrum of designs. The horizontal axis thus covers the full range of architectures from the serial computer on the left to the highly parallel computer on the right. It can also be argued that a computer only solves efficiently those problems with a vector length greater than its $n_{1/2}$. Thus the higher the value of $n_{1/2}$, the more limited is the set of problems that it may solve efficiently. In this sense therefore one can regard the $n_{1/2}$ axis as ranging from the most general-purpose on the left to the most specialised on the right.

1.3.5 Vector (SIMD) performance

We have considered so far only the maximum performance r_∞ of a processor on imagined vectors of infinite length. The generic timing formula (1.4a) can be used to define some other numbers that characterise the performance of a computer on actual programs with finite vector lengths. These are:

(a) *the average vector performance*

$$r = n/t = r_\infty/(1 + x^{-1}) = r_\infty \text{pipe}(x) \tag{1.9a}$$

or

$$r = \pi_0 n/(1 + x)$$

where $x = (n/n_{1/2})$, $\text{pipe}(x) = (1 + x^{-1})^{-1}$; and
(b) *the vector efficiency*

$$\eta = r/r_\infty = (1 + x^{-1})^{-1} = \text{pipe}(x). \tag{1.9b}$$

The relation between vector efficiency and vector length is shown in figure 1.16, in which we note from the definition that $\eta = 0.5$ when $n = n_{1/2}$, and that the efficiency asymptotically approaches unity as the vector length

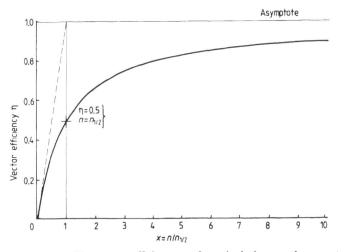

FIGURE 1.16 The vector efficiency η of a calculation on the generic computer as a function of vector length n. The actual processing rate is the maximum rate times the efficiency. This curve, $y = (1 + x^{-1})^{-1}$ is called the pipeline function, $\text{pipe}(x)$.

increases to infinity. It is important to realise that the approach to the asymptote is slow, for example a vector length of $n = 4n_{1/2}$ is required to reach 80% of the maximum performance, and $9n_{1/2}$ to reach 90%. The initial slope of the efficiency curve, which determines the short-vector performance, is given by the broken line. This joins the point $(n = n_{1/2}, \eta = 1)$ to the origin. The above functional form $y = (1 + x^{-1})^{-1}$ appears frequently in performance analyses to describe the way in which an asymptote is approached (see §1.3.6). Owing to its origin here in describing the performance of a pipeline as a function of vector length, we call it the *pipeline function*, pipe(x). Whenever this functional form occurs the performance can be described by two parameters, the asymptotic value (here r_∞) and the half-performance parameter (here $n_{1/2}$, but elsewhere $f_{1/2}$ (§1.3.6) or $s_{1/2}$ (§1.3.6)). For processor arrays, the actual efficiency curve is discontinuous as shown by the full line in figure 1.17. However the broken line shows how the generic approximation with $n_{1/2} = N/2$ gives a good average description of the efficiency of such a computer.

It is instructive to examine the performance of a computer on vectors that are both long and short compared with its half-performance length. Thus, let us consider the following two limiting cases.

(i) *The long-vector limit* $(n \gg n_{1/2}$ and $x \to \infty)$
In this limit, we have from equations (1.4a) and the first equation of (1.9a)

$$t = n/r_\infty; \qquad r = r_\infty. \tag{1.9c}$$

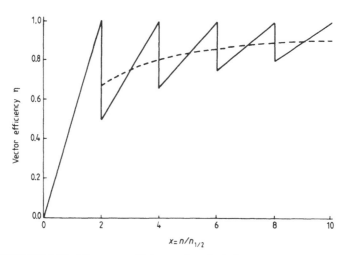

FIGURE 1.17   The vector efficiency of a calculation on a processory array (full line) compared with its approximate representation by the generic computer (broken line).

Thus the time of execution is proportional to vector length, and the performance is constant. Equation (1.9c) is the timing equation for serial computation, in which the time is simply the total arithmetic divided by the rate. Thus, in the long-vector limit any computer behaves like a serial computer even though it may have substantial parallelism and large $n_{1/2}$.

(ii) *The short-vector limit* ($n \ll n_{1/2}$ and $x \to 0$)
In this limit, it is appropriate to use the timing relation (1.4b) and the second equation of (1.9a), and we find

$$t = \pi_0^{-1}; \qquad r = \pi_0 n. \tag{1.9d}$$

Thus the time of execution is constant independent of the vector length, and the performance is proportional to vector length. This is the behaviour of an infinite array of processing elements, since there are, in this case, always enough processors to assign one to each of the vector elements. The computation can, therefore, always be completed in the time for one parallel operation of the array, independent of the length of the vector. Thus, in the short-vector limit all computers act like infinitely parallel arrays, even though their $n_{1/2}$ might be quite small.

We find therefore that r_∞ characterises the performance of a computer on long vectors, whilst π_0 characterises its performance on short vectors, hence the subscript zero.

Many parallel computers have a scalar unit with $n_{1/2} = 0$ and a maximum processing rate $r_{\infty s}$, as well as a vector processing array or pipeline with $n_{1/2} > 0$ and a maximum processing rate of $r_{\infty v}$. The vector breakeven length, n_b, is the vector length above which the vector processor takes less time to perform the operation on a vector than the scalar processor. Using the generic formula (1.4a) we obtain:

$$n_b = n_{1/2}/(R_\infty - 1), \tag{1.10}$$

where $R_\infty (= r_{\infty v}/r_{\infty s})$ is the ratio of maximum vector to maximum scalar processing rates, both measured in elements per second. The relationship (1.10) is plotted in figure 1.18. Generally speaking, it is desirable to have a small value of n_b, otherwise there will be few problems for which the vector processor will be useful. Equation (1.10) shows that this can be achieved by a small value of $n_{1/2}$ or a large ratio between the vector and scalar processing rates. The vector breakeven length and other parameters are given for a selection of computers in table 1.3.

Since the ratio of vector to scalar speed is usually substantial (of order 10), the overall performance of an actual program depends on the fraction v of the arithmetic operations between pairs of numbers (called *elemental operations*) that are performed by vector instructions compared to

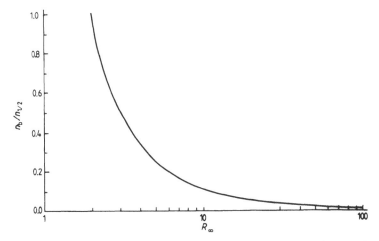

FIGURE 1.18   The influence of the maximum vector to scalar speed ratio R_∞ on the breakeven length n_b.

TABLE 1.3   The characteristic parameters of several parallel computers.

Computer†	$n_{1/2}$	r_∞ (Mflop/s)	R_∞	n_b
64′ CRAY-1	10–20	80	13	1.5–3
48′ BSP	25–150	50	20	1–8
2-pipe 64′ CDC CYBER 205	100	100	10	11
1-pipe 64′ TIASC	30	12	4	7
64′ CDC STAR 100	150	25	12	12
32′ (64 × 64) ICL DAP	2048	16	400	5

† Prime denotes bits.

those that are performed by scalar instructions. We shall refer to this ratio as the fraction of arithmetic vectorised. The average time per elemental operation is then

$$t = vt_v + (1 - v)t_s,$$ (1.11a)

where t_v and t_s are the average times required for an elemental operation when performed with, respectively, a vector or scalar instruction. The rate of execution is $r = t^{-1}$ and is a maximum r_v for complete vectorisation ($v = 1$, $t = t_v$); hence the fraction of the maximum realisable gain that is achieved with a fraction v of the arithmetic vectorised is

$$g = \frac{r}{r_v} = [R + v(1 - R)]^{-1},$$ (1.11b)

where $R = r_v/r_s = t_s/t_v = R_\infty \eta$ is the actual vector to scalar speed ratio for the problem concerned, and is a function of vector length through the efficiency η. Figure 1.19 shows g as a function of v for a variety of values of R from 2 to 1000. It is clear that for large R a very high proportion of the arithmetic must be vectorised if a worthwhile gain in performance is to be realised. Obtaining such levels of vectorisation may not be as difficult as it appears, because the introduction of one vector instruction of length n vectorises n elemental operations, where n may be very large. As a measure of the amount of vectorisation required, we define $v_{1/2}$ as the fraction of arithmetic that must be vectorised in order to obtain one-half of the maximum realisable gain. From equation (1.11b) we obtain

$$v_{1/2} = (R - 2)/(R - 1), \tag{1.12a}$$

which in the limit of large R becomes

$$v_{1/2} = 1 - R^{-1} \simeq \text{pipe}\,(R) \tag{1.12b}$$

The relationship (1.12a) is plotted in figure 1.20.

In the limit of an infinitely fast vector unit $(R \rightarrow \infty)$ we find from equation (1.11b),

$$r = r_{\infty s}/(1 - v). \tag{1.12c}$$

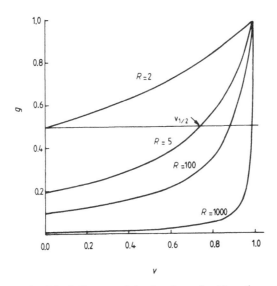

FIGURE 1.19 The influence of the fraction of arithmetic vectorised v on the fraction of the maximum gain that is realised g. R is the vector to scalar speed ratio, $v_{1/2}$ is the value of v required to realise half the maximum gain.

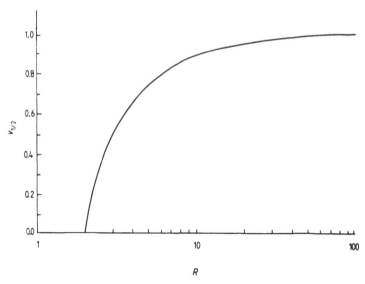

FIGURE 1.20   The fraction of arithmetic that must be vectorised to obtain half the maximum realisable gain, $v_{1/2}$, as a function of the actual vector to scalar speed ratio R.

That is to say, if the vector unit is much faster than the scalar unit, the speed of a vector computer is determined *only* by the speed of its *scalar* unit, $r_{\infty s}$, and the *scalar fraction* of the arithmetic $(1 - v)$. Put in another way, if a tortoise and a hare are in a relay race, the average speed of the pair is almost entirely determined by the speed of the tortoise and how far it has to travel. The speed of the hare is not important because the time taken by the hare is, in any case, negligible. It would not improve matters significantly, for example, if the hare were replaced by a cheetah. This effect is shown by the slow approach of $v_{1/2}$ to its asymptotic value in figure 1.20, as the speed of the vector unit increases. In a sentence, vector computers with slow scalar units are doomed, as can be seen from the history of the CDC STAR 100 (see §1.1.3). The above effect and equations (1.11) and (1.12) are collectively known as *Amdahl's law* (Amdahl 1967), and the steep rise in the curve of figure 1.19 as *Amdahl's wall*. In the design of their vector computer, IBM have concluded that it is not cost effective to build a vector unit that is more than four or five times faster than the scalar unit, although we have seen in table 1.2 that many of the most successful vector computers have values of R_{∞} significantly higher than this.

The Amdahl law arises whenever the total time of some activity is the sum of the time for a fast process and the time for a slow process. The same law therefore applies when a job is subdivided for execution on a MIMD multiprocessor. This is often called the problem of *multi-tasking* a job. If the number of processors is large, say N, then the speed of work that is subdivided for parallel execution on the N processors is N times the speed of execution of those parts of the job that must be executed sequentially on a single processor. Thus, for large N, the average speed of the job is primarily determined by the speed of a single processor (necessarily very slow if N is large) and the fraction of work that cannot be multi-tasked, because it must be processed sequentially, i.e. equation (1.12c) again (Larson 1984, 1985).

1.3.6 MIMD performance

It is important to realise that the adoption of multi-instruction stream programming *does* have a cost. Such MIMD programming incurs additional overheads and associated expenses, and furthermore the SIMD saving is lost. Whether or not a MIMD solution to a problem is advantageous obviously depends on the magnitude of these costs.

If, in a MIMD computer, p microprocessors are all executing the identical sequence of instructions (i.e. taking the same route through the same program) then p identical instructions are fetched from memory into p identical instruction processing units (IPUS), when only a single instruction fetch and IPU is really needed. In this case both memory bandwidth and silicon area are being wasted. A SIMD computer would be much more effective for such a job, because it would indeed only fetch and process a single instruction into its master control unit, and the p processing elements would perform identical operations under common control, as is required. This saving of instruction fetch and processing is known as the *SIMD saving*, and is the main reason why such computers were invented. The saved silicon area can be used to add more parallelism into the arithmetic units and hence increase their speed, and the saved memory bandwidth will reduce memory contention for data and associated delays.

It is highly likely that MIMD computers will be executing identical programs in all their microprocessors, for who is going to write, for example, a thousand different subroutines for a thousand-processor MIMD system? Of course, the key point is that identical instructions are not necessarily executed, even if the programs in the microprocessors are the same, because there may be data-dependent branches in the program which could cause different routes to be taken through the program by different microprocessors. A prime example is a Monte Carlo simulation where, by the choice of a different set of random numbers as data, different routes are deliberately taken through a common program every time it is executed. Such problems clearly require the multiple instruction streams of a MIMD computer for their solution, and are quite unsuited to solution on a SIMD computer.

There are, however, many cases where the natural subdivision of a problem onto a MIMD computer leads to the execution of an identical sequence of instructions by all the microprocessors. Consider, for example, the solution of p independent sets of tridiagonal equations, one set given to each of p processors, or the calculation of p independent fast Fourier transforms. Both these problems occur at different stages of the solution of partial differential equations by transform methods (see §5.6.2), and contain no data-dependent branches. They are therefore ideally suited to SIMD solution and do not require the multiple instruction streams of a MIMD computer.

The three problems (or overheads) associated with MIMD computing are:

(1) *scheduling* of work amongst the available processors (or instruction streams) in such a way as to reduce, preferably to zero, the time that processors are idle, waiting for others to finish;

(2) *synchronisation* of the processors so that the arithmetic operations take place in the correct sequence;

(3) *communication* of data between the processors so that the arithmetic is performed on the correct data.

The problem of communication of data from memory to the arithmetic units is present on all computers and causes the difference between peak performance rates of arithmetic pipelines usually quoted by manufacturers, and the average performance rates found for realistic problems. Scheduling and synchronisation are, however, new problems introduced by MIMD computation. Three parameters are used to quantify the problems, E_p for scheduling, $s_{1/2}$ for synchronisation and $f_{1/2}$ for communication (Hockney 1987b, c, d).

(i) Scheduling parameter: E_p

Scheduling is the most commonly studied problem in MIMD computation; indeed until recently it was the only problem of the three that had received much attention. Most of the literature on parallel algorithms is concerned with the problem of scheduling the work of different algorithms onto a p-processor system, on the assumption that the time taken for synchronisation and data communication can be ignored. For the moment we shall also make this assumption, because we wish to consider the latter two effects separately. Following the work of David Kuck's group at the University of Illinois (Kuck 1978 p 33) we introduce the *efficiency of scheduling*, E_p, of work amongst p processors as follows. If T_1 is the time to perform all the work on one processor, and T_p is the time to perform the work when it is shared amongst the p processors of the same type, then

$$E_p = \frac{T_1}{pT_p} \leqslant 1. \tag{1.13}$$

Perfect scheduling occurs when it is possible to give $(1/p)$th of the work to each processor, when $T_p = T_1/p$ and $E_p = 1$. If the work cannot be exactly balanced between the p processors, then some processors finish before others and become idle, making $T_p > T_1/p$ and $E_p < 1$. For this reason scheduling is sometimes referred to as *load balancing*.

If r_∞ is the maximum performance of the p-processor system, then each processor has a maximum performance of r_∞/p. Also, if we consider the processors to be serial computers ($n_{1/2} = 0$), as would be the case if they were microprocessors, and let the work be quantified as s floating-point arithmetic operations, then the time to perform all the work on one processor is

$$T_1 = \frac{s}{(r_\infty/p)} = \frac{sp}{r_\infty} \tag{1.14}$$

and the time for parallel execution on p processors is

$$T_p = \frac{T_1}{pE_p} = \frac{s}{r_\infty E_p}. \tag{1.15}$$

(ii) Synchronisation or grain size parameter: $s_{1/2}$
We introduce the synchronisation overhead into the above timing formula (1.15) by defining the concept of *work segment* as the basic unit out of which MIMD programs are built up. A work segment is a body of work lying between two synchronisation points, which is scheduled into p completely independent tasks, one of which is given to each of the p processors. 'Independent' means that there is no communication (i.e. data transfer) between the tasks of the work segment. The extent of the work segment is determined by the fact that it lies between two synchronisation points. That is to say all work started prior to the work segment must finish before the work segment is started, and all work within the segment must finish before the next part of the program proceeds. In scientific programming the amount of work is usually equated to the number of floating-point operations in the segment.

If the amount of work that is divided up for independent parallel execution on the p processors is large, we say that the *grain* of the parallelism is large, or that the program exhibits large-grain parallelism. Similarly, if the amount of work divided up is small, we speak of small-grain parallelism. The main difference between SIMD and MIMD programming is the grain size of the program parallelism. In SIMD computing synchronisation takes place by hardware after every vector operation and the grain of the parallelism is equal to the vector length and is often quite small. In MIMD computing the grain size is usually much greater and may involve a whole phase of an algorithm (e.g. Fourier transform over a whole three-dimensional mesh). We can quantify the idea of grain size by equating it to s, the amount of work in a work segment defined in the previous paragraph. Put slightly differently, the grain size of a SIMD program is the work within the innermost DO loops that are replaced by vector instructions. MIMD programs are more usually subdivided at the outermost DO loops, and the grain size equals large segments of the total program.

The above act of synchronisation will take a certain time, t_0, depending on the hardware facilities provided on the computer, and on the software facilities provided by the operating system and computer language. The *synchronisation parameter* $s_{1/2}$ is defined as the amount of arithmetic that could have been done (at the maximum rate of r_∞) during the time taken for synchronisation (Hockney 1985c). Thus, $s_{1/2} = r_\infty t_0$ is the amount of arithmetic that is, so to say, lost because of the need for synchronisation. It therefore measures the importance of the synchronisation overhead to the user, because it measures it in units of lost arithmetic which is the currency of the user's problem. If we add the synchronisation time to the time for the actual work previously computed, T_p, we get the time for a work segment to be

$$t_w = t_w t_0 + T_p = r_\infty^{-1}[(s/E_p) + s_{1/2}]. \tag{1.16}$$

For a problem that can be perfectly scheduled ($E_p = 1$), equation (1.16) for MIMD computation has the same form as equation (1.4a) for a SIMD vector instruction, with the amount of arithmetic, s, being analogous to the vector length, n, and $s_{1/2}$ being analogous to $n_{1/2}$. Thus $s_{1/2}$ is also the amount of arithmetic required in a work segment in order for the average computing rate to be half the maximum, i.e. $r_\infty/2$. In this situation, when $s = s_{1/2}$ only half the time is spent on useful arithmetic, the other half being spent on synchronising the work. The system is then being used with 50% effectiveness. The parameter $s_{1/2}$ may therefore be called the *half-performance grain size*. The average performance for other values of the amount of arithmetic s is given by

$$\bar{r} = \frac{s}{t_w} = \frac{r_\infty E_p}{1 + (s_{1/2} E_p/s)} = r_\infty E_p \, \text{pipe}(s/s_{1/2} E_p). \tag{1.17}$$

This expression is the same as equation (1.9a) and figure 1.16 for the average vector performance as a function of vector length, with r_∞ replaced by $r_\infty E_p$, and $n_{1/2}$ replaced by $s_{1/2} E_p$.

It is important to know the value of $s_{1/2}$ for a MIMD computer because it is a yardstick which can be used to judge the grain size of the program parallelism which can be effectively used. In other words, it tells the programmer how much arithmetic there must be in a work segment before it is worthwhile splitting the work between several processors. If we regard an effectiveness of less than 50% as unacceptable, then by definition the grain (or segment) size, s, must exceed $s_{1/2}$. If, on the other hand, we ask for the *break even grain size*, s_b, above which it is faster to suffer the synchronisation overhead and split the job between p processors than to avoid synchronisation altogether by using a single processor, we obtain, equating (1.14) and (1.16):

$$s_b = s_{1/2}/(p - E_p^{-1}), \qquad (1.18)$$

For perfect scheduling ($E_p = 1$) this result is exactly analogous to the vector break even length, n_b, calculated in equation (1.10) and plotted in figure 1.18, with the number of processors, p, taking the place of the ratio of vector to scalar processing rates R_∞.

Like $n_{1/2}$, the synchronisation parameter, $s_{1/2}$, should be considered as an experimental quantity to be measured on any MIMD system, and such measurements are reported for the two-CPU CRAY X-MP and IBM /CAP in Chapter 2 (Hockney 1985a, 1987d), and the Denelcor HEP in Chapter 3 (Hockney and Snelling 1984, Hockney 1985c). The programs used differ in detail according to the synchronisation and multi-tasking software provided by the system. However, the principle is as follows. As with the measurement of $n_{1/2}$ in program (1.5) an element-by-element vector multiply is chosen as the task to be subdivided amongst the instruction streams. When the vector length is a multiple of the number of instruction streams (or processors) the work can be perfectly scheduled, and for these vector lengths we measure the time, t, as a function of the amount of arithmetic, s, and fit the results to

$$t = r_\infty^{-1}(s + s_{1/2}). \qquad (1.19a)$$

and the average performance is given by the pipeline function

$$r = r_\infty \mathrm{pipe}(s/s_{1/2}). \qquad (1.19b)$$

The same subroutine is given to each instruction stream which consists of the DO 10 loop of program (1.5) with parameters specifying the starting and ending values of I. These are set by the master control program to different values in each instruction stream, to ensure that each stream performs a different section of the original vector multiply operation.

Different cases can be distinguished for the synchronisation measurement according to the different synchronisation primitives that are provided, and in a network system according to whether the required data is in the processors or must be transferred across the network. In the latter case one is also including the cost of communication in the measurement. To study synchronisation alone one would assume the data was available in the local memory of each processor. A study of the worst and best values for r_∞ and $s_{1/2}$ gives a span of values between which those applicable in actual problems can be expected to lie. In a tightly coupled system which has fast communication between the processors, the range between best and worst cases should be quite small; but in a loosely coupled system with slow communication between processors the range of values could be large. The variation of $s_{1/2}$ with communication delays is considered in the $(\hat{r}_\infty, \hat{s}_{1/2}, f_{1/2})$ benchmark described in §1.3.6, part (iv), below.

(iii) *Communication parameter*: $f_{1/2}$

In all the tests used above to measure $n_{1/2}$ and $s_{1/2}$ we have always taken element-by-element vector multiply as a standard problem, and distinguished different cases with different values for $(r_\infty, n_{1/2}, s_{1/2})$ depending on whether the vectors were stored in local registers or global main memory. If the processing rate of an arithmetic pipeline is faster than the rate at which data can be obtained from and returned to main memory, then there will be a memory access (or communication) bottleneck, and the effective processing rate of the computer will improve as the amount of arithmetic performed per memory access increases. This situation, which we refer to as inadequate memory bandwidth, is common in supercomputers because of their high-performance arithmetic pipelines. Our test case of vector multiply has only 1/3 of a floating-point operation per memory access, which is the worst possible case. Many calculations permit a much higher value and the following simple analysis attempts to take this into account.

Memory access may be approximately modelled (Hockney 1985d, Curington and Hockney 1986) by assuming that each vector memory access obeys the timing relation

$$t = (n + n_{1/2}^m)/r_\infty^m. \tag{1.20a}$$

and that vector operations in the arithmetic pipeline obey the timing relation

$$t = (n + n_{1/2}^a)/r_\infty^a. \tag{1.20b}$$

Introducing the important new variable f (the *computational intensity*), which is defined as the number of floating-point operations per memory reference, and expressing the average time per vector arithmetic operation as

$$\bar{t} = (n + \bar{n}_{1/2})/\bar{r}_\infty \tag{1.21a}$$

then, in the case that memory transfers and arithmetic cannot be overlapped, we find that

$$\bar{r}_\infty = \frac{r_\infty^a}{(1 + x^{-1})} = \hat{r}_\infty \ \text{pipe}(f/f_{1/2}) \tag{1.21b}$$

$$\bar{n}_{1/2} = \frac{n_{1/2}^m + n_{1/2}^a x}{(1 + x)} \tag{1.21c}$$

where

$$\hat{r}_\infty = r_\infty^a, \qquad f_{1/2} = r_\infty^a / r_\infty^m \qquad \text{and} \qquad x = f/f_{1/2}.$$

Thus the ratio x determines the extent to which the average performance of the combined arithmetic pipeline and memory approaches asymptotically that of the arithmetic pipeline alone. The peak performance $\hat{r}_\infty$ is defined as the performance in the limit $f \to \infty$, and is equal in this model to r_∞^a. Equation (1.21b) shows that the manner in which this asymptote is reached is identical to the way the average performance of a single pipeline, r, approaches its asymptote, r_∞, as a function of vector length n, namely like the pipeline function pipe(x) (see equation (1.9a) and figure 1.16). Hence, in analogy to $n_{1/2}$, we introduce another parameter $f_{1/2}$ (the *half-performance intensity*) which is the value of f required to achieve half the peak performance. The parameter $f_{1/2}$ is a property of the computer hardware alone and provides a yardstick to which the computational intensity, f (which is a function of the algorithm and application alone), should be compared in order to estimate the average performance, $\bar{r}_\infty$. If $f = f_{1/2}$, then the average performance is half the maximum possible (as with $n_{1/2}$); but if we require 90% of the maximum performance $(0.9r_\infty^a)$ then we need $f = 9\, f_{1/2}$. Equation (1.21c) shows how the $n_{1/2}$ of the combined memory and arithmetic pipeline varies from that of the memory for small f to that of the arithmetic pipeline for large f. Figures 1.21(a) and (b) show measurements of $f_{1/2}$ for an FPS 164 linked by a channel to an IBM 4381 host (Hockney 1987c). Similar measurements have also been reported for the FPS 5000 (Curington and Hockney 1986).

The value of $f_{1/2}$ is determined by plotting f/r_∞ versus f, which should be approximately linear. The inverse slope of the best straight line is $\hat{r}_\infty$ and the negative intercept with the f axis is $f_{1/2}$, because rearranging equation (1.21b) we have

$$f/r_\infty = \hat{r}_\infty^{-1}(f + f_{1/2}). \tag{1.21d}$$

If, on the other hand, memory transfers may take place at the same time as the arithmetic pipeline performs arithmetic, it is possible to overlap memory transfer with arithmetic. In this case, a different functional form arises which we define as the knee function. This is the truncated ramp function

$$\text{knee}(x) = \begin{cases} x & \text{if } x \leqslant 1 \\ 1 & \text{if } x \geqslant 1. \end{cases} \tag{1.22a}$$

The performance of the combined pipelines with memory transfer overlap is then

$$r_\infty = \hat{r}_\infty \, \text{knee}(0.5\, f/f_{1/2}) \tag{1.22b}$$

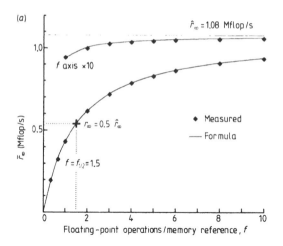

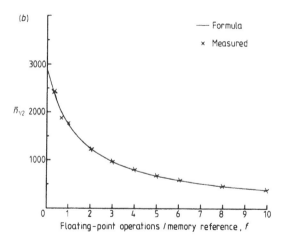

FIGURE 1.21 (a) Variation of asymptotic performance, $\bar{r}_\infty$, as a function of the number of floating-point operations per memory reference, f, for an FPS 164 linked via a channel to an IBM 4381 host. Data held in the host is transferred to the FPS 164 for computation, then returned. The measured values are fitted to equation (1.21b) with $\hat{r} = 1.08$ Mflop/s and $f_{1/2} = 1.5$. The computation is a dyadic vector multiply, $A(I) = B(I) * C(I)$; $I = 1, N$. (From Hockney (1987c).) (b) The average half-performance length, $\bar{n}_{1/2}$, as a function of floating-point operations per memory reference, f, for an FPS 164 linked via a channel to an IBM 4381 host. The measured values are fitted to equation (1.21c) with $f_{1/2} = 1.5$, $n_{1/2}^m = 2900$ and $n_{1/2}^a = 4$. (From Hockney (1987c).)

where

$$\hat{r}_\infty = r_\infty^a \qquad \text{as before, but now} \qquad f_{1/2} = 0.5\, r_\infty^a / r_\infty^m. \qquad (1.22c)$$

Thus, the effect of allowing memory transfer overlap is to halve the value of $f_{1/2}$, and therefore reduce the amount of arithmetic per memory reference that is required to achieve a given fraction of the peak performance.

In the above analysis, if the memory and arithmetic pipeline are in the same CPU, we have an analysis of a memory-bound single instruction stream/single data stream (SISD) or SIMD computer. If, on the other hand, the memory and arithmetic pipeline are in different CPUs, we have an analysis of the communication overhead in a MIMD computer. In both cases the key hardware parameter is $f_{1/2}$ which in this model of computation is proportional to the ratio of arithmetic performance to the rate of data transfer. Other analyses of performance degradation due to data communication problems are given by Lint and Agerwala (1981) and Lee, Abu-Sufah and Kuck (1984).

(iv) The $(\hat{r}_\infty, \hat{s}_{1/2}, f_{1/2})$ benchmark

In the above discussion we have considered the overheads of synchronisation for constant f (part (ii)), and the effects of communication delays without synchronisation (part (iii)). In this section we consider a more general benchmark which includes both effects, and in particular shows the variation of $s_{1/2}$ with f.

The vector A is to be computed as the element-by-element vector multiply of the vectors B and C, by dividing the work as equally as possible amongst p processors. The vectors B and C initially reside in the main system memory (wherever that might be), and the result vector A is to be assembled in the same memory. Thus each of the p processors must be sent a portion of the vectors B and C, perform the partial vector multiply, and return its portion of the result vector A to main memory. The job is finished only when all the processors have returned their partial vectors to main memory. That is to say, a synchronisation point must be programmed after the calculational part of the program (e.g. by including a barrier statement) and included in the timing measurement. In order to vary the ratio of computation to communication, the multiplication of the partial vectors B and C in each processor is repeated $3f$ times. This maintains the previous definition of f, since there are three I/O operations (input B, C; output A). Furthermore, when $f = 1/3$, the vector operation is performed once as described in part (ii).

As before we describe the overall timing of the benchmark as follows:

$$t = r_\infty^{-1}(s + s_{1/2}) \quad \text{giving} \quad r = r_\infty \, \text{pipe}(s/s_{1/2}). \tag{1.23}$$

However, now we separate the synchronisation, communication, and calculational parts of the time and consider the case that arithmetic and communication cannot be overlapped in time. Then the times add and

$$t = t_0(p) + t_t(p)m + t_a(p)s \tag{1.24a}$$

$$t = t(\text{start-up and synchronisation}) + t(\text{communication}) + t(\text{calculation}) \tag{1.24b}$$

where the three terms are identified respectively with start-up and synchronisation, communication and calculation as shown. The terms given in the equation are explained as follows. The term s is the total number of floating-point operations (flop) performed in all processors; m is the number of I/O data words in the work segment (see below); f equals s/m and gives the floating-point operations per I/O data word; $t_0(p)$ is the time for the null job $(s = m = 0)$; $t_t(p)$ is the time per I/O word on average; and $t_a(p)$ is the time per floating-point operation, on average.

The problem variable m quantifies the communication that a work segment has with the rest of the program. If, as is usual, the body of a work segment is written as a subroutine, it is the number of words contained in all the input and output variables and arrays of the subroutine, for all instantiations of the subroutine. In the case of the above benchmark, if the vector length is n, then $m = 3n$ (two input vectors and one output vector), and $s = 3nf$, because the arithmetic is repeated $3f$ times.

A little rearrangement shows that

$$r_\infty = \hat{r}_\infty \text{ pipe}(f/f_{1/2}); \qquad s_{1/2} = \hat{s}_{1/2} \text{ pipe}(f/f_{1/2}) \tag{1.25a}$$

and

$$r = r_\infty \text{ pipe}(s/s_{1/2}) = \hat{r}_\infty \text{ pipe}(f/f_{1/2}) \text{ pipe}(s/s_{1/2}) \tag{1.25b}$$

or

$$r = \frac{\hat{r}_\infty}{1 + (f_{1/2}/f) + (\hat{s}_{1/2}/s)} \tag{1.25c}$$

where

$$\hat{r}_\infty = t_a^{-1}, \qquad \hat{s}_{1/2} = t_0/t_a \qquad \text{and} \qquad f_{1/2} = t_t/t_a. \tag{1.25d}$$

The interpretation of equations (1.25) is as follows. The primary hardware parameters which determine the behaviour of the MIMD computer are the time parameters t_0, t_t and t_a. However, these only affect program performances through the ratios in equation (1.25d). The peak performance is $\hat{r}_\infty$, the inverse of the arithmetic time. The peak value of $s_{1/2}$, namely $\hat{s}_{1/2}$, is the ratio of synchronisation to arithmetic time, i.e. the maximum possible arithmetic lost during synchronisation. The communication parameter $f_{1/2}$ is the ratio of the communication time to the arithmetic time, i.e. the ratio of computation rate to communication rate (i.e. bandwidth). The overall performance parameters $(r_\infty, s_{1/2})$ are less than the peak hardware values because of the communication delays, and this variation is expressed by the pipeline function in equation (1.25a).

Having computed the overall parameters, the actual average performance is given in equation (1.25b), where the last expression shows how the peak rate is degraded by communication delays (first pipe function), and by inadequate grain size (second pipe function). In equation (1.25b) it must be remembered that $s_{1/2}$ is a function of f through equation (1.25a). The true variation of average rate with f and s is seen more clearly in equation (1.25c), where only the true hardware parameters are used. In equation (1.25c) the first term in the denominator comes from the arithmetic time, the second term comes from the communication time, and the third term from the synchronisation time.

Figures 1.22(a) and (b) show measurements of the overall performance parameters $(r_\infty, s_{1/2})$ of equation (1.23) for the $(\hat{r}_\infty, \hat{s}_{1/2}, f_{1/2})$ benchmark applied to the IBM /CAP parallel computer system (see §2.5.9), in which ten FPS 164 computers are attached by channels to an IBM 4381 host (Hockney 1987d). Since $f_{1/2} = 2$ the upper curves for $f = 100$ in these figures correspond to $\hat{r}_\infty$ and $\hat{s}_{1/2}$, and the lower curves show the performance degradation which occurs when f is small. The broken and chain curves (fit 1 and fit 2) are fits obtained using equations (1.25) with the following parameters:

$$\hat{r}_\infty = 1.08 \ p \quad \text{Mflop/s}$$

$$\hat{s}_{1/2} = 1.08 \ p \ (2500 + 4777 \ p) \quad \text{flop} \tag{1.26}$$

$$f_{1/2} = 2.02 \quad \text{flop/ref.}$$

These parameters tell us that in order to be suitable for solution on this computer system, a problem should have $f = 8$ flop/ref giving $f/f_{1/2} = 4$ and an asymptotic performance of 80% of the peak performance. The $s_{1/2}$ for ten processors would be about 400 kflop, and for 80% effectiveness one would need a grain size of $4 \times s_{1/2} \simeq 1.6$ Mflop. This is the numeric meaning of the statement that this system is loosely coupled and therefore only suitable for the solution of problems exhibiting large-grain parallelism (of the order 1 Mflop). Fortunately, many problems of physics do satisfy this criterion (Fox and Otto 1984).

(v) *Program performance*

The previous sections have considered only the timing of a single work segment which can be distributed amongst p processors. To consider a complete program comprising many work segments, one must take into account that

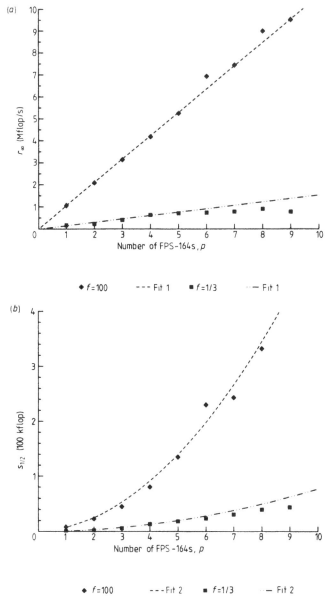

FIGURE 1.22   (a) The asymptotic performance, r_∞, as a function of the number of processors used, p, with $f = 100$ and $f = 1/3$, for the IBM /CAP. The top curve is effectively $\hat{r}_\infty(p)$. The broken and chain curves for fit 1 correspond to the values (1.26) in equation (1.25a). (b) The synchronisation parameter, $s_{1/2}$, as a function of the number of processors used, p, with $f = 100$ and $f = 1/3$, for the IBM /CAP. The top curve is effectively $\hat{s}_{1/2}(p)$. The broken and chain curves for fit 2 correspond to the values (1.26) in equation (1.25a).

some of these might contain code that has to be executed sequentially on a single processor in order to get the correct results. Let us therefore divide the time of execution of the original sequential program, T_1, into two parts

$$T_1 = t_{ser} + t_{par} \tag{1.27}$$

where t_{ser} is the time for the essentially serial part of the code, and t_{par} is the time, when executed sequentially, for the part that can be executed in parallel. When the program is parallelised the best parallel time will be

$$T_p = t_{ser} + \frac{t_{par}}{p} \tag{1.28}$$

where only the time of the second term is reduced by parallel execution. We have, of course, assumed that synchronisation and communication time are negligible. If we define the performance $R(p)$ of the program to be the inverse of the executing time, then

$$R(p) = T_p^{-1} = (t_{ser} + t_{par})^{-1} = R_\infty[1 + (p_{1/2}/p)]^{-1} \tag{1.29a}$$

which can also be written

$$R(p) = R_\infty \, \text{pipe}(p/p_{1/2}) \tag{1.29b}$$

where

$$R_\infty = t_{ser}^{-1} \qquad \text{and} \qquad p_{1/2} = t_{par}/t_{ser}. \tag{1.29c}$$

Thus the asymptotic rate R_∞ is the inverse of the time for the part of the program that could not be parallelised. This occurs in the model when the number of processsors $p \to \infty$ which reduces the time for the parallelisable part of the program to zero. We notice that the functional form of the approach to the asymptotic rate is again that of the pipeline function $\text{pipe}(x)$, and $p_{1/2}$, as usual, is the number of processors required to achieve half of the asymptotic performance.

The usual parameter that is used to compare the performance of algorithms is the *speed-up*, S_p (Kuck 1978), which is defined as

$$S_p = T_1/T_p = R(p)/R(1) = S_\infty[1 + (p_{1/2}/p)]^{-1} = S_\infty \, \text{pipe}(p/p_{1/2}) \tag{1.30a}$$

where T_1 and T_p are, respectively, the times for the algorithm to run on one or p processors. From which we conclude

$$S_\infty = R_\infty/R(1) = (1 + p_{1/2}) = T_1/t_{ser}$$

$$= (\text{fraction of original program time } not \text{ parallelised})^{-1}. \tag{1.30b}$$

Both the pipeline function equations (1.29b) and (1.30a) are expressions of Amdahl's law (Amdahl 1967), that the performance or speed-up cannot exceed that obtained if the vectorised or parallelised parts take zero time. The maximum rate is then determined by the time for the unvectorised or unparallelised part of the program.

Many parallelised programs are found to follow the functional dependence of equation (1.29b), at least for small values of p. In practice, however, as p becomes larger, synchronisation and other overheads usually increase rapidly with p, with the result that there is often a maximum in the program performance, and a subsequent reduction in performance as the number of processors further increases. This observed behaviour has been fitted to the function

$$R(p) = R_\infty \left\{ 1 + \frac{p_{1/2}}{p} \left[1 + \frac{1}{n-1} \left(\frac{p}{\tilde{p}} \right)^n \right] \right\}^{-1} \tag{1.31a}$$

In this formula the overhead of synchronisation is represented by the factor in braces, and the rapidity with which it increases with p is determined by n which we call the *index of synchronisation*. Clearly, it is desirable for it to be as small as possible.

The magnitude of the synchronisation overhead is determined by $\tilde{p}$, with larger values meaning a smaller overhead. The values of these parameters are likely to vary considerably between different computer software systems. The maximum performance occurs when $p = \tilde{p}$ and has the value

$$\tilde{R} = R_\infty \left[1 + \frac{n}{n-1} \left(\frac{p_{1/2}}{\tilde{p}} \right) \right]^{-1} \tag{1.31b}$$

As a practical example of parallelisation we have taken a large program for the simulation of electron flow in semiconductors, called FET6, and converted it for parallel execution on the IBM /CAP system at IBM ECSEC, Rome (Hockney 1987c). The program calculates new positions and velocities of approximately 10 000 electrons step-by-step in time, and is completely described in Hockney and Eastwood (1981), Chapter 10. Each time-step comprises three stages: *charge assignment*, *potential solution*, and *acceleration*. During the acceleration stage the motion of all electrons is independent, so that this part of the program can be parallelised by giving $1/p$ of the electrons to each of the p processors. Since this is the most time-consuming part of the time-step, the other parts of the time-step were not altered and thus form the sequential part of the program.

Figure 1.23 shows actual performance measurements of FET6 on multiple FPS 164s. The timing of the parallelised program is fitted well by equation (1.31a), in the range $1 \leqslant p \leqslant 10$, with the following parameters (equivalent speed-up ratios to $R(1)$ are given in parentheses):

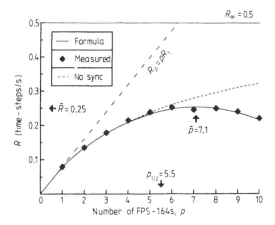

FIGURE 1.23 Absolute performance of the FET6 program on one to ten FPS 164s. Note that the performance reaches a peak of $\tilde{R}$ for $\tilde{p}$ processors. The measured values are fitted to equation (1.31a) with $R_\infty = 0.5$, $p_{1/2} = 5.5$, $\tilde{R} = 0.25$, $\tilde{p} = 7.1$ and $n = 5$. If the overhead of synchronisation could be eliminated the performance would follow the broken curve which asymptotes at R_∞.

$$R_\infty = 0.5 \text{ time-steps per second} \quad (S_\infty = 6.2)$$
$$p_{1/2} = 5.5 \qquad \tilde{p} = 7.1 \qquad n = 5. \tag{1.32}$$

The maximum value of the performance is then, from equation (1.31b),

$$\tilde{R} = 0.25 \text{ time-steps per second} \quad (\tilde{S} = 3.1). \tag{1.33}$$

We can interpret the significance of these parameters as follows. The value of $\tilde{p} = 7.1$ means that the parallelised program cannot usefully utilise more than seven processors, and in this sense is reasonably well matched to the ten-processor /CAP system. Values of $(\tilde{R}, \tilde{p}, n)$ depend on the synchronisation overhead. If it were possible to reduce this substantially by additional synchronisation hardware or improvements to the software, the effect would be represented by increasing $\tilde{p} \to \infty$ and $\tilde{R} \to R_\infty$. Since synchronisation takes place on the host, this could be achieved by providing a faster host computer on the /CAP system. The parameter R_∞ is thus the maximum possible performance for the program when synchronisation time is negligible. Ninety per cent of this performance can be achieved if the number of processors $p = 9p_{1/2} = 50$, which indicates that this particular program would benefit from the use of more processors than the ten that are available on the Rome /CAP system, provided the synchronisation overheads are substantially reduced.

In figure 1.23 ideally the performance would follow the claim line pR_1, along which the performance is p times that for a single processor (this is the ideal of so-called linear speed-up). However, because only part of the program is parallelised, Amdahl's law limits the performance to less than $R_\infty = 0.5$ time-step/s. Thus the Amdahl wall in this plot is an Amdahl ceiling.

In the absence of synchronisation overheads the performance would follow the pipeline function with $p_{1/2} = 5.5$ processors, and this is shown as the broken curve. However, after about five processors, synchronisation overheads rapidly reduce the performance, which is seen to peak at $\tilde{p} = 7.1$ processors when the performance is $\tilde{R} = 0.25$, a speed-up of only $\tilde{S}_7 = 3.1$ compared with the performance on a single processor is found.

This behaviour of program performance as a function of number of processors is typical of MIMD computer systems. One's concern is to produce hardware and software that make R_∞ and $\tilde{p}$ as large as possible, and $p_{1/2}$ as small as possible.

2 Pipelined Computers

2.1 SELECTION AND COMPARISON

In this chapter we describe in detail the architecture, technology and performance of several pipelined computers. First we discuss the latest models of the CRAY series of supercomputers, that is to say the CRAY X-MP and CRAY-2. These have been by far the most commercially successful, and some 88 of the original CRAY-1 computers, and its successors the CRAY-1S, -1M and X-MP had been installed by 1985. Three prototype one-CPU CRAY-2 computers were installed in test sites in 1984/5 and a total of 13 full 4-CPU systems had been installed by the end of 1987.

The second computer series described, the CDC CYBER 205, is now also well established with sales in the region of 30 by 1985. Its architecture has a long history going back to the CDC STAR 100 which was conceived in the mid-1960s, and is sufficiently different from that of the CRAY series to make a separate study worthwhile. The future of this line lies with the ETA[10], which can be described as eight CYBER 205 CPUs attached to a common memory, and implemented in CMOS technology cooled by liquid nitrogen. With this interesting development one can be assured that the CYBER 205 architecture will also be of lasting interest.

The three Japanese vector computers, the FUJITSU VP 100/200, HITACHI S-810/10 and 20, and the NEC SX1/SX2, appeared later and clearly owe much to the CRAY and CYBER 205 architectures. These are so similar that we treat them together in a comparative fashion.

The last series of pipelined architectures that we discuss is the Floating Point Systems 164 and 164/MAX. These computers have almost the same architecture as that of the earlier FPS-120B and FPS-100, and over 5500 of this series of computers had been installed by 1985. They have the advantage of being within the price range of a wide range of industries and universities, and when properly programmed can give exceptional price performance. Of particular interest is the availability of matrix accelerator, or MAX, boards, up to 15 of which can be attached to an FPS-164. Each of these is equivalent to two additional FPS-164 CPUs under single-instruction control. For the right matrix problems, performance in the supercomputer range can be obtained at minicomputer cost. A further interesting development is that by Clementi at IBM Kingston and Rome, and Wilson at Cornell University, which combines about ten FPS-164s to a host computer, making an economic MIMD system for the solution of large scientific problems. For these reasons, the FPS-164 architecture is also certain to be of lasting significance.

In the comparisons below we will describe: the *physical layout*, in order to give an idea of the size and appearance of the machine; the overall *architecture*, in order to appreciate the main storage areas, data paths, transfer rates and computing elements; the *technology*, in order to be aware of, for example, the level of integration; the *instruction* set, in order to give an idea of its richness; the *software*, in order to gauge the amount available and its breadth; and then finally the *performance* is judged by measurements of r_∞ and $n_{1/2}$ obtained where possible by running programs on the computers.

2.2   THE CRAY X-MP AND CRAY-2

The CRAY X-MP and CRAY-2 are both manufactured by Cray Research Inc.† in Chippewa Falls, Wisconsin, USA. They are derivatives of the CRAY-1 computer which was designed by Seymour Cray and first installed at the Los Alamos Scientific Laboratory in 1976 (Auerbach 1976b, Hockney 1977, Russell 1978, Dungworth 1979, Hockney and Jesshope 1981). The development of the CRAY X-MP, which is a multiprocessor version of the CRAY-1, is due to Steve Chen and his team (Chen 1984) in Chippewa Falls. A two-CPU version was announced in 1982, and a four-CPU version in 1984. Data on the CRAY X-MP is taken primarily from the CRAY X-MP Computer Systems reference manuals (Cray 1982, 1984a, 1984b). The next development in this series is the CRAY Y-MP, which was announced in 1987. This is anticipated to have a clock period of 4–5 ns, and to comprise of up to 16 CPUs and have 32 Mword of common memory, backed up with a secondary memory of 1 Gword.

The CRAY-2, on the other hand, has been developed as a separate project by Seymour Cray at his Chippewa Falls laboratory. It uses more advanced circuit technology and a new concept in cooling. This permits the use of the

---

† Corporate headquarters: 608 Second Avenue South, Minneapolis, Minnesota 55402, USA. UK Office: Cray Research (UK) Ltd, Cray House, London Road, Bracknell, Berks RG12 2SY.

fastest clock currently, circa 1988, found in production computers with a period of 4.1 ns. The CRAY-2 was announced as a produce in 1985 (Cray 1985).

2.2.1 Physical layout

The most striking feature of all CRAY computers is their small size. This is well illustrated in figure 2.1 which shows a typical CRAY X-MP installation. At the centre is the CRAY X-MP itself which is housed in the same three-quarter cylindrical mounting that has been a characteristic of the earlier CRAY-1 and CRAY-1S computers. To the left is a quarter-cylinder housing the solid state device (ssd) which is sometimes referred to as a solid state disc, and to the right is another quarter-cylinder housing the I/O subsystem (ios).

The CRAY X-MP comprises a central $4\frac{1}{2}$ ft diameter cylindrical column $6\frac{1}{2}$ ft high, surrounded by a circular seat bringing the diameter at floor level to about 9 ft. The central column is divided into three 90° segments each of

FIGURE 2.1 A typical CRAY X-MP installation, showing the computer in the centre, the I/O subsystem on the right and the solid state device (ssd) on the left. (Photograph courtesy of Cray Research Inc.)

which has four wedge-shaped columns holding up to 144 circuit modules. Each module comprises a pair of circuit boards mounted on opposite sides of a heavy copper heat transfer plate which makes a good thermal contact with the cast aluminium vertical cold bars as shown in figure 2.2. These cold bars are the prominent vertical features seen in the central column of figure 2.1. Freon is circulated through stainless steel pipes inserted in the cold bars which are thereby cooled to 21 °C. The copper plate temperature is maintained at 25 °C and the module temperature between 48 and 54 °C. The seats around the base of the CRAY X-MP enclose power supplies and some plumbing associated with the distribution of the freon cooling. Two 25-ton compressors are located externally to the computer room and complete the cooling system. A 175 kVA motor generator, also housed externally, supplies 208 V, 400 cycle three-phase primary power. The total power consumption of the machine is 128 kW, most of which must be removed by the freon cooling system. The CRAY X-MP occupies 45 square feet of floor

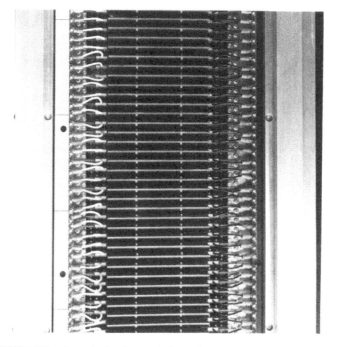

FIGURE 2.2   A vertical column of circuit boards on the CRAY X-MP. The boards are attached in pairs to copper coding plates which are clamped to the vertical freon-cooled aluminium cooling bars. (Photograph courtesy of Cray Research Inc.)

FIGURE 2.3   A single 6 inch × 8 inch circuit board of the CRAY X-MP
mounted with packaged integrated circuit chips. Each board has the
capacity for a maximum of 144 chips. (Photograph courtesy of Cray
Research Inc.)

space and weighs 5.25 tons. The ios and ssd each occupy 15 square feet of
floor space, and weigh 1.5 tons.

Figure 2.3 shows a double-layer module. Two circuit boards are attached
to each of two copper plates and rigidly fixed to form a three-dimensional
structure of four boards. As well as the connections within the circuit boards,
cross connections are made in the third dimension between the four printed
circuit boards. The four-board model then slides in grooves as shown in
figure 2.2.

The CRAY X-MP obtains it high performance, in part, from the compact
arrangement of its circuit boards that leads to short signal paths and
propagation delays. This can be seen in figure 2.4, which shows the layout

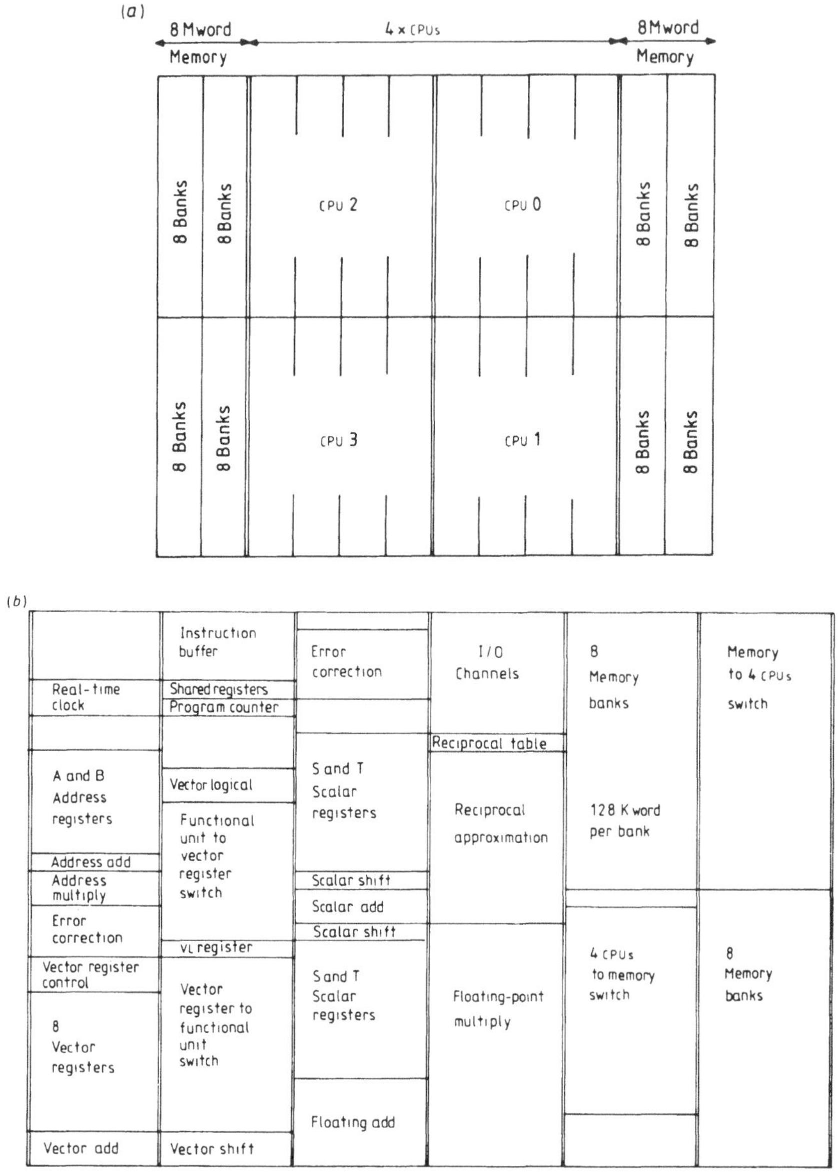

(a)

| 8 Mword Memory | 4 × CPUs | 8 Mword Memory |

8 Banks 8 Banks CPU 2   CPU 0   8 Banks 8 Banks

8 Banks 8 Banks CPU 3   CPU 1   8 Banks 8 Banks

(b)

	Instruction buffer	Error correction	I/O Channels	8 Memory banks	Memory to 4 CPUs switch
Real-time clock	Shared registers / Program counter				
			Reciprocal table		
A and B Address registers	Vector logical	S and T Scalar registers	Reciprocal approximation	128 K word per bank	
	Functional unit to vector register switch				
Address add					
Address multiply		Scalar shift			
Error correction		Scalar add			
	VL register	Scalar shift		4 CPUs to memory switch	8 Memory banks
Vector register control	Vector register to functional unit switch	S and T Scalar registers	Floating-point multiply		
8 Vector registers					
		Floating add			
Vector add	Vector shift				

FIGURE 2.4   (a) The layout of memory and logic within the central column of the CRAY X-MP/48 showing the location of the four CPUs and memory. (b) Allocation of circuit board positions to units in one quadrant (i.e. CPU 1).

of the modules in the X-MP/48. There are twelve columns in total, forming the 270° arc of the CRAY X-MP housing. The central four columns house the four CPUs and the left and right four columns the 8 Mword of main memory.

2.2.2   Architecture

The overall architecture of the CRAY X-MP can be described as one, two or four CRAY-1-like CPUs sharing a common memory of up to 8 Mwords. The original two-CPU model introduced in 1982 occupied a three-quarter cylinder of 12 columns, as shown in figure 2.1. However, in 1984 the use of higher density chips allowed the two-CPU and one-CPU models to be housed in six columns, and a four-CPU model was introduced using the full 12 columns. The one-CPU model is available with 1, 2 or 4 Mwords of 76 ns static MOS memory, arranged in 16 banks (1 and 2 Mwords) or 32 banks (4 Mwords); and the two-CPU model with 2 or 4 Mwords of 38 ns ECL bipolar memory, arranged in 16 and 32 banks respectively. The four-CPU model has 8 Mwords of ECL memory arranged in 64 banks. The latter computer is called the CRAY X-MP/48 where the first digit gives the number of CPUs and the second digit the number of megawords of shared memory.

The overall architectural block diagram in figure 2.5 is drawn for the CRAY X-MP/22 and X-MP/24. Each CPU has 13 or 14 independent functional units working to and from registers (there are two vector logical units in the model X-MP/48). As in the CRAY-1, there are eight 24-bit address registers (A0, ..., A7), eight 64-bit scalar registers (S0, ..., S7) and eight vector registers (V0, ..., V7), each holding up to 64 64-bit elements. The address registers have their own address integer add and integer multiply pipelines, and the scalar and vector registers each have their own integer add, shift, logical and population count pipelines. Floating-point arithmetic operations are performed in three pipelines, for multiply, add and reciprocal approximation (RA), which take their arguments from either the vector or scalar registers. Note that this means scalar and vector floating-point operations cannot be performed simultaneously, as they can in computers with separate scalar and vector units (e.g. CYBER 205). A 7-bit vector length (VL) register specifies the number of elements (up to a maximum of 64) that are involved in a vector operation, and the 64-bit vector mask (VM) register has one bit for each element of a vector register and may be used to mask off certain elements from action by a vector instruction.

In order to perform arithmetic operations, data must first be transferred from common memory to the registers. This may be done directly, or in the case of the S and A registers via block transfers to the T and B buffer registers, each of which holds 64 data items. The maximum data transfer to the S and A registers is one word every two clock periods, and to the T and B registers combined is 3 words per clock period. Transfers to and from the vector

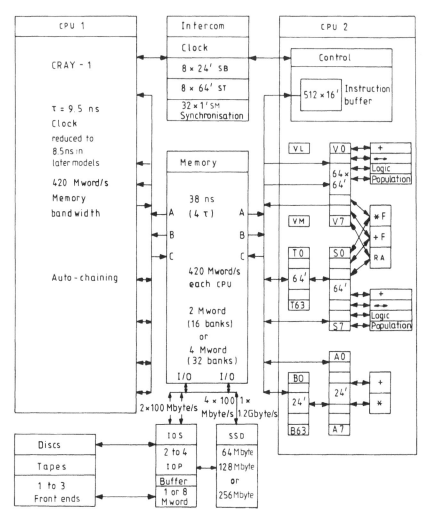

FIGURE 2.5 Architectural block diagram of a CRAY X-MP/2, showing the two CPUs, memory and principal data paths. The various letters are Cray's abbreviations for various registers in the machine (i.e. A, B, S, T and V). These are followed by a decimal number showing the number of registers. The 64′ represents the length of register in bits. SB, ST, VM and VL are other registers, see text. The ending 'F' designates floating-point operations.

registers must be made directly with common memory and three 64-bit data paths are provided for this purpose. Two of the paths transfer input arguments from common memory to the vector registers and the third transfers the result from the vector register to common memory. Because the data paths are separate, two input arguments and one result can be transferred per clock period, giving a data transfer rate of 315 Mword/s per CPU. A four-CPU machine has therefore a memory bandwidth of 1.2 Gword/s. If there is no contention for memory access (see next paragraph), the common memory can read and write data at this rate. Since, when full, each floating-point pipeline requires two new arguments and generates one result every clock period, the bandwidth to common memory is just sufficient to support *one* floating-point arithmetic pipeline from common memory. If other pipelines are to be simultaneously active, this is only possible if intermediate results are stored in the registers and not returned to common memory. The register-oriented architecture of the CRAY machines is designed to be used in this way. The advantage of providing registers is that, when properly programmed, there is less demand for common memory bandwidth and high rates of computation can be achieved for the inner-loops of calculations. The cost is the greater complexity of programming that is required to achieve the optimum use of the registers, whether this is done by a compiler or by hand coding in assembler.

There are four ports into common memory for each CPU, one of which (labelled I/O) is reserved for use by the input/output subsystem and the solid state device. The three other ports (labelled A, B, C) are available to support the data paths to the registers described in the last paragraph. All loads and stores to the A and S registers use port C. Block transfers from common memory to the B registers use port A, and to the T registers use port B. Block stores to common memory from the B and T registers use port C. Transfers from common memory and the vector registers use all three ports: ports A and B to load the registers from memory, and port C to store into memory. This means that three memory accesses can be made per clock period. With four ports from up to four processors accessing up to 64 memory banks, the memory interconnection system on the CRAY X-MP is complex. Cheung and Smith (1984) have analysed the system and suggested some improvements. For clarity, we describe now the X-MP/24 with two CPUs and 32 memory banks. The extension to more CPUs and more banks follows the same pattern.

The memory is divided into four sections, each with eight memory banks of 128 Kword. (The larger X-MP/48 has 16 banks per section, with each consecutive four banks in the same section.) There is an independent data path, called a *line*, from each CPU to each section of memory. The four lines on each CPU are connected by a crossbar switch to the A, B, C and I/O data paths coming from the functional units. Addressing is interleaved across the sections such that the two least significant bits of the address give the section number, and the five least significant bits give the bank number. When accessed a line is busy for one clock period, and a bank for four clock periods.

Memory contention on the X-MP can then take several forms. *Bank conflicts*, with which we are familiar on other systems, occur when a bank is accessed whilst it is still processing a previous memory reference. A *simultaneous conflict* arises when a bank is referenced simultaneously on the independent lines from different CPUs. This is resolved by alternating the priority of the CPUs every four clock periods. Finally there is a *line conflict* within a CPU, which occurs when two or more of the A, B, C and I/O data paths make a memory request to the same memory section (i.e. want the same line) in the same machine cycle. Line conflicts are resolved by giving priority to a vector reference with an odd stride over that with an even stride. If both references have the same parity the first reference to have been issued has priority. Memory references are resolved, and waits are initiated if conflicts arise, on an element-by-element basis during vector references. This means that the time interval between elements of a vector reference is not predictable or necessarily regular. Since these elements feed into the arithmetic pipelines, stages of a pipeline may become temporarily empty. These holes or 'bubbles' in the pipeline cause a degradation in average arithmetic performance, depending on the detailed pattern of memory referencing. Since contention for memory may depend on activity in CPUs that are not under the control of the programmer the degradation is likely to be unpredictable. However, in the very worst possible case of four CPUs each making three memory accesses per clock period to the same bank, each CPU would only satisfy an access every 16 clock periods, compared to a maximum rate of three accesses every clock period, giving a theoretical maximum degradation of memory bandwidth by a factor of 48. Cheung and Smith (1984) analyse more realistic patterns of memory access and conclude that performance is typically degraded by 2.5 to 7% on average due to memory contention, and in particularly bad cases by 20 to 33%.

The solid state device, or SSD, is a solid state MOS storage device ranging from 64 to 1024 Mbytes (i.e. up to 128 Mword arranged in 128 banks) which can be used in place of disc storage for the main data of very large user problems, and by the operating system for temporary storage. It is therefore often referred to as a solid state disc, and acts like a disc with an access time of less than 50 μs. It may be linked to an X-MP/48 via one or two 1250 Mbyte/s channels, and to an X-MP/2 by one such channel. The X-MP/1 may be connected by a 100 Mbyte/s channel.

Front-end computers, magnetic tape drives and disc storage units are always connected to the CRAY X-MP by an I/O subsystem (IOS). The first was introduced in 1979 to improve the I/O performance of the original CRAY-1 computer with both disc and magnetic tape storage. The enhanced version introduced in 1981 is an integral part of all CRAY X-MP installations, and comprises two to four 16-bit I/O processors (IOP) and an 8 Mword buffer memory. The first IOP handles communication with up to three front-end processors, and the second IOP handles communication between up to 16 disc units and the X-MP common memory via the buffer memory. The third and fourth IOPS are optional, and each may be used to attach a further 16 disc units. The DD-49 disc introduced in 1984 has a capacity of 1200 Mbyte and a 10 Mbyte/s transfer rate. The IOS communicates with the CRAY X-MP via two 100 Mbyte/s channels. There is also an optional direct path between the IOS and SSD with a transfer rate of 40 Mbyte/s.

All the CPUS of a CRAY X-MP share a single I/O section which communicates with the IOS and SSD. A 1250 Mbyte/s channel is provided for SSD data transfer, and two 100 Mbyte/s channels for IOS transfers. Four 6 Mbyte/s channels transfer I/O requests from the CPUS and other computers to the IOPS.

Each CPU has four instruction buffers, each holding 128 16-bit instruction parcels. On the X-MP/2 all eight memory ports are used for instruction fetch, which takes place at 8 words (32 parcels) per clock period. For this reason *all* other memory references on both processors are suspended during an instruction fetch from either processor. An exchange package of 16 64-bit words defines the status of a user job and may be exchanged in 380 ns.

The intercommunication section of the CRAY X-MP contains three clusters of common registers (five clusters on the X-MP/48) that may be accessed by all CPUS for communication and synchronisation purposes. There is also a common clock that allows program timing to be made to the nearest clock period. The clock cycles of all CPUS are synchronised. The common registers are eight 24-bit SB registers, eight 64-bit ST registers and 32 1-bit synchronisation or semaphore (SM) registers. Instructions are provided to transfer their contents to the A and S registers.

The 13 functional units take data from, and return data to, the A, S, V and vector mask registers only. The functional units, which may all operate concurrently, fall into four groups. The clock time, τ, of the first models of the CRAY X-MP was 9.5 ns. This was reduced to 8.5 ns on later models of the machine. We will use the value of 9.5 ns in the rest of this chapter.

Functional unit	(ns)	Unit time (clock periods) $= l$
Address units (24-bit)		
(1) integer addition	19	2
(2) integer multiplication	38	4

Functional unit	Unit time	
	(ns)	(clock periods) = l
Scalar units (64-bit)		
(3) integer addition	28.5	3
(4) shift	19 or 28.5	2 or 3
(5) logical	9.5	1
(6) population count	28.5 or 38	3 or 4
Floating-point units (64-bit)		
(7) addition	57	6
(8) multiplication	66.5	7
(9) reciprocal approximation (RA)	133	14
Vector units (64-bit)		
(10) integer addition	28.5	3
(11) shift	38	4
(12) logical	19	2
(13) population count	57	6
Main memory operations (64-bit)		
load scalar register from main memory	133	14
load vector register from main memory (64 elements)	769.5	81

All the functional units are pipelined and may accept a new set of arguments every clock period. In the above list the unit time is the length of the pipeline in nanoseconds or clock periods. In the latter units, it is therefore the variable l of §1.3.1. In the case of scalar instructions the above is the time from instruction issue to the time at which the scalar result is available (i.e. ready) in the result register for use by another instruction. In the case of a vector instruction an additional clock period is required to transfer each operand from a vector register to the top of a functional unit, and a further clock period is required to transfer each result element from the bottom of the pipeline to the result vector register. The timing formula for a vector instruction processing n elements is therefore

$$t = (l + 2 + n)\tau. \tag{2.1a}$$

Comparing equation (2.1a) with the general formulae (1.6), we see that the start-up time and half-performance length for register-to-register vector operations are respectively given by:

$$s = 3, \qquad n_{1/2} = s + l - 1 = l + 2. \tag{2.1b}$$

The address units are for address, index and other small-range calculations with integers less than about 16 million in two's-complement arithmetic. Data are taken from and returned to the A registers. The scalar units operate exclusively on 64-bit data in the S registers and, with the exception of population count, return results to the S registers. Integer addition is in two's complement arithmetic. Shifts may be performed on either the 64-bit contents of an S register (unit time 2 clock periods), or on the 128 bits of two concatenated S registers (unit time 3 clock periods). The logical unit performs bit-by-bit manipulation of 64-bit quantities and is an integral part of the modules containing the S registers. For this reason the data do not need to leave the S register modules and the function can be performed in one clock period. The population-count unit counts the number of bits having the value of one in the operand (unit time 4 clock periods) or the number of zeroes preceding the first one in the operand (unit time 3 clock periods). The resulting 7-bit count is returned to an A register.

The vector units perform operations on operands from a pair of V registers or from a V register and an S register. The result is returned to a V register or the vector mask register. Successive operand pairs are transmitted to the functional unit every clock period. The corresponding result emerges from the unit l clock periods later, where l is the functional unit time. After this, results emerge at the rate of one per clock period. The number of elements that are processed in such a vector operation is equal to the number stored in the 7-bit vector length register (VL). The elements used are the elements 0, 1, 2, ..., up to the number specified. Some vector operations are also controlled by the 64-bit vector mask register (VM). Bit n of the mask corresponds to element n of a vector register. The VM register is used in conjunction with instructions to pick out elements of a vector, or to merge two vectors into one (see §2.2.4). The vector integer addition unit performs 64-bit two's-complement integer element-by-element addition and sub-traction. The vector shift unit shifts the 64-bit contents of vector elements or the 128-bit contents of pairs of adjacent vector elements. Shift counts are stored in the instruction or an A register. The vector logical unit performs bit-by-bit logical operations on the elements of a vector and also creates 64-bit masks in the VM register.

The three floating-point functional units perform both scalar and vector floating-point operations. The arguments and results may therefore be either S or V registers. The 64-bit signed-magnitude floating-point number has a 48-bit mantissa giving a precision of about 14 decimal digits, and a 16-bit biased binary exponent giving a normalised decimal number range of approximately 10^{-2500} to 10^{+2500}. Separate units are provided for floating-point addition, multiplication and division. Division is performed in the reciprocal approximation and multiplication units by an iterative procedure. This approach makes possible the pipelining of the division operation.

The division algorithm to compute the ratio of two scalars S1/S2 is implemented by four machine instructions with the following actions:

RA unit $S3 = 1/S2$ reciprocal approximation   (2.2a)

$$
\begin{cases}
S4 = (2 - S3*S2) & \text{reciprocal iteration} & (2.2b) \\
S5 = S1*S3 & \text{multiplication by numerator} & (2.2c) \\
S6 = S4*S5 & \text{multiplication by correction} & (2.2d)
\end{cases}
$$

Multiplication unit

This can be recognised as the Newton iteration for the reciprocal of S2, because this problem is the same as finding the zero of the function

$$f(x) = S2 - x^{-1}, \tag{2.3a}$$

which has the derivative

$$f'(x) = x^{-2}. \tag{2.3b}$$

The Newton iteration is defined by

$$x^{(n+1)} = x^{(n)} - f(x^{(n)})/f'(x^{(n)}), \tag{2.3c}$$

therefore

$$x^{(n+1)} = (2 - S2*x^{(n)})x^{(n)} \tag{2.3d}$$

where n is the iteration number. In the above code, instruction (2.2a) forms an approximation to the reciprocal of S2 using the reciprocal approximation unit. This approximation is stored in S3 and corresponds to $x^{(n)}$ in equations (2.3c, d). The reciprocal iteration instruction (2.2b) computes with one instruction the contents of the bracket in equation (2.3d) and is executed by the floating-point multiplication unit. The multiplication instruction (2.2c) multiplies the initial approximation to $(S2)^{-1}$ by the numerator S1, and the multiplication instruction (2.2d) applies the correction specified by equation (2.3d). After one iteration the result is accurate to 47 bits, that is to say essentially the full precision of the 48-bit mantissa. The reason for multiplying by the numerator before applying the correction is that the instructions (2.2b) and (2.2c), although they use the same unit, may be started on successive clock periods; whereas instruction (2.2d), if it came first, would have to wait the completion of instruction (2.2b) because it requires the value of S4. With the ordering in equation (2.2) the division is complete in 29 clock periods corresponding to 2.8 Mflop/s. If the instructions (2.2) were replaced by vector instructions and placed in the order (2a, 2c, 2b, 2d), instructions

(2.2a) and (2.2c) would chain together (see three paragraphs ahead). The element-by-element division of two vectors can be accomplished at the asymptotic average rate of 27 Mflop/s.

Instructions on the CRAY X-MP are either one-parcel (16-bit) instructions or two-parcel (32-bit) instructions. Prior to execution, instructions reside in four instruction buffers, each of which can contain 128 one-parcel instructions or equivalent combinations of different length instructions. The buffers are filled cyclically from main memory. Whenever a required instruction is not present in the buffers, the next buffer in the cycle is completely filled by taking one 64-bit word (4 parcels) from each memory bank in parallel. There is a 22-bit program counter (P) containing the address of the next instruction to be executed, a 16-bit next instruction parcel (NIP) to hold the next instruction, a 16-bit current instruction parcel (CIP) to hold the instruction waiting to issue, and a 16-bit lower instruction parcel to hold the second parcel of a two-parcel current instruction.

An instruction can be issued (i.e. sent to the functional unit for execution) if the unit is not busy with a previous operation and if the required input and result registers are not reserved by any other instructions currently in execution. Different instructions place different reservations on the registers that they use, and the manual (Cray 1976) should be consulted for details. Broadly speaking a vector instruction reserves the output vector register for the duration of the operation and the input vector register until the last element has entered the top of the pipeline. However if a vector instruction uses a scalar register this is not reserved because a copy of the scalar is kept in the functional unit. The scalar register may therefore be altered in the clock period after the vector instruction is issued. Similarly the value of the vector length register is kept by the functional unit and the VL register can be changed immediately after instruction issue. Hence instructions with different vector lengths can be executing concurrently. In the case of scalar instructions only the result register is reserved by the instruction, in order to prevent it being read by other instructions before its value has been updated.

A special feature of the CRAY X-MP architecture is the ability to *chain together* a series of vector operations so that they operate together as one continuous pipeline (Johnson 1978). The difference between unchained and chained operations is illustrated in figure 2.6. The upper diagram shows the timing for three vector instructions if chaining does not take place.

The time for unchained operation of a sequence of m vector instructions is

$$t = \sum_{i=1}^{m} \tau[s_i + l_i + (n-1)]. \tag{2.4a}$$

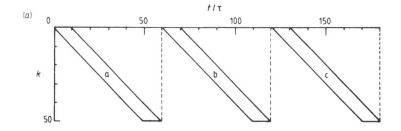

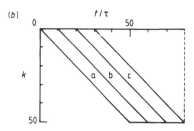

FIGURE 2.6   Timing diagram (a) for three unchained vector operations a, b and c and (b) for the same operations if they can be chained together. In this example $s + l = 10$ and $n = 50$. The time sequence for the processing of the kth element is given by a horizontal line drawn k units down from the time axis. An element passes through a pipeline for an operation when this line crosses the trapezoid for that operation.

Expressed as a time per vector operation this becomes

$$\frac{t}{m} = \frac{1}{\bar{r}_\infty}(\bar{n}_{1/2} + n), \tag{2.4b}$$

where the average asymptotic performance is

$$\bar{r}_\infty = \tau^{-1} \tag{2.4c}$$

and the average half-performance length is

$$\bar{n}_{1/2} = \frac{1}{m} \sum_{i=1}^{m} (s_i + l_i - 1) = (s + l - 1). \tag{2.4d}$$

In the case of m chained vector operations we have

$$t = \tau \left(\sum_{i=1}^{m} (s_i + l_i) + (n - 1) \right). \tag{2.5a}$$

The timing equation (2.4b) applies but now with

$$\bar{r}_\infty = m\tau^{-1}, \tag{2.5b}$$

$$\bar{n}_{1/2} = \sum_{i=1}^{m} (s_i + l_i) - 1 = m(s+l) - 1. \qquad (2.5c)$$

The second equalities in equations (2.4d) and (2.5c) are the results if all the pipes have the same start-up time and pipelength, as will frequently be approximately true. We see from the above that the average behaviour of a sequence of unchained vector operations is the same as the behaviour of a single operation. However, if m vector operations are chained together, both the asymptotic performance and the half-performance length are increased m-fold. This effect was noted previously in Chapter 1 as arising from the replication of processors. In the case of chaining it arises because the m pipelines are working concurrently which similarly increases both performance, r_∞, and the amount of parallelism, $n_{1/2}$. In the case of unchained operation the pipelines are working sequentially and there is consequently no increase in the average rate of computation or the parallelism.

The architecture of the CRAY X-MP can be conveniently summarised using the ASN notation of §1.2.4. A short description, ignoring I/O and the details of the register connections, but still presenting the essential computational features of the machine, would read:

$$C(\text{CRAY X-MP}/nm) = nC1(\text{CPU}) \times (8m)M1^{38}_{128K*64}(\text{common});$$

$$C1(\text{CPU}) = \text{Iv}^{9.5}_{16.32}[14\ \text{Ep} - M2 - \{\langle \tfrac{9.5}{64}, \langle \tfrac{9.5}{64}, \tfrac{9.5}{64} \rangle \rangle\}]_r;$$

$$M2(\text{registers}) = \{8M_{64*64}(\text{vector}), 72M_{1*64}(\text{scalar}),$$

$$72M_{1*24}(\text{address})\};$$

$$14\text{Ep} = \{3\text{Fp}_{64}, 5\text{Bp}_{64}, 4\text{Bp}_{64}, 2\text{Bp}_{24}\};$$

$$3\text{Fp}_{64}(\text{floating-point}) = \{\text{Fp}_{64}(+), \text{Fp}_{64}(*), \text{Fp}_{64}(\div)\};$$

$$5\text{Bp}_{64}(\text{vector}) = \{\text{Bp}_{64}(\text{integer} +), \text{Bp}_{64}(\text{shift}), 2\text{Bp}_{64}(\text{logical}),$$

$$\text{Bp}_{64}(\text{pop. count})\};$$

$$4\text{Bp}_{64}(\text{scalar}) = \{\text{Bp}_{64}(\text{integer} +), \text{Bp}_{64}(\text{shift}), \text{Bp}_{64}(\text{logical}),$$

$$\text{Bp}_{64}(\text{pop. count})\};$$

$$2\text{Bp}_{24}(\text{address}) = \{\text{Bp}_{24}(\text{integer} +), \text{Bp}_{24}(\text{integer} *)\}$$

2.2.3 Technology

The main 8 Mword memory of the CRAY X-MP/48 is composed of either bipolar or static MOS 64 Kbit VLSI chips with an access time of 38 ns or four clock periods. In contrast to the CRAY-1, which used 4/5 NAND gate chips, the logic circuits of the CRAY X-MP and the A and S registers are made from 16-gate array integrated circuits with a 300–400 ps propagation delay. As in the CRAY-1, the backwiring between modules is made by twisted pairs of wires.

2.2.4   Instruction set

The CRAY X-MP has approximately 128 instructions (a 7-bit operation code in most but not all instructions). The majority of the instructions are three-address instructions and specify the source of two input operands and the destination of the result. Since all functional units work register-to-register, only three bits are required to specify a source or destination, hence the operation code and three addresses just fit into a 16-bit single-parcel instruction. All but about 18 of the instructions are single-parcel. Instructions referring to a main-memory address (22 bits) for block transfers or branching require two parcels and are 32-bit. Instructions containing 22-bit constants for transfer to the A or S registers also require two parcels.

The instructions provide the range of logical, shift, branch, integer and floating-point arithmetic that one would expect from the nature of the 13 functional units. We give below examples in the notation of the CAL assembler language. The format of the instruction is

$$\|g|h|i|j|k\| \quad m \quad \| \qquad \text{op code is } g \text{ or } (g|h)$$

$$\|4|3|3|3|3\| \ 16 \text{ bits} \| \qquad\qquad\qquad (2.6)$$

$$\|\text{parcel 1} \ \| \text{parcel 2}\|$$

Read/Write:

Bjk, Ai	,A0	Read (Ai) words to B register jk from (A0)
,A0	Tjk,Ai	Store (Ai) words from T register jk to (A0)
Ai	exp,Ah	Read from ((Ah) + jkm) to Ai. (A0 = 0, $jkm = exp$)
exp,Ah	Si	Store (Si) to ((Ah) + jkm). (A0 = 0, $jkm = exp$)
Vi	,A0,Ak	Read (VL) words to Vi from (A0) incremented by (Ak)
,A0,Ak	Vj	Store (VL) words from Vj to (A0) incremented by (Ak)
Ai	Sj	Transmit (Sj) to Ai
Bjk	Ai	Transmit (Ai) to Bjk
Si	Tjk	Transmit (Tjk) to Si
SMjk	1	Set jkth semaphore register
Ai	SBj	Read jth shared B register

Control:

J	Bjk	Branch to (Bjk)

R	*exp*	Return branch to *ijkm* (= *exp*); set B00 to instruction counter
JSZ	*exp*	Branch to *ijkm* (= *exp*) if (S0) = 0
JAP	*exp*	Branch to *ijkm* (= *exp*) if (A0) positive

Logic:

Si	Sj & Sk	Logical product of (Sj) and (Sk) to Si
Vi	Sj!Vk	Logical sum of (Sj) and (Vk) to Vi
S0	Si < *exp*	Shift (Si) left *jk* (= *exp*) places to S0
Si	Si,Sj < Ak	Shift register pair (SiSj) left (Ak) places to Si

Arithmetic:

Si	*exp*	Transmit *jkm* (= *exp*) to Si
Ai	Aj + Ak	Integer sum of (Aj) and (Ak) to Ai
Si	Sj*Sk	Integer product of (Sj) and (Sk) to Si
Vi	Sj*FVk	Floating product of (Sj) and (Vk) to Vi
Si	/HSj	Floating reciprocal approximation of (Sj) to Si
Vi	Vj*IVk	2.0—product of (Vj) and (Vk) to Vi; reciprocal iteration

Miscellaneous:

VL	Ak	Transmit (Ak) to VL
Si	VM	Transmit (VM) to Si
Si	RT	Transmit (real-time clock) to Si
Ai	PSj	Population count of (Sj) to Ai
Ai	ZSj	Leading zero count of (Sj) to Ai

In the above, parentheses around a register name mean the contents of that register, and *exp* can be replaced by a simple arithmetic expression, the value of which is placed in the instruction.

Two instructions of the CRAY X-MP merit special explanation. These are:

Set Mask—the 64 bits of the vector mask (VM) register correspond one-for-one to the 64 elements of a vector register. If the element satisfies a condition the corresponding bit of VM is set to one, otherwise it is zeroed. The conditions are zero, non-zero, positive or zero, negative. Thus:

VM   V5, Z   set VM bit to 1 where V5 elements are zero,
VM   V7, P   set VM bit to 1 where V7 elements are positive or zero.

Vector Merge—the contents of two vector registers Vj and Vk are merged into a single result vector Vi according to the mask in the VM register. If the *l*th bit of VM is 1 the *l*th element of Vj becomes the *l*th element of the result register, otherwise the *l*th element of Vk becomes the *l*th element of the result register. Vj may alternatively be a scalar register. The value in the vector length register determines the number of elements that are merged.

Thus:

> Vi   Vj!Vk & VM   merge Vj and Vk into Vi according to the pattern in VM,
>
> V7   S2!V6 & VM   merge S2 and V6 into V7 according to the pattern in VM.

The purpose of the mask and merge instructions is to permit conditional evaluation with vector instructions. Consider, for example, the evaluation of

```
      DO 3 I=1,N
      IF (C(I)) 2,1,1
1     A(I) = EXPRESSION 1
      GO TO 3
2     A(I) = EXPRESSION 2                          (2.7)
3     CONTINUE
```

in which $A(I)$ is to take the value of *expression 1* if $C(I)$ is zero or positive, or the value of *expression 2* if $C(I)$ is negative. If we assume that $A(I)$ and $C(I)$ are stored in element $(I-1)$ of, respectively, vector registers V3 and V4, and VL contains the value of N (assumed for simplicity $\leqslant 64$) then the above code can be implemented using vector instructions as follows:

(1) evaluate *expression 1* for all elements and place in V1;
(2) evaluate *expression 2* for all elements and place in V2;
(3) VM   V4, P   set VM bit to 1 where V4 element $\geqslant 0$;
(4) V3   V1!V2 & VM   merge V1 and V2 according to VM.

We note that in this method *expression 1* and *expression 2* must be evaluated for all elements, even though only N of the $2N$ results will be used in the final merged vector V3. This is clearly wasteful of arithmetic but does allow the use of vector instructions throughout. Alternatively the FORTRAN code (2.7) can be implemented with scalar instructions, in which case one would examine the sign of $C(I)$ element-by-element and evaluate either *expression 1* or *expression 2* as indicated, but not both. In this case no unnecessary arithmetic is performed; however the scalar arithmetic used is slower than the vector arithmetic of the first method. Clearly there will be some breakeven vector length above which the vector merge method is faster than evaluation with scalar instructions.

Two situations cause particular problems to parallel computers. These can be described as the *scatter* and *gather* operations, which can be defined by the following FORTRAN code:

(1) Scatter

```
      DO 1 I=1,N
1     X(INDEX(I)) = Y(I)                           (2.8a)
```

(2) Gather

```
DO 1  I=1,N
1   Y(I) = X(INDEX(I))
```
                                                                (2.8b)

In either case the integer array INDEX contains a set of indices which may specify addresses that are scattered arbitrarily over the main memory of the computer. A scatter operation distributes an ordered set of elements $Y(I)$ throughout memory, according to the pattern of addresses in the array INDEX. Conversely the gather operation collects the scattered elements of X and sorts them into the ordered array Y. Such operations occur in sorting problems; in reordering as, for example, the unscrambling of the bit-reversed order of the fast Fourier transform (see §5.5.2); and in the charge assignment (scatter operation) and field interpolation (gather operation) steps of a simulation code using a particle-mesh model (see Hockney and Eastwood 1981).

CRAY X-MP models prior to the four-CPU announcement in 1984, and the earlier CRAY-1 and CRAY-1S computers, had no special hardware or instructions for implementation of scatter and gather operations. Such loops as (2.8) had to be executed by scalar instructions, and had a rather disappointing performance of about 2.5 Mop/s (Hockney and Jesshope 1981). The CRAY X-MP/4, however, provides vector scatter/gather instructions, and also related compress index instructions that permit these operations to execute at much higher vector speeds:

,A0, Vi	Vj	vector scatter Vj using indices in Vi
Vi	,A0, Vj	vector gather into Vi using indices in Vj
Vi, VM	Vj, Z	compress index of zero Vj, i.e. obtain indices of zero elements of Vj as a compressed vector in Vi, and set VM = 1 in corresponding bit positions.

The gather instruction is executed on either of the two read ports, and the scatter instruction in the write port. The compress index instruction is executed in the vector logical unit.

Using these instructions the scatter loop (2.8a) can be implemented by

V0	,A0, 1	vector load Y(I) into V0	
V1	,A0, 1	vector load INDEX(I) into V1	(2.8c)
,A0, V1	V0	vector scatter to X(INDEX(I))	

and the gather loop by

V0	,A0, 1	vector load INDEX(I) into V0	
V1	,A0, V0	gather X(INDEX(I)) into V1	(2.8d)
,A0, 1	V1	vector store Y to common memory	

The method of vectorising conditional expressions explained in connection with the vector merge instruction is inefficient because the arithmetic for both alternatives of the IF statement in (2.7) must be performed, even though only one alternative is needed—the unwanted results being discarded in the merge operation. If the number of elements that are to be changed by the IF statement is a small proportion of the total, a better strategy is to pick out the indices of the vector elements satisfying the condition using the compress index instruction, and subsequently gather the elements into a *compressed vector*. The arithmetic is then performed only on the compressed vector which is much shorter than the original. The results can be subsequently scattered back to their original location. As a trivial example illustrating this situation, consider the loop

$$\text{DO 1}\quad I = 1, N$$
$$1\quad \text{IF}(X(I).\text{NE}.0)\ Y(I) = Y(I) + X(I)$$

(2.8e)

in which we assume that the number of non-zero elements in X is small. This can be most effectively executed by the following code:

V0	,A0, 1	vector load $X(I)$
V1, VM	V0, N	put indices of non-zero values of $X(I)$ in V1,
V5	V1	V5 and V6; and set corresponding elements of
V6	V1	VM to 1
S1	VM	put vector mask in S1
A1	PS1	count population of ones in S1
VL	A1	set length of compressed vector in VL register

(2.8f)

V2	,A0, V1	vector gather $X(I)$
V3	,A0, V5	vector gather $Y(I)$
V4	V2 + FV3	$Y(I) = Y(I) + X(I)$ on compressed vectors
,A0, V6	V4	vector scatter of $Y(I)$ to original positions in memory

As in the previous case, the testing must be performed on all N elements of the vectors, however, the arithmetic (which could be substantial in a realistic example) is only performed on the compressed vectors. If the pattern of memory access avoids memory contention, the four final vector instructions of (2.8f) chain together and can execute at the full vector speed of one element per clock period.

2.2.5   Software

The principal new feature of the CRAY X-MP software compared to that previously described for the CRAY-1 is the provision of system routines to allow the multiple CPUs to cooperate in the solution of a single user problem. This facility is called *multi-tasking* on the CRAY X-MP and concerns provision for synchronising the operation of the multiple CPUs. In other computing systems, these may be called provisions for parallel, MIMD or multi-instruction programming, or for multiprocessing. CRAY reserve the term multi-tasking for parallel programming at the subroutine level, in which the unit of work that is handed for execution to a CPU is a subroutine, and preferably a large one. Such a unit is called a *task*. The software provides three basic facilities: first for initiating and waiting for such tasks (the fork/join or TASKS software), secondly for protecting critical sections of code that must be executed by only one CPU at a time (the LOCKS software), and thirdly for synchronising sequences of events occurring in different tasks (the EVENTS software). All facilities are invoked by calls to FORTRAN routines in the standard multi-tasking library.

A task is defined in the normal way as a subroutine and is initiated by

    CALL TSKSTART (*taskid, name, list*)

where *taskid* is the name of a three-element integer array identifying the task, *name* is the name of the subroutine, and *list* is a list of any arguments to the subroutine. A call to TSKSTART defines a new logical CPU which has a one-to-one correspondence with the task and puts it into a queue for execution. The actual physical CPU which executes the task is chosen by the multi-tasking software, and is not within the gift of the programmer. This has the advantage that multi-tasking programs written for a CRAY X-MP do not have to be aware of the number of CPUs available, and therefore will execute without change on a one-, two- or four-CPU model.

A multi-tasking program will always have a control program, which will make one or more CALLS to TSKSTART at a point in the program where time can be saved by giving independent pieces of work to other CPUs. Such a point is called a *fork* in the program. The companion facility is

    CALL TSKWAIT (*taskid*)

which makes the control program wait until the task identified by *taskid* has finished. If this is done for all the tasks that were initiated, one has produced a *join* in the program (or synchronisation point) which cannot be passed until all the parallel streams of instructions in the separate tasks have finished.

Very often the tasks that are initiated at a fork in the program are quite independent and require no intertask communication during their execution. Defining independent tasks is the clearest method of programming, and leads to the clearest programs and easiest program debugging. However, it may not always be possible. An example of independent tasks is the calculation of fast Fourier transforms on all 32 lines of a 32×32 mesh, during the solution of a partial differential equation (see §5.6.2, equation (5.133)). Half the transforms can be given to each CPU of an X-MP/2 with no inter-communication between them. All that is required is that all the Fourier transforms are complete before the next stage of the algorithm is performed. This may be achieved by a CALL to TSKWAIT.

In contrast, the separate tasks created at a fork may contain sections of code that could give wrong answers if they were executed simultaneously by more than one CPU. These are called *critical* sections of code, and an example is the updating of a shared common variable. Supposing two or more tasks are being executed, each of which includes the following code

COMMON S

$S = S + 1$

The intention is that the common variable S is incremented by one as each task reaches this part of the code. However, if CPU2 reads S while CPU1 is in the process of updating it but before the new value is stored, CPU2 will read the old value of S instead of the new. The updating by CPU1 will be lost, and the count in S will be incorrect. The only way to ensure that the updating is correct is to ensure that the statement $S = S + 1$ is executed to completion, including the store into S, by only *one task at a time*. This is the simplest example of a critical section of code that must be executed by only one CPU at a time. More complicated examples may involve substantial segments of code, but all have the feature of reading and writing into shared common variables. The failure to identify and protect critical sections of code is one of the most frequent programming errors, and the most difficult to detect because the program might work sometimes, depending on the detailed loading and timing of the computer.

The multi-tasking software provides *locks* to protect critical sections of code, which are manipulated by the following subroutines.

CALL LOCKASGN (*name*)

CALL LOCKON (*name*)

CALL LOCKOFF (*name*)

CALL LOCKREL (*name*)

where *name* is an integer variable assigned to and identifying the lock. A lock has only two states, either *locked* (also described as being *on* or *set*) or *unlocked* (also described as being *off, reset* or *cleared*). LOCKASGN creates the lock and sets it to off. LOCKON switches the lock on, if it was off, or suspends the task if it was on. In the latter case the task will wait in suspension until the lock is cleared by some other task. At this point all tasks which were waiting for the lock to be cleared continue execution. LOCKOFF switches the lock off unconditionally, and LOCKREL eliminates the lock and releases the identifier assigned to the lock, which may then be used for other purposes. Thus locks with the same *name* (i.e. using the same key) may be put around many critical sections of code in different tasks by writing

> LOCKON (*name*)
>
> ⟨code: e.g. S = S + 1⟩
>
> LOCKOFF (*name*)

but the important feature to remember is that the mechanisms work as though there is *only one key* for each lock *name*. This ensures that only one of the locked (or protected) segments may be executed at a time. In general such protected sections of code should be kept to a minimum, because locks force the code segments to execute sequentially. Some CPUs will become idle waiting for the key to the lock, and the whole point and advantage of parallel processing on multiple CPUs is lost.

Another cause for communication between tasks may be the need to establish a certain sequence of events between two interacting tasks. The following synchronisation primitives are provided by the multi-tasking software:

> CALL EVASGN (*name*)
>
> CALL EVWAIT (*name*)
>
> CALL EVPOST (*name*)
>
> CALL EVCLEAR (*name*)
>
> CALL EVREL (*name*)

where *name* is an integer variable associated with the occurrence of an event. EVASGN defines the existence of an event called *name* which has two possible states: *posted* (i.e. it has happened) or *cleared* (or not posted, i.e. it has not yet happened). EVWAIT checks the status of the event. If it is not posted the calling program is suspended and waits until it is, alternatively if it is posted the program proceeds. EVCLEAR clears the posted event and, since EVWAIT does not change the status of an event, it is normally called immediately after EVWAIT. EVREL eliminates the event, and releases the variable *name*. By appropriate definitions of events and calls to EVWAIT and EVPOST, any sequence of events can be forced between two interacting tasks. The LOCKS and EVENTS primitives have very similar facilities, and can clearly simulate each other. For clarity it is best to use locks for protecting critical sections and events for synchronising events.

As the CRAY manuals point out, multi-tasking has some cost. Calls to the TASKS, LOCKS and EVENTS subroutines take time, and are an overhead that has to be borne by any multi-tasked program, and which is not present in the same program if it is executed as a single task on one CPU. Put another way, multi-tasking does *not* reduce the number of CPU cycles required to complete a job, indeed it increases the number of cycles due to calls to the multi-tasking routines. What it does is to reduce the total elapsed wall-clock time for a job by sharing the CPU cycles amongst several CPUs. Multi-tasking will therefore only be worthwhile if the size of the tasks, known as the grain of the parallelism, is large enough to outweigh the multi-tasking overhead. It is important, therefore, to reduce calls to the multi-tasking library, and to exploit parallelism in the program at the highest possible level (i.e. between different instantiations of the outermost loops of the program). In order to enable a quantitative judgement to be made, the magnitude of the overhead associated with each of the three methods of synchronisation has been measured, and is given in the next section.

2.2.6   Performance

The performance of the original CRAY-1 computer has been fully discussed in the first edition of *Parallel Computers* (Hockney and Jesshope 1981). Here we present measurements made on the two-CPU CRAY X-MP/22 of the three parameters r_∞, $n_{1/2}$ and $s_{1/2}$ (see §§1.3.3 and 1.3.6, and Hockney (1985a)). First we give measurements of r_∞ and $n_{1/2}$ on a single CPU, and afterwards consider the overhead, $s_{1/2}$, of synchronising the two CPUs by different methods. Figure 2.7 shows the result of executing the equivalent of the code (1.5) for dyadic and two types of triadic operations. For each vector length N in the program (1.5), the measurement was repeated 100 times and the minimum value is plotted in figure 2.7. The result is obtained in the standard way by fitting the best straight line and recording its inverse slope as r_∞ and its negative intercept on the n-axis as $n_{1/2}$. The results are shown in table 2.1.

The first three cases in table 2.1 are measurements of vector instructions. The dyadic case uses only a single vector pipeline with all vectors stored in main memory, and is to be compared with values of $r_\infty = 22$ Mflop/s and $n_{1/2} = 18$ previously obtained on the CRAY-1 (Hockney and Jesshope 1981). We find a three-fold increase in r_∞ due, primarily, to the provision of three memory ports on the CRAY X-MP compared with one on the CRAY-1. The

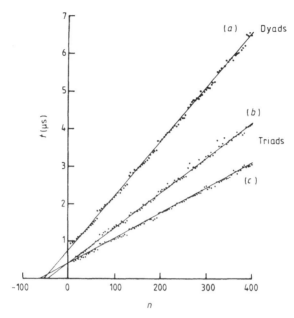

FIGURE 2.7   Measurements of r_∞ and $n_{1/2}$ on one CPU of the CRAY X-MP/22. Time, t, as a function of vector length, n, for a single vector operation. (a) Dyadic operations $A = B \times C$. (b) All vector triadic operations $A = D \times B + C$. (c) CYBER 205 triad $A = sB + C$. All vectors obtained from and returned to common memory. (Figure courtesy of North-Holland from *Parallel Computing.*)

TABLE 2.1   Measured values of r_∞ and $n_{1/2}$ on a single CPU of a CRAY X-MP for memory-to-memory operations. CRAY-1 values are in parentheses. The start-up time is $t_0 = n_{1/2}/r_\infty$.

Operation: statement 10 of code (1.5)	r_∞ (Mflop/s)	$n_{1/2}$ (flop)	t_0 (μs)
Dyadic $A(I) = B(I)*C(I)$	70	53	0.75
(CRAY-1 values)	(22)	(18)	(0.82)
All vector triad $A(I) = D(I)*B(I) + C(I)$	107	45	0.42
CYBER 205 triad $A(I) = s*B(I) + C(I)$	148	60	0.40
Scalar code $A(I) = B(I)*C(I)$	5	4	0.80

start-up time in microseconds, $t_0 = n_{1/2}/r_\infty$, has not changed significantly between the two machines: the extra complexity of memory access on the X-MP being compensated by a reduction in clock period. However, the importance of this overhead, which is what $n_{1/2}$ measures, is three times greater on the X-MP, because three times as much arithmetic could have been done during this time compared with the CRAY-1. The maximum rate at which one floating-point pipeline can deliver results is one result per clock period of 9.5 ns, that is to say 105 Mflop/s. The measured value of $r_\infty = 70$ Mflop/s is less than this because of the time taken to refill the vector registers from main memory. Because the vector registers hold 64 elements, this is an overhead that is incurred every 64 elements and is just visible (figure 2.7).

The second two cases are measurements for triadic vector operations, which involve the chaining of two vector instructions and the simultaneous use of both the floating-point multiply and add pipelines. At best we can expect a doubling of r_∞ which is achieved if one of the arguments is a scalar, but is not achieved in the all-vector case. The value of $n_{1/2}$ is not materially altered in the triad cases; however, the start-up time in microseconds is halved because a single start-up of 0.8 μs is shared between two instructions.

For comparison purposes, we have run the dyadic benchmark with instructions to the compiler to use only scalar instructions, and obtain $r_\infty = 5$ Mflop/s and $n_{1/2} = 4$. The start-up time t_0 remains at 0.8 μs but this is now of negligible importance because the arithmetic performance is about twenty times slower when scalar instructions are used.

Four methods of synchronising the operation of the two CPUs of a CRAY X-MP on a single job have been considered, and all the programs used are given in Hockney (1985a). They are the use of the TSKSTART and TSKWAIT primitives, which we refer to as the TASKS method; the use of the LOCKON and LOCKOFF primitives which we refer to as the LOCKS method; the use of the EVPOST and EVWAIT primitives which we refer to as the EVENTS method; and finally the use of a simplified LOCKS method written in CAL code. In all cases the programs were run on the computer in stand-alone mode, and timing was performed using the real-time clock function RTC(DUM). In this way we ensure that the second physical CPU is assigned to the second logical CPU in the programs, and that we are measuring the wall-clock time for the complete job. In the TASKS method (figure 2.8), after calling the real-time clock at the start of the measurement (T1), the second CPU is given a copy of the subroutine DOALL by the TSKSTART statement, and begins to execute it. The first CPU, which is performing the control program MULTI, then executes another copy of the subroutine DOALL in the CALL DOALL statement. The TSKWAIT

```
PROGRAM MULTI
COMMON/GLOBAL/A(400), B(400), C(400)
DIMENSION IDT(2)
EXTERNAL DOALL
DATA B/400*1.0/, C/400*1.0/

NMAX = 400
IDT(1) = 2

T1 = 9.5E-9*RTC(DUM)
T2 = 9.5E-9*RTC(DUM)
T0 = T2-T1

DO 20 N = 2, NMAX, 2
T1 = 9.5E-9*RTC(DUM)
NHALF = N/2,
NH1 = NHALF + 1

CALL TSKSTART (IDT,DOALL, NH1, N)
CALL DOALL (1, NHALF)
CALL TSKWAIT (IDT)

T2 = 9.5E-9*RTC(DUM)

T = T2-T1-T0
WRITE (6, 100) N, T
20   CONTINUE

100   FORMAT (' N: ', I4, 4X, 'TIME IN SECONDS:' F16.12)
STOP
END

SUBROUTINE DOALL (N1, N2)
COMMON/GLOBAL/A(400), B(400), C(400)

DO 10 I = N1, N2
10      A(I) = B(I)*C(I)

RETURN
END
```

FIGURE 2.8   Program for measuring r_∞ and $s_{1/2}$ when a job is split between the two CPUs of the CRAY X-MP/22 using the TASKS method of synchronisation.

statement ensures that both CPUs have finished their share of the work before the timer is called again to record the end of the measurement (T2). The parameters to DOALL are used to ensure that the two CPUs do different

TABLE 2.2   Measured values of r_∞ and $s_{1/2}$ when dyadic memory-to-memory operations are split between two CPUs on the CRAY X-MP22. The overhead is separately measured using TASKS, LOCKS, EVENTS and CAL code for synchronisation. The synchronisation overhead is $t_0 = s_{1/2}/r_\infty$ in microseconds.

Method	r_∞ (Mflop/s)	$s_{1/2}$ (flop)	t_0 (μs)	$\pi_0 = t_0^{-1}$ (k/s)
TASKS	130	5700	45	22
LOCKS	140	4000	28	36
EVENTS	140	2000	14	71
Simplified LOCKS CAL code	110	220	2	500

Note: results are deliberately rounded to two significant figures only. Greater precision would suggest spurious accuracy.

elemental operations ($N/2$ each) from the total of N operations. In this method the overhead of starting a new task occurs at every *fork* into a work segment that is divided between the two CPUs, and is therefore included in the measurement.

The measured time fits the formula

$$t = 45 + 3.2 \, s/400 \; \mu s$$

which leads to the values of r_∞ and $s_{1/2}$ given in table 2.2. The next least expensive method of synchronisation proves to be the LOCKS method. In this case we observe (table 2.2) $s_{1/2} = 4000$, about 2/3 of the value found for the TASKS method. On the other hand, we find the EVENTS method half as expensive as the LOCKS method with $s_{1/2} = 2000$. In order to determine the least overhead possible, a simplified form of the LOCKS method has been programmed in CAL by John Larson of Cray Research Inc. The overhead is thereby reduced by a factor of ten to $s_{1/2} = 220$. An examination of the CAL code shows that there is no wasted time, and it is unlikely that synchronisation can be achieved on the CRAY X-MP with less overhead. However, it must be said that in the CAL code, one CPU waits for the other to finish by continually testing one of the synchronisation registers. This prevents the waiting CPU from doing any other work during this time, and hence this code would hardly be acceptable as a general method of synchronisation.

2.2.7   CRAY-2 and CRAY-3

Figure 2.9 is a striking view of the CRAY-2 computer and its coolant reservoir

FIGURE 2.9   Overall view of the CRAY-2 computer, with the coolant
reservoir in the background.

(Cray 1985). The mainframe in the foreground, which weighs nearly three
tons, contains a foreground processor, four background processors, a large
common memory of 256 Mword, and all power supplies and backwiring.
This is all contained within a cylindrical cabinet 4 feet high and $4\frac{1}{2}$ feet in
diameter. The compression in size is remarkable—in effect a four-CPU CRAY
X-MP plus its I/O system and solid state device have been shrunk to a third
or a quarter of their present size, and placed in a single container. The power
generation of 195 kW is little changed from that of the CRAY X-MP, but
an entirely new cooling technology—*liquid immersion cooling*—is used to
remove it. In this method all the circuit boards and power supplies are totally
immersed in a bath of clear inert fluorocarbon liquid which is slowly circulated
(about one inch per second) and passed through a chilled water heat
exchanger to extract the heat. Cooling is particularly effective because the
coolant, which has a high dielectric constant and good insulating properties,
is in direct contact with the printed circuit boards and integrated circuit
packages. Figure 2.10 shows a closer view of the coolant circulating past the

FIGURE 2.10   Total immersion cooling technology of the CRAY-2. All circuit boards and backwiring are totally immersed in a bath of fluoro-carbon coolant, which is seen bubbling past the boards. (Photograph courtesy of Cray Research Inc.)

boards. The cooling system is closed and valveless; the collection of columns at the rear of figure 2.9 is a reservoir for the 200 gallons of coolant. If a circuit board needs to be replaced all the coolant must be pumped to the reservoir before the circuit boards can be reached. This can be accomplished in a few minutes.

The clock period on the CRAY-2 is 4.1 ns, and this necessitates short connecting wires between boards. This requirement has led to the develop-ment of three-dimensional pluggable modules, each comprising eight printed circuit boards rigidly held together as a unit with cross connections between the boards. Each module (see figure 2.11) forms an $8 \times 8 \times 12$ array of integrated circuit packages giving approximately 250 chip positions. The module measures approximately $1 \times 4 \times 8$ inches, weighs two pounds, and consumes 300 to 500 W of power. The 320 modules are mounted in 14 columns forming a 300° arc. There are 152 CPU modules and 128 memory modules, with approximately 240 000 chips, nearly 75 000 of which are

FIGURE 2.11   A CRAY-2 module comprising eight circuit boards. (Photograph courtesy of Cray Research Inc.)

memory. The logic chips use 16-gate arrays, as in the CRAY X-MP. As in the other CRAY computers, backwiring uses twisted pair wires between 2 inches and 25 inches in length. In all there are about 36 000 such connections with a total length of about six miles.

The overall architecture of the CRAY-2 is shown in figure 2.12, and can be described as four background processors accessing a large shared common memory, and under the control of a foreground processor. The common memory of 256 M 64-bit words is directly addressable by all the processors (32-bit addressing is used). It is arranged in four quadrants of 32 banks, giving a total of 128 banks with 2 Mword per bank. Dynamic MOS memory technology is used (256K bits per chip), which means that the access time (about 250 ns) is very long compared to the main ECL memory of the CRAY X-MP (38 ns). In this respect, the common memory is more correctly compared to the solid state device memory of an X-MP. Each memory bank has a functionally independent data path to each of four bidirectional memory ports, each of which connects to one background processor and one foreground communications channel. The total bandwidth to memory is therefore 1 Gword/s.

Access to common memory is phased, which means that each processor can have access to a particular quadrant only every four clock periods when its phase for accessing that quadrant comes around. There are four phases, one assigned to each of the background processors. The 32 memory banks in a quadrant share a data path to each common memory port; however, because of the phased access scheme, only one bank accesses the path in a

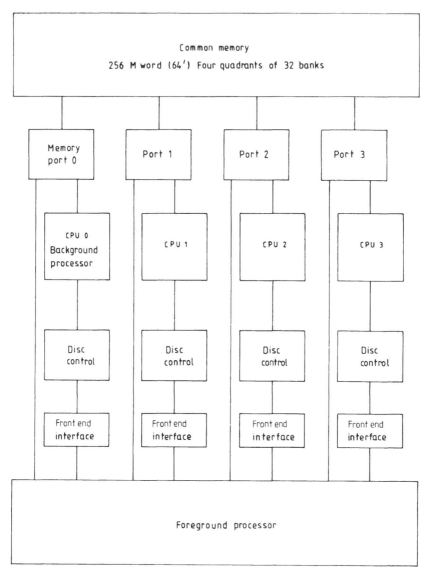

FIGURE 2.12   Overall architectural block diagram of the four-CPU CRAY-2.

given four-clock-period time slot. Therefore each bank, functionally, has an independent path to each of the four memory ports. The least significant two bits of the memory address select the quadrant, the next five bits select one of the 32 banks in a quadrant, and the remaining 25 bits of a 32-bit address specify the word within a bank (of course at present only 21 of these are needed to address the 2 Mword of a bank). Thus the elements of a contiguously stored vector are spread first across the quadrants, then across the banks, and finally across the words of a bank.

If references to successive elements of the vector occur every clock period, then a given quadrant is referenced every four clock periods, and a given bank every 128 clock periods (512 ns). The common memory access time of about 250 ns therefore avoids memory bank conflicts in this ideal case of reference to a contiguously stored vector. Bank conflicts will occur if the stride in memory address between successive elements is four or a larger power of two. In the worst case, when all the elements are stored in the same bank, the memory access rate is only 1/64 of the maximum for contiguous vectors. Consequently the performance of the CRAY-2 depends critically on the pattern of memory accesses to the common memory, and may vary widely for different jobs, and for differently programmed implementations of the same job.

The foreground processor supervises the background processors, common memory and peripheral controllers via four 4-Gbit/s communication channels. Each channel is a 16-bit data ring connecting one background processor, up to nine disc controllers, a front-end interface, a common memory port, and the foreground processor. The ring transfers a 16-bit data packet between stations every clock period. The foreground processor itself is a 32-bit integer computer with a 4 Kword (32-bit) local data memory and 32 Kbytes of instruction memory.

The architecture of a background processor is shown in figure 2.13, and can be described as a CRAY-1 architecture with the B and T intermediate registers replaced by 16 Kword of local memory. As in the CRAY-1, there is only a single bidirectional data path to the common memory, and there are eight 64-word vector registers and eight 64-bit scalar registers. The eight address registers have become 32-bit registers but there is no longer a data path directly from the common memory to the address registers. The functional units are also somewhat differently arranged, and reduced in number to nine. There is now no separate floating-point reciprocal approximation unit. This function now shares the floating-point multiply unit, which also provides the new function of hardware square root. The vector shift, population count, and integer arithmetic share a single vector integer unit; and the scalar integer add and population count are both performed in the scalar integer unit. Except for being 32-bit units, the address functional units are the same as on the CRAY-1. The 16 Kword local memory has an access time of four clock periods, and is intended for the temporary storage of scalars and vector segments during computation. The speed of this memory relative to the functional units is the same as the main memory of the CRAY X-MP, which of course is very much larger (up to 8 Mword).

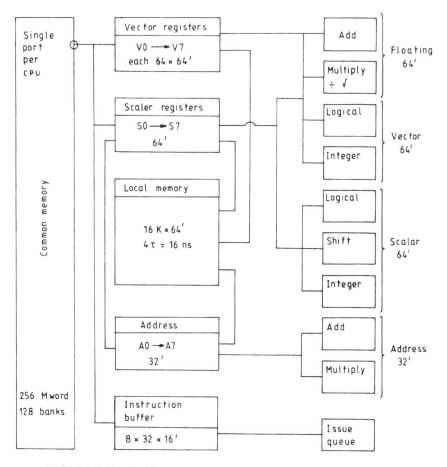

FIGURE 2.13   Architectural block diagram of a single CRAY-2 CPU.

Each background processor has its own 64-bit real-time clock which is advanced every clock period, and synchronised at start-up time with the clocks in the other background processors. There is a 32-bit program address counter, and common memory field protection is provided by a 32-bit base and limit registers. Eight 1-bit semaphore flags and a 32-bit status register are provided to control access to shared memory areas, and to synchronise the background processors. Eight instruction buffers hold 64 16-bit instruction parcels, and instruction issue takes place every other clock period. There are 128 basic instruction codes which include scatter/gather facilities, like the CRAY X-MP/4; however, the ability to chain together a succession of vector instructions, which is an important feature of the CRAY-1 and X-MP, has been lost and is not available on the CRAY-2.

The operating system on the CRAY-2 is based on the AT and T Unix†
system, which is rapidly becoming an industry standard (Ritchie and
Thompson 1974, Bourne 1982). This will provide the standard Unix facilities
and a C-language compiler. A new FORTRAN optimising (NFT) compiler
and a CAL-2 assembler are provided, together with a CRAY-1 to CRAY-2
assembler conversion utility. Extensions have been introduced to Unix to
enhance the I/O performance, multiprocessing facilities, and networking.

The peak performance of a background processor is two floating-point
results per clock period, or 500 Mflop/s, giving 2 Gflop/s for the four-CPU
computer. This could apply only if arguments and results are stored in the
registers and local memory (register-to-register operation), and can be
approached for favourable problems such as matrix multiply for which a rate
of 430 Mflop/s has been reported. The rates obtained for average FORTRAN
jobs which would use the slow common memory for storage will depend
greatly on the efficiency with which common memory transfers are buffered
by the compiler, and on the amount of arithmetic performed on average per
common memory reference. Values obtained for the $(r_\infty, n_{1/2})$ benchmark
(see §1.3.3) may be more indicative of the performance to be obtained from
average FORTRAN jobs working to and from data in common memory.

Table 2.3 shows the results on the CRAY-2 of the $(r_\infty, n_{1/2})$ benchmark
for a variety of simple kernels, using the first CIVIC compiler (1985) and an
improved compiler CFT77 (1.3) which appeared two years later. In these
tests, all the input data and results are stored in common memory, and no
use is made of the fast 16 Kword local memory. Also very little arithmetic
is performed per common memory reference ($f < 2$), so that the performance
is primarily determined by the speed of the common memory, and not by
the speed of the arithmetic pipelines. For these reasons the measured values
of r_∞ represent a worst case evaluation, but nevertheless one that may arise
from FORTRAN code that is not optimised to make use of the local memory,
by giving appropriate directives to the compiler. If the same loops were
executed with the input data and results stored in the vector registers of the
CPU, a performance approaching the theoretical peak performance of the
arithmetic pipelines alone ($\hat{r}_\infty$) should be attainable. This column therefore
represents the evaluation of the best possible case.

The first five kernels are vectorised by the compiler and show improved
performance as f increases, as is to be expected from a memory-limited
computation. The next two kernels, the recurrence and charge assignment,
do not vectorise and the performance is characteristic of scalar code.
The last two kernels concern data movement only. Transposition has a

† Unix is a trademark of AT and T Bell Laboratories.

TABLE 2.3   Results of the $(r_\infty, n_{1/2})$ benchmark for a variety of simple kernels on one CPU of the CRAY-2, using FORTRAN code and the CIVIC compiler (1985). The improved values in parentheses are for CFT77 (1.3), as of November 1987.

Operation: statement 10 program (1.5)	Stride	$\hat{r}_\infty$ (Mflop/s)	r_∞ (Mflop/s)	$n_{1/2}$
Dyad $A(I) = B(I)*C(I)$ $f = 1/3$	1 8	244 244	32 (56) 9 (10)	53 (83) 10 (0.5)
All vector triad $A(I) = B(I)*C(I) + D(I)$ $f = 1/2$	1 8	488 488	54 (65) 14 (17)	31 (28) 7 (2.5)
All vector 4-op $A(I) = B(I)*C(I) + D(I)*E(I) + F(I)$ $f = 2/3$	1	488	73 (100)	33 (33)
Matrix multiply middle product (§5.3.2) $f = 2$	1	488	75 (72)	75 (59)
Inner product $S = S + B(I)*C(I)$ $f = 1$	1	488	73 (119)	283 (236)
First-order recurrence $A(I) = B(I)*A(I - 1) + D(I)$ $f = 1/2$	1	488	3.1 (12)	4.1 (15)
Charge assignment $A(J(I)) = A(J(I)) + S$ $f = 1/2$	1	244	1.4 (2.8)	2.6 (10)
Transposition $A(I,J) = A(J,I)$	1	—	2.1 (3.4)	1.6 (6.2)
Random scatter; gather $A(J(I)) = B(I); A(I) = B(J(I))$	—	—	19 (29)	26 (45)

performance similar to scalar operation. However, scatter/gather is quite efficiently implemented at a rate near vector speeds. The results also show a severe degradation of performance when non-contiguous vectors are used (here chosen to have a stride of eight). The large ratio between the theoretical peak rates, $\hat{r}_\infty$, and the measured performance r_∞ confirms the statement that the performance of the CRAY-2 depends critically on the care with which problems are programmed, in particular on minimising common memory references and maximising the use of the CPU local memory for intermediate results.

The CRAY-3 is an implementation of the CRAY-2 in gallium arsenide (GaAs) technology which should allow a clock period of 1 ns. The chips are manufactured by Gigabit Logic Inc., California (Alexander 1985) and Harris Microwave (McCrone 1985). The CRAY-3 is also likely to have more processors, perhaps 8 to 16 and a common memory of one gigaword. It is scheduled for 1988/9.

2.3   THE CDC CYBER 205 AND ETA[10]

The CYBER 205 is manufactured by Control Data Corporation† in Saint Paul, Minnesota, USA. The machine was announced in 1980, and the first delivery to a customer site was to the UK Meteorological Office, Bracknell, England in 1981. By 1985 about 27 CYBER 205 computers had been installed. The CYBER 205 represents the culmination of a long programme of research and development that began with the design and delivery of the CDC STAR 100 computer in the period 1965–75 (see §1.1.3). A very interesting account of the technology and design trade-offs that were made during the creation of this computer is given by Neil Lincoln (1982) who was in charge of the design team. Other details are to be found in the CDC CYBER 200 Model 205 Computer System Hardware Reference Manual (CDC 1983). In 1983 the CYBER 205 series 600 was introduced which replaced the 4 Mword bipolar main memory of the original computer (now called the series 400) with a larger 16 Mword static MOS memory. Both these machines are described in this section. Since they differ only in the technology and packaging of the memory, we base the description below on the two-pipe 4 Mword series 400 machine, and point out only when necessary differences between it and the series 600 machines. The CYBER 205 architecture is important because it is being continued into the next generation of supercomputers. The ETA[10] computer which appeared in 1986 may be described as eight up-rated CYBER 205 computers working from a large shared memory. This is described in §2.3.7.

2.3.1   Physical layout

A photograph of the CYBER 205 and a block diagram of the different parts of the machine are shown in figures 2.14 and 2.15. The photograph is taken

† Corporate Headquarters: PO Box 0, Minneapolis, Minnesota 55440, USA. UK Office: Control Data House, 179/199 Shaftesbury Avenue, London WC2H 8AR.

FIGURE 2.14   View of the CDC CYBER 205, showing the vector stream and string units on the left and the floating-point pipes on the right. (Photograph courtesy of Control Data Corporation.)

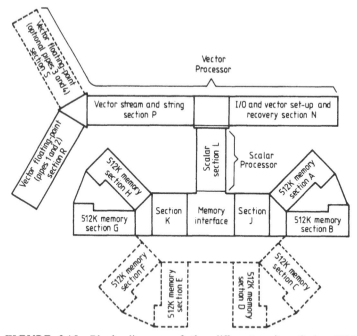

FIGURE 2.15   Block diagram of the different units of the CDC CYBER 205 series 400. The series 600 differs only in the memory section, which is smaller and rectangular in plan. (Diagram courtesy of Control Data Corporation.)

from the top left of figure 2.15 and is of a two-pipe machine. The main memory comprises four or eight memory sections. In the series 400 machine, these are housed in two or four wedge-shaped cabinets, each containing one million 64-bit words in two memory sections. The scalar section which contains the instruction processing unit, forms the central core of the machine. To one end is attached the memory through the memory interface unit, and to the other end is attached the vector processor. The latter comprises one, two or four vector floating-point arithmetic pipelines, a vector stream and string section, and an I/O and vector set-up and recovery section. Overall a 4 Mword computer occupies a floor area of about 23 ft × 19 ft. Cooling for the basic central computer with 1 Mword of memory consists of two 30-ton water-cooled condensing units which are housed separately, and power is supplied by one 250 kVA motor-generator set. The heat dissipation is about 118 kW. An additional 80 kVA motor-generator set and 90 tons of cooling are needed for the 4 Mword system, and a standby 250 kVA generator is provided. The CYBER 205 is designed to be attached to a front-end system, typically a CYBER 180, CDC 6000, IBM or VAX computer.

The series 600 computer differs from the above description only in the memory cabinets, which fit to the end and side of sections J and K forming a rectangular plan. There are still four or eight memory sections, but these may now contain 0.5, 1, 1.5, or 2 Mwords each, giving configurations with 1, 2, 4, 8, 12 or 16 Mwords of memory.

2.3.2 Architecture

The principle units and data paths in the CYBER 205 are shown in figure 2.16. Memory comprises eight *sections* (A to H), each divided into eight memory stacks (called memory modules on the series 600). Each memory stack (or module) is divided into eight memory banks and has an independent 32-bit data path to the memory interface unit. Each bank contains 16K 39-bit half-words on the series 400 (32 data bits plus 7 SECDED bits). On the series 600, however, a bank may contain 16K, 32K, 48K or 64K half-words, depending on the number of ranks of chips mounted on the memory board (see §2.3.3).

The memory is organised into pairs of sections (A/H, B/G, C/F, D/E) each of which therefore has 16 stacks and a 512-bit data path to the interface unit. This data width is known as a superword or *sword*. It is equivalent to eight 64-bit words or 16 32-bit half-words, and is the unit of access to memory for vectors. Successive addresses in a sword are stored in different memory stacks, so that a sword may be accessed in parallel by taking a half-word from each stack of the 16 stacks in a double section of memory. The access and cycle time of memory is 80 ns. However, provided successive references

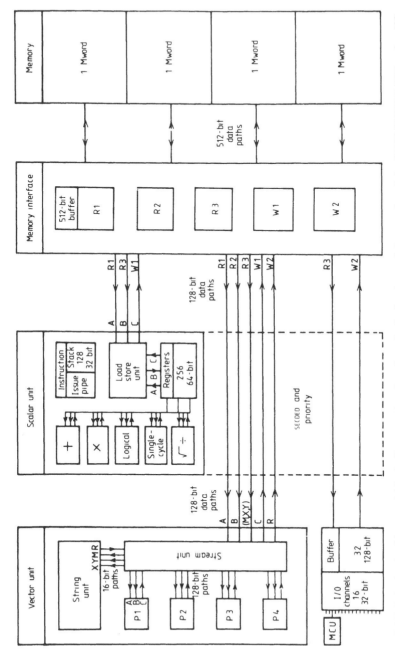

FIGURE 2.16   Architectural block diagram showing the principal units and data paths of the CDC CYBER 205 series 400. The series 600 increases the memory up to 16 Mwords.

are to different banks, a fresh sword can be referenced every clock period of 20 ns from each double section. This is a bandwidth of 400 Mword/s per double memory section. However, not all this bandwidth is used by the memory interface unit (see below).

Although memory may be addressed by the bit, byte (8 bits), half-word (32 bits) or word (64 bits), access to memory is by the sword (512 bits) for vectors, by the word or half-word otherwise. The memory interface unit organises memory requests at each 20 ns interval into swords, words or half-words, then delivers or assembles this data via 128-bit wide paths to the scalar and vector sections. Communication with the rest of the computer is in terms of three read paths and two write paths, and the memory interface unit has a one-sword buffer associated with each path (R1, R2, R3 for read and W1, W2 for write). The memory interface unit connects to the scalar, vector and I/O sections via 10 128-bit data paths, each of which has a maximum transfer rate of 128 bits every clock period, giving a maximum total transfer rate of 1000 Mword/s.

On a two-pipe machine the memory interface unit operates as described above, and has a maximum read throughput in vector mode of one sword or 8 words per clock period. As each pipe requires two new input arguments per clock period, the capacity of the data paths and interface unit just matches the needs of the arithmetic pipelines. A four-pipe machine, however, requires twice the above amount of data per clock period, and four-pipe machines have the R1, R2, and W1 buffers increased to 1024 bits, and the corresponding data paths to the vector unit increased to 256 bits. The interface unit then makes simultaneous reference to a double-sword (1024 bits) of data, by simultaneously accessing a half-word from 32 memory stacks spread over four memory sections. Clearly this is still not using the full bandwidth of the memory itself, which is capable of supplying a half-word simultaneously from each of its 64 memory stacks (eight stacks in each of eight memory sections).

The scalar section reads from buffers R1 and R3, and writes to buffer W1. All data paths to the vector section also pass through the scalar section where SECDED checking and priority determination for memory requests take place. The inclusion of SECDED checking in the architecture substantially extends the mean time between failures. The scalar section contains the instruction issue pipeline, which has a maximum issue rate of one instruction every 20 ns clock period (τ). The three address instructions are drawn from a stack which may hold up to 128 32-bit instructions or 64 64-bit instructions, or mixtures with equivalent total length. Both vector and scalar instructions are decoded in the instruction issue pipeline, which dispatches the decoded vector instructions to the vector unit for execution. The issue of decoded scalar instructions to the scalar arithmetic functional units is controlled on a reservation system, the principal conditions of which are:

(1) *Source operand conflict*—an instruction requiring the result of a previous instruction as input must wait until the operand is available;

(2) *Output operand conflict*—an instruction whose result refers to the same result register as a previously issued instruction must wait until the previous instruction is complete.

Sixteen result address registers hold the register file addresses for the output operands of previously issued instructions, and these are checked against the operands of any instruction awaiting issue until no conflicts arise.

The arithmetic portion of the scalar section comprises a load/store unit and five independent arithmetic functional units which take data from and return data to a file of 256 64-bit registers. The load/store unit moves data between the register file and main memory. The clock time $\tau = 20$ ns and the unit times are:

Functional unit	Unit time (ns)	Unit time (clock periods)
Load/Store	300	15
Addition/Subtraction	100	5
Multiplication	100	5
Logical	60	3
Single Cycle	20	1
Division, Square root, Conversion		
for 64 bits	1080	54
for 32 bits	600	30

The above unit times are the total times required to compute either a 32-bit or a 64-bit result; however all units except the last are pipelined and may take a new set of arguments every clock period. However note that the register file can supply at most one new pair of arguments per clock period and this is the factor most likely to limit the processing rate of the scalar section as a whole. However, the division, square root and conversion unit is not pipelined in this way, and new arguments are only accepted every 54 clock periods. The result from any of the above units can be passed directly to the input of any unit, in a process called *shortstopping*. This process, when applicable, eliminates the time needed to write results to a register and retrieve them for use in the next arithmetic operation. The unit times above assume that shortstopping takes place and do not include the time to write results to registers or memory. The register file may supply at most two operands for the current instruction and store one result from the previous instruction concurrently during every clock period. This is found to be sufficient to support a scalar performance of 45 Mflop/s out of a peak potential of one instruction every 20 ns or 50 Mflop/s.

Access to main memory is controlled by the load/store unit which acts as a pipeline and may accept one read (load) from memory every clock period or one write (store) to memory every two clock periods. A buffer is provided in the unit for up to six read and three write requests. A randomly accessed word can be read from memory and loaded into the register file in 300 ns provided memory is not busy. If it is, up to a further 80 ns is added to the time.

Operations on vectors of numbers or strings of characters are performed in the vector section which comprises either one, two or four *floating-point pipelines* and a *string unit* that are fed with streams of data by a *stream unit*. Unlike the CRAY X-MP there are no vector registers, and all vector operations are main-memory to main-memory operations, necessitating data to travel a round trip of about 50 ft, compared to less than 6 ft on the CRAY X-MP. This difference partly explains the much longer vector start-up time on the CYBER 205. A vector may comprise up to 65 535 consecutively addressed elements. If the required data is not consecutively stored the required elements can be selected by a *control vector* of bits, one bit for each word of the vector. The operation is then only performed for elements for which the corresponding control bit is one. However all elements of the consecutively stored vector must be read from memory even though only a small fraction may be operated upon. Alternatively if the control vector is sparse in ones, the specified elements of a long vector may be selected by a *compress* operation and re-stored consecutively. Subsequent operations may then be performed with better efficiency on the new compressed vector. In addition efficient scatter/gather instructions are implemented by microcode in the stream unit. These reference memory either randomly according to an index list, or periodically (i.e. at equal intervals).

Data is received from main memory in three input streams: A and B for the two streams of floating-point numbers, and (M, X, Y) for control vectors and character strings. There are two output streams: C for floating-point numbers and R for character strings. Each of these streams is 128 bits wide and is distributed by the stream unit into 128-bit data streams for use by the floating-point pipelines and 16-bit streams for use by the string unit. Each of the identical floating-point pipelines (P1 to P4 in figure 2.16) comprises five separate pipelined functional units for addition, multiplication, shifting operations and delay, connected via a data interchange (see figure 2.17).

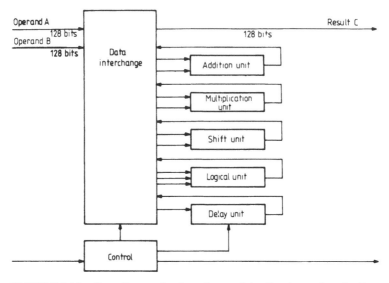

FIGURE 2.17   Overall organisation of one of the floating-point pipelines on the CDC CYBER 205. The machine may contain one, two or four such identical general-purpose pipelines. (Diagram courtesy of Control Data Corporation.)

Division and square root are performed in the multiplication unit. Each unit is attached to the data interchange by three 128-bit data paths (two input paths A and B, and the output path C). These paths can therefore support a rate of 100 million 64-bit results per second (Mr/s) per unit. The units themselves however are only capable of generating results at half this rate (50 Mr/s for 64-bit operation and 100 Mr/s for 32-bit operation). For simple vector instructions using a single unit the data interchange connects the two input streams A and B and the output stream C to the appropriate functional unit, leading to an asymptotic operation rate of 50 Mflop/s (64-bit operation) and 100 Mflop/s (32-bit operation) per pipeline. If the two successive vector instructions use different units, contain one operand that is a scalar and are preceded by the select-link instruction, then 'linkage' takes place. The output stream from the first unit used is fed by the data interchange to the input of the second unit. In this way the two units operate concurrently and the two instructions act as a single vector instruction with no intermediate reference to main memory. Examples of such linked triadic operations† are:

$$\text{vector} + \text{scalar} * \text{vector}, \qquad (\text{vector} + \text{scalar}) * \text{vector},$$

† A triadic operation is one involving three input arguments, e.g. $A + B \times C$; a dyadic operation is one involving two input arguments, e.g. $A + B$.

which occur frequently in matrix problems (for example the inner product of two vectors by the middle-product method, see §5.3.2). This facility plays the same role as chaining does on the CRAY X-MP, but is more restrictive. On the CYBER 205 two operations at most may be linked together and one of the operands must be a scalar. For such linked triads the asymptotic performance of a floating-point pipeline is doubled to 100 and 200 Mflop/s for 64-bit and 32-bit arithmetic respectively. The maximum asymptotic performance on the CYBER 205 is therefore 800 Mflop/s for linked triads in 32-bit arithmetic on a 4-pipe machine.

Figure 2.18 shows in more detail the overall organisation of the addition and multiplication pipelines. The addition operation (figure 2.18(a)) is seen to be divided into seven principal suboperations. A backward connection (or shortstop) is provided around the ADD segment to allow the unnormalised addition result of one element of a vector operation to be added to the next element of the vector. This facility is used in the interval instruction which forms the vector $C_{i+1} = C_i + B$; $C_0 = A$. Another shortstop takes the normalised result C of the addition pipe back to become the B operand input. The result arrives back at B eight clock periods after the operands contributing

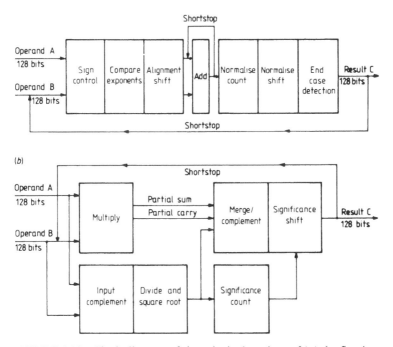

FIGURE 2.18   Block diagram of the principal sections of (a) the floating-point addition and (b) multiplication pipelined units on the CDC CYBER 205. (Diagram courtesy of Control Data Corporation.)

to it entered the pipe, hence one can accumulate $C_{i+8} = C_i + A_{i+8}$. This facility is used in summing all the elements of a vector, and in the dot- or inner-product instructions. The multiplication pipeline (figure 2.18(b)) has a similar shortstop (7 clock period delay) that is used to find the product of all elements of a vector. The asymptotic rate for multiplication is 50 Mflop/s (64-bit arithmetic) or 100 Mflop/s (32-bit arithmetic) per pipeline, and for division is 4 Mflop/s (64-bit arithmetic) or 15.35 Mflop/s (32-bit arithmetic) per pipeline. A divide enhancement is available as an option which doubles the asymptotic rate for division.

The string unit performs all bit logical and character string operations on strings (i.e. vectors) of bits and bytes. It also processes the control vector associated with the masking of floating-point operations. All data paths to the string unit are 16 bits wide. There are two paths, X and Y, for data input and one path, M, for a mask of control bits. The output data stream is R. The result rate on bit logical operations is 800 Mbit/s on any CYBER 205.

The basic CYBER 205 provides eight I/O ports, each 32 bits wide and capable of transferring 200 Mbit/s. A second set of eight I/O ports may be optionally added leading to a total maximum I/O bandwidth of 3200 Mbit/s. Each I/O channel contains a 4096-bit buffer register. All channels share a further I/O buffer register which connects to the memory interface unit via a 128-bit read data bus (R3) and 128-bit write data bus (W2). The full I/O rate is available at any level of vector or scalar usage in the CPU. Any I/O channel can be utilised by the maintenance control unit (MCU) which provides the interface to the operator for maintenance, system control (including initial start-up) and monitoring. The MCU consists of a control unit, line printer, disc drive and channel interface. Working in off-line mode the MCU loads diagnostic routines from disc and displays their results; working on-line the MCU performs real-time monitoring of the CPU and displays its status.

The architecture of the CYBER 205 can be summarised using the ASN notation of §1.2.4 as follows:

$$C(\text{CYBER } 205) = \text{Ipv}_{32,64}^{20}[\{\text{Ep, P2, P3}\}_{\overline{10 \cdot 128}}\text{M2}_{\overline{4 \cdot 512}}4\text{M1}_{1\text{M} \cdot 64}^{80}]$$

$$\text{M1(main)} = \{_{\overline{512}}16\text{M3}_{128\text{K} \cdot 32}\}; \qquad \text{M3(stack)} = \{_{\overline{32}}8\text{M}_{16\text{K} \cdot 32}^{80}\}$$

$$\text{Ep(vector)} = \{4\bar{\text{F}}\text{p, Bp3}\} - \text{Bp4}_{\overline{5 \cdot 128}}; \qquad \text{M2(buffer)} = 5\text{M}_{1 \cdot 512}$$

$$\text{Fp} = \text{Fp}_{32,64}(+, *, \div, \sqrt{})$$

$$\text{Bp3(string)} = \text{B}_{16}(\text{bit, byte}); \qquad \text{Bp4(stream)} = \text{Bp(scatter/gather)}$$

$$\text{P2(scalar)} = 5\text{Epl}_{\overline{3 \cdot 64}}256\text{M}_{1 \cdot 64 \, \overline{3 \cdot 128}}$$

$$5\text{Ep}1 = \{\text{Fp}_{64}(+), \text{Fp}_{64}(*), \text{Ep}(1\text{-cycle}), \text{Bp}_{64}(\text{logical}), \text{Fp}_{64}(\sqrt{}, \div)\}$$

$$P3(I/O) = 16\{D_{\overline{32}}IO_{\overline{32}}M_{\overline{32}}\}_{\overline{32}}M_{32\bullet128}(I/O\ \text{buffer})_{\overline{2\bullet128}}$$

2.3.3 Technology

The main interest in the technology of the CYBER 205 centres on the use of a novel LSI circuit, packaging and cooling technique for the logic of the computer, based on LSI bipolar ECL gate-array logic. As an example we show in figure 2.19 an overall view of a 15-layer arithmetic circuit board which can hold an array of 10×15 LSI chips. The main feature is the $\frac{1}{4}$ in $\times \frac{1}{4}$ in coolant pipe carrying freon that passes horizontally across the board 10 times. The LSI chips are mounted on ceramic holders approximately 0.5 in $\times$ 0.7 in $\times$ 0.1 in and clamped directly onto the cooling pipes which maintain a chip temperature of $55 \pm 1\,^{\circ}$C. In figure 2.20 a technician is shown

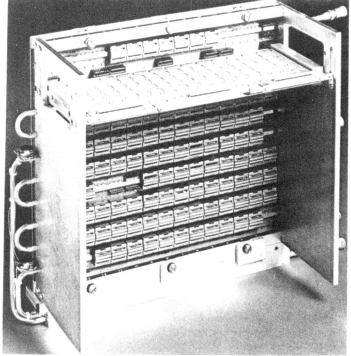

FIGURE 2.19   An arithmetic circuit board of the CDC CYBER 205. Freon coolant flows through $\frac{1}{4}$ in $\times \frac{1}{4}$ in pipes to which LSI chips are clamped. Each 15-layer board may hold up to 150 LSI chips. (Photograph courtesy of Control Data Corporation.)

FIGURE 2.20   A close-up of a CDC CYBER 205 circuit board, showing
a technician replacing a 168-switch LSI chip mounted on its ceramic switch
holder. The horizontal bars are the coolant pipes. (Photograph courtesy
of Control Data Corporation.)

replacing an LSI chip and numerous empty clamps awaiting chips are shown
attached to the coolant pipes. In figure 2.21 the various connectors and
clamps that hold the ceramic-mounted chip to the pipe are shown, and
figure 2.22 shows how these are assembled. The copper pad on the ceramic
mount makes direct thermal contact with the cooling pipe. The LSI chip has
52 external connections that are brought out to the side and bottom of the
ceramic mount (called an LSI array, figure 2.22 *left*). Two connectors, one for
each side of the ceramic mount and each containing 26 pins, are plugged
into the circuit board (figure 2.21 *centre* and figure 2.22 *right*) and a plastic
retainer mounted over them. The LSI array is dropped into the retainer and
clamped in place by a metal spring clip. To give an idea of the compression
in size achieved by LSI, we note that the entire scalar section L (see figures 2.14
and 2.15) is housed on 16 LSI boards in a cabinet about seven feet long. The
LSI chips use bipolar transistors with emitter-coupled logic (ECL) circuitry.
An array of 168 ECL switches (equivalent to about 300 logic gates) is packed
onto each LSI chip. Subnanosecond gate delays are achieved with a power ×
gate-delay product of about 6 pJ. Only 29 different LSI chips are used in the
whole computer. The use of LSI in the CPU reduces the power consumption
and makes the machine easier to maintain. Reliability is also improved by
reducing the number of connections external to the chip by a factor of about
six. The power consumed by each 168-switch LSI chip is smaller by a factor
of 10 than the power required by the previous SSI technique.

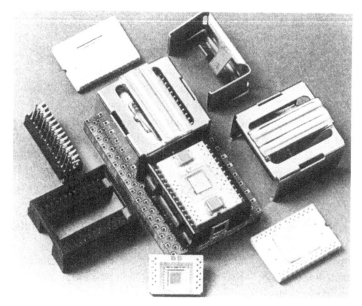

FIGURE 2.21   The various holders, connectors and clamps that are
used to attach the 0.5 in × 0.7 in ceramic-mounted LSI chip to the coolant
pipes. In the centre two chips are shown mounted on a section of circuit
board. (Photograph courtesy of Control Data Corporation.)

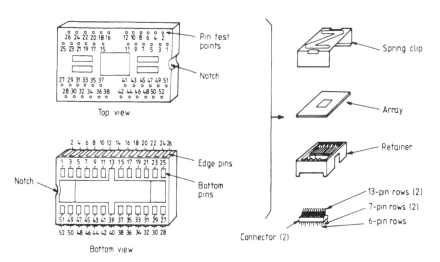

FIGURE 2.22   Left: top and bottom views of the LSI chip mounted on its ceramic
holder. The mounted chip is called an LSI array. Right: exploded view of how the LSI
array is attached to the coolant pipe and circuit board. (Diagrams courtesy of Control
Data Corporation.)

The main memory of the CYBER 205 series 400 uses 4K bipolar memory chips with a cycle and access time of 80 ns. Auxiliary logic uses emitter-coupled circuitry with ECL 100K chips. Each one million 64-bit words of memory is housed in two memory sections, each of which holds eight memory stacks, as is shown in figure 2.23. Figure 2.24 shows a close-up of one stack which stores 128K 32-bit half-words in eight independent memory banks. A stack contains two input boards, one output board and 16 memory boards. Cooling is provided by freon cooling plates that lie between the boards. A memory board, which is shown in figure 2.25, provides 20 or 19 bits of a word in parallel from groups of four 4K memory chips. A pair of memory boards forms a memory bank that accesses in parallel 39 bits (32 data bits and 7 SECDED bits), one bit from each of 39 memory chips. The memory address specifies which of the 16K bits from a group of four 4K-bit chips is accessed.

Although the organisation of the CYBER 205 series 600 memory is identical to the series 400 from the user's point of view, the technology and packaging are quite different. Static MOS 16-Kbit chips are used, and the greater level of integration enables the memory to be offered as either 1, 2, 4, 8, 12 or

FIGURE 2.23   The racks associated with the storage of one million 64-bit words in the CYBER 205. There are two sections of memory on the extreme left and right, and two smaller cabinets of the memory interface in the centre. In the memory section on the right eight memory stacks holding a total of half a million words can be seen. The CYBER 205 may contain 1, 2 or 4 such million-word assemblies.

FIGURE 2.24   A memory stack from the CYBER 205. The stack has
two input boards, one output board and 16 memory boards. A pair of
memory boards comprise one memory bank of 16K 32-bit half-words.
There are eight banks in a stack which gives storage for 128K half-words.
(Photographs courtesy of Control Data Corporation.)

16 Mwords. In the maximum 16 Mword machine, each of the eight memory
sections contains 2 Mwords, organised in eight memory modules of 512K
32-bit half-words. Each module plays the same role as a memory stack in
the series 400, and comprises a control board with ECL 100K series logic
chips and a 12 inch × 15 inch storage board assembly, on which is mounted
a 4 × 4 array of 3.25 inch × 2.25 inch ministorage boards. These latter boards
play the role of the 16 memory boards of the series 400 stack, and each
contains 20 16-Kbit MOS chips, ten on each side. One layer (called a rank)
of ministorage boards therefore gives a capacity of 128K half-words, the
equivalent of the maximum 4 Mword series 400 machine. Up to three
additional ranks of ministorage boards may be attached piggy-back fashion
to the first layer. Each extra rank provides an additional 4 Mwords of
consecutive storage locations to form the 8, 12 and 16 Mword memories.

FIGURE 2.25 A memory board from the CYBER 205 that provides 20 bits in parallel. The 4K-bit memory chips are mounted centrally in two 4 × 10 arrays. Four chips are associated with each bit position of the word, giving 16K memory addresses. Two such boards form a memory bank, and provide the 39 bits of a half-word (32 data bits and 7 SECDED check bits). (Photograph courtesy of Control Data Corporation.)

The memory access time of the series 600 is unchanged at 80 ns, as are all other features of the CYBER 205.

The operation of the CYBER 205 scalar and vector sections is controlled by microcode that is stored in memories housed on auxiliary logic boards that may hold up to 90 ECL 100K circuit chips. One such board is shown lying horizontally at the top of the LSI board in figure 2.19. These memories, the 128 32-bit instruction stack, and the 256 64-bit register file are all assembled on auxiliary boards from the ECL 100K chips. These memory elements have a read/write cycle time of 10 ns which includes the 1.0 ns ECL 100K gate delay time.

2.3.4   Instruction set

The instruction set of the CYBER 205 is particularly rich in facilities, but it is necessary first to discuss the addressing system and the arithmetic formats. The CYBER 205 has a virtual addressing storage system. The virtual address field in an instruction has 48 bits and is an address to an individual bit in virtual memory. Thus one may address up to 2.8×10^{14} bits, 3.5×10^{13} bytes, 8.8×10^{12} 32-bit half-words or 4.4×10^{12} 64-bit words of virtual memory. The top half of virtual storage is reserved for the operating system and vector temporaries, leaving 2.2×10^{12} 64-bit words of virtual address space for user programs and data. On the other hand, the physical main memory has a maximum of 4.2×10^6 64-bit words. The operating system transfers programs and data into main memory as either short pages (512, 2K or 8K 64-bit words long) or long pages of 64K 64-bit words. The translation between the virtual memory address and the physical memory address is performed in the scalar section using 16 associative registers. The registers hold associative words which contain the virtual and corresponding physical addresses of the 16 most recently used pages. They may all be compared in one clock period for a match between a virtual address in the instruction being processed and the virtual page addresses in the associative word. If there is no match, the comparison continues into the space table which is the extension of the list of associative words into main memory. If there is a match the virtual address is translated into the physical address, and the program execution continues. If the page is not found, the program state is automatically retained in memory and the monitor program is entered in order to transfer control to another job.

Floating-point arithmetic may be performed on either 32-bit half-words or 64-bit full-words. Numbers are expressed as $C \times 2^E$ where the coefficient C and the exponent E are two's complement integers. In the 32-bit format E has 8 bits and C has 24 bits, allowing a number range from $\pm 10^{-27}$ to $\pm 10^{+40}$. In 64-bit format E has 16 bits and C has 48 bits, giving a number range from $\pm 10^{-8616}$ to $\pm 10^{+8644}$. The binary points of both the exponent and the coefficient are on the extreme right of their bit fields, since they are both integers, and the sign bit is on the extreme left of the field. A number is normalised when the sign bit of the coefficient is different from the bit immediately to its right. Double-precision results are stored as two numbers in the same format as the single-precision result, and referred to as the upper and lower results. They may be operated upon separately. Another feature is the provision of significance arithmetic. In this mode the result of a floating-point operation is shifted in such a way that the number of significant digits in the result is equal to the number of significant digits of the least significant operand.

Instructions in the CYBER 205 are three-address and may be either 32 or 64 bits long with 12 possible formats. There are 219 different instructions which may be divided into the following categories (the number of instructions in each category is given in parentheses):

Register (60)   Vector macro (15)

Index (9)	String (1)
Branch (29)	Logical string (8)
Vector (28)	Non-typical (51)
Sparse vector (11)	Monitor (7)

In this short review we cannot attempt to describe all the instructions, but we do attempt to give the flavour of the instruction set by giving examples of the more interesting instructions.

Register instructions manipulate data in the 256 64-bit register file, either as 32-bit half-words or as 64-bit full-words, depending on the instruction. R, S and T stand for 8-bit register numbers. The bits in a word are numbered from left to right starting at zero. Examples of register instructions are:

ADDX R, S, T Add address part (i.e. bits 16 to 63) of register R to register S and store in register T.

EXPH R, T Take the half-word exponent from register R and place in least significant bit positions of register T.

Index instructions load and manipulate 16, 24 or 48 portions of registers:

IS $R, I16$ Increase the rightmost 48 bits of register R by the 16-bit operand I16 in bits 16 to 31 of the 32-bit instruction.

Branch instructions can be used to compare or examine single bits, 48-bit indices, 32- or 64-bit floating-point operands. The results of the comparison determine whether the program continues with the next sequential instruction or branches to a different instruction sequence:

CFPEQ $A, X, [B, Y]$ Compare for equality the 64-bit floating-point numbers in registers A and X. Jump to location specified by the contents of $[B, Y]$ if comparison is successful.

Vector instructions perform operations on ordered sets of numbers that are stored in consecutive storage locations. The designated operation is performed element-by-element and the result stored in a consecutive set of storage locations. The maximum length of such a vector is 65 535 elements. A vector is specified in an instruction by giving the numbers (8 bits each) of a pair of registers, for example $[A, X]$ or $[01, 02]$. The first register contains the base address (48 bits) and field length (16 bits) of the vector, and the second register contains a 16-bit offset to the base address identifying the start of the vector. A vector instruction also includes the number of the register (8 bits) that contains the address of the start of the control vector

(48 bits). The control vector is a bit vector containing one bit position for each element of the vector operands. It is used to control (or mask) the storage of the result of the vector operation. One may, for example, require that storage only takes place for elements for which the corresponding control bit is a one (or alternatively a zero). In the following A, B, C, X, Y, Z stand for register numbers in the range 00 to FF in hexadecimal:

MPYUV $[A, X]$, $[B, Y]$, C, Z Multiply the vector specified by registers $[A, X]$ by the vector specified by registers $[B, Y]$ under the control vector specified by register Z. The output vector is specified by register C. The offset for the output vector is in register $C + 1$ by convention.

Vector instructions include add, subtract, multiply, divide (using either the upper or lower portions of the double-length results, normalised or significance arithmetic), the ceiling and floor functions of APL (Iverson 1962), contraction and expansion of numbers between 64-bit and 32-bit floating-point formats, square root, truncation, and the packing and unpacking of the coefficient and exponent of floating-point numbers.

A special data format is provided for use if a vector contains many zero or near-zero elements. Such a vector may be compressed into a sparse vector which is specified by an order vector and data vector. The order vector is a bit vector with a bit position for every element of the full vector. The presence of a non-zero element is identified by a one bit and the presence of a zero-element by a zero bit. With the positions of all the non-zero elements identified by the order vector, the data vector need only store the values of the non-zero elements in the order that they appear in the full vector. A sparse vector is specified in an instruction by giving the numbers of a pair of registers. The first register of the pair contains the base address of the data vector and the second register contains the base address and field length of the order vector. Addition, subtraction, multiplication and division operations can be performed on such sparse vectors, for example:

ADDNS $[A, X]$, $[B, Y]$, $[C, Z]$ Add normalised the sparse vector specified by registers $[A, X]$ to the sparse vector specified by registers $[B, Y]$, storing the result as a sparse vector specified by registers $[C, Z]$.

Sparse vectors are formed by vector compare instructions, for example:

CMPEQ $[A, X], [B, Y], Z$ Compare and form order vectors: if $A_n = B_n$, bit Z_n† equals 1 otherwise $Z_n = 0$,

followed by a vector compress operation, for example:

CPSV A, C, Z Compress vector A into vector C, controlled by the order vector Z.

A sparse vector can be expanded to a full vector, for example, by:

MRGV A, B, C, Z Merge vector A with vector B under control of order vector Z. If B is a vector of all zeroes then the result vector C is the expanded form of the sparse vector A:
if $Z_n = 1$, $C_n =$ next element of A;
if $Z_n = 0$, $C_n =$ next element of B.

A similar instruction that does not expand the vectors is the mask instruction:

MASK A, B, C, Z Form output vector C from corresponding elements of A or B, depending on the value of the corresponding bit in the control vector Z:
if $Z_n = 1$, $C_n = A_n$;
if $Z_n = 0$, $C_n = B_n$.

The vector compare, compress, merge and mask instructions are examples of non-typical instructions. Other instructions in this category include: read real-time clock, count ones in field, simulate fault, find maximum or minimum element in a vector, merge bit and byte strings, scan for a given byte.

Another set of instructions of particular interest are the vector macro instructions, which perform with one instruction some of the more frequent operations in numerical analysis. They are implemented in microcode and perform operations that would normally require a subroutine of instructions. In all cases the elements involved in the operation may be selected by a control vector Z. Examples are:

ADJMEAN $[A, X], C, Z$ Adjacent mean:
$$C_n = (A_{n+1} + A_n)/2$$
AVG $[A, X], [B, Y], C, Z$ Average: $C_n = (A_n + B_n)/2$

† A_n, B_n, C_n mean the nth element of the floating-point vector specified by the words in registers A, B, C respectively. Z_n means the nth bit of the order or control vector specified by the word in register Z.

DELTA $[A, X], C, Z$ Delta or numerical differentiation: $C_n = (A_{n+1} - A_n)$

DOTV $[A, X], [B, Y], C, Z$ Double-length dot product $(\sum A_n B_n)$ stored in registers C and $C + 1$.

SUM $[A, X], C, Z$ Sum of elements of A (double-length) to registers C and $C + 1$.

VREVV $[A, X], C, Z$ Transmit vector A to C with its elements in reverse order.

The operations of periodic or random scatter and gather (see §2.2.4) are performed by single instructions on the CYBER 205. They apply to items or groups moved to or from either main memory or the register file. Examples are:

VTOVX $[A, X], B, C$ Vector to indexed vector transmission $B \rightarrow C$ indexed by A

VXTOV $[A, X], B, C$ Indexed vector to vector transmission B indexed by $A \rightarrow C$

Index lists that are used above may be generated in any convenient way; however a special search instruction is provided that can be useful for this purpose, for example:

SRCHEQ A, B, C, Z Search for equality and form index list. A_n is compared with all elements of B until equality is found. The number of unsuccessful comparisons before the 'hit' is entered into C_n. This is repeated for all elements of A. The counts C_n are in fact the indices of the elements satisfying the condition of equality. The comparison may be limited to certain elements by the control vector Z.

Another form of the search instruction provides a single index, for example:

SELLT $[A, X], [B, Y], C, Z$ Select on less than conditions: the corresponding elements of A and B are compared in turn starting with the first element. The index of the first pair to satisfy the condition (here $A_n < B_n$) is placed in the register C. Element pairs are skipped or included according to the bits in the control vector Z.

String instructions perform operations on strings of data in the form of 8-bit bytes. These represent characters from a possible set of 256, including for example the ASCII and EBCDIC standards. In contrast with the CYBER 203, the CYBER 205 implements only one such byte-oriented instruction:

MOVL $[A, X], [C, Y], I8$ The bytes in A are moved left to become the bytes of C. Repeats of byte I8 are brought in from the right as required.

The logical string operations, on the other hand, perform logical operations on strings of single bits. Eight logical operations are provided in separate instructions, for example:

AND $[A, X], [B, Y], [C, Z]$ The logical bit-by-bit AND of the bits in A with the bits in B form the bits of C.

NOR $[A, X], [B, Y], [C, Z]$ The NOT of the inclusive OR of the bits of A with the bits of B form the bits of C.

The monitor instructions function only when the machine is in monitor mode. They are used by the operating system to load and store associative words to absolute addresses in memory in order to manage the virtual memory system and to deal with interrupts. These instructions cannot be executed in a user program without causing a program fault.

2.3.5 Software

The software used on the CYBER 205 is a development of that written for the CDC STAR 100, and has been in operational use on the STAR 100, CYBER 203 or CYBER 205 since 1974. The principal items of software are:

(1) CYBER 205-OS—a batch and interactive operating system;

(2) CYBER 200 FORTRAN—a vectorising compiler for the main high-level language;

(3) CYBER 200 META—the assembler language that gives access to all the hardware features of the machine;

(4) CYBER utilities—including a loader and file editing and maintenance facilities.

The CYBER 200 operating system is designed to handle batch and interactive access, either locally or from remote sites, via a front-end computer such as a CYBER 180 series, IBM or VAX. The mass storage for user files is on CDC 819 disc units (capacity 4800 Mbit, average data rate

36.8 Mbit/s, average positioning time 50 ms) attached to data channels of the CYBER 205. Each user may program using memory of up to 2.2×10^{12} 64-bit words. Such programs are stored on the mass storage system and transferred into the physical store of the computer (maximum 4×10^6 64-bit words) in pages. It is the job of the operating system to allocate the physical memory into appropriately sized pages and distribute these amongst the several user programs that may be executing in a multiprogramming environment. The operating system achieves this by using monitor mode instructions to change the associative words in the associative registers and page table. The operating system is highly modular, and communication between different parts is by messages. The bulk of the operating system is stored in virtual memory and is paged into main memory when required. Only the *kernel* and *pager* are resident in the memory of the CYBER 205 at all times. The kernel handles the allocation of time to active jobs and the communication of messages to different parts of the software. The pager handles the allocation of memory and page swapping.

Another part of the operating system, called the virtual system, enters jobs into the operating system from the batch or interactive terminals, and removes jobs from memory if they finish or become inactive. Messages from active jobs for I/O are processed by the virtual system and sent to the loosely coupled network (LCN) for execution. Accounting tasks are also performed. An *operator* program is provided for interactive communication with the operator who may: display user jobs and accounting information; terminate, suspend and resume jobs; look at system tables; and generally control the flow of jobs through the computer.

The CYBER 200 FORTRAN compilers include both the ANSI standards X3.9–1966 and 1978, with extensions to permit the user to make use of the vector and string hardware facilities of the CYBER 205. Some features have been added that comply with ANSI standard X3.9–1978, and that have become a Control Data standard in earlier CDC compilers (NAMELIST I/O, ENCODE/DECODE, BUFFER IN and BUFFER OUT). The compiler has an automatic *vectoriser* that substitutes either vector instructions or appropriate STACKLIB routines (see p. 178) for DO loops, where such replacement cannot alter the logic of the problem. In addition the CYBER 200 FORTRAN compiler provides an *optimiser* that reschedules the order of scalar instructions so as to make optimum use of the scalar registers and pipelined scalar functional units, thus providing the maximum of functional concurrency without specific user intervention.

Vector operations are specified by descriptors which define vectors. These translate directly into the machine instruction format described in §2.3.4. A vector is specified by an array name, a starting index and a length. Thus given an array A, $A(10;100)$ means the vector which starts at location $A(10)$ and which is 100 elements long. Descriptors can be used implicitly within expressions, or by dynamic declaration. Examples are given below.

```
        DIMENSION A(1000),B(1000)
        DESCRIPTOR AD,BD

C  (1)  BY DECLARATION
        ASSIGN AD, A(1;999)
        ASSIGN BD, B(2;999)
        AD = BD*2.0

C  (2)  IMPLICIT USE OF DESCRIPTORS
        A(1;999) = B(2;999)*2.0
```

In both cases the array of elements $B(2), \ldots, B(1000)$ is multiplied by two and stored in locations $A(1), \ldots, A(999)$. In case (1) the descriptors are used in place of the array names in the arithmetic statement, and in case (2) the array elements to be used are specified in the arithmetic statement.

Access to all instructions of the CYBER 205 may be obtained through special subroutine calls of the form CALL Q8ADDX(R, S, T) which, for example, generates the single machine instruction ADDX R, S, T. The mnemonics for other machine instructions can similarly be prefixed with the reserved letters Q8 and used as a subroutine call in order to generate a single instruction in the place in the FORTRAN code where the subroutine call is made. Alternatively user-supplied assembler code can be incorporated in a FORTRAN program by linking a subprogram generated by the CYBER 200 assembler itself as an external reference to the FORTRAN program during the loading of the program parts.

Some frequently occurring dyadic and triadic operations on vectors, including recursive operations, have been efficiently programmed and are available via special subroutine calls to the STACKLIB routines. Examples from the 25 general forms, are:

(1) Add Recursive V1

```
        CALL Q8AD10(A(2),B(2),A(1),N-1)
```

being the equivalent of

```
        DO 1 I=2,N
    1   A(I) = B(I)+A(I-1)
```

(2) Multiply Add

```
        CALL Q8MA400(A(2),C,B(2),D(2),N-1)
```

being the equivalent of

```
        DO 1 I=2,N
    1   A(I) = C*B(I)+D(I)
```

(3) Subtract Multiply, Recursive V1, Reverse Order

```
CALL  Q8SM013(A(N-2),B(N-2),C(N-2),A(N-1),N-1)
```

being the equivalent of

```
DO 1 I=2,N
  J = (N+1)-I
1   A(J) = B(J) - (C(J)*A(J+1))
```

In the above the letters following Q8 identify the type of arithmetic operations involved, and the numerical code indicates whether the operands are scalar or vector and which operands are recursive.

The META assembler program for the CYBER 205 generates relocatable binary code from mnemonic machine instructions, procedures, functions and miscellaneous directives. Access is thereby provided to all the hardware facilities of the machine. Directives allow the programmer to control the process of assembly. Some features of the assembler are: conditional assembly capability for selective assembly, generation of re-entrant code that can be used simultaneously by several users without duplication of the code, ability to redefine all or any of the instruction mnemonics, ability to define a symbol for a set (or list) of data, the attributes of such sets (type and number of elements etc) may be assigned and referenced by the programmer. The assembly process takes place in two passes. In the first pass, all statements are interpreted, values are assigned to symbols, and locations are assigned to each statement. In the second pass, external and forward references are satisfied, data generation is accomplished, and the binary output and assembly listings are produced. Assembler programs are modular in form and may consist of several subprograms that are linked together by the LOADER program.

The LOADER program is one of the operating system utilities. It takes relocatable binary code produced by the FORTRAN compiler of the META assembler, links these with any requested library routines, and produces an executable program file. The user has control over the characteristics of the program file and may, for example, specify that certain routines be loaded as a group in either a small or large virtual page. Source files of the CYBER 205 software system, including the compiler and assembler, and user programs are all stored as card images in program files that may be created, edited and maintained on a card-by-card basis by the utility program UPDATE. Object binary files may be edited with the object library editor.

The above file maintenance activities, job preparation and input/output are performed on the front-end computer, thereby leaving the CYBER 205 for its main task of large-scale calculation. The hardware link between the front-end computer and the CYBER 205 is controlled by link software which permits multiple front-end computers to operate concurrently.

2.3.6   Performance

We consider first the performance of the CYBER 205 in the best case, when successive elements of all vectors are stored contiguously in memory. The performance of non-contiguously stored vectors is considered on page 184. Table 2.4 gives the expected performance of a selection of such contiguous vector operations in a 64-bit floating-point arithmetic on a two-pipe CYBER 205. We notice immediately that the half-performance length $n_{1/2}$ is, with the exception of scatter and gather, close to 100; that is to say at least twice as long as the CRAY X-MP. Since the value of $n_{1/2}$ determines the best algorithm to use (see Chapter 5), it may be the case that different algorithms should be used on the two machines, even though they are both in the general category of pipelined vector computers. For most instructions, the asymptotic operation rate r_∞ is 100 Mop/s or Mflop/s, compared with 70 Mflop/s for such dyadic operations on the CRAY X-MP. This rate is reduced to 40–50 Mop/s for the scatter, gather, max/min and product of element instructions.

TABLE 2.4   Expected vector performance of a two-pipe CYBER 205 for a selection of instructions (64-bit working), interpreted in terms of r_∞ and $n_{1/2}$. The actual performance in a multiprogramming environment may differ somewhat from these values. N = number of elements in the output vector, I = number of elements in the input vector or vectors. All vectors are contiguously stored in memory.

Instruction	Time (clock periods)	r_∞ (Mop/s)	$n_{1/2}$
Full vector addition	$51 + 0.5N$	100	102
Full vector multiplication	$52 + 0.5N$	100	104
Sparse vector addition	$88 + I/16 + 7N/16$	$\leqslant 89$†	$\leqslant 156$
Sparse vector multiplication	$88 + I/16 + 7N/16$	$\leqslant 73$‡	$\leqslant 128$
Dot product	$116 + I$	100	116
Product of elements	$126 + I$	50	126
Max or min of elements	$86 + I$	50	86
Compress	$52 + 0.5I$	100	104
Mask or merge	$56 + 0.5N$	100	112
Random scatter	$83 + 1.25N$	40	66
Random gather	$69 + 1.25N$	40	55
Vector relational	$56 + 0.5N$	100	112

† $I \leqslant 2N$
‡ $I \leqslant 4N$

In the last paragraph we have compared the performance of the two-pipe CYBER 205 and the CRAY X-MP when doing arithmetic to the same precision namely 64 bits. There are two ways in which the performance of the CYBER 205 can be increased. One option is to decrease the precision of the arithmetic to 32-bit floating-point, and the other is to increase the number of pipes from 2 to 4. The actions of halving the precision or doubling the number of pipes both have the effect of doubling the number of results that are produced in a given time. Performing both actions, i.e. going to 32-bit arithmetic on a four-pipe machine, produces four times the number of results in the same physical time. If the time to produce n results on the two-pipe machine in 64-bit arithmetic is

$$t = r_\infty^{-1}(n + n_{1/2}) \tag{2.9}$$

then this is also the time for operating on a vector of length $n' = cn$, where $c = 2$ or 4 in the above cases. Substituting in equation (2.9), we have

$$t = r_\infty^{-1}(n'/c + n_{1/2}) \tag{2.10a}$$

$$= (cr_\infty)^{-1}(n' + cn_{1/2}). \tag{2.10b}$$

In the new situation (indicated by a prime) we have, by definition

$$t = (r'_\infty)^{-1}(n' + n'_{1/2}), \tag{2.10c}$$

and hence by comparison with equation (2.10b) we have

$$r'_\infty = cr_\infty, \qquad n'_{1/2} = cn_{1/2}. \tag{2.10d}$$

Therefore doubling or quadrupling the vector length processed in a given time, doubles or quadruples both r_∞ and $n_{1/2}$. This effect can be seen in the data given by Kascic (1979) in figure 2.26, which shows the timing curve for dyadic vector addition or element-by-element multiplication of the form $\mathbf{C} = \mathbf{A} \text{ op } \mathbf{B}$ on a two-pipe CYBER 205 in 64-bit arithmetic (curve A:$r_\infty = 100$ Mflop/s, $n_{1/2} = 100$) with the same operation on a 2-pipe CYBER 205 in 32-bit arithmetic or on a 4-pipe CYBER 205 in 64-bit arithmetic (curve B:$r_\infty = 200$ Mflop/s, $n_{1/2} = 200$).

Figure 2.26 also shows the timing curve C for a linked triadic operation such as

$$\mathbf{D} = \mathbf{A} + \mathbf{B}*c, \tag{2.11}$$

in which vector $\mathbf{B}$ is multiplied by the scalar c, and then added element-by-element to the vector $\mathbf{A}$. Since most of the overhead that contributes to $n_{1/2}$ is associated with the reading and writing of numbers to main memory, roughly doubling the length of the pipeline does not significantly alter $n_{1/2}$.

However two operations must be credited for every result returned to memory and r_∞ is thereby doubled. Put another way: let the timing equation for either a multiplication or addition vector operation be

$$t = \tau(n + s + l - 1), \qquad (2.12a)$$

where s is the time to read and write to memory and l is the arithmetic pipe length. In the CYBER 205 $s \gg l > 1$, hence

$$r_\infty = \tau^{-1}, \qquad n_{1/2} = s + l - 1 \approx s. \qquad (2.12b)$$

If two such operations are linked together, then the timing equation per vector operation becomes:

$$t = \tau(n + s + 2l - 1)/2, \qquad (2.13a)$$

and hence

$$r_\infty = 2\tau^{-1} \qquad \text{and} \qquad n_{1/2} = s + 2l - 1 \approx s. \qquad (2.13b)$$

Thus, as previously stated, r_∞ is doubled and $n_{1/2}$ is approximately unchanged. This is shown by curves B and C in figure 2.26. In fact $n_{1/2}$ is increased by about 50% due to the time for the select-link instruction, which has been ignored in the above analysis, but which must be executed just prior to the

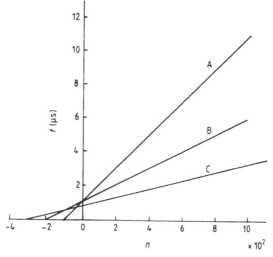

FIGURE 2.26   Timing curves used in the determination of r_∞ and $n_{1/2}$ for the CDC CYBER 205, taken from data given by Kascic (1979). A, 2-pipe CYBER 205 in 64-bit arithmetic on dyadic operations; B, 4-pipe CYBER 205 in 64-bit arithmetic or a 2-pipe CYBER 205 in 32-bit arithmetic also on dyadic operations; C, 4-pipe CYBER 205 in 64-bit arithmetic on triadic operations.

vector instructions that are to be linked together. This increase in $n_{1/2}$ can be seen in curve C of figure 2.26.

The results of the above measurements of r_∞ and $n_{1/2}$ are summarised in table 2.5 and compared with the previous results for a single CPU of the CRAY X-MP. The specific performance π_0 measures the short vector performance (see §1.3.5), hence one can see immediately that the short vector performance of the X-MP/1 is always greater than that of the CYBER 205, even the four-pipe machine. On the other hand—except for the one-pipe 205 in 32-bit mode—the long vector performance of the CYBER 205, which is measured by r_∞, is always greater than that of the X-MP/1. It follows that there must be a vector length, say $\hat{n}$, above which the CYBER 205 is faster and below which the CRAY X-MP is faster.

The value of $\hat{n}$ can be obtained by equating the performance of the two machines. If we use a superscript (2) for the CYBER 205 and the superscript (1) for the CRAY X-MP, one obtains

$$(\hat{n} + n_{1/2}^{(2)})/r_\infty^{(2)} = (\hat{n} + n_{1/2}^{(1)})/r_\infty^{(1)} \tag{2.14a}$$

whence

$$\hat{n} = n_{1/2}^{(2)}(1 - \gamma)/(\alpha - 1) \tag{2.14b}$$

where $\alpha = r_\infty^{(2)}/r_\infty^{(1)}$ and $\gamma = \pi_0^{(2)}/\pi_0^{(1)}$ are the ratio of asymptotic performance and the ratio of specific performance respectively. Equation (2.14b) was used to calculate the values of $\hat{n}$ in table 2.5.

TABLE 2.5 The asymptotic performance r_∞, half performance length $n_{1/2}$ and the specific performance $\pi_0 = r_\infty/n_{1/2}$ for contiguous memory-to-memory operations on the CYBER 205 and a one-CPU CRAY X-MP. The term $\hat{n}$ is the vector length above which the CYBER 205 has a high average performance than the CRAY X-MP/1. The CRAY X-MP/1 has the higher performance for vector lengths less than $\hat{n}$.

Computer	Bits	r_∞ (Mflop/s)		$n_{1/2}$		π_0 (M/s)		$\hat{n}$
		Dyad	Triad	Dyad	Triad	Dyad	Triad	Dyad
CYBER 205	64	50	100	50	75	1	1.3	∞
one-pipe	32	100	200	100	150	1	1.3	58
CYBER 205	64	100	200	100	150	1	1.3	58
two-pipe	32	200	400	200	300	1	1.3	26
CYBER 205	64	200	400	200	300	1	1.3	26
four-pipe	32	400	800	400	600	1	1.3	20
CRAY X-MP/1	64	70	148	53	60	1.3	2.4	—

The performance parameters given in tables 2.4 and 2.5 apply only if successive elements of the vectors involved are stored in successive memory addresses. Such vectors are said to be *contiguous*, and the memory of all computers is usually organised to access such vectors without memory-bank or memory-data-path conflicts. The *stride* of a vector is the interval in memory address between successive elements of the vector. A contiguous vector is therefore a vector with a stride of one, and any other vector is a *non-contiguous* vector. In general, vectors may have other constant strides. For example, if the elements of an $(n \times n)$ matrix are stored contiguously column by column (normal FORTRAN columnar storage), the rows of the matrix form vectors with a constant stride of n. Such vectors are sometimes described as being *periodic*. Other vectors may have elements whose location is specified by a list of addresses which may have arbitrary values. Such vectors are referred to as *random* vectors, and are accessed by using the scatter/gather (or indirect addressing) instructions of a computer.

Since computers are normally optimised for rapid access to elements of a contiguous vector, their performance is usually degraded (sometimes dramatically) if *non-contiguous* vectors are involved. This is particularly true in the case of the CYBER 205 and, as an example, we consider the timing of a dyadic operation $X = Y*Z$ between random vectors. Since the only vector instructions available on the CYBER 205 are between contiguous vectors, this non-contiguous vector operation must be performed in several stages: first the two input vectors Y and Z must be 'gathered' into two temporary contiguous vectors; then a contiguous vector operation can be performed, producing a temporary contiguous result; and finally the contiguous result is 'scattered' to the random locations of the vector Z. We can calculate the timing for this non-contiguous operation by using the timing formulae in table 2.4:

$$\text{gather } Y \qquad = (69 + 1.25n)\tau$$

$$\text{gather } Z \qquad = (69 + 1.25n)\tau$$

$$\text{contiguous multiply} = (52 + 0.5n)\tau \qquad\qquad (2.15)$$

$$\text{scatter } X \qquad = (83 + 1.25n)\tau$$

$$\text{total time} \qquad = (273 + 4.25n)\tau$$

$$= (4.25/\tau^{-1})(n + 64)$$

Since τ^{-1} corresponds to 50 Mflop/s, we find that the effective parameters describing a non-contiguous dyadic operation are

$$r_\infty = \tau^{-1}/4.25 = 12 \text{ Mflop/s}; \qquad n_{1/2} = 64. \qquad (2.16)$$

Thus we find that the use of non-contiguous vectors has degraded the performance by almost a factor 10 from the asymptotic contiguous performance of 100 Mflop/s. Because the time for the non-contiguous operation is dominated by the time for the scatter/gather operations which are not speeded up by increasing the number of vector pipelines, the above performance of approximately 10 Mflop/s maximum is virtually unchanged if the number of vector pipelines is increased.

It would, of course, be absurd to program the CYBER 205 entirely with non-contiguous vector operations of the kind discussed in the last paragraph. First, all problems should be structured so that the number of non-contiguous operations is reduced to a minimum, possibly even zero; and secondly if non-contiguous operations are unavoidable, it is desirable to group them so that many contiguous operations (rather than one in the above example) are performed on the temporary contiguous vectors. In this way the overhead of the scatter/gather operations is amortised over many vector operations. Even so the contiguous and non-contiguous performance for a dyadic operation are the best and worst possible cases, and actual performance on a particular problem will lie between the two. The fact that the range of performance between the worst and best case on the CYBER 205 is so large is indicative that considerable program restructuring may be necessary to get the best performance out of this computer.

2.3.7 The ETA[10]

The ETA[10] is the first product of ETA Systems Inc.†, which was formed in August 1983 as an off-shoot of Control Data Corporation, to shorten the development cycle of new supercomputers and, in particular, to continue the development of the CYBER 205 line of supercomputers. This 10 Gflop/s computer was announced in 1986, with first customer deliveries in early 1987. The system is to be developed into a 30 Gflop/s computer, the ETA[30], by 1992, possibly using gallium arsenide technology. ETA[10] computers have been ordered for the National Advanced Scientific Computing Centers at Princeton, Minnesota and Florida State Universities, and for the Supercomputer Applications Laboratory (SAL) at the University of Georgia, Athens. Orders outside the USA have been announced for the German Weather Service in Offenbach, West Germany, and for the Atmospheric and Environmental Service of Canada. The first delivery was to Florida State University, Tallahasse, in January 1987.

The overall architecture of the ETA[10] is shown in figure 2.27(a) and comprises 2, 4, 6 or 8 CPUs and 2 to 18 I/O units working from a large shared

† 1450 Energy Park Drive, St Paul, MN 55108, USA.

(a)

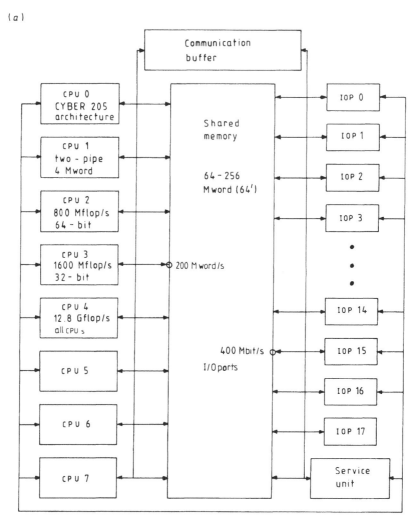

FIGURE 2.27   (a) Overall architectural block diagram of the ETA[10] computer.

memory of 64, 128, 192 or 256 Mwords (64-bit). Each CPU is architecturally
the same as a two-pipe CYBER 205 with 4 Mwords (64-bit) of local memory,
as described in §§2.3.1 to 2.3.5. The installation at Tallahasse is shown in
figure 2.27(b). The two low cabinets at the front each house four CPUs with
their local memories, and the taller cabinet behind houses the shared memory
and I/O units. Although the memory is hierarchical, programmers will see
a uniform virtual address space, addressed by 48 bits, as in the CYBER 205.
The enhanced ETA[30] is expected to have 16 Mwords of local memory per
processor and 1 Gword of shared memory. The target peak performance of

(*b*)

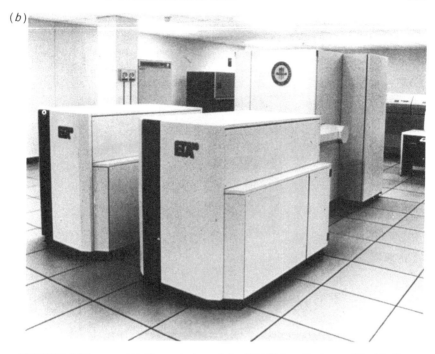

FIGURE 2.27 *cont.* (*b*) General view of the ETA[10] installation at Florida State University, Tallahasse, which was installed in January 1987. Each of the two low cabinets at the front holds four CPUs, each with 4 Mword of local memory. The large shared memory of up to 256 Mword and the I/O units are contained in the taller cabinet behind.

each ETA[10] CPU on triadic operations is 800 Mflop/s in 64-bit working and 1600 Mflop/s in 32-bit working, which corresponds to a clock period of 5 ns. This is a four-fold increase in performance over the CYBER 205 with its 20 ns clock. The peak performance of the complete eight-CPU system is, therefore, 12.8 Gflop/s in 32-bit mode, thus reaching the company's goal. The first machines will, however, have a clock period of 7 ns.

Each CPU is connected to the shared memory by a separate high-speed CPU port with a bandwidth of one 64-bit word per clock period (200 Mword/s or 12.8 Mbit/s). This is 1/6 of the data rate required to drive the two arithmetic pipelines of the CPU directly from the shared memory, hence it is assumed that substantial calculations are performed within a CPU using its local memory, before results are returned to shared memory. The Communication Buffer is a high-speed, one million word memory used to communicate, coordinate and synchronise the multiprocessor activity. Input and output are performed via 18 slower I/O ports with a bandwidth of 400 Mbit/s. Each I/O port contains up to eight functional units, each

(c)

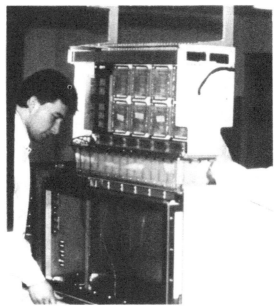

FIGURE 2.27 *cont.* (c) (Top) A single 19 in × 24 in CPU board of the ETA[10], which contains all the logic of a two-pipe CYBER 205. Above this is the 4 Mword of local memory. (Bottom) The CPU board being lowered into the tank of the liquid nitrogen cryostat. The CPU board is totally immersed in the liquid nitrogen, but the local memory above is air-cooled. A thick layer of insulation can be seen separating the two.

containing a 68020 microprocessor to support standard peripherals (such as discs and tapes), and networks (such as Ethernet, Hyperchannel and LCNS) via multiple I/O channels.

As the internal speed of logic devices within a chip has increased, the relative importance of interconnection delay, experienced when logic signals pass between chips, has increased. A major aim of supercomputer designers is therefore the reduction of this interconnection delay by placing more and more logic gates on each chip, thus reducing the number of chips and their interconnections. Unfortunately, increasing the level of VLSI in this way leads to the use of slower technologies which are normally unacceptable in a high-performance computer: for example, a much higher density of gates can be achieved using CMOS than by using the faster ECL technology. ETA Systems has resolved this dilemma by adopting the slower CMOS technology because of its higher gate density, and regaining the lost speed by operating the chips at liquid nitrogen temperatures, as we now see.

The principal innovation in the ETA[10] computer is technological, since the architecture of the CPU has been around since the CDC STAR 100 in the mid-1960s. ETA Systems has adopted high-density CMOS gate-arrays for the computer logic, with 20 000 gates on a 1×1 cm² chip. These were developed by Honeywell for the US Defense Department's VHSIC (very high-speed integrated circuit) program, and use a 1.25 μm feature size. Because of the complexity of the chips, self-testing logic is incorporated on the chip. The technology is voltage-driven and only uses power when it changes state. Consequently, the chip generates only about 2 W, which is to be compared with 4 W for the 250-gate ECL chip used on the CYBER 205.

By using the above CMOS chips, a reduction in the number of chips of about 80 times can be achieved compared to the CYBER 205. This technology, along with very dense 42-layer printed circuit boards and electronic computer-aided design tools, has enabled a single CPU of a two-pipe CYBER 205, to be placed on a single 19 inch × 24 inch board containing 250 CMOS chips. The 4 Mword local memory is mounted separately above. Thus the whole 4 Mword CYBER 205 illustrated in figures 2.14 and 2.15, and occupying 23 ft × 19 ft, has been reduced to one logic board and a compact set of memory boards occupying about two square feet (figure 2.27(c), top).

At room temperature, CMOS is not as fast as the bipolar ECL technology which is usually used for supercomputer logic circuits. However, if the CMOS chips are cooled to liquid nitrogen temperatures (77 K), circuit speeds are increased about two-fold, and speeds equivalent to room temperature ECL can be achieved. The life of the circuit boards is also found to be increased. The CPU of the ETA[10] will therefore be totally immersed in liquid nitrogen, which is part of a closed-loop system including a cryogenerator. The local and shared memory, however, will be air cooled. The ETA[10] will therefore to be the first commercially produced *cryogenic computer*. Figure 2.27(c) (bottom) shows the CPU boards being lowered into the cryogenic tank, and the thick insulating layer that separates the liquid-nitrogen-cooled logic boards from the air-cooled local memory.

The 4 Mword local memory in each CPU is made from 64K-bit CMOS static RAM chips, and the shared memory uses 256K-bit dynamic RAM chips. The use of the above density of VLSI enables the maximum eight-CPU 256 Mword ETA[10] system to be contained in a single cabinet occupying only 7 ft × 10 ft of floor space (figure 2.27(b)).

The instruction set of the ETA[10] is upward compatible with that of the CYBER 205: that is to say all CYBER 205 machine instructions are included. However, some instructions have been added to manipulate the communication buffer and permit MIMD programming, i.e. the synchronisation of the multiple CPUs when working together to solve a single problem (multi-tasking on the CRAY X-MP).

A large software program has been mounted to support the ETA[10]. The VOS (virtual operating system) will be provided with user interfaces to maintain compatibility with the CYBER 205 virtual operating system, and UNIX will be provided for compatibility with a large range of workstations and minicomputer front-ends. VOS design emphasises direct interactive communication between the user and the ETA[10], with support for both high-speed and local area networks, thus eliminating the requirements for a larger general purpose front-end computer. Although FORTRAN 77 is available, the main programming language is expected to be FORTRAN 8X, which anticipates the ANSI 8X standard (see Chapter 4) and provides structures for expressing program parallelism.

In addition to vectorising compilers for the above languages, the KAP preprocessor developed by Kuck and Associates, and based on Kuck's parafrase system, automatically identifies parallelism in programs and restructures them to enhance the level of subsequent vectorisation (Kuck 1981). The operating system will also contain a multi-tasking library for parallel processing. The multi-tasking tools allow access to a shareable data set from each processor.

2.4 JAPANESE VECTOR COMPUTERS

In 1983/4 three Japanese manufacturers (Fujitsu, Hitachi and NEC) announced pipelined vector computers that combined the best features from the CRAY-1 and CYBER 205 machines. All the computers had separate scalar and vector units like the CYBER 205, and vector registers like the CRAY-1; indeed a single-block architectural diagram could be used for all three machines. The computers are the Fujitsu FACOM VP-100 and VP-200 with an advertised peak performance of 266 and 533 Mflop/s respectively; the Hitachi HITAC S810 models 10 and 20 with a peak performance of 315 and 630 Mflop/s; and the NEC SX1 and SX2 with a peak performance of 570 and 1300 Mflop/s. Although similar in overall architecture, the three computers differ significantly in detail, particularly in the cooling technology. We will now describe the computers in more detail.

2.4.1 The Fujitsu FACOM VP-100/200/400

The FACOM VP-100/200 was the first of the Japanese vector computers to appear (Motegi, Uchida and Tsuchimoto 1984, Miura and Uchida 1984, Tamura, Kamiya and Ishigai 1985). The project was started by Fujitsu in 1978 and the first machine was operational in 1982. The first customer delivery of a VP-100 was made to the Institute of Plasma Physics at Nagoya University in December 1983, and the second delivery, also of a VP-100, was made to Kyoto University in April 1984. The VP-200 model shown in figure 2.28 was taken at the Fujitsu Ltd Numazu Production Plant in the foothills of Mount Fuji, where all the vector processor manufacture and software development takes place. The VP-100 and VP-200 are marketed in the USA and UK by Amdahl Corporation as the Amdahl VP1100 and VP1200 and, in Germany, the Fujitsu vector processors are marketed by Siemens who installed a VP-200 in their Munich support centre in February 1985. Siemens first sale of a VP-200 was to the IABG Institute in Munich. A VP-400 with twice the performance of the VP-200 was installed at the Japanese National Aerospace Laboratory in 1985. Its theoretical peak performance is 1 Gflop/s.

The overall view of the VP-200 in figure 2.28 shows three cabinets in the foreground for, from left to right, memory, the scalar unit, and the channel processors. At the rear are two further cabinets for the vector unit and a second block of memory. The machine shown occupies about 20 ft × 15 ft of floor space, weighs about 7 tons and consumes about 80 kVA of power.

Unlike the CRAY and CYBER 205 computers which use freon refrigeration, the Fujitsu vector processors use forced air cooling. This is achieved by attaching cooling fins to the chip mounting, as shown in figure 2.29. Figure 2.29(a) shows a logic chip which uses bipolar ECL LSI with 400 gates per chip. Register files are contained on similar chips with 1300 gates per chip. In both cases the gate delay time is 350 ps. Figure 2.29(b) shows a high-speed memory module comprising four 1K-bit bipolar RAMs with an access time of 5.5 ns. In this case four fins are used, one for each chip. This memory module is used for buffer memory and control store. Both the assemblies shown in

FIGURE 2.28   Overall view of the Fujitsu VP-200. The three cabinets on the right are for memory, scalar unit, and channel processors; the two on the left are the vector unit and a second cabinet of memory.

figure 2.29 are about 0.75 inch × 0.75 inch × 0.75 inch, and are mounted on multichip carriers (MCCs) holding up to 121 assemblies in an 11 × 11 array, as shown in figure 2.30(a). The MCC is a 31 × 31 cm² 14-layer printed circuit board. Thirteen such MCCs are then mounted horizontally in a 5 cm cube called a stack. This is shown in figure 2.30(b). The main memory of the VP-100/200 is made from 64K-bit static MOS chips with an access time of 55 ns. These chips do not require fins to dissipate the heat and are mounted as flat packs on 24 × 38 cm² 6-layer printed circuit memory boards (not illustrated, but like any other such board). Each board contains a 4 × 32 array of 64K-bit data chips or 1 Mbyte of memory.

The overall architecture of the FACOM vector processors is given in figure 2.31. The architecture of the vector unit is CRAY-like, in the sense that multiple functional units (for floating-point add, multiply and divide) work from a vector register memory (64 Kbyte). However, there is a separate scalar unit, as in the CYBER 205, with 64 Kbyte of buffer storage (5.5 ns

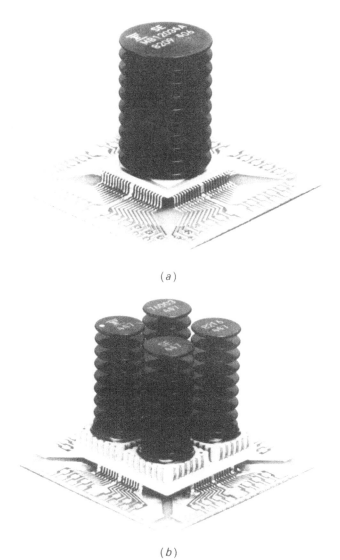

(a)

(b)

FIGURE 2.29   Air-cooled chips of the Fujitsu vector processors, (a)
An ECL LSI logic chip with 400 gates. (b) Four 1 Kbit memory
chips mounted together on a module.

access). The main memory (55 ns access) of 256 Mbyte is arranged as
256 banks, and is connected to the vector registers by two load/store pipelines.
The above numbers are for the VP-200; the main memory and vector register
storage is halved in the VP-100. There are also 256 mask registers, each of

(a)

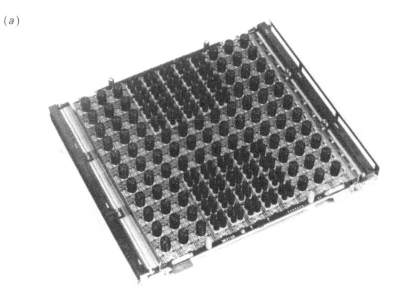

(b)

32 bits (16 bits on the VP-100), which are used to store mask vectors which control conditional vector operations and vector editing operations.

A unique feature of the computer is that the vector register storage may be dynamically reconfigured under program control either as 256 vector registers of 32 64-bit elements, 128 registers of 64 elements, ... etc, or as eight registers of 1024 elements. The length of the vector registers is specified by a special register, and can be altered by a program instruction.

The clock period of the scalar unit is 15 ns, which is called the major cycle. The vector unit, however, works on a clock period of 7.5 ns, the minor cycle. On the VP-200 the floating-point add and multiply pipelines can deliver two 64-bit results per clock period, leading to a peak performance of 267 Mflop/s for register-to-register dyadic operations using one pipeline, and 533 Mflop/s for register-to-register triadic operations which use both the add and multiply pipelines simultaneously. These rates are halved on the VP-100. The divide pipeline is slower and has a peak performance of 38 Mflop/s. On the VP-200 each load/store pipeline can deliver four 64-bit words every 15 ns, or equivalently a bandwidth of 267 Mword/s (133 Mword/s on the VP-100). These rates are 2/3 of the bandwidth which is required to support a dyadic operation with arguments and results stored in main memory. Thus, unlike the CRAY X-MP and CYBER 205, the Fujitsu VP has insufficient memory bandwidth to support such memory-to-memory operations. This puts a heavier burden on the compiler to make effective use of the vector registers for intermediate results in order to limit transfers to and from main memory.

The instruction set of the Fujitsu VP is identical to the IBM 370, with the addition of vector instructions; indeed IBM-370-generated load modules will run on the VP without change. Vector instructions include conditional evaluation of a vector arithmetic operation controlled by a mask with one bit per element of the vector (as on the CYBER 205, §2.3.4); compress and expand vectors according to a condition; and vector indirect addressing, that is to say a random scatter/gather instruction as described for the CRAY X-MP in §2.2.4. This instruction can gather four elements every 15 ns.

It is anticipated that most users will write their programs in FORTRAN, and an extensive interactive software system is being developed for the interactive optimisation and vectorisation of such programs (Kamiya, Isobe, Takashima and Takiuchi 1983, Matsuura, Miura and Makino 1985). For example, the vectorisation of IF statements presents a particular problem,

◁

FIGURE 2.30 Logic technology of the Fujitsu vector processor. (a) A multichip carrier (MCC) with space for 121 LSI chips. (b) A stack of 13 horizontally mounted MCCs.

and the FORTRAN 77/VP vectorising compiler selects the best of three possible methods. These are: (a) conditional evaluation using a bit-mask; (b) selection of the participating elements into a compressed vector before performing the arithmetic; and (c) the use of the vector indirect addressing to select the participating elements. The compiler compares the three methods, based on the relative frequency of load/store operations in the DO loop, and the fraction of the vector elements which are participating (the true ratio). If the true ratio is medium to high, a masked arithmetic operation is best; otherwise the compress method is best when the frequency of load/store operations is low, and indirect addressing is best when the frequency is high. Interaction takes the form of suggestions to the programmer on how to restructure his program to improve the level of vectorisation.

2.4.2   The Hitachi HITAC S-810

The Hitachi HITAC S-810 was the second of the Japanese vector processors to appear, the first customer delivery being to the University of Tokyo in

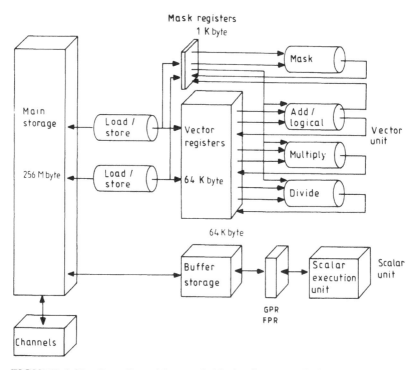

FIGURE 2.31   Overall architectural block diagram of the Fujitsu vector processor. (FPR denotes floating-point registers and GPR denotes general-purpose registers.)

1984. Other machines have been installed for internal company use. The Hitachi S-180 model 10 and model 20 vector computers are similar in overall architecture to the Fujitsu machines, as can be seen by comparing figures 2.31 and 2.32. The principal difference is that the S-810 has more pipelines (Nagashima, Inagami, Odaka and Kawabe 1984). There are three load and one load/store pipelines on the S-810 compared to only two load/store pipelines on the VP. This means that memory-to-memory dyadic operations can be supported on the S-810 at the full rate. The main memory size is 256 Mbytes (40 ns access time) and there is 64 Kbytes of vector register storage. Like the Fujitsu VP, this register storage can be reconfigured dynamically to hold vectors of different lengths.

The model 20 has 12 floating-point arithmetic pipelines (four add, two multiply/divide followed by add, and two multiply followed by add). The clock period for both models is 14 ns, which corresponds to a theoretical peak performance of 71.4 Mflop/s per pipeline for register-to-register operation, or 857 Mflop/s for the 12 pipelines. However, if one takes into account the time to load the vector registers from main memory this is reduced to a realistic maximum performance of 630 Mflop/s if all the pipelines are used. The design is optimised to evaluate expressions such as $A = (B + C)*D$ which require three vector loads and one vector store and thus use the four load and store pipelines. The model 10 has a 6 floating-point

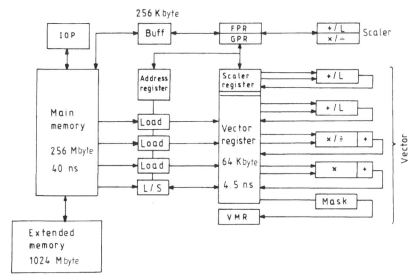

FIGURE 2.32   Overall architectural block diagram of the Hitachi S-810 model 10. (VMR denotes the vector mask register and L denotes logical operations.)

pipelines and half the vector register and main memory size. Its peak performance is quoted as 315 Mflop/s.

As with any of the computers discussed, the observed performance on actual problems will be less than the above peak rates, because of problems of memory access. As a simple test, the $(r_\infty, n_{1/2})$ benchmark described in §1.3.3 has been executed for a number of vectorised DO loops (statement 10 of program segment (1.5)). The results are given in table 2.6 for the S-810 models 10 and 20 in both 32- and 64-bit precision. For the model 10 we observe a maximum performance of approximately 240 Mflop/s for the four-ops case, which is an expression that makes maximum use of the hardware. In this case, there are three input vectors and one output vector which occupy all the four pipelines to memory. In addition, the expression uses the two add and the two multiply pipelines. Since the pipelines have a clock period of 14 ns this corresponds to a maximum rate of 71 Mflop/s per pipeline, giving a maximum expected performance of 284 Mflop/s. The measured value of 240 Mflop/s is less than this, due to the time required to load the vector registers.

If there are more than three input vectors and one output vector, the memory bandwidth is insufficient to feed the arithmetic pipelines with data at the rate

TABLE 2.6   Results for the $(r_\infty, n_{1/2})$ benchmark on the Hitachi S-810/10 with figures for the model 20 in parentheses. Upper case variables are vectors, and lower case are scalars. (Data courtesy of M Yasumura, Hitachi Central Research Laboratory, Tokyo.)

Operation: statement 10 program (1.5)	Stride	Precision bits	r_∞ (Mflop/s)	$n_{1/2}$
Dyad	1	32	60(97)	60(130)
		64	62(119)	73(143)
A = B + C	8	32	31(61)	46(108)
		64	56(110)	92(208)
Triad	1	32	118(180)	71(126)
		64	121(238)	73(152)
A = B + e*C	8	32	43(85)	27(61)
		64	66(131)	41(108)
Four-ops	1	32	238(345)	91(157)
		64	231(489)	88(190)
A = B + (e*C + f*D)	8	32	85(163)	24(58)
		64	134(263)	49(111)

required. The performance is degraded, and the expression must then be treated as a combination of simpler dyadic and triadic operations, whose performance we consider next. Simpler expressions such as dyads and triads cannot make use of all the arithmetic pipelines and we observe rates of approximately 60 and 120 Mflop/s respectively for contiguous vectors with a stride of unity. This rate may be degraded to a half or a third due to memory-bank conflicts if the vectors are non-contiguous, in our case with a stride of eight. In summary, one could say that, for contiguous vectors, $r_\infty = 60$ Mflop/s per arithmetic operation in the expression (up to a maximum of four), with a degradation of up to a factor of three for unfavourably stored vectors. These results seem to be unaffected by the precision of the arithmetic. Thus a performance in the range 30 to 240 Mflop/s is to be expected from the model 10 depending on the circumstances. Values of $n_{1/2}$ range from 40 to 90.

The results for the model 20 are shown in parentheses in table 2.6. This model has twice the number of arithmetic pipes and we observe almost exactly double the r_∞ and also double the value of $n_{1/2}$, leading to a performance between approximately 60 and 480 Mflop/s depending on the circumstances. This is to be compared with a peak performance of 630 Mflop/s quoted by the manufacture. Because $n_{1/2}$ is also doubled, vectors twice as long are required to achieve the same fraction of the maximum performance.

The Hitachi S-810 is air-cooled, like the Fujitsu VP, and uses ECL LSI with 550 gates per chip (350 ps gate delay) or 1500 gates per chip (450 ps gate delay) for its logic. Vector register storage uses bipolar 1K-bit chips with 4.5 ns access time, and the main storage uses 16 K-bit CMOS static RAMs. Each plug-in board holds up to 40 chips and has 14 interconnection layers.

As with the Fujitsu VP, the instruction set is an extension of IBM 370 and a vectorising compiler for IBM FORTRAN 77 is seen as the principal input language. The techniques used for vectorisation are described by Yasamura et al (1984). IBM-generated load modules will run without change.

2.4.3 The NEC SX1/SX2

The Nippon Electric Corporation's SX1 and SX2 computers were the last of the three Japanese vector computers to appear, and the SX2 has the highest theoretical peak performance of over 1 Gflop/s (figure 2.33). The first two deliveries of SX2 computers were made in 1985 to Osaka University and the Sumito Trading Company. European marketing is by Mitsui and Company Europe Group.

The SX computers use current mode logic (CML) bipolar gates with a density of 1000 gates per LSI chip and a gate delay time of 250 ps. Cache and vector registers are made from 1K-bit bipolar RAMs with a 3.5 ns access time.

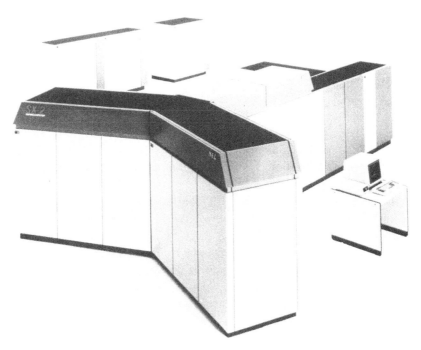

FIGURE 2.33   General view of the Nippon Electric Company SX2 computer.

These chips (36 of them) are packaged on a 10 cm square ceramic base to form a multichip package (figure 2.34(a)), which in turn is placed in a liquid cooling module (figure 2.34(b)) through which water circulates. The main memory of 256 Mbyte is composed of 64K-bit MOS static RAMS with a 40 ns access time. The large extended memory of 2 Gbyte is composed of dynamic MOS chips.

The overall architecture of the NEC SX2 is shown in figure 2.35 (Watanabe 1984). There are four general-purpose vector pipelines, each of which computes every fourth element of a vector operation and has inside it a combined floating-point multiply/divide pipeline and an add pipeline. Thus the elements of a vector operation are spread across the available vector pipelines, as in the CYBER 205. The clock period is 6 ns, so that when all eight floating-point pipelines are working simultaneously a maximum $r_{\infty} = 1333$ Mflop/s is asymptotically possible. The above figures refer to the SX2 model. The SX1 computer has a 7 ns clock period and half the number of pipelines, giving a maximum $r_{\infty} = 570$ Mflop/s. The data transfer rate between memory and vector registers is eight numbers per clock period.

As with the other Japanese machines the input language is FORTRAN 77.

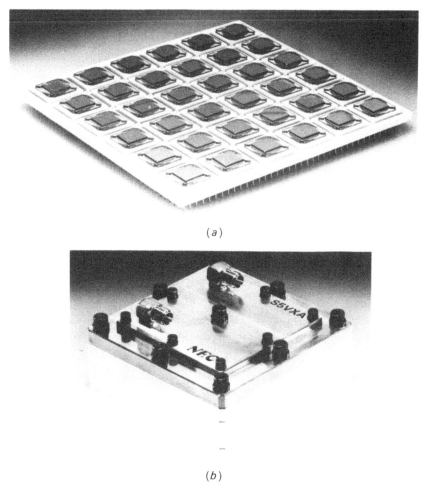

(a)

(b)

FIGURE 2.34   Water-cooled technology of NEC SX1/SX2. (a) A 10 cm² multichip package, containing 36 LSI chips, each of which has 1000 logic gates. (b) The liquid cooling module.

There is an automatic vectoriser, analyser and optimiser to assist in restructuring the FORTRAN to obtain a better level of vectorisation. Unlike the other machines, the instruction set is not compatible with IBM 370, and IBM-generated load modules will not run on the machine.

2.4.4   Performance Comparisons

A substantial number of benchmark comparisons have been made between the vector computers that we have described, and some of these are given

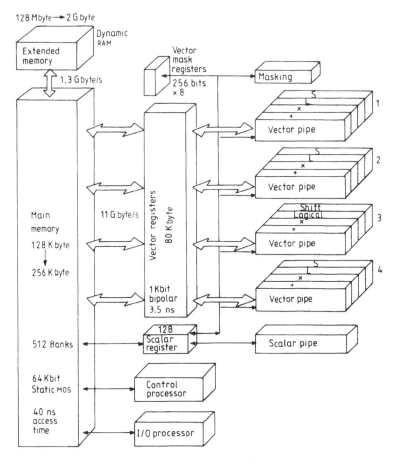

FIGURE 2.35   Overall architectural block diagram of the NEC SX2.

in table 2.7. The top three rows give the average performance in Mflop/s for three of the so-called *Livermore loops* (McMahon 1972, Arnold 1982, Riganati and Schneck 1984). Fourteen such DO loops were selected by the Lawrence Livermore Laboratory as being typical of their computer-intensive work. We have selected three which generally exhibit both the best and the worst performance of a computer. Loop 3, the inner product, is at the centre of most linear algebra routines. Most vector computers make special provision for this loop, and the highest vector performance is usually observed. The performance in loops 6 and 14 is usually more characteristic of the scalar performance because the DO loops involve recurrences and the opportunity for vectorisation is reduced or eliminated. These problems are the solution of a tridiagonal system of equations and the movement of particles in a computer simulation of a hot gas plasma.

The Livermore loops are called kernel benchmarks because they consist of small program segments, rather than the solution to a complete problem. We can see in the comparison between Livermore loop 3 and loops 6 or 14 that the kernel performance ranges over at least a factor of 10. Hence it is also of interest to see what performance is found on a complete problem which would involve both vectorisable and non-vectorisable loops, together with unavoidable scalar code. To address this point, Dongarra (1985) has compiled a set of benchmark measurements from the timings of the well known LINPACK routines for the solution of a full set of 100 linear equations by lower/upper triangular (LU) decomposition and back substitution (Dongarra et al 1979). Rows four and five of table 2.7 give the performance for FORTRAN code and for specially optimised code for the same problem using assembler routines for the key inner loops. The solution of 100 linear equations is not a large enough problem to demonstrate the full capabilities of a large supercomputer, particularly one with multiple CPUs. Hence in row 6 we show the best performance that has been obtained for solving 300 linear equations, using the matrix–vector method (see next paragraph). The performance obtained on this benchmark is indicative of what should be obtainable from carefully optimised code.

A glance at table 2.7 shows that it is not possible to see a clear distinction between the performance of the computers. All of the computers are of similar performance and can be expected to work at a few tens of Mflop/s on non-optimised FORTRAN code, and a few hundreds of Mflop/s on carefully optimised code (possibly needing assembler code). Which computer has the best performance on any particular problem is likely to depend on the care with which the program is optimised. For example, substantially improved performance can be obtained on the CRAY X-MP if the LU decomposition is expressed in terms of matrix–vector operations, instead of vector–vector operations (see row 6). The work of a matrix–vector operation may then be multi-tasked very efficiently across the multiple CPUs (Dongarra and Eisenstat 1984, Chen et al 1984). The best performance is achieved on the CRAY X-MP if three rows are simultaneously eliminated (Dongarra and Hewitt 1985). In this case, a performance of 718 Mflop/s is obtained for the solution of a set of 1000 equations on a four-CPU CRAY X-MP. Other detailed comparisons of the performance of the above supercomputers are given by Bucher (1984), Lubeck et al (1985), and Bucher and Simmons (1986).

TABLE 2.7 Some benchmark comparisons between pipelined vector computers. The average performance is given in Mflop/s for the stated problem using 64-bit arithmetic.

Problem	CRAY X-MP[d]	CYBER 205[f]	IBM[f] 3090VF[d]	Fujitsu[n] VP200(2) VP400(4)	Hitachi S-810/20	NEC SX2	CRAY-2[d,l]	ETA[10]
Theoretical peak performance	210(1) 420(2) 840(4)	100(1) 200(2) 400(4)	108(1) 216(2) 432(4)	533(2) 1142(4)	840	1300	488(1) 976(2) 1951(4)	1250(1) 5000(4) 10000(8)
$(r_\infty, n_{1/2})$	—	—	—	—	(119, 143)	—	(56, 83)(1)	—
FORTRAN (§1.3.3)								
Livermore 3 inner product	96(1)[e]	86(1)[e]	—	331(2)[e] 359(4)[g]	212[e] 332[i]	561[g]	63(1) 91(1)	—
Livermore 6 tridiagonal	8(1)	5(1)[e]	—	10(2)[e] 10(4)[g]	5[e] 32[i]	13[g]	4(1) 8(1)	— —
Livermore 14 particle pusher	7(1)[e]	5(1)[e]	—	14(2)[e] 15(4)[g]	9[e] 11[i]	24[g]	6(1) 9(1)	— —
LINPACK[a] FORTRAN[m] n = 100	24(1)	20(2)	12(1)	18(2)[k]	17	46[j]	15(1)[k]	—
Assembler[b] inner loop n = 100	44(1)	25(2)	—	—	—	—	—	—

Matrix–vector[c] best assembler $n = 300$							
171(1)	31(2)	18(1)[h]	183(2)[h]	158[h]	309	93(1)	
257(2)		27(1)	230(2)				
480(4)							

Notes

a Solution of linear equation using DGEFA and DGESL for matrices of order n (Dongarra *et al* 1979).

b Basic linear algebra subroutines (BLAS) optimised in assembler (Dongarra 1985).

c Matrix–vector method of Dongarra and Eisenstat (1984). Matrix order 300. Best reported assembler (Dongarra 1985).

d Number of CPUS used is shown in parentheses

e Fuss and Hollenberg 1984.

f Number of pipelines is shown in parentheses.

g Van der Steen 1986.

h All FORTRAN.

i After optimisation (Nagashima *et al* 1984).

j Dongarra 1986.

k Dongarra and Sorensen 1987.

l Single-processor CRAY-2 prototype (Bruijnes 1985).

m FORTRAN code with single-statement BLAS loop (Lawson *et al* 1979, Dongarra 1985). FORTRAN directives allowed.

n Number in parentheses indicates the model in the table below.

2.5   THE FPS AP-120B AND DERIVATIVES

The FPS AP-120B and its derivatives—the AP 190L, the FPS-100, 164, 164/MAX, 264 and the FPS 5000—are all members of a single family of computers based on a common architecture, namely that of the FPS AP-120B. These computers have been renamed as follows: the original MSI version of the FPS-164 (now discontinued) is called the M140, the later VLSI version (first called the FPS-364) is called the M30, the FPS-164/MAX is called the M145, and the FPS-264 is now the M60. They are all manufactured by Floating Point Systems Inc.† in Beaverton near Portland, Oregon, USA. The company was founded in 1970 by C N Winningstad to manufacture low-cost yet high-performance floating-point units to boost the performance of minicomputers, particularly for signal processing applications. Starting in 1971, the company produced floating-point units for inclusion in other manufacturers' machines (e.g. Data General). The first machine marketed under the company's name, the AP-120B, was co-designed by George O'Leary and Alan Charlesworth, and had a peak performance of 12 Mflop/s. Deliveries began in 1976, and by 1985 approximately 4400 machines had been delivered. The FPS-100 is a cheaper version of the AP-120B, made for inclusion as a part of other computer systems. The AP-120B was designed for attachment to minicomputers, and a version with more memory, called the AP-190L, was introduced for attachment to larger mainframe computers such as the IBM 370 series.

In 1980 the concept of the AP-120B was broadened from rather specialised signal processing applications to general scientific computing by increasing the word length from 38 to 64 bits, and the addressing capability from 16 to 24 bits. The memory capacity was also greatly increased, first to 1 Mword then to 7.25 Mword. The new machine which evolved was the FPS-164 which was first delivered in 1981. By 1985 about 180 FPS-164s had been sold. Although capable of solving much larger problems than the AP-120B, the FPS-164 was no faster at arithmetic—indeed its peak performance of 11 Mflop/s was 1 Mflop/s less than that of the AP-120B. The first improvement in arithmetic speed came in 1984 with an enhancement to the architecture called the matrix accelerator (MAX) board. Each such MAX board can be regarded computationally as the equivalent of two additional FPS-164 CPUs, so that a machine with the maximum of 15 MAX boards has a theoretical peak performance of 31 FPS-164 CPUs or 341 Mflop/s. The AP-120B and the FPS-164 are both implemented in low-power (and therefore low-speed) transistor–transistor logic (TTL), and the next improvement

† Headquarters: PO Box 23489, Portland, Oregon 97223, USA. UK Office: Apex House, London Road, Bracknell, Berkshire, UK.

in performance came in 1985 with the announcement of an ECL version of the FPS-164, called the FPS-264, which had a peak performance of 38 Mflop/s. The FPS-164 was designed in 1979 with the then current MSI of between 10 and 100 gates per chip. The machine has now been re-engineered in CMOS VLSI and is marketed as the FPS M30 computer, which may be attached to microVAX and Sun workstations. The architecture is the same as the FPS-164 which we describe here.

The FPS-164/MAX is a SIMD computer because the 31 CPUs in a full configuration operate in lock-step in response to a single stream of instructions. Another development of the AP-120B architecture has, however, been towards multi-instruction stream computing (MIMD). The FPS-5000 series of computers, announced in 1983, have a control processor, up to three arithmetic coprocessors and an I/O processor attached via a common bus to a common memory and a host computer. Each arithmetic coprocessor (AC) has its own control unit, and may be executing a different subroutine from the other ACs, thus providing a MIMD capability. According to the classification given in §1.2.6 the FPS-5000 is a bus-connected, shared-memory MIMD computer. The control processor of the FPS-5000 is either an AP-120B or FPS-100 computer. The XP-32 arithmetic coprocessor is of a new design due to Pincus and Kallio, but follows the same general pattern as the AP-120B.

All the above computers are called array processors because they are designed to process arrays of numbers efficiently. Architecturally, however, they are all pipelined computers with a small number of pipelined arithmetic units working from a common memory and registers. In this respect their architecture is CRAY-like. It is important to realise that although the above computers are called array processors, they are not arrays of processors like the ICL distributed array processor DAP (see Chapter 3). This ambiguity in the meaning of the expression 'array processor' has led us to avoid its use in this book. However, the term is commonly used for the computers described in this section and by the manufacturers. Some, however, prefer to interpret the initials AP to mean *attached processor*, since most require a host computer.

We start by describing in detail the father of the family, namely the AP-120B (§2.5.1 to §2.5.6), then follow with separate sections on the special features introduced in the FPS-164 and 264 (§2.5.7), the FPS 164/MAX (§2.5.8) and the FPS-5000 (§2.5.10). The use of multiple FPS-164s attached to a host, to form a loosely coupled array of processors (/CAP) is considered in §2.5.9.

2.5.1   FPS AP-120B

Apart from company documentation, the principal references describing the AP-120B are by Wittmayer (1978), Harte (1979) and Charlesworth (1981).

FIGURE 2.36   An overall view of an FPS AP-120B installation. The AP-120B occupies only 29 inches of rack space, and is attached to a PDP 11/34 with two disc units, a tape reader and output printer. The control teletype or VDU is not shown. (Photograph courtesy of D Head and Floating Point Systems, S A Ltd.)

The FPS AP-120B occupies 29 in of rack space on a standard EIA 19 in wide rack. Figure 2.36 shows a small installation comprising a PDP 11/34 host in the lower portion of the rack below the AP-120B, two disc units, a magnetic tape reader and output printer. The control teletype or VDU is not shown. The machine weighs about 160 pounds and consumes less than 1.3 kW (compared with about 115 kW for the CRAY X-MP). Cooling is by forced air driven by push–pull fans that are part of the AP-120B chassis. The machine can therefore be carried by one man, and plugged into a standard 13-amp domestic power socket. In contrast to the CRAY X-MP and CYBER 205, no auxiliary coolant plant or motor–generator sets are required to drive the machine.

Figure 2.37 shows a rear view of an AP-120B with a circuit board partially withdrawn. The cooling fans which can be seen at the top, blow air over the

FIGURE 2.37   Rear view of the FPS AP-120B showing the vertically mounted 10 in × 15 in circuit boards and fan cooling. (Photograph courtesy of D Head and Floating Point Systems, S A Ltd.)

vertically mounted 15 in × 10 in circuit boards. The chassis has capacity for 28 etched-circuit boards which are chosen to suit the particular requirements for memory and input/output. Two of the circuit boards are shown in greater detail in figure 2.38. The circuit boards are six-layered, comprising three signal layers and three power supply layers for ± 5 V and + 12 V. The power supplies, which are not shown in figure 2.37, are mounted on a separate power panel that occupies the rear part of the 29 in rack space behind the circuit boards.

2.5.2   Architecture

The overall architecture of the AP-120B is shown in figure 2.39. It is based on multiple special-purpose memories feeding two floating-point pipelined arithmetic units via multiple data paths. The machine is driven synchronously from a single clock with a period of 167 ns. This means that the state of the machine after a sequence of operations is always known and reproducible.

FIGURE 2.38 Two circuit boards from the FPS AP-120B. Left: the control buffer logic board from the control unit which decodes instructions; right: a board from the program memory which stores instructions. (Photograph courtesy of D Head and Floating Point Systems, S A Ltd.)

The operation of the machine can therefore be exactly simulated, clock period by clock period, and the machine does not suffer from the delicate timing uncertainties that had plagued some earlier computers which had separate clocks driving several independent units. Multiple data paths are provided between the memories and the pipelines, in order to minimise the delays and contentions that can occur if a single data path is shared between many units.

Starting at the top of figure 2.39 the memories are: a *program memory* of up to 4K 64-bit words for storing the program (cycle time 50 ns); a *scratch-pad* (S-pad) memory of 16 16-bit registers for storing addresses and indices; a *table memory* (167 ns cycle time, either read only or read/write memory) of up to 64K 38-bit words for storing frequently used constants, such as the sine and cosine tables for use in calculating a Fourier transform; two sets (*data pad X* and *data pad Y*) of 32 38-bit registers for storing temporary floating-point results; and a *main data memory* for 38-bit words (plus three parity bits), directly addressable to 64 Kwords but, with an additional 4-bit page address, expandable up to 1 Mword. Separate 38-bit data paths are provided to each of the two inputs to the floating-point adder and multiplier. These four independent paths may be fed from the main data memory, the data pads or from table memory. Three further 38-bit data paths feed results from the two pipelines back to their own inputs, or to the data pads or main data memory. These multiple paths allow an operand to be read from each data pad and a result written to each data pad during one machine cycle.

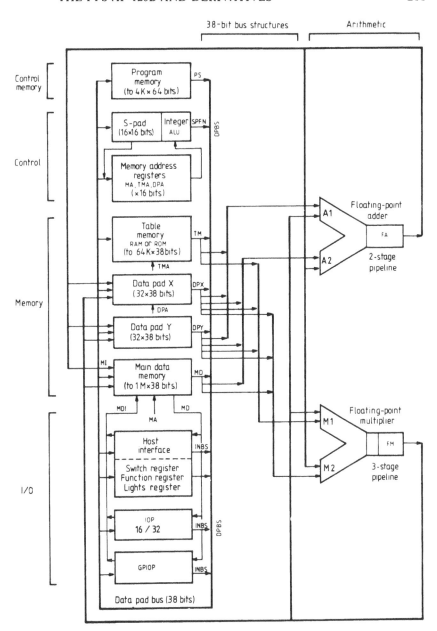

FIGURE 2.39   Overall architecture of the FPS AP-120B, showing the multiple memories, arithmetic pipelines and data paths. (Diagram courtesy of Floating Point Systems Inc.)

The address within both data pads is given by the contents of the data pad address register (DPA). Relative addressing (−4 to +3) with respect to this address is available separately for each data pad within the instruction in the data pad index fields XR, YR, XW, YW (see §2.4.4).

The main data memory is available in 8K modules (or 32K modules depending on the chip type) which are each organised as a pair of independent memory banks, one bank for the odd addresses and the other bank for the even addresses. The standard memory has an access/cycle time of 500 ns and the optional fast memory has a cycle time of 333 ns. Successive references to the same memory bank (e.g. all even addresses less than 8K) must be separated by at least three clock periods with standard memory or two clock periods with fast memory. Two successive references to different memory banks (e.g. two neighbouring addresses which are from odd and even banks, or two even addresses separated by 8K and therefore in different modules) may however be made on successive clock periods. Alternating references to the odd and even memory banks, as would occur when accessing sequential elements of a long vector, can occur at one reference per clock period for fast memory, giving an access to the same bank every 333 ns (matching the capability of the memory chip), and an effective minimum cycle time between requests to memory as a whole of 167 ns. For standard memory this rate must be halved, giving an effective memory cycle time for such optimal sequential access of 333 ns. If repeatedly accessing the same bank, a cycle time of 500 ns (three clock periods) applies. The memory is therefore described by the manufacturer as having an *interleaved* 'cycle' time of 167 ns for fast memory or 333 ns for standard memory, even though the memory chips have a physical cycle time of 333 and 500 ns respectively. However, it should be remembered, when making comparisons with other machines, that we have previously quoted the cycle time of the memory chips as a measure of the quality of the memory (e.g. 38 ns main memory of the CRAY X-MP, although this is organised into banks so as to give an interleaved 'cycle' time of 9.5 ns). If a memory reference to a part of the memory that is busy occurs, the machine stops execution until the memory becomes quiet.

Instructions on the AP-120B are 64 bits wide, and each instruction controls the operation of all units in the machine. Thus there is, in this sense, only one instruction in the instruction set (see §2.5.4) with fields which control each of the 10 functions, although some fields overlap and thus exclude certain combinations of functions. This arrangement of control is referred to as 'horizontal microcode'. Instructions are processed at the maximum rate of one per clock period, i.e. 6 million instructions per second, but since each instruction controls many operations this is equivalent to a higher rate on a conventional machine whose instructions only control one unit.

The S pad contains an independent 16-bit integer arithmetic and logical unit (ALU) for computing addresses, loop counts and indices in its set of 16 16-bit registers. The operations provided are: integer addition, subtraction; logical AND, OR, and equivalent; and shifts and bit reversal for use in Fourier transformation (see §5.5). All these operations take one clock period. There is no provision for multiplication. These integer operations are performed in parallel with floating-point calculations in the pipelines.

Both the addition and multiplication floating-point pipelines may take a new pair of operands every clock period and deliver a result every clock period. Thus the maximum rate of generation of results is 6 Mflop/s per pipeline, or 12 Mflop/s if both pipelines can be kept supplied with data. The performance on actual problems is more likely to be in the range 4–8 Mflop/s (see §2.5.6). The length of the multiplication pipeline is three clock periods (500 ns) and of the addition pipeline is two clock periods (333 ns). The addition pipeline also performs the logical operations of AND, OR and equivalence, absolute value, scale and number conversions between sign–magnitude and two's complement formats. The operation of the addition and multiplication pipelines is clarified by figures 2.40 and 2.41 which show the possible source and destinations of operands, the two input registers (A1, A2 or M1, M2), the partial operations performed at each stage of the pipelines, and the buffer registers for holding the intermediate partial results. The AP-120B does not provide a hardware divider, and floating-point division is accomplished in software by evaluating an approximating polynomial.

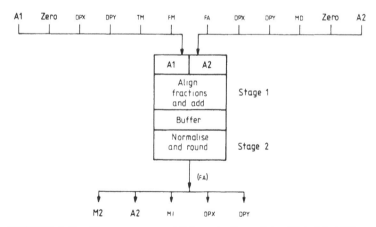

FIGURE 2.40   The two-stage addition pipeline of the FPS AP-120B. Note the possible sources of operands and destinations for results, using the notation of figure 2.39. Buffers are provided between each stage to store intermediate results. (Diagram courtesy of Floating Point Systems Inc.)

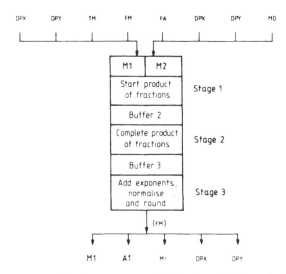

FIGURE 2.41   The three-stage multiplication pipeline of the FPS
AP-120B. Notation as used in figure 2.39. (Diagram courtesy of Floating
Point Systems Inc.)

Floating-point arithmetic is performed in a 38-bit format (10-bit radix-2
exponent and a 28-bit two's complement mantissa). This representation has
a dynamic range $10^{\pm 153}$ and a precision of eight decimal places. It provides
significantly more precision than IBM 32-bit radix-16 format ($10^{\pm 78}$ and six
decimal places). Extra guard digits are also kept during the arithmetic
operation in order to minimise loss of precision. The conversion of numbers
between the format used by the host computer and the internal 38-bit format
of the AP-120B takes place as the numbers are transferred between the
machines.

Input and output to the AP-120B is performed by an I/O port (IOP) or a
general programmable I/O port (GPIOP). The IOP provides either 16- or 38-bit
direct memory access to the main data memory of the AP-120B, by 'stealing'
memory cycles as required. Data transfer rates of 1.5 Mword/s into the AP
and 1.3 Mword/s out of the AP are obtained. The 16-bit wide port is used
for analog-to-digital input, display outputs and standard peripherals. The
38-bit wide port also contains a full adder and can be wired for a variety of
data format conversions. As well as normal I/O this port can be used to link
one AP-120B to another. The IOP occupies one circuit board of the
AP-120B, and can accommodate up to 256 external devices. The GPIOP is a
programmable I/O channel which provides up to 3 Mword/s continuous
transfer with 'in-flight' format conversion including fixed-point to floating-
point. It contains two 18 Mips microprocessors and occupies three circuit

boards. It is used typically to interface with discs, real-time displays, video cameras and other computers.

The architecture of the AP-120B with a 64K standard main data memory can be expressed in the ASN notation of §1.2.4 as follows:

$$C(AP\text{-}120B) = I1[\{Fp_{38}(+), Fp_{38}(*)\}_{\overline{7 \times 38}}$$

$$\{P1, M1-a, M2, M3, P2\}]_h$$

$$I1 = \{I_{64}^{167} - M4\}$$

$$P1(S\text{-pad}) = B_{16} - M_{16 \bullet 16}$$

$$M1(\text{main data}) = 8\{2M_{4K \bullet 38}^{500}(MOS)\}$$

$$M2(\text{table}) = M_{64K \bullet 38}$$

$$M3(X, Y \text{ data pads}) = 2M_{32 \bullet 38}$$

$$M4(\text{program}) = M_{4K \bullet 64}^{50}(\text{bipolar})$$

$$P2(IOP) = 256D - IO_{16} - a$$

2.5.3 Technology

The AP-120B is designed for reliability and therefore uses only well proven components and technologies, under conditions well clear of any operating limits. As a result mean time between failure (MTBF) of the hardware is typically several months to a year. The logic of the computer is made from low-power Schottky bipolar TTL (transistor–transitor logic) chips with a level of integration varying from a few gates per chip to a few hundred gates per chip. Typical gate delays in this logic technology are 3–5 ns. Various registers in the computer are also made in this logic technology. These are the S-pad and data-pad registers, and the subroutine return stack. The 50 ns program source memory and the 167 ns table memory both use 1K Schottky bipolar memory chips, whereas the slower and larger main data memory uses either 4K or 16K MOS memory chips.

It is interesting to compare the CRAY X-MP with the AP-120B from the point of view of technology, speed and power consumption, as they represent opposite extremes. The CRAY X-MP uses high-speed and high-power bipolar ECL technology with sub-nanosecond gate delays and a clock period of 9.5 ns, leading to the need for a large freon cooling system to dissipate a total of about 115 kW. The AP-120B on the other hand uses mostly low-power technology and consequently has a much longer clock period of 167 ns. However this permits the use of air cooling and limits the total power consumption to about 1.3 kW.

2.5.4 Instruction set

There are no vector instructions *per se* on the FPS AP-120B. Instead a 64-bit instruction is issued every clock period that has fields which control the operation of all units in the computer during that clock period. As an example, if the field FM (see below) controlling the multiplication pipeline is activated (i.e. bit 51 of the instruction is one), then all data in the multiplication pipeline are advanced to the next stage. The pipeline must therefore be activated three times to 'push' one pair of arguments completely through the three-stage pipeline and thereby complete one multiplication operation. A further activation of the pipeline is necessary for every subsequent element of a vector operation. This would most likely be set within a loop. The format of the single universal instruction is shown in figure 2.42. The notation for the data sources and destinations corresponds with figure 2.39. For a complete description of the instruction the reader is referred to the AP-120B Processor Handbook (FPS 1976a). In order to indicate the operation of the instruction, we give below examples of the uses of the data fields:

(1) S-pad group (control of 16-bit integer ALU and 16-bit registers)

> SOP   specifies dyadic S-pad operation, e.g. ADD, SUB, MOV, AND, OR, EQUIV; operands are SPS and SPD registers; result goes to SPD.

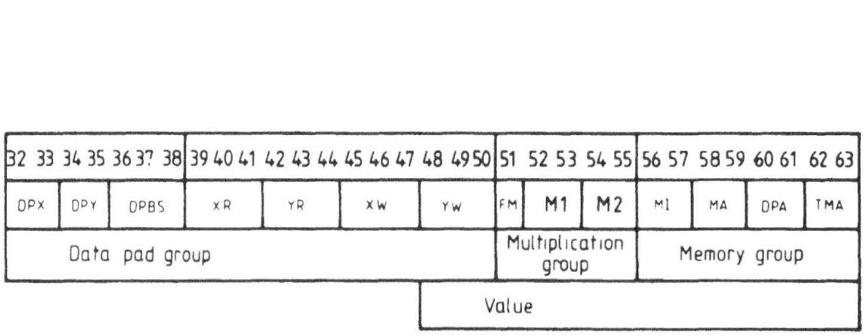

0 1 2	3 4 5	6 7	8 9 10 11 12 13	14 15 16 17 18	19 20 21 22	23 24 25 26 27	28 29 30 31
B	SOP	SH	SPS · SPD	FADD	A1 · A2	COND	DISP
	S-pad group			Adder group		Branch group	
	SOP1			FADD1			
	SPEC operation			I/O			

32 33 34 35 36 37 38	39 40 41 42 43 44 45 46 47 48 49 50	51 52 53 54 55	56 57 58 59 60 61 62 63
DPX · DPY · DPBS · XR · YR · XW · YW		FM M1 M2	MI MA DPA TMA
Data pad group		Multiplication group	Memory group
	Value		

FIGURE 2.42 The data fields in the 64-bit instruction of the FPS AP-120B. This single instruction controls the operation of all units in the computer at every clock period. (Diagram courtesy of Floating Point Systems Inc.)

SPS   number (4 bits) of 16-bit source register.

B   if 1, reverses bits of source register before performing operation.

SPD   number of 16-bit destination register.

SH   specifies single left or right shift, or double right shift after operation.

SOP1   specifies monadic operations on data in SPD register when SOP = 0, e.g. increment ($+1$), decrement (-1) or complement SPD register.

SPEC   control, conditional and branch instructions.

(2) Floating-point adder group

FADD   specifies dyadic operations, e.g. FADD, FSUB, AND, OR, EQUIV using data in A1 and A2. Intermediate results are moved one segment down the pipeline to the next buffer register.

A1   source of data to be loaded into first adder input register, e.g. FM (multiplier output), DPX, TM.

A2   source of data to be loaded into second adder input.

FADD1   specifies monadic operations on data in A2 when FADD = 0, e.g. convert A2 to integer, sign–magnitude or two's complement; take absolute value.

(3) I/O group (controls I/O and transfers to and from buses), e.g.

DPBS → SPD   data-pad bus contents to S-pad destination.

DPBS → TMA   data-pad bus contents to table memory address register.

SPFN → PNLBS   output of S-pad to panel bus.

INTA   interrupt acknowledge; device address put in DPBS.

(4) Branch group

COND   condition for branch, e.g. always, on flag, on arithmetic error, return jump from subroutine, on FA or SPFN =, $\neq$, $\geqslant$, > 0.0.

DISP   if branch true next address is current address + DISP − 16, relative jump of −16 to +15.

(5) Data-pad group (controls transfers to and from data pads X and Y)

DPA   current data-pad address.

DPX   load data pad X from DPBS, FA or FM.

DPY   load data pad Y from DPBS, FA or FM.

DPBS   selects DPX, DPY, MD, SPFN or TM to be sent to data-pad bus.

XR   data-pad register with address $DPA + XR - 4$ is sent to DPX.

YR   data-pad register with address $DPA + YR - 4$ is sent to DPY.

XW   DPX is sent to data-pad register with address $DPA + XW - 4$.

YW   DPY is sent to data-pad register with address $DPA + YW - 4$.

(6) Floating-point multiplier group

FM   multiply or no operation. Intermediate results are moved one stage down the pipeline to the next buffer register.

M1   loaded from FM, DPX, DPY or TM.

M2   loaded from FA, DPX, DPY or MD.

(7) Memory group (controls transfers to and from main data and table memories)

MI   load memory input register from FA, FM or DPBS.

MA   increment or decrement memory-address register by one, or read from SPFN, and initiate data memory cycle.

DPA   increment or decrement data-pad address by one or set address from SPFN.

TMA   as DPA but for table memory.

2.5.5   Software

Software for the AP-120B, except for device drivers, is written in FORTRAN, so that it may be compiled to run on a variety of host computers. It may be subdivided into the following categories:

(1) operating system;
(2) program development software;
(3) application libraries.

The operating system consists of an executive APEX and a set of diagnostic routines APTEST. The executive controls transfers of data between the host and the AP-120B, transfers AP programs from the host to the AP program source (PS) memory and initiates the execution of programs in the AP. The operation of APEX is illustrated in figure 2.43. Most user programs will be FORTRAN programs that call either upon AP-120B maths library programs

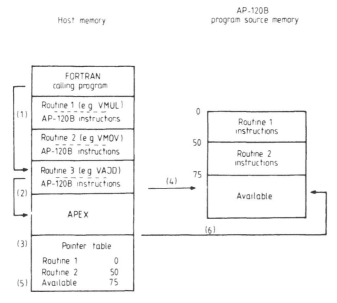

FIGURE 2.43   Diagram showing the action of the APEX executive during program execution (courtesy of Floating Point Systems Inc). Both AP-120B instructions and data are transferred between the host computer and the AP-120B under the control of APEX.

or user-supplied subroutines of AP-120B instructions in order to manipulate arrays in the AP-120B. APEX itself is a subroutine that is linked as part of the compiled FORTRAN program and runs on the host. It contains a table recording the location and contents of all AP routines that have already been loaded in the AP program source memory. The AP-120B instructions for each AP routine are stored as part of each subroutine in the host memory. Assuming that subroutines 1 and 2 have already been called, the following sequence of events takes place as the FORTRAN program executes in the host computer:

(1) FORTRAN program calls on AP using routine 3 (VADD);
(2) routine 3 calls APEX;
(3) PS memory table searched: routine 3 not in PS memory;
(4) APEX transfers AP-120B instructions from host to PS memory;
(5) PS memory table updated;
(6) APEX initiates execution of routine 3 in the AP-120B.

Data is transferred from the host to the AP-120B by calling the subroutine APPUT and results are transferred back with the subroutine APGET. Both these routines also call upon APEX to control the transfer. Once the execution of a program on the AP has begun, APEX returns control to the FORTRAN calling program on the host, which may then proceed with other calculations that do not use the AP. If a call to another AP routine is met before the first has finished, APEX will wait for the first call to be completed.

The program development software comprises:

(1) *Mathematics library*—over 250 FORTRAN callable subroutines for the manipulation of arrays. The subroutines are written in assembler code and carefully optimised. Examples from the standard library are:

VADD, VMUL	element-by-element vector addition and multiplication;
SVE, DOTPR	sum of vector elements and dot product;
MMUL, MATINV	matrix multiplication and inverse;
CFFT, ACORT	complex fast Fourier transform and autocorrelation.

In addition there exists the advanced maths library which contains routines for function generation, binary search, tridiagonalisation, diagonalisation, solution of real and complex sparse systems of equations, and the solution of ordinary differential equations by Runge–Kutta integration.

(2) *APAL*—the assembler for AP-120B code which assembles programs on the host for subsequent execution on the AP.

(3) *APLOAD*—links separate APAL object modules into a single module for execution on the AP.

(4) *APSIM, APDBUG*—simulates an AP program and allows debugging on the host or AP respectively.

(5) *VFC*—vector function chainer. Consolidates multiple calls to maths library routines into a single call and thereby reduces calling overhead.

(6) AP FORTRAN—a compiler running on the host that accepts FORTRAN IV and produces code for execution on the AP.

The above software allows programs to be prepared in a variety of ways. In the order of decreasing speed in execution and increasing ease of preparation, these methods are likely to be: APAL assembler, vector function chainer, maths library routines and AP FORTRAN.

Application libraries are available for:

(1) *SIGLIB*—signal processing library with routines for histograms, hanning windows, correlations, transfer and coherence functions etc.

(2) *IMP*—image processing library with routines for 2D fast Fourier transform and convolution, image filtering etc.

(3) *AMLIB*—advanced maths library with routines for function generation, Runge–Kutta integration, sparse matrix solution and matrix eigenvalues.

2.5.6 Performance

The AP-120B does not include a real-time clock and it is therefore impossible to time the execution of programs accurately. Attempts to time programs by using the clock on the host computer are usually imprecise and variable because of the effect of the host operating system. This is particularly the case if a time-sharing system is in use. In estimating the performance of the AP-120B we are therefore forced to rely on the timing formula given in the AP-120B maths library documents. The document (FPS 1976b) which we use gives timing formulae that may be related directly to formula (1.4a) defining r_∞ and $n_{1/2}$. The minor differences from later documents (FPS 1979a, b) are unimportant. Because of the synchronous nature of these machines theoretical timings should be reliable; however the absence of a clock makes the optimisation of large programs very difficult. The detailed timing of large programs soon becomes tedious and error prone. An alternative is to simulate the execution of the AP program on the host computer using the program APSIM (see §2.5.5). This program produces the theoretical program timing but again may be impractical for the timing of large programs because it runs about 1000 times slower than the program would execute on the AP-120B itself.

Using the maths library documents (FPS 1976b) we give the timing formulae for a selection of simple vector operations, and derive from them estimates for $n_{1/2}$ and r_∞. We quote the timing formulae for the standard memory (500 ns chip cycle time) and give the improved values of r_∞ for the fast memory (333 ns chip cycle time) in parentheses. Where there is a small timing variation because of the choice of odd or even memory locations for the vectors, we have taken the minimum timing. None of these minor timing alternatives substantially change the character of the machine, and they can largely be ignored.

(1) Vector move   CALL VMOV (A, I, C, K, N)

$$C_{mK} \leftarrow A_{mI}, \qquad m = 0, 1, \ldots, N - 1$$

Memory-to-memory move of vector **A** to **C**. K and I are the memory increments between successive elements of **A** and **C** respectively. The time for N operations is:

$$t = \tfrac{2}{3}(N + 1) \ \mu s,$$

hence

$$r_\infty = 1.5(3.0) \ \text{Mop/s}, \qquad n_{1/2} = 1.$$

We note that this operation is memory bound and the transfer rate doubles for the fast memory. However $n_{1/2}$ is unaffected by the memory type.

(2) Vector addition   CALL VADD (A, I, B, J, C, K, N)

$$C_{mK} \leftarrow B_{mJ} + A_{mI}, \qquad m = 0, 1, \ldots, N - 1.$$

The time for N operations is:

$$t = N + 1 \; \mu s,$$

therefore

$$r_\infty = 1(2) \text{ Mflop/s}, \qquad n_{1/2} = 1.$$

(3) Vector multiplication   CALL VMUL (A, I, B, J, C, K, N)

$$C_{mK} \leftarrow B_{mJ} * A_{mI}, \qquad m = 0, 1, \ldots, N - 1.$$

The time for N operations is:

$$t = N + 2 \; \mu s,$$

therefore

$$r_\infty = 1(2) \text{ Mflop/s}, \qquad n_{1/2} = 2.$$

(4) Vector division   CALL VDIV (A, I, B, J, C, K, N)

$$C_{mK} = B_{mJ}/A_{mI}, \qquad m = 0, 1, \ldots, N - 1.$$

The time for N operations is:

$$t = 1.83(N + 3) \; \mu s,$$

therefore

$$r_\infty = 0.55(0.55) \text{ Mflop/s}, \qquad n_{1/2} = 3,$$

and we see that, because the calculation is dominated by arithmetic, the faster memory does not increase the performance.

(5) Vector exponential   CALL VEXP (A, I, C, K, N)

$$C_{mK} \leftarrow \exp(A_{mI}), \qquad m = 0, 1, \ldots, N - 1.$$

The time for N exponentials is:

$$t = 4.87(N + 0.3) \; \mu s,$$

therefore

$$r_\infty = 0.2 \text{ Mflop/s}, \qquad n_{1/2} = 0.3.$$

(6) Dot product   CALL DOTPR (A, I, B, J, C, N)

$$C \leftarrow \sum_{m=0}^{N-1} A_{mI} * B_{mJ}.$$

The time for $2N$ operations is:

$$t = \tfrac{2}{3}(N+2) \ \mu s,$$

therefore

$$r_\infty = 3(6) \ \text{Mflop/s}, \qquad n_{1/2} = 2.$$

The above selection of timings and performances, which may be considered typical for simple memory-to-memory vector operations, shows that only a small fraction of the potential performance of 12 Mflop/s can be realised for such operations. This is because the memory bandwidth is insufficient to support a memory-to-memory processing rate of this magnitude. A single vector operation has two input vector operands and one output result vector. Therefore a memory-to-memory processing rate of 12 Mflop/s requires a memory bandwidth of 36 Mword/s. The standard memory on the AP-120B has a bandwidth varying between 2 and 3 Mword/s depending on whether references are all to the same memory bank or alternate between different memory banks. The fast memory has similarly a bandwidth varying between 3 and 6 Mword/s. Thus it is evident that the memory bandwidth is only about one-tenth of that required to sustain both the addition and multiplication pipelines working with data from main memory.

The most likely constraint on the realisation of fast processing rates on the AP-120B is therefore the low memory bandwidth. In order that memory bandwidth does not become a bottleneck, it is necessary for the *computational intensity* (see p106), f, to be at least two floating-point operations per memory reference (flop/ref) with fast memory, or at least 4 flop/ref with standard memory. These conditions are realised for more complicated algorithms and large enough problem sizes, n: e.g. matrix multiplication, $f = 2n/3$ flop/ref; fast Fourier transform (FFT), $f = 1.25 \log_2 n$ flop/ref.

The FFT algorithm is worthy of closer examination because it is the main algorithm around which machines like the AP-120B were designed. Throughout the algorithm operations are of the 'butterfly' type (see equation (5.87)):

$$c = a + wb, \tag{2.17a}$$

$$d = a - wb, \tag{2.17b}$$

where a and b are complex elements from main memory, and c and d are complex results to be returned to main memory. The complex constant w may reside in a register. Equations (2.17) require one complex multiplication $(s = w*b)$ and two complex additions $(a+s, \ a-s)$ for four memory references to complex numbers. This is the equivalent of 10 real arithmetic operations for eight real memory references, or $f = 1.25$ flop/ref. The above considerations are for the radix-2 transform and show that this algorithm does not have a high enough computational intensity to keep the arithmetic pipes busy. By combining two levels of the FFT together, we obtain the radix-4 algorithm and increase the computational intensity to 2.5 flop/ref, which is a figure satisfying the conditions given above for the fast memory. We find that the maths library subroutine CFFT does use the radix-4 algorithm and gives a performance of 8 Mflop/s.

The values of half-performance length found above are in the range $n_{1/2} = 1-3$, showing that the AP-120B, although it has many parallel features, actually behaves very similarly to a serial computer. In this respect the computer is similar to the CRAY-1 ($n_{1/2} \sim 10$) and quite different from the CYBER 205 ($n_{1/2} \sim 100$) or ICL DAP ($n_{1/2} \sim 1000$). The selection of the best algorithm is normally determined by the value of $n_{1/2}$ (see Chapter 5) and we would expect algorithms optimised to perform well on a serial computer also to perform well on the AP-120B. However, as has been emphasised above, the performance of a program may be more dependent on the management of memory references than on the questions of vector length that are addressed by the value of $n_{1/2}$.

2.5.7   FPS-164 (renamed M140 and M30) and 264 (renamed M60)

As can be seen from figure 2.44 the FPS-164 is substantially larger than the AP-120B, being about 5.5 ft high and occupying about 2.5 ft × 7 ft of floor space, principally because of the need to accommodate a much larger memory. The same cabinet is also used for the FPS-164/MAX and FPS-264. The principal improvements introduced in the FPS-164 (compared with the AP-120B) are:

(a) 64-bit floating-point arithmetic compared with 38-bit;
(b) 32-bit integer arithmetic compared with 16-bit;
(c) 24-bit addressing to 16 Mword compared to 16-bit addressing to 64 Kword only;
(d) 64-bit X- and Y-pad data registers compared with 32-bit;
(e) 64 32-bit S-pad address registers compared with 16 16-bit;
(f) 1024 64-bit instruction cache replacing program memory;
(g) 256 32-bit subroutine return address register stack;
(h) main memory expandable from 0.25 to 7.25 Mword with memory protection;
(i) table memory of 32 Kwords RAM;
(j) a clock for timing programs—sadly lacking on the AP-120B.

The increase in arithmetic precision and addressing range generally lift the

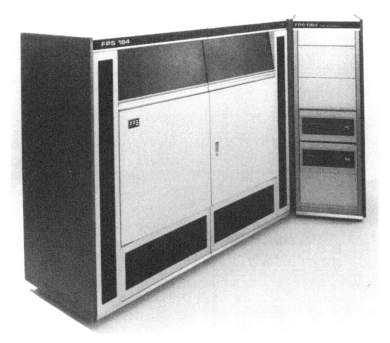

FIGURE 2.44   Overall view of the FPS-164.

specification of the computer up from that of a minicomputer to that of
current large mainframe computers. The way these features fit into the overall
architecture can be seen in figure 2.45. In contrast to the AP-120B, both
instructions and data share the same main memory, and instructions are
automatically read into the instruction cache as needed. This cache memory
therefore replaces the separate program memory of the AP-120B. Subroutine
referencing is made more efficient on the FPS-164 by the inclusion of the
subroutine stack which provides storage for 256 32-bit subroutine return
addresses. The table memory, also called auxiliary memory, has become
primarily a random access memory for use as temporary register storage for
intermediate results. The first 8K of this memory is however reserved for
read-only constants of which about 5K are assigned, the next 16K or 32K
(depending on the option purchased) may be used as random access memory
by user programs. The main memory is organised into modules, each with
even and odd memory banks, as in the AP-120B. The original FPS-164 used
16K-bit dynamic NMOS chips and had 12 memory modules occupying 24
memory boards (each board a bank), giving a capacity of 1.5 Mword.
Subsequently, the use of 64K-bit memory chips has enabled the memory

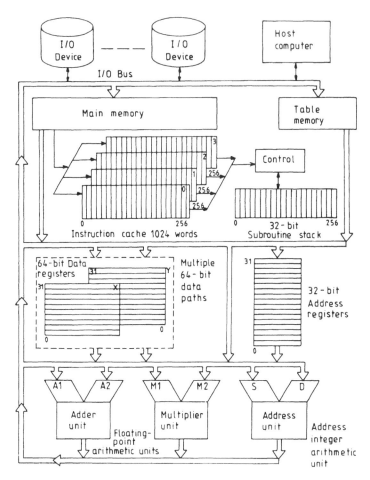

FIGURE 2.45   Overall architecture of the FPS-164.

capacity to be increased to 7.25 Mword. Main memory acts as a three-stage pipeline, and successive requests to the same bank may occur every other clock period. Table memory acts as a two-stage pipeline, and successive requests may occur on successive clock cycles. A new feature on main memory is memory mapping and protection. If this is available, the contents of the MDBASE register are added to the requested address on line MA (see figure 2.39) to form a physical address. If this exceeds the contents of the MDLIMIT register no memory access takes place: a read produces zeros and a write is ignored.

The FPS-164 is said to have a nominal clock period of 167 ns like the AP-120B. However, the actual clock period used is slightly longer at 182 ns, which leads to an asymptotic pipeline rate of 11 Mflop/s. The machine is also provided with a programmable real-time clock and a CPU timer that increments every clock period of the computer. This enables accurate timings to be made within user programs.

The theoretical arithmetic performance of the FPS-164 is $11/12$ of the theoretical figures given for the AP-120B. With the availability of the CPU timer, accurate benchmarking is possible and we quote results obtained by Thompson (private communication). The result of the $(r_\infty, n_{1/2})$ benchmark for memory-to-memory arithmetic operations is given in table 2.8. The results for straight FORTRAN code (top number of pair), and for the use of the optimised mathematics library routines (bottom numbers) are compared. As expected from the theoretical timings for the AP-120B, a performance of about 1 Mflop/s is to be expected for dyadic operations. Except for the dyadic add operation, there is nothing to recommend the use of the library routines, because the same performance can be obtained using FORTRAN, and the overhead of the operation is less using FORTRAN as can be seen from the smaller value of $n_{1/2}$. A dyadic operation is the worst case for performance because there is no opportunity to store intermediate results in fast access registers, and the number of floating-point operations per memory reference, f, is $1/3$ only. The last two cases of the triad and three-ops raise f to $2/3$

TABLE 2.8 Benchmark results of $(r_\infty, n_{1/2})$ for memory-to-memory operations on the FPS-164. Upper case variables are vectors, lower case are scalars. (Data courtesy of Bill Thompson, TUCC.) (Upper numbers FORTRAN, lower numbers library routine or higher optimisation level, OPTC.)

Operation: statement 10 program (1.5)	r_∞ (Mflop/s)	$n_{1/2}$
$A = B + C$	0.88	5
CALL VADD	1.06	16
$A = B * C$	1.07	5
CALL VMUL	1.04	17
$A = B/C$	0.30	7
$A = b*(C - D)$		
OPTC = 1	0.8	—
OPTC = 3	3.4	—
$A = B + C*(D - E)$		
OPTC = 1	1.0	—
OPTC = 3	3.2	—

and 3/4 respectively. With the compiler optimisation level (OPTC) equal to one, which performs local optimisation only, the dyadic performance of about 1 Mflop/s is obtained. However, if OPTC = 3, software pipelining is employed to overlap operations and the performance is raised three-fold.

Performance on more substantial benchmarks, the Livermore loops and LINPACK, are given in table 2.9. Broadly speaking, a performance on actual problems between 1 and 5 Mflop/s is typical, depending on the care with which the problem is programmed.

The FPS-264 is an ECL logic implementation of the FPS-164 architecture, which allows the clock period to be reduced from 182 ns to 53 ns, a ratio of 3.4. Other improvements to the instruction cache and main data memory lead to a claimed performance ratio of between four and five times the FPS-164. Logic uses air-cooled custom Fairchild 100K ECL chips (compared with Schottky TTL on the FPS-164), and the main memory uses 64K-bit static NMOS chips (compared with dynamic NMOS on the FPS-164). The latter allows the memory to be packaged with 0.5 Mword per memory board, divided into two banks. A maximum main data memory of 4.5 Mword divided into 18 banks was available on the first machines. The instruction cache is divided into two interleaved banks of 4 Kword, giving a total of 8 Kword, compared with 1 Kword on the FPS-164. The FPS-264 has the same external appearance as the FPS-164, being packaged in the same cabinet (see figure 2.44). The reported performance of the FPS-264 on the Livermore loops and the LINPACK benchmark is given in table 2.9, and generally supports the assertion that the FPS-264 can be expected to perform about four times faster than the FPS-164.

2.5.8   FPS-164/MAX (renamed M145)

A novel enhancement to the FPS-164 was announced in 1984, the FPS-164/MAX which stands for matrix algebra accelerator (Charlesworth and Gustafson 1986). This machine has a standard FPS-164 as a master, and may add up to 15 MAX boards, each of which is equivalent to two 164-CPUs with the addition of four vector registers of 2048 elements in each CPU. In total there is therefore the equivalent of 31 164-CPUs or 341 Mflop/s.

The architecture of the MAX board is shown in figure 2.46. The board contains two CPUs, each with an eight-stage 64-bit floating-point multiplier which feeds its results into an eight-stage floating-point adder. One input to every multiplier in the machine (and there are 31 multipliers in a 15-board machine) is broadcast from main data memory, whilst the other input comes from the local scalar or vector registers on the MAX board. Similarly, the second input to the adder comes from the local registers. In the broadcast operation the same number is sent simultaneously from main data memory

TABLE 2.9 Performance of a selection from the Livermore and LINPACK benchmarks on the FPS-164, 264 and FPS-164/MAX. Figures are Mflop/s for 64-bit arithmetic.

Problem	FPS-164	FPS-264	FPS-164/MAX[f]
Theoretical peak	—	—	33(1)
performance	—	—	99(4)
	11	38	341(15)
$(r_\infty, n_{1/2})$ FORTRAN §1.3.3	(1.07, 5)	—	—
Livermore 3 inner product	3.0[e]	—	—
Livermore 6 tridiagonal	1.1[e]	—	—
Livermore 14 particle pusher	1.5[e]	—	—
LINPACK[a] FORTRAN[b] $n = 100$	1.4	4.7	—
Assembler[c] inner loop $n = 100$	— 2.9	— 10	6(1)[g] 20(15)[g]
Matrix–vector[d] best assembler $n = 300$	— — 8.7	— — 33	15(1)[h] 26(4)[h] —

Notes

a   Solution of linear equation using DGEFA and DGESL for matrices of order 100 (Dongarra et al 1979).
b   All FORTRAN code (Dongarra 1985).
c   BLAS routines optimised in assembler (Dongarra 1985).
d   FORTRAN matrix–vector method of Dongarra and Eisenstat (1984). Matrix order 300. Best reported assembler (Dongarra 1985).
e   Gustafson 1985.
f   Number of MAX boards used in parentheses.
g   FPS 1985b.
h   Dongarra 1986.

to all the multipliers. The clock period of the MAX board is the same as the FPS-164 main CPU, namely 182 ns, hence each board has a peak performance of 22 Mflop/s. The logic of the FPS-164/MAX is implemented in CMOS VLSI,

Broadcast to all MAX boards
Matrix data or vector register indices

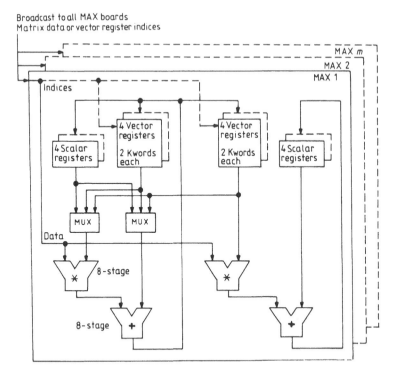

FIGURE 2.46   Architecture of a matrix accelerator (MAX) board.

and the arithmetic pipelines of the MAX board (shown in figure 2.47) use the pipelined WEITEK arithmetic chips.

MAX boards occupy memory board positions of the FPS-164, and it is possible to upgrade an existing FPS-164 to an FPS-164/MAX. The MAX boards look to the host computer to be the top 1 Mword of the 16 Mword of its address space, leaving a maximum addressable normal memory of 15 Mword. The FPS-164/MAX uses the same memory board as the FPS-264, with 0.5 Mword per board of static NMOS chips. A full FPS-164/MAX has 29 memory board slots of which 14 are used to hold the 7 Mword of physical main data memory, and 15 slots are used for the 15 MAX boards. The availability of 256 K-bit static NMOS chips will allow 1 Mword per memory board and a physical memory size of 15 Mword, to match the full addressing capability.

The idea of the MAX board is to speed up the arithmetic in a nest of two or three DO loops, such as one finds in many matrix operations, in particular in the code for matrix multiply. In this example the 31 164-CPUs of a full system would be used to simultaneously calculate the 31 inner products that are required to produce 31 elements in a column of the product matrix. The FPS architecture is already optimised for the efficient calculation of inner

FIGURE 2.47   A MAX board.

products, and the only problem is to ensure that the required data is available
to the pipelines. One of the vector registers of each of the 31 CPUs receives
one of the 31 rows of the first matrix, whilst the elements of the column of
the second matrix are broadcast, one-by-one, to all CPUs as the 31 inner
products are accumulated, as is illustrated in figure 2.48. This produces 31
elements in the corresponding column of the result matrix. All other columns
for the same 31 rows can be computed without altering the contents of the
vector registers, and it is only when this saving in data movement is possible
that the FPS 164/MAX can approach its maximum performance. Fortunately
many important problems in linear algebra (solving equations, eigenvalues
etc) can be formulated to satisfy this condition. At this stage in the calculation
of the matrix product, 31 rows of the product matrix have been computed.
The vector registers must now be refilled in order to compute the next 31
rows, until the product matrix is completed.

The kernel of the above algorithm is the multiplication of the (31×2048)
matrix **A** by the (2048×2048) matrix **B**, to give the (31×2048) product
matrix **C**. We assume **C** is initially cleared and transferred to main data
memory after the execution of the following code

$$
\begin{aligned}
&\text{DO 1}\quad \text{J} = 1,\ 2048 \\
&\text{DO 1}\quad \text{K} = 1,\ 2048 \\
&\text{DO 1}\quad \text{I} = 1,\ 31 \\
&\quad 1\quad \text{C}(\text{I,J}) = \text{C}(\text{I,J}) + \text{A}(\text{I,K}) * \text{B}(\text{B,J})
\end{aligned}
\qquad (2.18)
$$

where I is the CPU-number.

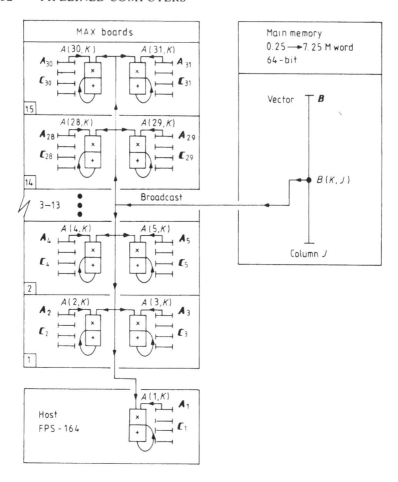

FIGURE 2.48   The simultaneous calculation of 31 inner products, $(A_i.B; i = 1, 31)$ using 15 MAX boards. The elements of B are broadcast one-by-one to all boards, and each board accumulates two inner products $(S_i = S_i + A_{ik}B_{k,j}; k = 1, 2048)$.

The DO-I loop is implemented by the broadcast of the scalar quantity $B(K,J)$ to all 31 CPUs, multiplying it by the $A(1,K)$ which is an element taken from the local vector A_i, and accumulating the inner product in $C(I,J)$ which is taken from the local vector C_i. In these operations all the 31 CPUs work in lockstep (i.e. in unison) but on their own individual data (I is the CPU-number), that is to say the control is SIMD. The DO-K loop accumulates the inner product, and the DO-J loop moves from column to column. All the three DO loops in (2.18) can be executed without transferring data between main data memory and the MAX board registers, and this is the essential requirement for efficient MAX board utilisation. This is expressed by saying that there must be re-use of data in both space and time if the peak speed is to be reached. Re-use in space refers to the broadcast of a single quantity $B(K,J)$ simultaneously to the 31 CPUs in the DO-I loop, and re-use in time refers to the fact that $C(I,J)$ and $A(I,K)$ are continually re-used from local memory in the DO-K and DO-J loops.

The purpose of the re-use facility is to limit the need for transfers between the main data memory and the MAX boards, and to perform the maximum amount of arithmetic between such transfer so as to dilute the penalty of loading the registers. We have expressed this before by the computational intensity, f, which is the number of floating-point operations per memory reference. In the above matrix multiply example we have two floating-point operations per execution of statement 1, and the references are the read of **A** and **B**, and the store of **C** (there is hardware provision for the initial clearing of **C**). Thus

$$f = (2 \times 31 \times 2048 \times 2048)/(2110 \times 2048) = 60. \qquad (2.19)$$

This is to be compared with the hardware parameter, $f_{1/2}$, which is half the ratio of asymptotic arithmetic performance to memory bandwidth in the relevant case of memory transfer overlap which occurs on the FPS-164 (see §1.3.6 and equation (1.20)). The memory bandwidth to the MAX boards, r_∞^m, is one word per clock period, and the maximum arithmetic rate, r_∞^a, is 62 arithmetic operations per clock period, hence

$$f_{1/2} = 31. \qquad (2.20)$$

The expected average performance can be estimated from (1.22b) as

$$\bar{r}_\infty = r_\infty^a \ knee \ (0.5f/f_{1/2}) \qquad x = f/f_{1/2} \qquad (2.21a)$$

whence

$$x = 1.9 \qquad \bar{r}_\infty = 0.97 r_\infty^a. \qquad (2.21b)$$

Thus we find that when the conditions for re-use in space and time are satisfied, a performance within 3% of the maximum peak performance is possible.

The MAX boards can only execute a limited number of instructions of the type that are given in table 2.10. The operation of the MAX boards and their registers are memory-mapped onto the top Mword of the addressable 16 Mword, as shown in figure 2.49. That is to say that the boards are operated simply by writing and reading to appropriate parts of the upper Mword of

TABLE 2.10   The instructions that may be executed by a MAX board, and the peak performance in Mflop/s for 1 and 15 boards. The same performance applies to both real and complex arithmetic. Full vector operations use $J(I) = I$.

Name	FORTRAN program line	MAX boards	
		1	15
Dot product	$S = S + A(I) * B(J(I))$	22	341
Complex dot product	$S = S + A(I) * B(I)$	22	341
VSMA	$A(J(I)) = S * B(J(I)) + C(I)$	11	167
VMSA	$A(J(I)) = B(J(I)) * C(I) + S$	11	167
Vector mult	$A(J(I)) = B(I) * C(J(I))$	5.5	83
Vector add	$A(J(I)) = B(I) + C(J(I))$	5.5	83

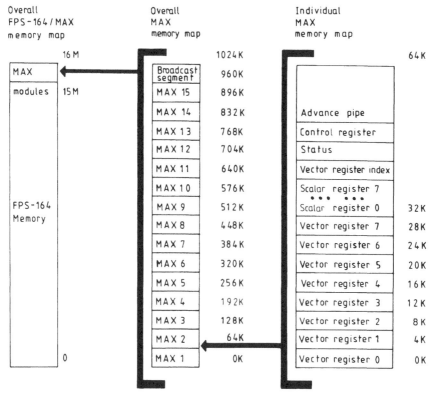

FIGURE 2.49   The memory mapping of MAX boards onto 16 Mword address space of an FPS-164.

the FPS-164 address space. This is divided into 16 individual MAX memory maps of 64 Kword each. The first 15 of these maps operate the 15 MAX boards individually, and the last is the broadcast segment which operates all the boards in unison. The first 32 Kword of the memory map addresses is the vector register storage which allows for eight registers of 4K elements each. The first MAX implementation, however, limits the vector length to 2K elements. Next are the eight scalar registers and the vector index registers. The MAX board is operated by placing appropriate words in the 'advance pipe' section.

The nature of the software that is available for driving the FPS-164/MAX can be seen from the following FORTRAN code that implements the matrix multiply discussed above

```
CALL SYS$AVAILMAX(NUMMAX)

MAXVEC = 8 * NUMMAX + 4

NUMVEC = MAXVEC

DO 10 I = 1, N, MAXVEC
   NUMVEC = MIN(NUMVEC, N − I + 1)                    (2.22)

IF(NUMVEC .LE. 0) GOTO 10

   CALL PLOADD(A(I,1),N, 1, NUMVEC, ITMA, 1, IERR)

   DO 20 J = 1, N

      CALL PDOT(B(1,J),1, N, C(I,J), 1, NUMVEC, ITMA, 1, 0, IERR)

20    CONTINUE

10  CONTINUE
```

The maximum performance is achieved by using as many vector registers as possible. The first statement in the above code obtains in NUMMAX the number of MAX boards that are mounted on the system. Since each contains eight vector registers, and there are four vector registers on the host FPS-164, NUMVEC is the number of vector registers. The DO-I loop loads NUMVEC rows of the matrix **A** into the available vector registers in preparation for accumulating NUMVEC inner products. The CALL PDOT forms the inner product with the Jth row, and this is repeated for all the rows by the DO-J loop.

2.5.9 IBM /CAP and Cornell systems

Both IBM and Cornell University are developing replicated MIMD computing systems, based on linking together multiple FPS processors. The IBM loosely coupled array of processors (/CAP) is the brainchild of Enrico Clementi and is installed at the IBM Kingston laboratory. In 1985 a similar configuration was installed at the IBM Scientific Center, Rome, to be the first computational heart of the newly set up 'European Center for Scientific and Engineering Computing' (ECSEC).

A simplified drawing of the IBM /CAP computer system is shown in figure 2.50. Ten FPS-164 computers, each with 4 Mbytes of main data memory are connected by 2–3 Mbyte/s channels to IBM host computers (Berney 1984). Seven are connected to an IBM 4381 for computational work and three to an IBM 4341 for program development, although all ten can be switched to the IBM 4381, making a system with a theoretical peak performance of 110 Mflop/s. The actual performance on quantum chemical problems for the ten-FPS-164 configuration is reported to be about the same as a CRAY-1S, or about 60 Mflop/s (Clementi *et al* 1984). Possible enhancements to the initial configuration involve the addition of two MAX boards to the ten FPS-164s, which gives a peak performance of 550 Mflop/s. If the maximum number of 15 boards were added to each FPS-164 the peak performance would be raised to 3.4 Gflop/s.

The above computer system is described as a loosely coupled array because in the initial configuration there was no direct connection between the computing elements (the FPS-164s), and because the connection to the host is by slow channels. Consequently, only problems which exhibit a very large grain of parallelism can be effectively computed. That is to say that a very

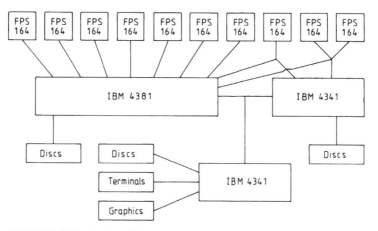

FIGURE 2.50   A simplified block diagram of the IBM experimental loosely coupled array of processors (/CAP) at Kingston, with ten FPS-164 computers connected via IBM host computers.

large amount of work must be performed on data within the FPS-164 before the results are transferred over the slow channels to the host, or via the host to other FPS-164s. A fast 22 Mbyte/s FPS bus (called FPSBUS, directly connecting the FPS-164s, has subsequently been developed by FPS which substantially reduces the overhead of transferring data between the FPS-164s.

A similar project to the above has been developed for some years at Cornell University's Theory Center under the direction of Professor Kenneth Wilson. Initially, this comprised eight FPS-100 processors connected by a custom 24 Mbyte/s bus. The work is now part of the Cornell Advanced Scientific Computing Center which is sponsored by NSF, IBM and FPS. It is envisaged that up to 4000 MAX boards could be interconnected to give a peak performance rate of about 40 Gflop/s.

In order to quantify the synchronisation and communication delays on the /CAP, measurements have been made of the performance parameters $(\hat{r}_\infty, \hat{s}_{1/2}, f_{1/2})$ and these are discussed in §1.3.6, part (iv). The benchmark has been conducted using either the channels or the FPS BUS for communication, and gives rise to the following total timing equations (Hockney 1987d)

$$t_{\text{channel}} = 2500 + 4777p + (1.87m/p) + t_a(p)s \quad \mu s \qquad (2.23a)$$

$$t_{\text{BUS}} = 18926 + 8181p + 0.191m + t_a(p)s \quad \mu s \qquad (2.23b)$$

where m is the number of I/O words and s the number of floating-point operations in the work segment (see §1.3.6, part (iv)). The first two terms in equations (2.23) are a fit to the synchronisation time, the third term is the time spent on communication, and the last term is the time spent on calculation. The fact that the communication term is inversely proportional to the number of processors, p, in the case of the channels, shows that the channels, although slow, are working in parallel. On the other hand, we see that the communication time does not depend on p in the case of the FPSBUS showing that the bus, although much faster, is working serially.

In order to compare the use of the channels with the use of the FPSBUS we equate the two timing formulae (2.23a) and (2.23b) and obtain the equation for the equal performance line (EPL):

$$m = p \frac{(16426 + 3404p)}{(1.87 - 0.191p)}. \qquad (2.24)$$

This relationship is plotted on the (p, m) phase plane in figure 2.51. Given a number of processors p and the number of I/O words m, a point is specified on this plane. Its location in the plane determines whether bus or channel communication should be used. There is an infinity in the relationship (2.24) at $p = 9.78$ (broken line) showing that the channels will always be faster if

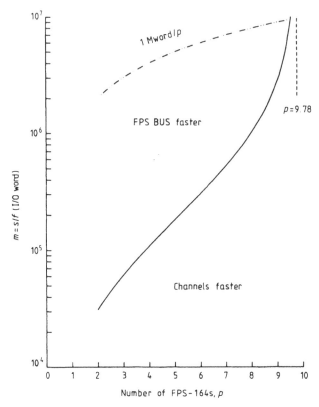

FIGURE 2.51   Phase diagram comparing the use of the FPS BUS with the use of the channels. For any number of processors chosen, p, either the FPS BUS or the channels is faster depending on the number of I/O words m. For more than about 10 processors the channels are always faster.

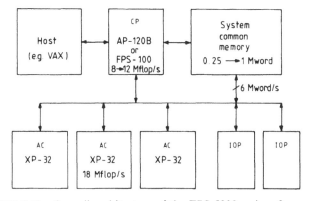

FIGURE 2.52   Overall architecture of the FPS-5000 series of computers.

there are ten processors or more. This is because the channels have a smaller start-up and synchronisation overhead than the bus, and a faster asymptotic rate if there are ten or more working in parallel. That is to say, more than ten 2–3 Mbyte/s channels working in parallel are faster than a 22 Mbyte/s bus working serially. We also show in figure 2.51 the line corresponding to 1 Mword per processor, which is a typical main memory size for an FPS-164 installation. Values of m above this line are inaccessible in such an installation, because problems requiring such a magnitude of I/O would not fit into the memory of the installation. However, the memory size can be increased to 28 Mword/FPS-164 (using 1 Mword memory boards) corresponding to a line off the top of the diagram.

2.5.10   FPS-5000 series

The FPS-5000 series of computers is interesting because it forms one of the few commercially available MIMD computer systems. It is the latest offering of so-called 'array processors' from FPS, and is a multicomputer development of AP-120B type of architecture. In our classification of MIMD computers (§1.2.6 and figure 1.9), it falls in the class of bus-connected, shared-memory, switched MIMD systems.

The overall architecture of an FPS-5000 series computer system is shown in figure 2.52 (FPS 1984a). The system is connected to its host computer by the *control processor* (CP) which is either an FPS AP-120B with a 167 ns clock or the slower FPS-100 with a 250 ns clock. Both CPs have the same architecture but differ in technology and packaging. The CP may perform useful arithmetic itself at peak rates, respectively, of 12 and 8 Mflop/s. More importantly, however, it controls the rest of the system which comprises up to three *arithmetic coprocessors* (ACs) and a number of *I/O processors* (IOPs) which share a common bus connection to a *system common memory* (SCM) of between 0.25 and 1 Mword. The ACs are FPS XP-32 computers with a peak performance of 18 Mflop/s. The largest system announced in 1983 used an FPS-100 CP (8 Mflop/s) and three XP-32 ACs, giving a theoretical peak performance of 62 Mflop/s. Continuity with previous systems is maintained because all software prepared for the AP-120B will run on the control processor, and the computing power can be enhanced in stages by adding arithmetic coprocessors as required.

The progression of technology in the 1970s and early 1980s is exemplified by the progression from the AP-120B (1975), through the FPS-100 (1978) to the XP-32 (1983). The basic processor of the AP-120B used Schottky TTL chips operating at 8 MHz and occupied 20 boards. Of these, three boards were required for the multiplier and three for the adder. The FPS-100 was able to compress the identical architecture onto ten boards by using low-

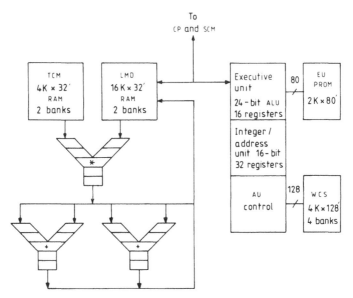

FIGURE 2.53   Internal architecture of the FPS XP-32 arithmetic coprocessor.

power Schottky TTL chips, at the cost of a slower clock (4 MHz). The multiplier and adder were both compressed to one board each. The XP-32, on the other hand, fits the whole processor onto a single board and, furthermore, includes a second floating-point adder. This is achieved by using fast Schottky VLSI chips with a 6 MHz clock. The multiplier is now reduced to a single chip (from the three boards in the AP-120B), namely the 32-bit WEITEK WTL-1032. Similarly the floating-point adders each use the WEITEK WTL-1033 floating-point ALU chip. The rest of the logic uses the Advanced Micro Devices 29500 series of VLSI circuits, and INMOS IMS-1040 static RAMs. The use of standard WEITEK floating-point chips has meant a change to the floating-point number representation. The FPS AP-120B and FPS-100 use a 38-bit format as described in §2.5.2. This has a dynamic number range from 10^{-155} to 10^{+153}, and a precision of 28 bits. The WEITEK chips, however, have adopted the IEEE 754 32-bit floating-point standard (IEEE 1983), which has a much reduced dynamic range of 10^{-38} to 10^{+38}, but a more precise mantissa equivalent to 33 bits.

The architecture of the XP-32 coprocessor (figure 2.53) is similar in general concept to that of the AP-120B but differs in detail (FPS 1984c). A five-stage floating-point multiplier pipeline and two five-stage floating-point adder pipelines are connected by multiple data paths to a *local main data* (LMD) and a *table coefficient memory* (TCM). The LMD stores 16K 32-bit words arranged in two banks, and the TCM has 4K 32-bit words also arranged in two banks. Overall control of the XP-32 is exercised by the *executive unit* (EU) which can operate simultaneously with the *arithmetic unit* (AU), thereby providing for the parallel execution of I/O and address calculation with floating-point arithmetic. The EU performs all communication of programs and data between the AC and the CP and SCM. The AU performs arithmetic only on data in the local TCM and LMD memories. Microcode programs for the EU reside in EU PROM, which contains 2K 80-bit microcode instructions. Similarly, microcode programs for the AU reside in a *writable control store* (WCS) of 4K 128-bit microcode instructions, arranged in four banks.

There is no direct connection between the arithmetic coprocessors. The coprocessors can only take data from and return results to the SCM, so that any communication between the ACs is by shared data in the SCM. The SCM acts as the main data memory of the control processor. In the case of the AP-120B control processor, it operates as described in §2.5.2 for the fast memory (333 ns access) with a 167 ns clock period. In the case of the FPS-100 control processor, the memory works on the slower 250 ns clock period. The arithmetic coprocessors may also have direct memory access (DMA) to the SCM by taking turns with the CP with the available memory cycles, according to a priority scheme. In the case of access by the ACs, the SCM may either read or write one word per clock period (but not both at the same time), giving a total SCM memory bandwidth of either 6 Mword/s (24 Mbyte/s) or 4 Mword/s (16 Mbyte/s). However, the memory is organised such that any individual XP-32 coprocessor may only use half this bandwidth, thereby allowing two ACs on an FPS-5000 system before the memory bus restricts its performance. This is achieved by limiting memory requests from any particular AC to every other memory cycle.

The FPS-5000 may be programmed entirely by calls to library programs in a subset of FORTRAN 77 called CPFORTRAN. In most cases company documentation gives timing formulae from which values of r_∞ and $n_{1/2}$ can be derived (FPS 1984a, b, 1985a). We give below a selection indicating the facilities provided.

(i) Host interface routines
The following routines run on the host computer, load programs and data into the FPS-5000, and start the CPFORTRAN program running on the CP.

      CPOPEN   Open a CPFORTRAN program file.

      CPLOAD   Load a CPFORTRAN program file from host to CP.

CPRUN   Start CPFORTRAN program running on CP.

EXPUT   Start data transfer from host to FPS-5000.

EXGET   Start data transfer from FPS-5000 to host.

APWAIT   Wait for data transfer and CP program to stop.

APWD   Wait for data transfer to stop.

APWR   Wait for CP program to stop.

(ii) *Synchronisation of the ACS by the CP*
The following routines run on the CP, and control the ACS.

XPSEL   Select the XP-32 for subsequent XPWAIT.

XPRUN   Start program running in selected XP-32.

XPWAIT   Wait for selected XP-32 to finish.

XPSTAT   Obtain status of XP-32.

(iii) *Data transfer to and from SCM*
The following routines run on the XP-32s, and transfer data prior to and after calculation.

XPDMAR   Transfer data between SCM and LMD.

XTMDMA   Transfer data between SCM and TCM.

$r_\infty = 2\,\mathrm{Mop/s}$

XPISNC   Wait for transfer (or arithmetic) to finish.

(iv) *Arithmetic within the XP-32*
The following XPMLIB routines run on the XP-32 and perform arithmetic on data in the LMD of the XP-32.

ZVMUL(IA, IB, IC, N)   Element-by-element vector dyadic multiply of $A*B$ to C, N elements ($r_\infty = 4$ Mflop/s, $n_{1/2} = 33$).

ZVDIV(IA, IB, IC, N)   Element-by-element vector divide ($r_\infty = 0.5$ Mflop/s, $n_{1/2} = 9$).

ZVSASM(IA, IB, ID, IC, N)   One-vector triad, vector scalar add scalar multiply: $C = (A + b)*d$ ($r_\infty = 12$ Mflop/s, $n_{1/2} = 56$).

ZVASM(IA, IB, ID, IC, N)   CDC 205-type two-vector triad,
vector add scalar multiply: $C = (A + B)*d$
($r_\infty = 8$ Mflop/s, $n_{1/2} = 37$).

ZVAM(IA, IB, ID, IC, N)   All-vector triad, vector add multiply:
$C = (A + B)*D$ ($r_\infty = 6$ Mflop/s, $n_{1/2} = 28$).

The above performance figures are for dyadic and triadic operations within
a single XP-32 processor and they do not take into account the time taken
to synchronise the multiple ACS of an FPS-5000 (i.e. the MIMD performance),
or the time to transfer data from the SCM to LMD prior to performing the
calculations. These have been separately measured by Curington and
Hockney (1986) and interpreted in terms of the parameters $n_{1/2}$, $s_{1/2}$ and
$f_{1/2}$ (§1.3.6). The results are given in tables 2.11 and 2.12.

The FPS-5320A computer which was used for the measurements comprises
a control processor and either one or two XP-32 arithmetic coprocessors.

TABLE 2.11   Measurements of $(r_\infty, n_{1/2}, s_{1/2})$ on a multiprocessor FPS-5320A
with a control processor (CP), and one or two XP-32 coprocessors, operating
on data in their respective local memories.

Operation	Configuration	r_∞ (Mflop/s)	$n_{1/2}$† or $s_{1/2}$
Dyad	CP only	1.5	14†
$A_i = B_i * C_i$	One XP-32	4.0	470
VMUL or ZVMUL	Two XP-32	8.0	1320
	CP + two XP-32	9.2	1545
Triad	CP only	3.9	40†
$A_i = (B_i + s)*c$	One XP-32	12.0	1490
VSASM or ZVSASM	Two XP-32	24.0	4200
	CP + two XP-32	27.7	4820

TABLE 2.12   Values of peak performance, $\hat{r}_\infty$, and $f_{1/2}$ for a single
FPS XP-32 arithmetic coprocessor when performing triadic ZVSASM
operations on data originating in system common memory.

Case	$\hat{r}_\infty$ (Mflop/s)	$f_{1/2}$
Sequential I/O	12.5	4.2
Overlapped I/O	12.6	2.2

The measurements with the CP alone in table 2.11 involve no synchronisation and hence are of $n_{1/2}$, whereas those using multiprocessors include the synchronisation time and are therefore of $s_{1/2}$. The values of the latter indicate the minimum amount of arithmetic that is worth dividing amongst the XP-32 coprocessors. Both the CP and ACS work on a 6 MHz clock, and have arithmetic pipelines with a peak performance of 6 Mflop/s for dyadic operations which use only one pipeline, or 12 Mflop/s for triadic operations which use two pipelines. The peak performance for the maximum configuration of the control processor and two XP-32s is therefore 18 Mflop/s for dyads and 36 Mflop/s for triads. These peak rates are achieved in the case of the XP-32 executing triads. However, in the other cases inadequate memory bandwidth prevents the peak rates being realised, although the XP-32 suffers much less than the CP in this respect.

The measurements in table 2.11 are for operations performed on data residing in the local memories (LMD) of the XP-32s, which it is assumed has already been loaded. However, in actual use the data for a problem will be stored in SCM, and will have to be transferred to the LMD before calculation can take place. The overall rate of computation will critically depend, therefore, on the amount of arithmetic performed in the XP-32 per data transfer between SCM and LMD (i.e. the variable f defined in §1.3.6) and can be characterised by the parameter $f_{1/2}$. The FPS-5000 is an ideal vehicle for studying this dependence because f can easily be varied and the value required to achieve half the peak performance can be obtained. This is the performance parameter $f_{1/2}$, and is given in table 2.12. The observed peak rate of 12 Mflop/s is as expected for the ZVSASM operation. The test proceeds by transferring a vector of n data from SCM to LMD using XPDMAR, and then performing f ZVSASM operations upon the vector. Two cases are considered, first when the I/O transfers between SCM and LMD take place sequentially with the arithmetic operation, and secondly when the I/O takes place simultaneously with the arithmetic (i.e. is overlapped with it). We find that the effect of overlapping the I/O is to halve the value of $f_{1/2}$.

3 Multiprocessors and processor arrays

3.1 THE LIMITATIONS OF PIPELINING

There are only two techniques for introducing parallelism into computer hardware, replication and pipelining; pipelining can be considered as replication which has been made possible through sequence, as each component of replication in a pipeline follows another in time. Pipelined operations are performed by overlapping their simpler component operations using parallel hardware. This is performed such that at any given time, component parts of a sequence of operations are being processed in the pipeline. In this way a single operation will share the pipeline with a number of other operations as it progresses through the various stages.

The fundamental difference between pipelining and spatial replication is that the parallel component operations of a pipeline are quite likely to perform different tasks, which when performed in sequence make up the operation required. There is obviously a limit to the parallelism available by splitting an operation into subtasks in this way, unless the operations are extremely complex. Although complete programs are complex, and the use of pipes of concurrent tasks as a style of programming can be very attractive, such large pipelines are very application-specific. Thus for general use pipelining can only provide a limited degree of parallelism, by exploiting commonly used complex operations such as floating-point arithmetic.

Pipelining is, however, the most attractive form of parallelism available, because pipelining does not create the same communications problems found using spatial replication. A pipeline is designed to reflect the natural data flow of the operation being performed, whereas spatial replication will utilise either a fixed network or a programmable connection network. In the first case the network may not necessarily reflect the data flow required in the operations being performed; in the latter case, although it is more general, the costs are large.

Although it is attractive, pipelining alone will not achieve the goal of unlimited computing power, which has motivated the use of parallelism in computer systems. This will only come about through the use of spatial replication, or indeed the use of both pipelining and spatial replication. This thesis seems already to have been proved, as is evident in the evolution of the pipelined computers described in Chapter 2. The CRAY X-MP has evolved by the replication of a faster but functionally identical computational section to that found in its predecessor, the CRAY-1. Another good example is found in the CDC CYBER 205, which has evolved from the CDC STAR by the replication of the STAR's re-engineered pipes. Indeed, this same architecture has a more highly replicated future in the ETA[10], where a VLSI implementation of a two-pipe CYBER 205 will be replicated up to eight times to obtain a maximum performance of 10 Gflop/s.

These architectures based on the limited replication of very fast processors are very expensive to engineer. They rely on fast clock periods, which mandate the use of high-speed, high-power circuits such as ECL, which is not a very dense circuit technology. Moreover, in order to match the high-speed circuits, the packaging must keep the circuits as close to each other as is physically possible. In this way the clock rates are not limited by signal propagation times. This in turn exacerbates the problems of removing the power dissipated by these 'hot' circuits. The ETA[10] is an exception to this use of ECL circuits in supercomputers, as it uses high-packing-density CMOS circuits. In order to obtain fast clock speeds in this VLSI technology, low voltage levels and very cold operating temperatures are required (liquid nitrogen temperatures).

Although the ETA[10] uses VLSI circuits, it still does not exploit the full economies of scale that can be obtained from VLSI. Many different custom chips, using the gate array design technique, are required. The alternative of using mass-produced VLSI circuits implies massive replication in the system. It is only recently (*circa* 1985) that components suitable for replication on this scale have become available. This chapter explores the alternatives of replication in computer architecture; the technological issues are discussed further in Chapter 6.

3.2   THE ALTERNATIVE OF REPLICATION

3.2.1   A matter of scale

In replicated systems, the most important issues are a matter of scale; whereas small numbers of processors may communicate efficiently using bussed systems or by shared memory, in larger numbers the inherent sequentiality of these communication methods produces bottlenecks. The designer is then forced to consider systems in which data is switched between processors. This switching can take place either in a fixed network, in which distances grow with the number of processors, or in a programmable connection network, in which the costs follow a square law. (The latter can also have considerable wiring problems.)

Figure 3.1 shows schematically a generic spatially replicated system. It is an attempt to distil the common features found in a wide variety of such systems, and in doing so perhaps reflects none. However, it does illustrate the major features of such architectures. These are: a parallel memory. from which many words may be accessed simultaneously; an array of processors; and the all-important switching circuits. The interprocessor switch provides a set of connections between ports on the processors, which may provide a small fixed set of permutations, or may indeed provide the complete set of all $n!$ permutations. The processor–memory switch provides data paths between the various banks of memory and the processors. There are two sensible alternatives for this switch, one allows only the identity permutation and the other allows a given processor to access all memory banks. It should be noted that it may not be desirable to have an equal number of processors and memory banks in this latter case.

The differences between the many and diverse parallel architectures that may be found in the literature can be quantified by the following metrics:

(a) the first quantity is the size of the array and the power of the individual

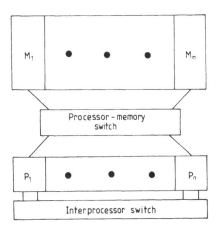

FIGURE 3.1   Diagram showing a generic spatially replicated system, where P_1 to P_n are the processors and M_1 to M_m are the memory modules.

processor, the product of the two to some extent determining the power of the overall system;

(b) the complexity of the switching networks, which will determine the flexibility of the system and hence whether the power obtained by replication can be utilised by a large class of problems;

(c) the distribution of the control to the system, i.e. whether the whole array is controlled by a central control processor, or whether each processor has its own controller; and

(d) the form of the control to the system, which may be derived from the flow of control through a predefined instruction sequence or a control structure more suited to declarative programming styles, such as data flow or reduction.

It should be noted that in this model the processors may not have a simple atomic structure, but may themselves contain replication or concurrency, as in the case of the replicated pipeline structures referred to above.

3.2.2   The control organisation

Of all the issues raised above, it is perhaps control which provides the major differences between parallel architectures. This is certainly true when considering the overall form of the control (i.e. control flow, reduction, data flow etc) and much controversy surrounds the choice of control for the next generation of computers, the fifth generation (Uchida 1983).

Both data flow (Dennis 1980, Chamber *et al* 1984) and reduction control strategies (Turner 1982, Chamber *et al* 1984) can be instrumental in relieving the programmer of the task of explicitly sequencing instructions. This has a profound effect on software engineering (see Chapter 4), as well as providing a framework for a machine execution strategy. In these functional and logical languages instruction sequencing arises from the decomposition of a description of the problem, rather than from an algorithm.

(i) Data flow

In data flow, an instruction does not execute under the influence of a program counter, but instead is able to execute if and only if all of its operands are available. A dataflow program can therefore be considered as a directed graph, along which data tokens flow, with the output from an operation being connected with an arc of the graph to the operations that consume its result. Instructions in a physical machine would generally be represented as packets containing operations, operands (as data or references) and tag fields giving meaning to this data. The latter is required because the state of the machine can no longer provide a context from which the interpretation of the data may be derived. During execution these tags would be matched with result packets, containing tag and data fields, and when all operands to a given instruction have been matched, the instruction can be queued for immediate execution.

To illustrate this consider the program and execution for the expression given below

$$(A + B)*(C + D)$$

The program, comprising a number of instruction packets, is 'loaded' into the system by injecting those packets into a program memory. The program execution would be initiated by injecting data packets, containing values for A, B, C and D into the system. One implementation of the variables A to D would be to assign unique tags to them which would be stored in the instruction packets. The data would then need to be similarly tagged. A matching unit could then match the tags of the data to the corresponding instructions as the data circulated through the system. For example, the two addition operations would attract the associated data and then become available for execution, possibly in parallel. There is scope for pipelining and replication in data flow computers.

Pipelining can be introduced in the flow of data packets (sequence of operations) through the system. Replication can be achieved by sharing the instruction packets between processors. If this latter form of parallelism is exploited, then some equitable means of sharing the program packets and associated tags between the processors is required.

Only when these two addition operations have been completed, generating values for the bracketed sub-expressions, will the multiplication operation be able to execute. It can be seen that the execution strategy is data driven and commences from the innermost level of nesting of an expression and proceeds outwards. Obviously, in a realistic program, the data flow graph or program will be very much more complex than this simple example. However, this example is sufficient to illustrate the notion of asynchronous parallelism being totally controlled by the data-driven mechanism. Because all data dependencies have been resolved within the graph description of the program, no explicit parallel declarations are required to allow data flow programs to run on multiple processors. Programs must be decomposed, however, in ways from which parallelism may be extracted. A simple list recursion, for example, would produce a sequential algorithm, whereas a recursive dividing of the list, expressing the function to both halves, would generate an algorithm containing parallelism. This recursive halving is the basis for many common algorithms, for example quick sort, and is a classical expression for generating implicit parallelism.

A data flow machine based on this concept of tagged instruction packets would comprise one or more processors containing the following units:

(a) one or more execution units;
(b) one or more data stores;
(c) a program store, containing the program graph;
(d) a matching unit, to complete unmatched tokens.

Such a machine has been constructed by a research group at Manchester University (Gurd and Watson 1980, and Watson and Gurd 1982). For further reading on data flow see Chamber *et al* (1984) and Hwang and Briggs (1984).

(ii) Reduction

Reduction as a means of computer control can also yield parallelism without explicit control. Reduction is based on the mathematics of functions and lambda calculus, and using this formalism programs can be considered as expression strings or as parse trees. For example, a program could be represented by the following expression, either as a string, or in structured form as a parse tree for the expression, with the operator '*' at the root and two subtrees containing ' + ' operators and the operands A and B and C and D respectively.

$$((A + B)*(C + D))$$

Whereas in data flow execution is data driven, in a reduction strategy execution is demand driven. Thus if this program was entered into the system, or activated by a request for its result from a larger program, then a series of rewrites would take place, reducing this expression to its component operations. A rewrite is the procedure of taking an expression or tree and reducing that expression, performing the operation if leaves are known, or by activating its sub-expression or subtrees if they are not. The term 'reduction' is perhaps misleading, for early operations in this sequence will generate more programs for execution, as new subtrees or sub-expressions are activated.

Each subtree may of course be distributed on concurrent hardware for evaluation in parallel. At some later stage the program has been reduced to its component operations, which in a similar manner to data flow systems can be represented by tagged packets. It can be seen that reduction approaches the evaluation of an expression from the outermost nesting and works inwards, generating work as it proceeds.

A research group at Imperial College has been building an architecture for the implementation of functional languages by reduction (Darlington and Reeve 1981). This machine is based on transputers, and a few prototype machines were delivered by ICL, one to Imperial College in 1986. Another practical implementation of a reduction architecture is being funded by the Alvey program at University College, London (Peyton-Jones 1987a, b). For further reading concerning recent practical work on data flow, reduction and other advances in parallel processing see Chamber *et al* (1984) and Jesshope (1987b).

Despite the fact that both of the above methods of computer control generate parallelism without explicit command, they can also suffer from inefficiencies when compared with the more exploited control flow strategy. In both cases a substantial amount of computation can be required in organisation. There may be tens or even hundreds of instructions executed for every useful instruction (e.g. floating-point operation in a number-crunching application). This is very inefficient when compared to the highly optimised control flow computers which have been developed over the last three to four decades of von Neumann computing. However, these architectures attempt to increase the level of abstraction of the programming model towards one in which the computer executes the specification of a problem. An analogy to this situation would be to compare programming in assembler and a high-level language, where the latter should not be compared with the former in terms of efficiency, unless programming efficiency is also considered. In this case, there have been shifts towards architectures for executing high-level languages (Organick 1973) which have minimised any loss of efficiency paid for the higher level of abstraction.

In the same way, as research continues in the field of declarative systems, architectures will become more efficient as refinements are made in the hardware implementations and in compiler technology to exploit these improvements. Indeed, in recent presentations on data flow research at Manchester (Gurd 1987), results were presented which show the Manchester data flow machine comparing very favourably with conventional architectures. Functional language implementations are likely to follow this development path but are currently some five years behind the development stage of data flow.

One fundamental limitation in these architectures is that of communication bandwidth between processors, but this is shared with all replicated systems. This problem grows with the size of the replicated system and because data flow and reduction machines effectively require the distribution and communication of programs as well as data, the communications bandwidth requirements are greater and are likely to be more of a limitation if these strategies are adopted. For example, data packets of around 100 bits are common in such architectures, even for 16-bit operations. In control flow only one of a pair of operands need ever be communicated through a communication network. In data flow and reduction, many of these large packets may need to be communicated in order to obtain a single useful operation. Latency or pipelining is an effective tool in combating communication complexity, but for an effective system there must be a balance of load between communication and processing and, as indicated above, this load balancing is biased against communication as the degree of replication increases.

Generally the most efficient replicated systems either communicate data or programs between processors, whichever requires the least bandwidth. For example, if two processors need to enter into a sustained communication with each other, it may well be appropriate for the code of each processor to gravitate to each other, rather than for data to be passed between them. In SIMD machines of course it is only necessary to pass data.

(iii) Control flow

Considering just control flow architectures, there is still a wide choice in the implementation of control strategy in a replicated system. For example, if we use a single central controller for all processors, then we have a SIMD machine. The 'alternative' is to make each processor obey its own instruction stream by giving it an autonomous controller. If this is the case, then we have the MIMD machine in Flynn's categorisation. However, between these two extremes lies a whole range of control strategies.

Consider the following example: at one level of a given architecture we may find an array of processors under MIMD control, but each 'processor' may also contain an array of processors under SIMD control. This structure has been proposed by a number of designers (Pease 1977, Jesshope 1986a, b, c, Lea 1986).

Alternatively, consider the case where we have a single central controller which provides most of each control word for a SIMD array, but where each of the processors may also provide local control information, depending on local conditions. Most SIMD computers provide at least a rudimentary amount of control at the processor level, although this is usually restricted to an on/off switch. In other arrays more control fields have been devolved to the processors, while maintaining overall synchronisation and sequencing centrally. If such an array has a significant amount of local control it is said to be adaptive. An example of an adaptive array is considered in §3.5.4.

The control of an array may also have a number of levels; indeed this technique is often used to combat complexity in control unit design. It is possible that at one level the array may be synchronised, under the control of a single instruction stream, but that at lower levels (microcode for example) different instruction sequences may be obeyed. Thus, depending on local state, the local microcontroller may take autonomous action, albeit in response to an identical broadcast instruction. Such a system was proposed by Pease (1977) and implemented in Baba (1987).

In reality these decisions related to control and illustrated by the examples above are merely engineering decisions based on the implementation technologies. Unlike the distinction between declarative and control flow, these decisions are not decisions of philosophy, for in practice any form of control may be simulated by any other. However, a penalty in efficiency will be paid if the control structure does not match the algorithm. Running SIMD programs on MIMD machines is wasteful in hardware, and may pay a large penalty in synchronisation overhead, whereas the opposite would require the SIMD machine to run an interpreter, probably very inefficiently, which could then interpret a processor's data as instructions.

Whatever the choices then, decisions will be made based on technological restrictions and likely applications. For example, if implementing a replicated system to perform binary image processing, a single-bit serial processor would be a good choice of processor. However, a MIMD construction of single-bit processors would not be efficiently implemented because of the large overhead added by the controller and program store. A SIMD machine would therefore be most appropriate. Unfortunately, when entering the realm of replicated systems, there is no universal solution, or at least one has not yet been discovered.

3.2.3   The distribution of processing power

One of the major and most controversial issues concerned with replication is the distribution of processing power. Designs have been produced in which the number of processors varies from a few tens to many thousands. Obviously for a given cost, the more complex the processor, the fewer can be combined in a single array. This is a classic trade-off problem, between the power of the individual processor and the size of the processor array. What then are the factors influencing this decision?

One major factor, which has already been discussed in Chapter 1, is the applicability of a parallel computer to a given set of problems. The more parallel a computer is, the more specialised it becomes. This is really an efficiency argument, for if the problem is less parallel than the computer, then the efficiency of the computer suffers. However, this argument may be countered for a large range of applications, as for these, parallel algorithms exist; see Chapter 5. Quantifying the amount of parallelism is obviously impossible, unless the size and type of the problem is known. However, for a large class of problems, the amount of parallelism which may be exploited is far greater than the largest arrays being built c1988. Moreover, the size of these problems is likely to increase with the processing power available and, for a processor of fixed complexity, this will vary with the amount of replication.

However, for a certain class of problems a small amount of parallelism may be desirable and this need will be met by small arrays of very powerful processors, employing relatively expensive floating-point hardware, or indeed by pipelined vector computers. For other problems, parallelism is feasible on a large scale and other considerations may be taken into account.

The simplest of all processors use the bit-serial approach and many thousands of these may be combined very cheaply, by exploiting VLSI technologies. In this way the processing power is spread very thinly over a single bit slice of the data, which must be highly parallel. It can be seen that given the parallelism, this approach provides very efficient use of the hardware for all forms of data: boolean, character, integer and floating-point. It can also be shown that, for a given number of logic elements, this approach provides the maximum computational power at least for simple operations.

Consider as a simple problem the addition of a number of pairs of b-bit numbers (N say, where $N \geqslant b$), given b 1-bit full adders with delay time τ_{FA}. Figure 3.2 gives the truth table for a full adder and also shows an implementation.

One solution to the problem is to link the full adders in a chain, which then adds all bits within a word. This is sometimes called a parallel adder

A	B	C_{in}	P	G	S	C_{out}
0	0	0	0	0	0	0
0	0	1	0	0	1	0
0	1	0	1	0	1	0
0	1	1	1	0	0	1
1	0	0	1	0	1	0
1	0	1	1	0	0	1
1	1	0	0	1	0	1
1	1	1	0	1	1	1

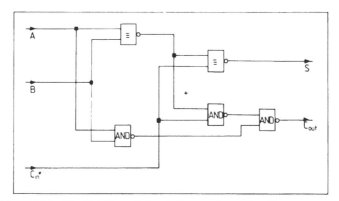

FIGURE 3.2   Truth table and circuit diagram for a full adder, constructed from exclusive-OR and NAND gates. The inputs are A, B and a carry-in signal C_{in} and the outputs are S (the sum) and C_{out} the carry-out. The truth table also shows the signals P and G, functions of A and B only, indicating that a carry is propagated at this position or generated at this position.

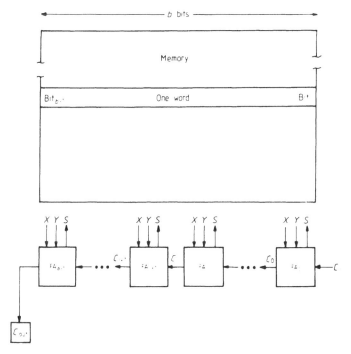

FIGURE 3.3   Parallel or ripple carry adder.

and is shown in figure 3.3. It can be seen however, that the process is not parallel at all, it is sequential. The carry out from the first 1-bit adder is only defined after time τ_{FA} from A, B and C_{in} all being available, where τ_{FA} is the delay introduced by two NAND gates and an exclusive OR gate. This carry is then required as input to the second adder in the chain. A more apt description of this adder is given by its alternative name, the ripple carry adder. The total delay, t_r, for the b-bit ripple addition is given by

$$t_r = \tau_{FA} + 2(b-1)\tau_d \tag{3.1}$$

where τ_d is the delay introduced by one NAND gate. Therefore, ignoring memory access time, the solution to our problem will be given by N times equation (3.1). We call this time T_r:

$$T_r = N[\tau_{FA} + 2(b-1)\tau_d]. \tag{3.2}$$

The bit-slice approach uses each 1-bit adder as an independent unit, which adds pairs from different words in parallel. The carry out from one step must be held in a register, for input at the next step. This is illustrated in figure 3.4 and the time to add each bit slice, t_b, is given simply by:

$$t_b = \tau_{FA} + \tau_l \tag{3.3}$$

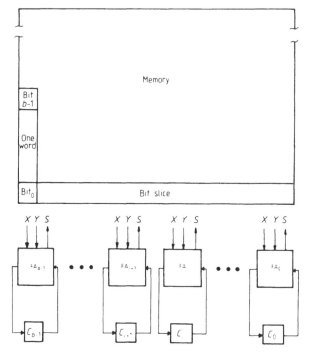

FIGURE 3.4   Parallel bit-slice addition.

where τ_1 is the time taken to catch the carry signal. The total time to complete the problem using the bit-slice approach is given by

$$T_b = \lceil N/b \rceil b(\tau_{FA} + \tau_1) \qquad (3.4)$$

where $\lceil \ldots \rceil$ denotes the integer ceiling function.

It can be seen that the time required in the bit-slice approach grows with N, whereas the ripple-carry approach grows with N/b.

It is instructive to take into account the memory access times; for example, by adding the memory access time, τ_M, to both equations (3.1) and (3.3), the following ratio is obtained for the solution to this problem:

$$\frac{T_r}{T_b} = \frac{N[\tau_M + \tau_{FA} + 2(b-1)\tau_d]}{[N/b]b(\tau_M + \tau_{FA} + \tau_1)}.$$

This simplifies, if N is an integral multiple of b, to give:

$$\frac{T_r}{T_b} = \frac{\tau_M + \tau_{FA} + 2(b-1)\tau_d}{\tau_M + \tau_{FA} + \tau_1}. \qquad (3.5)$$

Obviously if $\tau_M \gg \tau_d$ there is very little to be gained by the bit-slice approach. It seems very likely that this is the reason why the machines of the early

1950s followed the bit-parallel, word-serial line of evolution. It is interesting to speculate how the history of computing would have progressed, if there had been less of a disparity between the memory and logic technologies of the 1950s.

Today it is economic to build memory and logic from the same technology, even on the same VLSI chip. Thus the ratio in equation (3.5) is significantly greater than 1, for even modest word sizes. This is not the end of this discussion however, as seemingly sequential processes may often be speeded up by new techniques. This is true for example in recursion, as is shown in Chapter 5, which describes parallel algorithms. The addition of two b-bit numbers is a classic case of a recursive algorithm and may be computed in $\log_2 b$ steps. This has been shown to be a lower bound (Winograd 1967).

How then can this be accomplished using logic elements? The key lies in the definition of the two states in a full adder called propagate and generate, and are labelled P and G respectively on figure 3.2. Based on the two input values to be summed, they define the two states: 'carry is generated' and 'carry will be propagated'. Given a pair of such states, it is simple to define new generate and propagate states for the combined pair. For example, given P_0, P_1 and G_0, G_1, two new states P' and G' are defined by:

$$G' = G_1 \vee (P_1 \wedge G_0)$$

and   (3.6)

$$P' = (P_1 \wedge P_0).$$

Given these, a carry-out state may also be defined using the carry in:

$$C' = G' \vee (C \wedge P').$$ (3.7)

Figure 3.5 shows the layout of a carry look-ahead unit based on equations (3.6) and (3.7). This unit may be incorporated into a tree-like structure, to give the carry look-ahead adder. This is illustrated for an eight-bit word in figure 3.6. It can be seen that in general, b full adders and $b-1$ carry look-ahead units are required. The delay for this circuit for $b > 2$ is given by the time to obtain the most significant sum bit, which is given by:

$$t_c = 2\tau_{FA} + (\log_2 b - 1)\tau_{CL}$$ (3.8)

where τ_{CL} is the delay in obtaining C' in a single carry look-ahead unit (figure 3.5). This gives a great improvement over equation (3.1) for typical values of b, at the expense of doubling the number of gates. However, for the problem of adding N b-bit pairs of numbers this is still slower than using the bit-slice technique, as can be seen in equation (3.9):

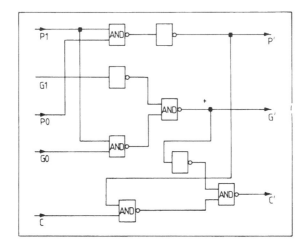

FIGURE 3.5   A carry look-ahead circuit, constructed from NAND gates and inverters, which has as inputs the propagate and generate signals from two bit positions (P0, G0 and P1, G1); which a carry-in signal C, has as outputs a modified propagate and generate signal (which can be used to cascade the circuit) and a carry-out signal C'.

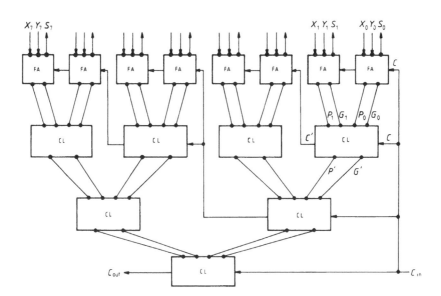

FIGURE 3.6   Eight-bit carry look-ahead adder. CL, carry look-ahead units; FA, full adders (see also figures 3.2 and 3.5).

$$\frac{T_c}{T_b} = \frac{\tau_M + (\log_2 b - 1)\tau_{FA}}{\tau_M + \tau_{FA} + \tau_I}. \tag{3.9}$$

Here the delays in the carry look-ahead unit (τ_{CL}) have been equated to the full adder delays τ_{FA}.

Similar techniques can be applied to obtain fast multiplication (Waser 1978) but again there are penalties to pay in the increased number of logic gates required. Thus we have shown techniques that are available to build faster but more complex functional units, although the relative cost increases and the efficiency can never match the bit-slice approach (unless $\tau_M \gg \tau_{FA}$).

This does not imply that all hardware should employ the bit-slice approach, as for single scalar operations the more complex hardware must be used. Thus the distribution of processing power will depend on several factors, of which the cost–performance ratio and expected parallelism of the workload are most prominent.

3.3 SWITCHING NETWORKS

3.3.1 Introduction

The theory and construction of switching networks are fundamental to the success of large-scale parallelism, which has now become feasible through the exploitation of VLSI technology. Replication on a large scale, as described in §3.2.1, is not viable unless connections can be established, either between processors or between processors and memory, in a programmable manner. Such connections can be established using switching networks—a collection of switches, with a given connection topology. Much of the early theory concerning switching networks was motivated by the needs of the telephone industry (Clos 1953, Benes 1965) but the convergence of this and the computer industry, in computer networks, digital telephone exchanges and now in parallel computers, has led to much more interest in this area. This section explores switching network architecture, with particular emphasis on large-scale parallel processing. Further reading in this same area can be found in a recent book by Siegel (1985).

Switching networks provide a set of interconnections or mappings between two sets of nodes, the inputs and the outputs. For N inputs and M outputs there are N^M well defined mappings from inputs to outputs, where by well defined we mean that each output is defined in terms of one and only one input. Figure 3.7 illustrates this by giving all well defined mappings between 3 inputs and 2 outputs. A network performing all N^M such mappings we will call a generalised connection network (GCN).

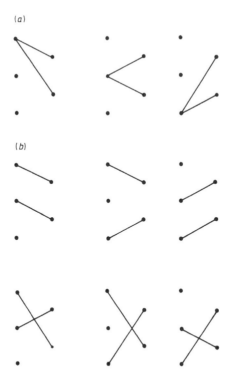

FIGURE 3.7   All possible mappings from three inputs to two outputs:
(*a*) one-to-many; (*b*) one-to-one.

If we limit the mappings only to the class of one-to-one mappings, then
$N!$ such mappings are well defined. Figure 3.7 also shows the subset of
one-to-one mappings: they do not broadcast data from one input to more
than one output. For this class of mapping to be non-empty, there must be
at least as many inputs as outputs. A network which performs these $N!$
mappings we will call a connection network (CN).

The obvious way of implementing a generalised connection network is
with a full or complete crossbar network, where each input can be switched
to any output. This is illustrated in figure 3.8, using two different
representations. In the first (figure 3.8(a)), the edges represent the input and
output sets and the nodes represent the crosspoints or switches. An alternative
representation is given by the bipartite graph (figure 3.8(b)), where the nodes
represent the input and output sets and the edges represent the crosspoints.
The full crossbar is the most general form of switching network, but the
number of crosspoints required is NM since there is one crosspoint between
each input and output. This network will therefore become very expensive
for large N. In general, a practical limit for N is around 2^7, whereas the

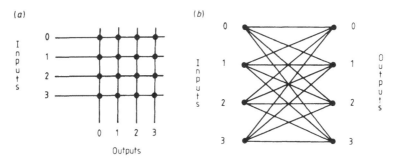

FIGURE 3.8   Two representations of the crossbar switch from four inputs to four outputs.

number of processors to be switched may be as high as 2^{14} or greater. Networks of this kind were used in the Burroughs BSP, one with $N = 16$ and $M = 17$ and another with $M = 16$ and $N = 17$.

Other switching networks are either based on many small interconnected crossbar networks, or on incomplete crossbar networks. These will be discussed later, but first we must introduce a notation and define some permutations which will simplify the discussion considerably.

3.3.2   Some fundamental permutations

A permutation on an ordered set of N nodes can be defined by a function $\pi(x)$, where x and $\pi(x)$ are integers in the range $0 \leqslant x$, $\pi(x) \leqslant N - 1$. The function $\pi(x)$ must also be one-to-one. For example, the function

$$\varepsilon_{(k)}(x) = \left\lfloor \frac{x}{2^k} \right\rfloor 2^k + \|x\|_{2^k} + 2^{k-1}|_{2^k} \tag{3.10}$$

defines an exchange permutation (figure 3.9).

However, for many permutations it is often found that a simpler way of defining the permutation can be obtained by looking at the binary representation of x. Thus

$$x = \{b_n, b_{n-1}, \ldots, b_1\} = b_n 2^{n-1} + b_{n-1} 2^{n-2} + \ldots + b_1 \qquad 0 \leqslant b_i \leqslant 1$$

represents the binary address of an element in the set. Permutations of the set of inputs can now be defined by operations or permutations on their binary address. This notation has been used to good effect in several papers concerning data manipulation and routing networks (Flanders 1982, Parker 1980, Nassimi and Sahni 1980).

(i) Exchange permutation

The exchange permutation defined above can now be more simply defined in terms of the binary representation of x.

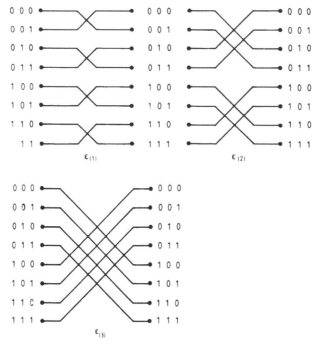

FIGURE 3.9   Exchange permutations.

$$\varepsilon_{(k)}(x) = \{b_n, \ldots, \bar{b}_k, \ldots, b_1\}. \tag{3.11}$$

The bar denotes the complement of a given bit. Thus the kth exchange permutation can be defined by complementing the kth bit of the binary representation of x.

(ii) Perfect shuffle permutation
The perfect shuffle is so called as it can be performed by cutting the set in two and interleaving the two sets obtained, as in the perfect card shuffle (see figure 3.10). This permutation corresponds to a unit circular left shift of the binary representation of x:

$$\sigma(x) = \{b_{n-1}, b_{n-2}, \ldots, b_1, b_n\}. \tag{3.12}$$

The kth subshuffle $\sigma_{(k)}$ and kth supershuffle $\sigma^{(k)}$ can also be defined by cyclic left shifts on the least and most significant k bits respectively:

$$\sigma_{(k)}(x) = \{b_n, \ldots, b_{k+1}, b_{k-1}, \ldots, b_1, b_k\}, \tag{3.13}$$

$$\sigma^{(k)}(x) = \{b_{n-1}, \ldots, b_{n-k+1}, b_n, b_{n-k}, \ldots, b_1\}. \tag{3.14}$$

These are also illustrated in figure 3.14 for $n = 3$ and $k = 2$. Clearly

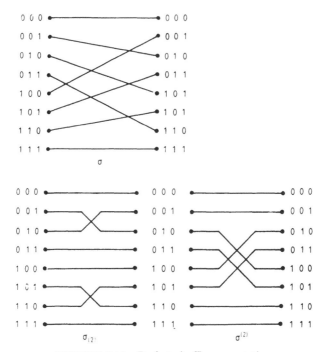

FIGURE 3.10   Perfect shuffle permutations.

$$\sigma^{(n)}(x) = \sigma_{(n)}(x) = \sigma(x)$$

and

$$\sigma^{(1)}(x) = \sigma_{(1)}(x) = x.$$

It can also be seen from figure 3.10 that the subshuffles (least significant bits) treat the set as a number of subsets, performing the perfect shuffle on each. The supershuffles, however, shuffle the whole set, but increase the width of the data shuffled.

(iii) Butterfly permutation
The butterfly permutation (figure 3.11) is defined over the binary representation of x by exchanging the first and last bits:

$$\beta(x) = \{b_1, b_{n-1}, \ldots, b_2, b_n\}. \tag{3.15}$$

As with the shuffle, we can also define the kth sub-butterfly and kth superbutterfly of x. The kth sub-butterfly exchanges kth and first bits, and the kth superbutterfly the nth and $(n - k + 1)$th bits:

$$\beta_{(k)}(x) = \{b_n, \ldots, b_{k+1}, b_1, b_{k-1}, \ldots, b_k\}, \tag{3.16}$$

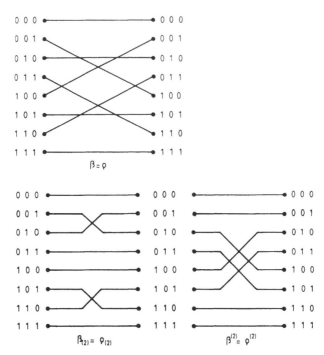

FIGURE 3.11   Butterfly and bit reversal permutations.

$$\beta^{(k)}(x) = \{b_{n-k+1}, \ldots, b_{n-k+2}, b_n, b_{n-k}, \ldots, b_1\}. \tag{3.17}$$

The sub- and superbutterfly are also illustrated in figure 3.11, for $n = 3$ and $k = 2$. Again

$$\beta^{(n)}(x) = \beta_{(n)}(x) = \beta(x)$$

and

$$\beta^{(1)}(x) = \beta_{(1)}(x) = x.$$

(iv) *Bit reversal permutation*
The bit reversal permutation, as its name suggests, is defined over the binary representation of x by reversing the order of the bits:

$$\rho(x) = \{b_1, b_2, \ldots, b_n\}. \tag{3.18}$$

One application where this permutation occurs is in the fast Fourier transform algorithm (see §5.5).

As with the previous two permutations, the kth sub bit reversal and kth super bit reversal can also be defined over the least and most significant k bits of x:

$$\rho_{(k)}(x) = \{b_n, \ldots, b_{k+1}, b_1, b_2, \ldots, b_k\}, \tag{3.19}$$

$$\rho^{(k)}(x) = \{b_{n-k+1}, b_{n-k+2}, \ldots, b_n, b_{n-k}, \ldots, b_1\}. \tag{3.20}$$

Figure 3.11 also illustrates the bit reversal for $n = 3$. However, the two permutations are not always equivalent, as will be shown later when we consider the algebra of permutations.

(v) Shift permutation

This last permutation is in fact more easily described without resorting to the bit representation of x:

$$\alpha(x) = |x + 1|_{2^n}. \tag{3.21}$$

Again we define sub- and supershifts on the least and most significant k bits of the binary representation of x. In terms of the bit representation of x, equation (3.21) defines the binary addition over the n-bit field, ignoring overflow. Therefore the sub- and supershifts can be defined as follows:

$$\alpha_{(k)}(x) = |x + 1|_{2^k} + \left\lfloor \frac{x}{2^k} \right\rfloor 2^k \tag{3.22}$$

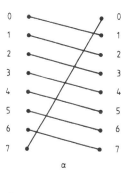

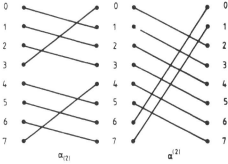

FIGURE 3.12   Shift permutations.

$$\alpha^{(k)}(x) = |x + 2^{n-k}|_{2^n}. \tag{3.23}$$

These are illustrated in figure 3.12 for $n = 3$ and $k = 2$.

3.3.3 The algebra of permutations

In order to establish more complex permutations and to establish equivalences, we will wish to combine and manipulate the simple permutations described above. This is really described by the algebra of functions. Thus we combine two functions by writing for example $\rho(\sigma(x))$, which describes the permutation defined by a perfect shuffle followed by a bit reversal. In terms of the bipartite graphs, this is equivalent to attaching the outputs of the shuffle to the inputs of the bit reversal. In order to simplify expressions, and at the same time preserve this feeling of left to right data flow in bipartite graphs, the expression above will be abbreviated to $\sigma\rho$. There is precedence for this form of functional product in the syntax of the combination of operators in APL (Iverson 1979). Also, $\sigma\sigma$ will be abbreviated to σ^2.

As in any algebra, we will require a unit or identity permutation, which we denote by i. This permutation preserves the order of the input set:

$$i(x) = x. \tag{3.24}$$

Having defined an identity permutation, we can now define the inverse of a mapping. For example the inverse of the perfect shuffle is denoted by σ^{-1} and $\sigma\sigma^{-1} = \sigma^{-1}\sigma = i$. The inverse mapping can be understood by reading the bipartite graphs from right to left.

Using this permutation algebra, we will now establish some important identities. In most cases these identities may be verified by considering the binary representation of x.

The first identity concerns the relationships between the kth sub- and superpermutations, where for the general permutation π,

$$\pi^{(k)} = \sigma^k \pi_{(k)} \sigma^{-k}. \tag{3.25}$$

Now we give a group of identities which concern the inverses of the permutations defined above:

$$\varepsilon_{(k)}^{-1} = \varepsilon_{(k)}, \tag{3.26a}$$

$$\rho_{(k)}^{-1} = \rho_{(k)}, \tag{3.26b}$$

$$\beta_{(k)}^{-1} = \beta_{(k)}, \tag{3.26c}$$

$$\alpha_{(k)}^{-1} = \alpha_{(k)}^{k-1}, \tag{3.26d}$$

$$\sigma_{(k)}^{-1} = \sigma_{(k)}^{k-1}. \tag{3.26e}$$

Another group of identities concerns the relationships which can be defined between the different permutations:

$$\varepsilon_{(k)} = \sigma_{(k)}\alpha_{(1)}\sigma_{(k)}^{-1}, \tag{3.27a}$$

$$\sigma_{(k)} = \beta_{(1)}\ldots\beta_{(k)}, \tag{3.27b}$$

$$\rho_{(k)} = \sigma_{(1)}\ldots\sigma_{(k)}. \tag{3.27c}$$

The final group of identities concerns the first sub-permutations only:

$$\sigma_{(1)} = i, \tag{3.28a}$$

$$\rho_{(1)} = i, \tag{3.28b}$$

$$\beta_{(1)} = i. \tag{3.28c}$$

As an example of the use of this algebra and the identities defined, let us establish what was apparent from figure 3.11, that for $n = 3$, $\rho = \beta$. From equations (3.27b) and (3.27c)

$$\rho_{(3)} = \beta_{(1)}\beta_{(1)}\beta_{(2)}\beta_{(1)}\beta_{(2)}\beta_{(3)}.$$

Since $\beta_{(1)}$ is the identity permutation (equation 3.28c) and $\beta_{(k)}$ is its own inverse (equation 3.26c),

$$\rho_{(3)} = \beta_{(3)}.$$

3.3.4   Single-stage networks

Single-stage networks consist of a fixed or single stage of switches. Thus in a P-processor array, the network would consist of a single array of P multiway selectors. This form of network is easily partitioned, with the selectors being a part of the processor hardware. Thus this form of network is usually associated with the interprocessor switch in figure 3.1.

Single-stage networks normally establish only a limited number of permutations directly. Other more general permutations must be built up iteratively. Thus, for example, if the network provides the shift permutation α, then to perform $\alpha^{(n-k)}$, where $\alpha^{n-k} = \alpha^{2^k}$, would require 2^k iterations through the network.

We will now describe a number of single-stage networks, defining their switches in terms of the set of permutations which they can generate. Thus

$$SW = \{\pi_1, \pi_2, \ldots, \pi_m\}$$

defines the switch SW, which performs the m permutations, π_1 or π_2 through to π_m. In all the discussion that follows we will assume that $P = 2^p$, unless otherwise stated.

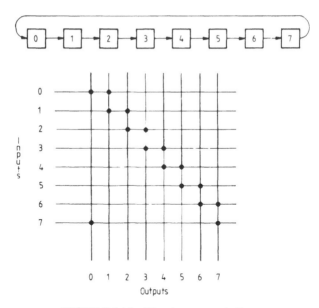

FIGURE 3.13   The ring network $R_{(1)}$.

(i) *The ring network*
This is the simplest of all networks; it consists of a ring of processors, with a undirectional flow of information. This is illustrated in figure 3.13 and like all single-stage networks is an incomplete crossbar switch, which is defined as:

$$R = \{i, \alpha_{(p)}\}.$$

(ii) *Nearest-neighbour networks*
This is a simple extension of the ring network, which allows a bidirectional flow of information. It is illustrated in figure 3.14 and defined as:

$$NN = \{\alpha_{(p)}^{-1}, i, \alpha_{(p)}\}.$$

This network is simple and cheap to construct but is not very suitable for a large number of processors. However, it may be generalised to more than one dimension. For example, if $P = Q^k$ and $Q = 2^q$, then the k-dimensional nearest-neighbour network can be defined as follows:

$$NN_{(k)} = \{\sigma_{(lq)}^q \alpha_{(q)}^{-1} \sigma_{(lq)}^q, i, \sigma_{(lq)}^q \alpha_{(q)} \sigma_{(lq)}^q : \text{for } l = 1, \dots, k\}.$$

The ICL DAP uses this switching network with $k = 2$ and $q = 6$.

$$S_{DAP} = \{\sigma_{(12)}^6 \alpha_{(6)}^{-1} \sigma_{(12)}^{-6}, \alpha_{(6)}^{-1}, i, \alpha_{(6)}, \sigma_{(12)}^6 \alpha_{(6)} \sigma_{(12)}^{-6}\}.$$

The permutations here can be considered as shifting north, south, east and west over a two-dimensional grid, with wraparound at the edges. The

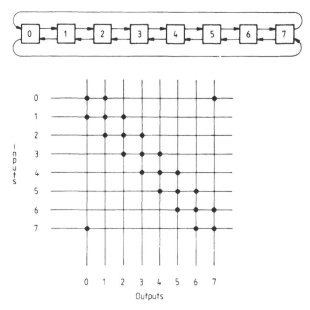

FIGURE 3.14   The nearest-neighbour network $NN_{(1)}$.

ILLIAC IV also has a two-dimensional nearest-neighbour switching network, with $q = 3$, although here the periodicity is defined somewhat differently:

$$S_{14} = \{\alpha_{(6)}^3 \alpha_{(3)}^{-1} \sigma_{(6)}^{-3}, \alpha_{(6)}^{-1}, i, \alpha_{(6)}, \sigma_{(6)}^3 \alpha_{(3)} \sigma_{(6)}^{-6}\}.$$

It is interesting to note what happens when we take the limiting case of the nearest-neighbour network, with $k = p = \log_2 P$. Here $q = 1$ and from (3.27a)

$$NN_{(p)} = \{\varepsilon_{(l)}, i: \text{for } l = 1, \ldots, k\}.$$

This describes the binary hypercube, whose structure consists of a hypercube of p dimensions with two processors per side. This is illustrated in figure 3.15 for $p = 3$, where it can be seen that the kth exchange is equivalent to exchanging information across the kth plane of symmetry ($k = 1$ plane shown).

(iii) Perfect shuffle networks

The perfect shuffle has had strong support as a permutation on which to base a switching network (Stone 1971, Lang and Stone 1976, Lang 1976). The perfect shuffle switch is defined below and illustrated in figure 3.16:

$$PS = \{\sigma_{(p)}^{-1}, i, \sigma_{(p)}\}.$$

It can be seen that this is not a very satisfactory switch, as it leaves four disconnected subsets of processors. Because of this the perfect shuffle exchange

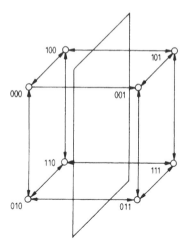

FIGURE 3.15   Illustration of the binary hypercube with eight nodes. The first plane of symmetry is shown.

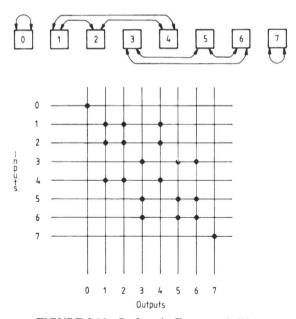

FIGURE 3.16   Perfect shuffle network $PS_{(1)}$.

(Stone 1971) and perfect shuffle nearest-neighbour networks (Grosch 1979) have been proposed. These are defined below and are illustrated in figure 3.17 for $N = 8$.

$$PSE = \{\sigma_{(p)}^{-1}, i, \varepsilon_{(1)}, \sigma_{(p)}\},$$

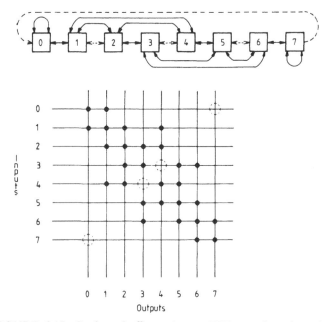

FIGURE 3.17   Perfect shuffle exchange $PSE_{(1)}$ and perfect shuffle nearest-neighbour $PSNN_{(1)}$ networks; broken lines convert $PSE_{(1)}$ to $PSNN_{(1)}$.

$$PSSN = \left\{ \sigma_{(p)}^{-1},\ \alpha_{(p)}^{-1},\ i,\ \alpha_{(p)},\ \sigma_{(p)} \right\}.$$

Both of these switches can be generalised to more than one dimension, as for the nearest-neighbour switch. However in the limiting case $k = p$, both are equivalent to the binary hypercube.

3.3.5   Some properties of single-stage networks

Single-stage networks or switches can be used iteratively to simulate a connection network. By iteration we mean that after one pass through the switch, the output at a given processor will be passed on as input to the same or another switch setting. Thus, for a given network, it is important to know how many iterations will be required to establish a mapping. This may affect the speed of the computation, especially if a large number of iterations are required.

To compare the networks discussed above, we introduce the concept of distance through the network. For example, the distance between two nodes i and j in the network is defined as the number of iterations required to establish the mapping $i \rightarrow j$. Having defined this we introduce two distance measures, which are a function of a given network and which give an indication of the effectiveness of the network. These measures were first defined by Jesshope (1980b, c) and have been used to establish timing estimates for the manipulation of data in processor arrays (Jesshope 1980a, c). A more complete study of the simulation of one network by another can be found in Siegel (1985).

(i) Maximum distance measure

The maximum distance measure $D_{(k)}$, for a k-dimensional network, gives the maximum distance between any two nodes in the network. It gives a measure of the effectiveness of the network for problems which involve global communication. In one dimension of an R switch, the processor that is farthest from processor i is processor $i - 1$ and for the NN switch it is processor $i + N/2$. An inductive argument is given by Jesshope (1980b), which affirms the scaling by k as given in table 3.1. For the perfect shuffle networks it is simpler to consider the binary representation of the address of the nodes. Thus for the PSE switch, node $i = \{b_q, b_{q-1}, \ldots, b_1\}$ is most distant from node $j = \{\bar{b}_q, \bar{b}_{q-1}, \ldots, \bar{b}_1\}$. This mapping can therefore be performed in $q - 1$ perfect shuffles (shift bits left) and q first-order exchanges (complement first bit). For the PSNN switch, the nodes which are most distant are i and $i + j$, where $j = \{1\,0\,1\,0\,1\ldots\}$. This again requires $q - 1$ perfect shuffles but only $q/2$ shifts (add one to i). The maximum distance in the PS switch is of course infinity. All of these results are summarised in table 3.1.

(ii) Fan out, fan in measure

The fan out, fan in measure gives a measure of the distance required to propagate one piece of information to all nodes in the network, or its inverse, the reduction (collection) of information. This is an important measure for many divide and conquer algorithms (see Chapter 5). This mapping will require a number of parallel permutations (possibly under boolean control). For example, in one dimension of the NN or R switch, an algorithm to perform this is given in figure 3.18. It can be seen that the total number of shift permutations for one dimension is $Q - 1 = 1 + 2 + 4 + \ldots + Q/2$. For

TABLE 3.1   Properties of single-stage networks.

Property	Network				
	$R_{(k)}$	$NN_{(k)}$	$PS_{(k)}$	$PSE_{(k)}$	$PSNN_{(k)}$
Maximum distance $D_{(k)}$	$k(Q-1)$	$kQ/2$	∞	$k(2q-1)$	$k(q+[q/2]-1)$
Fan out $F_{(k)}$	$k(Q-1)$	$k(Q-1)$	∞	$k(2q-1)$	$k(2q-1)$

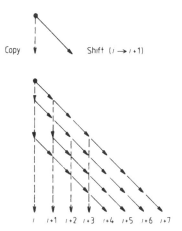

FIGURE 3.18   Illustration of the algorithm to fan out a single item of
data to all nodes in a ring $R_{(1)}$ or nearest-neighbour network $NN_{(1)}$.

the PSE and PSNN networks, however, the algorithm is best considered in
terms of the binary representation of the address or identifier of each
processor. The algorithm is illustrated schematically in figure 3.19, showing
data being propagated to all processors from processor number 0. Consider
first the PSE network: the shuffle and exchange permutations are described
in §3.3.2 by equations (3.11) and (3.12). These equations complement the
least significant bit of the address of a processor and perform a circular left
shift on the address of a processor respectively. Thus the problem can be
more easily specified as one of generating the addresses of all processors from
the source address by complementing the least significant bit of the address
and left-circular-shifting the bits of the address.

An intuitive lower bound can be derived for the number of steps that are
required to complete this operation. With the operations that are available,
the address that is 'farthest' from the source address is that which is its
complement. To complement an n-bit address requires n complement
operations (least significant bit only) and $n - 1$ shift operations, giving a total
of $2n - 1$ operations. Figure 3.19 illustrates that by suitable masking all
addresses can be generated while complementing the source address.

Implementing this algorithm on the PSE network, only one half of the
exchanged data is ever actually used, either the left shift or right shift,
depending on the parity of the source of the data. Therefore, exactly the same
algorithm can be used in the PSNN network. The odd and even masked
exchanges can be simulated in the same time using the left or right shifts of
the NN network. The algorithm is illustrated in a different form in figure 3.20
and the above results for the fan out process are summarised in table 3.1.

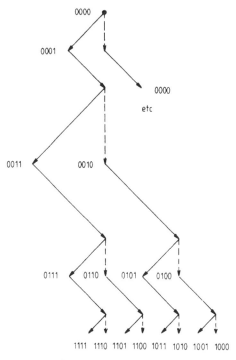

FIGURE 3.19   Illustration of the algorithm to fan out a single item of data to all nodes in a perfect shuffle exchange network $PSE_{(1)}$.

To compare the effectiveness of the network switches described, we will look at all possible configurations of a 4096 processor array. In order to do this we also need a rough estimate of the cost of a given configuration. Thus we assume a cost function C, given by the number of permutations the switch can perform, excluding the identity permutation (i.e. $C = m - 1$). $D_{(k)}$ and C are evaluated in table 3.2 for the 4096 processor example.

It can be seen from table 3.2 that the PSE and PSNN networks are the most cost effective for $k = 1$, but that the R and NN networks improve markedly for $k = 2$. Beyond this the returns in the R and NN networks diminish for increasing dimensionality. The PSE and PSNN networks show no benefits from increased dimensionality until $k = p$, when it can be seen

```
1 1 1 1 1 1 1 1   0 0 0 0   0 0   0 0
1 1 1 1 0 0 0 0   1 1 1 1   0 0   0 0   Binary
1 1 0 0 1 1 0 0   1 1 0 0   1 1   0 0
1 0 1 0 1 0 1 0   1 0 1 0   1 0   1 0

15 14 13 12 11 10 9 8   7 6 5 4   3 2   1 0   Decimal

                                        X   Exchange
                                     X  X   Shuffle

                                  X  X  X   Exchange
                            X  X  X  X  X   Shuffle

                   X  X  X  X  X  X  X  X   Exchange
      X     X     X     X  X  X  X  X  X  X   Shuffle

X X X X X X X X   X X X X   X X   X X   Exchange
X X X X X X X X   X X X X   X X   X X   Completed
```

FIGURE 3.20   A diagram showing the fan out operation using the shuffle–exchange network. The binary and decimal addresses of the processors are shown at the top of the table and the operations performed are shown on the right of the table. The crosses represent the propagation of the information from address 0.

TABLE 3.2   Maximum distance $D_{(k)}$ in a 4096 processor array. The cost function C is shown in parentheses.

k	Q	$R_{(k)}$	$NN_{(k)}$	$PSE_{(k)}$	$PSNN_{(k)}$
1	4096	4095 (1)	2048 (2)	23 (3)	17 (4)
2	64	126 (2)	64 (4)	22 (6)	16 (8)
3	16	45 (3)	24 (6)	21 (9)	15 (12)
4	8	28 (4)	16 (8)	20 (12)	16 (16)
6	4	18 (5)	12 (12)	18 (18)	12 (24)
12	2	12 (6)	12 (12)	12 (12)	12 (12)

that no network is any better than any other. Indeed it can easily be shown, using the algebra of permutations in §3.3.3, that all four networks are equivalent; the difference in cost reflects the increased bandwidth from multiple equivalent permutations.

It is perhaps not too surprising that the most common network found in processor arrays is the two-dimensional NN network as in the ILLIAC IV (McIntyre 1970), ICL DAP (Flanders et al 1977), Goodyear MPP (Batcher 1980), GEC GRID (Robinson and Moore 1982), NTT AAP (Komdo et al 1983). Linkoping University LIPP (Ericsson and Danielson 1983), NCR

GAPP (NCR 1984) and Southampton University RPA (Jesshope *et al* 1986). A recent exception is the connection machine (Hillis 1985), which uses, at least at the VLSI level, the limiting case of any of these networks, with $k = p$. This is often called the binary hypercube network or just the cube network (see figure 3.15).

One may ask why the $PSE_{(1)}$ network has not found favour, since it provides a very cost effective solution to long-range communication. There are perhaps several answers concerning both its use and implementation. Many algorithms require local communication, and usually in some nearest-neighbour form, and yet it requires $2\log_2 (N-1)$ iterations for the PSE network to simulate a two-dimensional NN network (Siegel 1985). To put this in perspective, this is also the maximum distance through the network. This can be easily verified by considering the fact that 0 and N are adjacent in any NN switch and yet are the farthest separated in the PSE switch. The PSNN switch overcomes some of these limitations and the two-dimensional PSNN has been considered in many paper designs; it has the two-dimensional nearest-neighbour connections used in many algorithms, but also has additional long-range communication properties. The reason that this network has not to our knowledge been used in any implementation, lies in the implementation itself.

Unlike the NN networks, networks based on the perfect shuffle permutation cannot be partitioned. If you slice a $NN_{(1)}$ network in half, it can be connected by only two wires; even a divided $NN_{(2)}$ network requires only $N^{1/2}$. What is more important, each half of the NN network will also function as an NN network of reduced size. However, if any network containing the perfect shuffle permutation is divided in two, then $N/2$ wires are required to connect each half and neither half has any function on its own. There are three implications to this: the first is in design, as if the network cannot be partitioned, then it must be designed as a whole, there being little regularity to submodules; the second is related and concerns reliability, this is achieved through redundancy, but with no regularity, which is very expensive; the third concerns the number of wires between submodules, for in modern system wires are more expensive and detrimental to performance than logic gates. All three issues are increasingly important in the current era of VLSI devices (see Chapter 6).

3.3.6 Multistage networks
In some situations it is required to connect one set of resources to another, so that any member of one set may access every member of the other set. One such example is found when providing switching between processors and memory banks, as in figure 3.1, where we wish to create a full access, shared-memory, multiprocessor system. Here it may be required that a given

processor needs access to every memory bank in the system. To provide such access a full connection network is required.

We have already met the full crossbar network (figure 3.8), which is a full connection network, and also encountered its major disadvantage—that the number of gates required to implement it grows as the square of the number of inputs. To put this in perspective, consider it as an alternative to the single-stage networks considered in the previous section to connect 4096 processors. This would require at least 16 million transistors but probably many times more than this for a switch of reasonable performance. The two-dimensional NN switch on the other hand could be implemented in a mere 25 000 transistors.

Multistage networks can also provide a cheaper alternative to the complete crossbar switch, when a full connection network is required. These networks are based on a number of interconnected crossbar switches, the most common being built up from the 2×2 crossbar switch. This switch is illustrated in figure 3.21 and can generate two permutations and a further two broadcast mappings.

If one bit is used to control this switch then only the permutations in figure $3.21(a, b)$ will be selected. Using an array of $N/2$ such switches we can define the kth-order exchange switch, a single-stage network, which requires $N/2$ control bits as:

$$E_{(k)} = \{\varepsilon_{(k)}, i\} = \sigma_{(k)} E_{(1)} \sigma_{(k)}^{-1}. \tag{3.29}$$

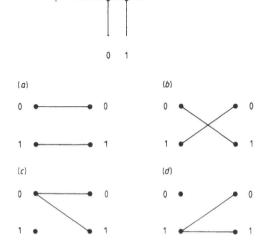

FIGURE 3.21 2×2 switching element and mappings.

(a)

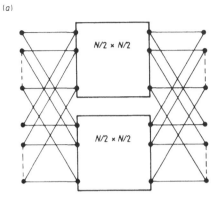

(b)

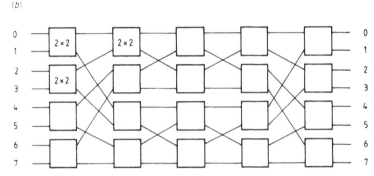

FIGURE 3.22 Benes network: (a) reduction of an N by N crossbar switch to two $N/2$ by $N/2$ crossbar switches and two exchange switches; (b) binary Benes network with full reduction and simplification ($N = 8$).

Alternatively if two control bits are used for each 2×2 crossbar, then all four mappings in figure 3.21 can be generated. Thus we can extend equation (3.29) to give the kth-order generalised exchange switch, which requires N control bits:

$$\mathrm{GE}_{(k)} = \{\varepsilon_{(k)}, 1_{(k)}, u_{(k)}, i\} = \sigma_{(k)}\mathrm{GE}_{(1)}\sigma_{(k)}^{-1}. \tag{3.30}$$

Here $1_{(k)}$ and $u_{(k)}$ are the upper and lower broadcast mappings which are defined by:

$$1_{(k)}(x) = \{b_n, \dots, b_{k+1}, 1, b_{k-1}, \dots, b_1\},$$
$$u_{(k)}(x) = \{b_n, \dots, b_{k+1}, 0, b_{k-1}, \dots, b_1\}.$$

(i) *Connection networks*

It was shown by Benes (1965) that the complete $N \times N$ crossbar switch could be reduced to two $N/2 \times N/2$ crossbar switches together with two N-input exchange switches as described above. This is illustrated in figure 3.22(a) and if we define an $N \times N$ crossbar switch as $X_{(n)}$, where $N = 2^n$ them this reduction can be formally described by:

$$X_{(n)} = E_{(n)}X_{(n-1)}E_{(n)}. \tag{3.31}$$

Obviously this reduction can be continued recursively giving:

$$X_{(n)} = E_{(n)}E_{(n-1)}E_{(n-2)}\ldots E_{(2)}X_{(1)}E_{(2)}\ldots E_{(n)}. \tag{3.32}$$

This can be modularised using the relationship in equation (3.29) giving:

$$X_{(n)} = \sigma_{(n)}E_{(1)}\sigma_{(n)}^{-1}\ldots\sigma_{(2)}^{-1}E_{(1)}\sigma_{(2)}\ldots\sigma_{(n)}E_{(1)}\sigma_{(n)}^{-1}.$$

Finally this can be simplified, by noting that the pre- and post-permutation of the inputs and outputs of a switch only redefine the order of these sets. Thus we obtain an expression for the binary Benes network (Lenfant 1978):

$$B_{(n)} = X_{(n)} = E_{(1)}\sigma_n^{-1}E_{(1)}\sigma_{(n-1)}^{-1}\ldots\sigma_{(2)}^{-1}E_{(1)}\sigma_{(2)}\ldots\sigma_{(n)}E_{(1)}. \tag{3.33}$$

This is illustrated in figure 3.22(b). By definition this is a full connection network giving $N!$ possible permutations; moreover, if we generalise this using the $GE_{(1)}$ switch, then all N^N well defined mappings can be established.

(ii) Shuffle exchange networks

Another class of networks which are not full connection networks are the so called shuffle exchange networks. Four such networks will be described here, the omega network (Lawrie 1975), the indirect binary n-cube network (Pease 1977), the banyan network (Goke and Lipovski 1973) and the R network (Parker 1980). These are identified below by Ω, C, Y and R respectively:

$$\Omega_{(n)} = (\sigma_{(n)}E_{(1)})^n,$$

$$C_{(n)} = E_{(1)}\beta_{(2)}E_{(1)}\beta_{(3)}\ldots\beta_{(p)}E_{(1)}\sigma_p^{-1},$$

$$Y_{(n)} = E_{(1)}\beta_{(2)}E_{(1)}\beta_{(3)}\ldots\beta_{(p)}E_{(1)},$$

$$R_{(n)} = E_{(1)}\sigma_{(n)}^{-1}E_{(1)}\sigma_{(n-1)}^{-1}\ldots\sigma_{(2)}^{-1}E_{(1)}\sigma_{(n)}.$$

It can be seen that the banyan switch is only the binary n-cube without the final unshuffle and that the R network is derived from the first half of the binary Benes switch, followed by a shuffle. Another interesting point which has been proved by Parker (1980) is the relationship between Ω, C and R, which gives the following identity:

$$\Omega_{(n)}^{-1} = C_{(n)} = R_{(n)} = Y_{(n)}\sigma_{(n)}.$$

The omega and binary n-cube networks are illustrated in figure 3.23.

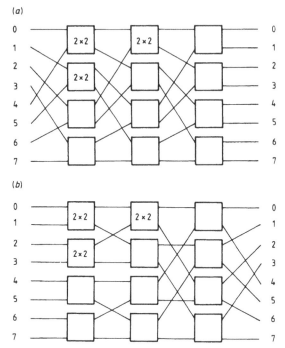

FIGURE 3.23   Shuffle exchange networks: (a) omega network; (b) indirect binary n-cube network.

Although these switches are not full connection networks, they do provide a very rich class of permutations, suitable for many applications on multiprocessor systems. Apart from mesh and matrix manipulation, they are highly suited to the FFT and related algorithms (Pease 1969) and to sorting techniques (Batcher 1968). For more information concerning multistage networks see Siegel (1985).

3.3.7   Network control

Up until now no account has been taken of how the setting of the individual switches or selectors in both the single- or multistage networks is obtained. For a single-stage network, which supports m permutations, over P processors, then $P\log_2 m$ bits of control are required to fully control the switch. However, it is common for single-stage networks to be controlled by a single instruction stream. In this case a single $\log_2 m$ bit control field is broadcast to all processors in the array. There are some exceptions to this, and it has been shown using the RPA (Jesshope et al 1986) to be a very powerful extension to the SIMD model to devolve the control of the switch to the processor, so that it may be set from local state. The RPA is described in some detail in §3.5.4.

For the crossbar switch, with N inputs, N^2 control bits are required, one for each crosspoint. To establish a connection between input i and output j, it is sufficient to assert the bit in position i, j in the matrix of crosspoints. If only one-to-one mappings are to be allowed, then for each input row, only one output column may be connected to it. Thus the number of control bits may be reduced to $N\log_2 N$, by coding the column numbers.

The situation is not so simple and perhaps still not so well understood, when considering multistage networks. These also require $O(N\log_2 N)$ control bits, with the constant depending on topology and whether this network is generalised to perform broadcast mappings. To store control words for even a limited subset of the available mappings will not be feasible. For example to store the control to compute only shift permutations will require $O(N^2\log_2 N)$ bits, which is already unreasonable for a large array of processors.

The alternative is to compute the control words for a given permutation. The method of Waksman (1968) can calculate the control word for any given permutation in $O(N\log_2 N)$ steps using a sequential algorithm, which can be reduced to $O(N)$ steps using a parallel algorithm (Thompson 1977). However, this is again unacceptable as it would dominate the switching delay time.

All is not lost, however, as there do exist algorithms which can compute the control required but only for a limited number of permutations. Such algorithms can be found in Lenfant (1978) and Parker (1980).

So far we have considered the control solution, which establishes a complete circuit for the permutation, until a new control word is loaded. An alternative strategy is to use the network as a packet switch. This involves using the network statistically, so that a given connection is only established in order to let a small packet of data through. In this case packets of data need to carry the address of their destination and possibly some sequencing information if large packets are sent in small fixed sizes. Thus the packets contain their own control and can be routed by intelligent switch nodes and if they arrive out of sequence can be reassembled at their destination.

The routing algorithm at each switch node must decide in which direction the switch must be set in order to send the packet closer to its destination. Because the network is being used statistically, more packets may arrive than the switch can handle in any time slot. In this case provision must be made either to refuse a packet or to buffer unwanted packets. In some cases the algorithm for refusal may send a packet further away from its intended destination.

The implementation of a packet-switched network, if efficient, can provide a very powerful basis for many parallel computational models. It provides a structure which has distributed control, and therefore provides a very dynamic environment.

A good analogy to illustrate the differences between a circuit switch and a packet switch is to consider the services of a telephone company and the post office. The post office forwards discrete packages from one branch to another, until delivered, while the telephone company will establish a direct link between you and another subscriber when you dial his number. In fact these days you may not actually have exclusive use of the circuits used, but for the bandwidth you require this is apparently what you perceive. Indeed, some experiments have been made with packet-switched voice communications, but this just illustrates the power of packet switching as a routing technique.

3.3.8   Data access and alignment

A major feature in replicated systems, especially SIMD arrays, is the way in which memory is organised. How the memory is connected to the processors and how it is addressed are two important features of this organisation.

Let us consider first the situation where each processor is connected to its own memory partition. This is the simplest configuration and the only consideration here is how the memory is addressed. An instruction issued from the control unit would normally supply the same address to all processors. A major issue is whether to allow the individual processor to modify this address, using its own index register.

This facility was provided on the ILLIAC IV, one of the first processor arrays to be built. In this computer the programmer may set an index register in each processor, based on information available locally, and then use this to select a non-planar slice of the array memory. This is important in many classes of problems as it often saves the manipulation of data (Jesshope 1980c).

The ICL DAP, on the other hand, does not provide this facility but then this machine has bit-serial arithmetic; to modify the DAP address would require a 14-bit addition (16K memory). Consequently it would require many processor cycles to modify the address. More sophisticated hardware would be required to obtain address modification in a realistic time. Roberts (1980) has proposed VLSI processor designs to accomplish this. These designs use carry look-ahead adders for address modification, while maintaining a bit-serial approach to arithmetic, but is this a realistic use of the additional complexity made available by the use of VLSI?

With a switch between the processors and memory, a richer set of mappings is required. Likewise it becomes more important that each memory bank is addressed independently.

Memory bank conflicts are one of the major drawbacks of any parallel memory organisation. This is true of the overlapped or interleaved organisation found in vector processors as well as the parallel organisation found in processor arrays. To illustrate this figure 3.24 shows a 4×4 array stored in

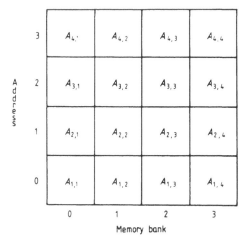

FIGURE 3.24   Storage scheme giving conflict-free access to rows but not columns of the 4 × 4 matrix.

four banks of parallel memory. It can be seen that the rows of **A** may be accessed without conflict but that data in a column of **A** all reside within the same memory bank. There is no way therefore that this data may be accessed in parallel.

With a switching network that can connect all memory banks to a given processor, it is very desirable to have a memory or data structure in which rows, columns and other principle substructures from arrays may be accessed without conflict. One method of achieving this is through skewed storage schemes. Figure 3.25 shows one such scheme for our 4 × 4 matrix. This allows

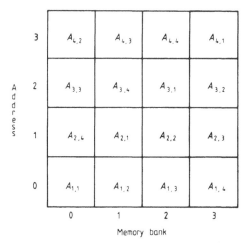

FIGURE 3.25   Skewed storage scheme giving conflict-free access to both rows and columns of the 4 × 4 matrix.

the conflict-free access of rows and columns however the diagonals of **A** remain in conflict. The question of more general skewing schemes has been investigated by Budnik and Kuck (1971) and Shapiro (1978), who consider the case of a prime number of memory banks, or alternatively a number which is not a power of 2. More recently Deb (1980) presents an example of a skewing scheme (4 × 4) which allows conflict-free access to rows, columns and both major diagonals. It does not however give conflict-free access to all circulant diagonals (Jesshope 1980b).

Burroughs, in their ill-fated BSP design (see §3.4.3), have devised a memory organisation based on the skewing schemes described by Budnik and Kuck (1971). The principle feature of this organisation is a memory with a prime number of memory banks, which allows conflict-free access to all linear subarrays defined by a start address and a skip distance. The exception to this is where the skip distance and the number of memory banks have a common factor. This occurs only when the skip distance is a multiple of the number of memory banks, which is prime.

We will illustrate this organisation with our 4 × 4 array as an example. Instead of having four memory banks for our four-processor array, we would choose the next highest prime number, five in this case. In general, to obtain parallelism N from a memory organisation choose M to be a prime number such that $M \geqslant N$. The address of each element of an array is then given by equation (3.34), where a is the corresponding linear address of the element.

$$
\begin{aligned}
\text{Memory bank:} \quad & \mu = |a|_M, \\
\text{Address within bank:} \quad & i = \lfloor a/N \rfloor, \tag{3.34}
\end{aligned}
$$

where $\lfloor f \rfloor$ gives the integer floor function of f and $|f|_g$ is the value of f modulo g. Table 3.3 gives these address mappings for our 4 × 4 example and figure 3.26 illustrates this. It is perhaps a bad example as one of the principle sets of subarrays, the forward diagonals, causes memory conflict. For an $N \times N$ matrix the forward diagonal has a skip distance of $N + 1$, which in our example is 5, the same as the number of memory banks. However, all other linear subarrays can be accessed without conflict; these include rows, columns and backward diagonals. As an example consider the access of row 2 of this matrix. The start address is 1 and the skip distance is 4, which define the linear address of each element of this row. Thus:

$$\mathbf{a} = (1, 5, 9, 13),$$

and using (3.34) gives

$$\boldsymbol{\mu} = (1, 0, 4, 3)$$

and correspondingly

TABLE 3.3   The mapping of a 4 × 4 array.

Element of **A**	Linear address	Memory bank number	Address in bank
1,1	0	0	0
2,1	1	1	0
3,1	2	2	0
4,1	3	3	0
1,2	4	4	1
2,2	5	0	1
3,2	6	1	1
4,2	7	2	1
1,3	8	3	2
2,3	9	4	2
3,3	10	0	2
4,3	11	1	2
1,4	12	2	3
2,4	13	3	3
3,4	14	4	3
4,4	15	0	3

$$\mathbf{i} = (0, 1, 2, 3).$$

In accessing a subarray, μ defines a mapping vector or permutation which can be used to set the switching network to align the memory banks to the correct processors. The indexing vector $\mathbf{i}$ is used to modify the base address. It should be noted that index components of $\mathbf{i}$ correspond to the memory banks given in μ and must therefore be permuted according to μ to give the index in memory bank order. A glance at figure 3.26 will confirm that the vectors above do in fact select row 2 of our matrix.

3.4   AN HISTORICAL PERSPECTIVE

3.4.1   A chequered history

Parallelism in computer design can be applied to either the bits within a word or across a number of words, or indeed to both. This trade-off has already been illustrated in §3.2.3. The choice of bit-parallel/word-serial designs, pioneered by von Neumann and others in the 1950s, has set a trend in computer architectures that has only recently been questioned seriously. This choice correctly reflected the engineering and technological constraints

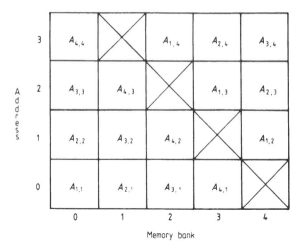

FIGURE 3.26   Storage scheme using a prime number of memory banks (five), which gives conflict-free access to all linear subarrays (rows, columns etc) of the 4 × 4 matrix, provided that successive elements are not separated by five.

that were current at that time, when logic was relatively very fast and expensive, and storage was relatively slow and inexpensive. These constraints resulted in the word-organised processor–memory structure (the so called von Neumann architecture), which kept the expensive parts busy (i.e. the processors built from vacuum tubes).

In modern computer design, both memory and logic are now cheap and the critical component has become the wire or interconnect. Many hundreds of thousands of transistors are routinely designed into custom VLSI circuits and will probably be pushing into the millions in the near future. Certainly circuits with in excess of one million transistors have already been fabricated (see Chapter 6). Because memory and logic can now be made using the same technology, the speed and costs are comparable; indeed it is very desirable to integrate memory and logic onto the same chip. These new technological factors require architectures with an equal balance between processing and memory and, moreover, require the components separated in the von Neumann architecture to be much more closely integrated.

The processor array, and to some extent the multiprocessor system, has taken an alternative evolutionary path to the conventional von Neumann processor in that many processors cooperate on a single problem, each accessing data from its own memory. It should be noted that with a full permutation switch between processors and memories, the notion of ownership may take on a very temporary nature! The grid-connected SIMD computer has perhaps been the primary contender in this architectural development, as will be seen later. However, to date not one of these systems has been a commercial success. Some have been technically successful and some have been commercial, but user preference for computers which execute standard serial FORTRAN has dominated the supercomputer market. However, with the increasing awareness of parallelism, the further development of parallel languages (see Chapter 4) and the growing cost/performance ratio that can be obtained with highly replicated VLSI circuits, processor arrays and multiprocessor systems are beginning to come of age. It remains to be seen whether any manufacturer can grasp the bit and produce a success story at last.

(i) An early start

The idea for connected computers, which could be applied to spatial problems, was conceived as early as 1958 in the paper by Unger (1958). In this machine a two-dimensional array of neighbour-connected logical modules was conceived as being controlled by a single master controller. Each module (these would be called processing elements or PES today) consisted of an accumulator, a boolean processor and about six bits of RAM. Data was input to a module either by connecting the modules in a shift register, or directly from the master controller. The controller acted as a conventional von Neumannn sequencer would have, with one exception—the ability to make a branch decision based on the logical sum of data from all modules in the array (the accumulators supplied this data). This exception was one of the most important innovations in Unger's design and it can still be found in nearly all processor arrays designed today.

This feature provided the only data-dependent branch possible in Unger's design, which translates to the parallel construct:

> FOR ALL PROCESSORS
>
>   IF ANY (ACCUMULATOR = TRUE)
>
>   THEN ACTION 1
>
>   ELSE ACTION 2

Of course with a suitable change in the sense of the logic, a branch on the 'all accumulators true' condition could also be obtained.

This architecture was never built, because of the 'alarming figures' that Unger estimated for the number of logic gates required. The actual figures called for around tens of thousands of logic gates, which may easily be implemented on a single chip today.

(ii) *SOLOMON*

The grid-connected computer was further consolidated in the SOLOMON computer design (Slotnick *et al* 1962, Gregory and McReynolds 1963) which was a 32 × 32 array of PES, and although this was also never built, it was the precursor to the ILLIAC (University of Illinois Advanced Computer) range of designs, which culminated in the ILLIAC IV and Burroughs BSP designs (see §§1.1.4 and 3.4.3). The SOLOMON design also had a major influence on the ICL DAP (see §3.4.2) and was perhaps notable for introducing another control concept into the SIMD processor array, that of mode control (or activity as it would be called today). The mode of a PE was a single-bit flag, which could be set with reference to local data and then used to determine the action of later instructions. In particular it could be used to inhibit storage of results in the array where not set, and thus provide a local conditional operation.

This mode or activity control is equivalent to the following construct:

    FOR ANY PROCESSOR

    IF MODE

    THEN ACTION 1

This parallel IF–THEN construct is sometimes written

    WHERE MODE

    THEN ACTION 1

(iii) *The ILLIAC IV*

The ILLIAC IV was really the first technically successful array processor to be built and will be fondly remembered by at least one of the authors of this volume, as it was used in anger at a very early stage of a program of research into the use of parallel computers for solving problems in science and engineering (Jesshope and Craigie 1980). For many researchers the ILLIAC IV, delivered to NASA Ames in 1972, was offering a full service via ARPANET from 1975.

The ILLIAC IV consisted of 64 complex processors arranged in a two-dimensional grid. The processors were 64 bits wide, although these could be reconfigured to operate on 8-bit data giving an array size of 256. The original design comprised four quadrants each the size of the array actually built and the whole, using four controllers, should have been capable of in excess of 1 Gflop/s (10^9). One of the pioneering features of the ILLIAC IV, which was only feasible due to its complex processor design, was the provision of local indexing operations. An X register in each processor could be used to modify the broadcast address it received from the central controller. Therefore each processor could access a different location in its own memory.

This facility provides the parallel construct:

FOR ALL PROCESSORS {LABELLED I = [0 FOR 64]}

A[I]: = MEMORY[X[I]]

As was shown in §3.3.8, this is a very powerful feature, especially when accessing arrays in multiple dimensions.

Although the ILLIAC IV was delivered late, was over budget, never really worked to full specification and was very unreliable, it provided the research community in the area of parallel machine construction and use with a tremendous stimulus. One has only to look at the development of parallel languages for the ILLIAC to see this. For example, the ALGOL-like TRANQUIL (Abel *et al* 1969) and GLYPNIR (Lawrie *et al* 1975), the FORTRAN-like CFD (Stevens 1975) and the PASCAL-like ACTUS (Perrott 1979) were all developed originally for the ILLIAC IV.

(iv) The real beginning

The 1970s also saw the development of many other processor array designs. In Britain two developments resulted in machines being fabricated, the CLIP project at University College, London, which was a design specifically built for image processing applications (Fountain and Goetcherian 1980) and the ICL DAP, which is described in more detail in §3.4.2. The first commercial ICL DAP was delivered in 1979 to Queen Mary College, London (QMC), and despite being well favoured in the research communities, was not a commercial success for ICL. It seemed that its successor, an LSI DAP would have a similar fate; however, at the time of writing, ICL had spun off its commercial DAP development into a newly created start-up company called Active Memory Technology Ltd (AMT Ltd). It remains to be seen whether this new company will be able to compete with the growing number of transputer products.

Of the six main-fram DAPs built, one went to QMC, two others went to Edinburgh University, one to the Hydraulics Research station at Wallingford, one to the National Physical Laboratory at Teddington and the last was used internally at ICL for integrated circuit CAD. In the US, STARAN and PEPE were other array processors built in the 1970s. The STARAN was an associative processor built by Goodyear Aerospace and PEPE was a system designed for tracking many ballistic missiles in real time.

Most of the systems mentioned above have had successors, many of which are described in the later sections of this chapter. For example, the ill-fated BSP, built by Burroughs, followed directly from their experience as contractors on the ILLIAC IV project. Although only one BSP machine was ever built, it is described in §3.4.3 as it represented the state-of-the-art in array processors design at that time. However, it had to compete commercially with established vector processors such as the CRAY-1. To do this it should have been able to outperform the CRAY-1 on maximum performance, but it gave only a fraction of the CRAY's peak performance. The fact that, unlike the CRAY, the BSP was designed to give a high fraction of its maximum performance on a wide range of problems when programmed in FORTRAN, did not seem to cut any ice with the end users or their representatives. There is a moral here perhaps.

3.4.2   The ICL DAP

Design of the pilot DAP (distributed array processor) was started in 1974 and was similar to the SOLOMON design (Gregory and McReynolds 1963), which consisted of a two-dimensional array of 1024 1-bit processors. However the DAP design introduced two new contributions to the SOLOMON formula. The first of these was a hardware feature, which effectively slices the array in two orthogonal directions. A number of registers are provided in the master control unit (MCU), which match and can be aligned with either dimension of the DAP array. This is achieved by the use of two orthogonal data highways, which thread rows and columns of processing elements (PES). These highways have one bit for each bit in the MCU register, which terminates in either a row or column of the DAP array. Thus $PE_{i,j}$ will have a one-bit highway directly to bits i and j of the MCU register. These highways collect and broadcast data to slices of the DAP array and provide the DAP with much of its flexibility in manipulating data.

The second contribution made by the DAP design concerns the manner in which the DAP is integrated into a complete system. The DAP is designed to emulate a memory module for an ICL main-frame computer and also to process data autonomously in a highly parallel manner. This concept has given the DAP its name, as processing power is distributed throughout the memory of a conventional computer.

The pilot DAP hardware was completed less than two years after its inception, as an array of 32×32 PEs having 1 Kbit of memory each. Some six years later in 1980, the first three production models were delivered to their respective customer sites. These main-frame machines consist of a large array of 4096 PES arranged in the same two-dimensional geometry, each having 4 Kbits of memory per PE. This gives a total of 2 Mbytes of memory (later increased to 8 Mbytes) which was attached to one of the top-end machines in the ICL 2900 range. There are some minor differences between pilot and main-frame machines and later differences between this and the

LSI DAPs. However, we will describe the first production version here. For the interested reader the pilot machine was first described in Reddaway (1973) and was evaluated on several applications in Flanders *et al* (1977), and the LSI DAPs are considered later in this section.

The main-frame DAP was constructed in units of 16 processors and associated memory on one 12 in × 7 in circuit board. This contained around 80 16-pin TTL integrated circuits, with typical levels of integration being 10–40 gates per chip. A single gate delay of 5 ns gives an overall clock cycle of 200 ns, including memory access. All memory for one processor is provided by a single chip, initially a 4K static MOS device. Thus 256 boards comprise the array section of the DAP, which together with control unit boards and host store access control are housed in a single air-cooled cabinet, occupying some 20–30 ft² of floor space. This is illustrated in figure 3.27. The cost of the complete DAP system was about £500 000, in addition to the cost of the 2900 host computer. However, it should be noted that the extra memory provided by the DAP would have cost a significant part of this figure in any case.

Figure 3.28 illustrates a typical 2900 system, which consists of an order code processor and a store access controller, both cross connected with a number

FIGURE 3.27   The DAP array memory and access unit (courtesy of ICL).

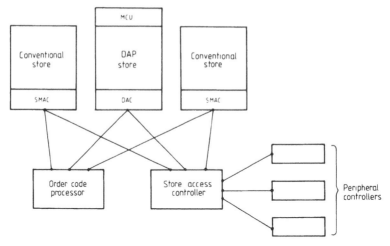

FIGURE 3.28   Relationship of the DAP to the 2900 main-frame computer.
SMAC, store multiple access controller. DAC, DAP access controller; MCU, master control unit.

of memory units. One or more of these memory units may be a DAP, which provides memory in the conventional way and may also be instructed by the order code processor to execute its own DAP code. The memory may still be used while the DAP is processing, by stealing unused memory cycles. Protection of DAP code and data is obtained by giving various access permissions. For example, read only and read execute segments may be defined in the DAP to protect data and code.

The store access controller provides memory access to peripherals and also provides a block transfer facility between memory units. Thus if the DAP is considered as the number crunching core of the system, then conventional store can be considered as fast backing store to the DAP, with the facility for pre- and post-processing by the order code processor. Figure 3.29 identifies the major components and data highways in the DAP. Interface to the 2900 system is provided by the DAP access controller and the column highway, which has one bit for each column of processors in the DAP array. Thus one 2900 64-bit word corresponds to a row across the DAP memory. Incrementing the 2900 address first increments down the columns of the DAP array and then through the 4K DAP address space. The column highway also provides a path between rows of the DAP array and the MCU registers, which can be used for data and/or instruction modification. Finally, the column highway provides the path for the MCU to fetch DAP instructions from the DAP store. DAP instructions are stored two per store row and one row is fetched from memory in one clock cycle.

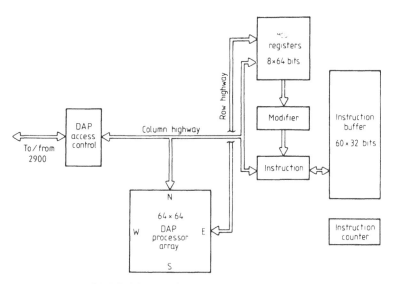

FIGURE 3.29 Major components of the DAP.

Under certain conditions, instructions can be stored in the instruction buffer for repeated execution. More details of this are given later.

It can also be seen from figure 3.29 that the row highway connects the DAP array to the MCU registers in the orthogonal direction. The row highway has one bit for each row of processors in the array and is used exclusively for transmitting data to and from the MCU registers. Figure 3.30 illustrates the various components and data paths which comprise a single processing element. The array forms a two-dimensional grid, with each processing element having four neighbours. The undefined connections at the edge of the array are defined by the instruction being executed. The geometry of an instruction is either planar or cyclic in rows and columns. Planar geometry defines a zero input at the edges, whereas cyclic geometry gives periodic connections in rows or columns of the array. The geometry of rows and columns can be set independently.

Within the processor are three 1-bit registers, two multiplexers and a 1-bit full adder. The A register provides programmable control over the action of the processing element, as certain store instructions are only enabled if this activity register is set. The A register also has a gated input for the rapid combination of the control masks by anding the contents of the register with the input. The other registers and accumulator (Q) and carry store (C).

The adder adds Q, C and the input to the processing element, giving sum and carry outputs, which can be stored in the Q and C registers respectively. One exception to this is when an add to store instruction is being executed.

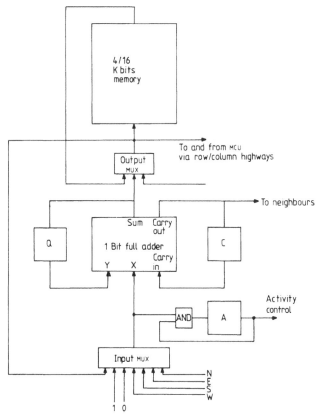

FIGURE 3.30 A simplified diagram of the ICL DAP processing element.

This family of instructions takes 1.5 cycles to execute and during this time reads an operand and writes the sum bit back to the same location (most instructions take only one cycle to execute, see below), thus saving half a cycle over an accumulator addition followed by a store accumulator instruction.

Parity processing elements (not illustrated in the figures) are incorporated into the design. These check both memory and logical functions (Hunt 1978). Also when acting as 2900 memory, a full Hamming error code is maintained, which gives single error correction and double error detection on every 64 bits of data read.

Instructions in the DAP are executed in two phases, the fetch and execute cycles. Each of these cycles is 200 ns in the main-frame DAP. However when an instruction appears after and within the scope of a special hardware DO loop instruction, then the first phase which fetches the instruction will only be performed once for all *N* passes of the scope of the loop. The DO loop instruction has two data fields, a length field which indicates the scope of the loop and a count field which may be modified and gives the number of times the loop is to be executed. The maximum length of the loop is 60 instructions and the maximum loop count is 254. Within a loop, instructions may have their addresses incremented or decremented by 1 on each pass. The DO loop is essential for building up software to operate on words of data. The rate of instruction execution is asymptotically one every clock period in the loop, compared with one every 1.5 clock periods when instructions have to be fetched for each execution (two instructions are fetched in one fetch cycle).

Most DAP instructions have fields as illustrated in figure 3.31(*a*). The operation code and inversion field effectively specify the instruction. The inversion bit creates pairs of instructions which are identical, with the exception that one of the inputs to the instructions is inverted. For example QA and QAN are the related pair which load the Q register from the contents of the A register. QAN inverts the input. Many DAP instructions have such complementary pairs. The two other 1-bit fields specify whether the DAP instruction is to have its address incremented or decremented within a DO loop.

The two 3-bit fields specify MCU registers. The first is set if a data register is required and the second if modification of the instruction is required. The two remaining fields give either a store address or an effective shift address in shift instructions. These fields can be modified by the MCU register specified in the modifier field.

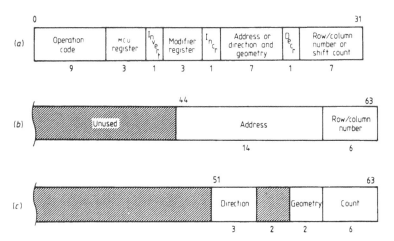

FIGURE 3.31 Instruction and modifier formats: (*a*) instruction; (*b*) address modifier; (*c*) shift modifier.

An instruction which references memory contains two 7-bit address fields. One specifies a row or column number and the other a 7-bit offset in memory. These addresses are added to the contents of the modifier register (figure 3.31(b)) to form the absolute address. A carry across into the memory address is only allowed in instructions which reference rows. For instructions which reference columns the column number is truncated to 6 bits to give a value modulo 64.

In instructions which shift data through the PE array, the two address fields specify a relative address across the array. One field is used to specify the geometry and direction of the shift, and the other is used as a shift count. Both may be modified with the contents of an MCU register having the format given in figure 3.31(c). The possible values of direction are self, N, S, E and W. Geometry has four possibilities, giving row and column geometries which may be set independently to plane or cyclic.

With either form of addressing, no modification is performed if the modifier field is set to zero. Thus register zero cannot be used as a modifier.

We will not attempt to describe all instruction types here, the interested reader should consult the DAP APAL assembly language manual (ICL 1979c).

TABLE 3.4   Summary of DAP instructions.

Register–register	Register–memory
1-bit addition Full or half add, sum to Q, carry to C	*1-bit addition to store* Full or half add, sum to store, carry to C
Vector addition Ripply carry add $Q \leftarrow Q + C$	
Transmit Register-to-register within PE, includes shift instructions	*Load/store* Load and store Q and A registers (rows, columns or entire store plane)
MCU/array Load MCU registers, broadcast or write selectively to/from Q and A	*MCU/store* Load MCU registers, broadcast or write selectively to/from store
MCU only Control, logic and arithmetic on MCU registers	

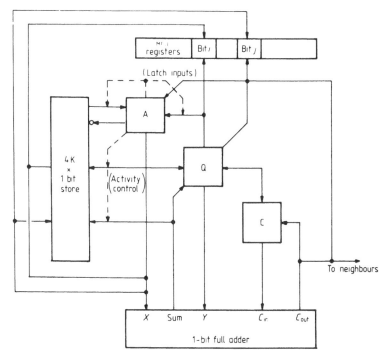

FIGURE 3.32   Conceptual data paths in DAP processing element.

What we do attempt, however, is to summarise the instructions into groups
in table 3.4 and indicate the conceptual data paths created by these
instructions in figure 3.32. In this figure the broken lines indicate control
exercised by the activity register.

The normal mode of processing on the DAP is bit-serial on 4096 words
in parallel. The microcode for addition will therefore use the hardware DO
loop facility to access consecutive bits of data, which must be stored
contiguously in memory.

Alternatively the DAP processing elements may be configured to form a
parallel or ripple carry adder (figure 3.3) (the PES may be linked in any
direction). Thus 64-bit words may be processed in parallel 64 at a time. The
ripple carry is propagated at better than one bit position in each clock period,
with at least four bit positions being guaranteed.

Figure 3.33 illustrates the data mappings of these two modes of operation,
matrix mode and vector mode. Processing word data in vector mode is always
performed one work per row in the DAP, although this is wasteful of
processing elements for short data (less than 64 in the main-frame DAP).
The RPA described in §3.5.4 is designed to make use of these unused elements
to provide a more flexible architecture. It should be noted that unless a data

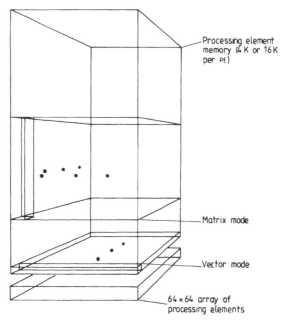

FIGURE 3.33   A diagram of the DAP processor array and memory, showing the two storage modes available; vector mode, in which the word is stored horizontally over a row of PES; and matrix mode, where the word is stored vertically over a single PE.

remapping is performed prior to the operation when processing data which does not map onto the DAP, then storage as well as processing power is lost.

At the system level on the main-frame DAP, which is closely integrated into the host system (ICL 2900), all DAP code appears to the user as a part of a mixed language system, with DAP FORTRAN subroutines being called from within the FORTRAN program executing on the 2900. As an example, consider the typical application illustrated in figure 3.34.

It can be seen that normal systems services are provided by the host system, including compilation, data set-up and analysis. A single entry point in the host FORTRAN program would then transfer control to the DAP entry subroutine and its siblings. Communications between DAP and host are by *common blocks*, which when accessed by the DAP will be loaded into the 2900 store corresponding to the DAP memory, where they are accessible to both host and DAP systems.

The language DAP FORTRAN, a parallel FORTRAN-based language, is described in §4.4.2. A DAP assembly language APAL is also provided, with interfaces to both FORTRAN and DAP FORTRAN. When using floating-point arithmetic and the standard system routines (which are highly

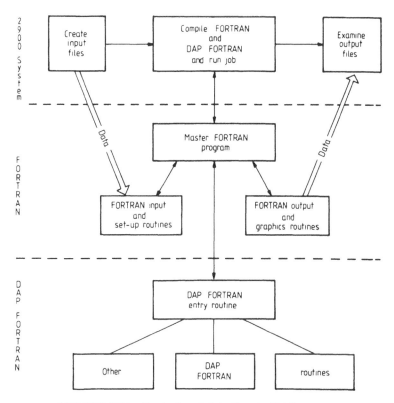

FIGURE 3.34   Control and data flow in DAP programs.

optimised), there is little benefit to be gained from using APAL. However, if algorithms can be found which exploit the bit nature of the processing elements, then orders of magnitude performance improvements may be made by coding at the assembler level (Eastwood and Jesshope 1977).

As the DAP consists of bit-serial processing elements, all arithmetic must be built up in software. For the mode of processing on the DAP matrix arithmetic will be based on sequences of single-bit operations. This means that there will be a strong performance dependency on the word length of the representation of the data. For integer arithmetic the proportionality is by word length and word length squared for addition and multiplication respectively. For floating-point calculations these proportionalities are masked somewhat by the overheads of exponent and mantissa manipulation. However in both cases, this word-length dependency is both the good news and the bad news concerning DAP performance. The bad news is that applications requiring extreme accuracy do rather poorly on the DAP; the good news is complementary, with some applications which require only short word lengths giving many hundreds of millions of operations per second.

For most problems, 32-bit floating-point calculation gives acceptable accuracy and we have therefore given timings for this precision in table 3.5. These are taken from Reddaway (1979) and represent the timings of the optimised system routines provided for arithmetic. These figures may also be used to give fairly accurate timing estimates of DAP FORTRAN programs, as in the DAP the overheads from high-level language manipulation are minimal. This is due to the rapid rate at which the DAP can perform data manipulation and is highlighted by the timing for assignment in table 3.5.

Some of these arithmetic routines have been timed in DAP FORTRAN programs, to establish how large the overhead is for a high-level language. These measured timings are given in table 3.6. It can be seen that the overheads range from less than 10% for single matrix operations (4096 elements) and increase to a little over 20% for DO loops performing 10 matrix operations (40 960 elements).

Great savings can sometimes be made if bit-level algorithms can be used. Some examples of this are illustrated in table 3.5, where some operations give seemingly contradictory execution rates. Perhaps the most noticeable is the sum over a matrix of 4096 elements. One could expect this result to take about 12 floating-point additions ($\log_2 4096 = 12$); however it takes less than two! Other examples are square root and logarithm, which normally

TABLE 3.5   DAP arithmetic routines (32-bit precision). X, Y and Z are real (4096 elements); IX, IY and IZ are integer (4096 elements); S is a real scalar.

Operation	Time (μs)	Processing rate (Mop/s)
$Z \leftarrow X$	17	241
$Z \leftarrow X*S$	40–130	32–102
$Z \leftarrow X**2$	125	33
$Z \leftarrow X + Y$	150	27
$Z \leftarrow SQRT(X)$	170	24
$Z \leftarrow X*Y$	250	16
$Z \leftarrow LOG(X)$	285	14
$Z \leftarrow X/Y$	330	12
$Z \leftarrow MAX(X,Y)$	33	124
$Z \leftarrow MOD(Z)$	1	4096
$IZ \leftarrow IX + IY$	22	186
$S \leftarrow SUM(X)$	280	175
$S \leftarrow MAX(X)$	48	85

TABLE 3.6 Arithmetic time in DAP FORTRAN (32-bit precision). X, Y and Z are real (4096 or 40 960 elements); S is a real scalar.

Operation	4096 elements		40 960 elements	
	Time (μs)	Rate (Mflop/s)	Time (μs)	Rate (Mflop/s)
Z ← X + Y	152	27	1848	22
Z ← X∗Y	272	15	3048	13
Z ← X∗S†	112–200	20–37	1368–2272	18–30
Z ← X∗∗2	152	27	1816	23
Z ← SQRT(X)	192	21	2208	19
Z ← X/Y	376	11	4080	10

† Uses S = 2 and S = e for minimum and typical figures respectively.

take many floating-point multiplications when performed iteratively. Some of the algorithms used for these functions are described in Flanders *et al* (1977) and Gostick (1979).

3.4.3 The Burroughs BSP

Unlike the DAP, the BSP is an array of complex processors, capable of performing floating-point operations on 48-bit words. It thus draws heavily on Burroughs previous experience in its contractor role when designing and constructing the ILLIAC IV. Although the design has many interesting features, it suffered from a reorganisation within the then ailing Burroughs company and never went into production. This was probably a fair decision, as the machine did not compete in performance with the then established CRAY-1. Indeed it did not even compete with its predecessor the ILLIAC IV, with a maximum processing rate of only 50 Mflop/s compared with the 80–100 possible on the ILLIAC IV. Despite this disappointing maximum performance, the design tackled a number of major deficiencies in its predecessors and is of sufficient interest to be included here.

One of the major problems tackled by Burroughs was that of maintaining a large computation data base, which was matched in performance to give continuous use of the processor array. This was a severe limitation on the ILLIAC IV (Feierbach and Stevenson 1979a), which had only 128 Kwords of random-access memory backed up with a large and unreliable disc memory. In contrast the BSP has from 1 to 8 Mwords of random-access memory, backed up by a fast electronic file memory. In addition to its size, Burroughs have organised the central BSP memory so that many regular array subsets can be accessed in parallel without conflict.

Other lessons that Burroughs have learnt from their ILLIAC IV experiences concern the organisation of the control processor, another weak point in the design of the ILLIAC. Whereas the ILLIAC IV had a small buffer memory and only limited processing power, the BSP provides a 256 Kword memory and a more complex scalar processor in addition to the control processor. This memory is used for code and data.

The details of the BSP described here are taken from the pre-production prototype (Burroughs 1977a–d, Austin 1979), which was built and tested by 1980. Figure 3.35 shows the overall configuration of a BSP system.

The interface between the system manager and the BSP provides two data highways, one slow and one fast. The slow highway (500 Kbyte/s) interfaces the I/O processor directly with the BSP control unit and is used for passing messages and control between the two systems. The second fast data highway ($\frac{1}{4}$ Mwords/s) interfaces the I/O processor with the file memory controller and is used for passing code and data files for processing on the BSP. The file memory provides a buffered interface between front-end and back-end systems for high bandwidth communication.

The file memory is one of the three major components of the BSP. The

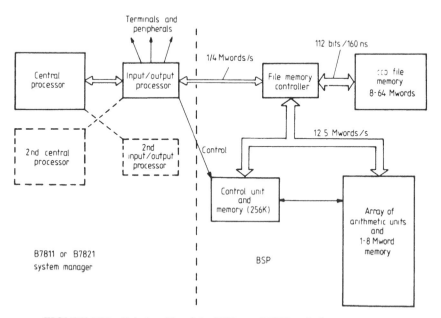

FIGURE 3.35   Relationship of the BSP to a B7800 main-frame computer.

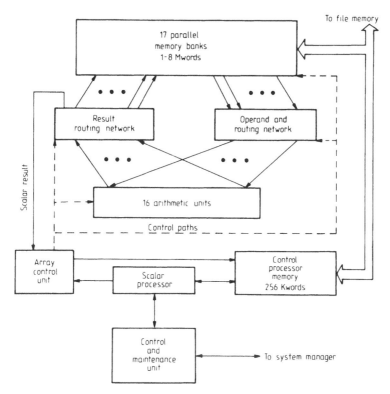

FIGURE 3.36   BSP control processor and array section, showing major components and data paths.

other two, the control processor and processor array are illustrated in figure 3.36 and described below.

The control processor portion of the BSP contained four asynchronous units, which between them provided array control, job scheduling, I/O and file memory management, error management and finally communication of commands between the system manager and BSP.

The control processor had 256 Kwords of 160 ns cycle MOS memory. This memory held both scalar and vector instructions and scalar data and had a data highway to the file memory.

The scalar processing unit was a conventional register oriented processor, which used identical hardware to that found in the 16 arithmetic units (AUS) of the array. However it differed from the AUs in that it had its own instruction processor, which read and decoded instructions stored in the control processor memory.

The scalar processor had 16 48-bit general-purpose registers and was clocked with a cycle of 80 ns. It performed both numeric and non-numeric operations, performing floating-point arithmetic at a maximum rate of 1.5 Mflop/s. One of the major tasks of the scalar processor is to pre-process vector instructions. This includes optimisations, checking for vector hazards and the insertion of various data fields into the vector instruction format. These operations were performed using sixteen 120-bit vector data buffers, which each held a vector instruction descriptor. Completed instructions were passed onto the array control unit, which queued instructions, finally decoding and broadcasting microcode control to the array section of the BSP.

It should be noted that the control processor was completely overlapped with the array section of the BSP. When initialised, the queue of instructions in the array control unit would keep the array section processing continuously. The scalar processor independently processed scalar instructions and pre-processed vector instructions. This approach was pioneered in the ILLIAC IV design, where careful assembler coding was required to exploit the overlap. In the BSP the scalar processor and array control unit hardware provided the necessary run-time operations.

The array section of the BSP contained four units which taken cyclically form a five-element pipeline. The operations or tasks performed by this pipeline are as follows:

  (a) read parallel memory for operands;
  (b) align operand data with arithmetic units;
  (c) perform operation;
  (d) align result data with memory banks;
  (e) store results in parallel memory.

These operations are overlapped with successive sets of 16 elements taken from the vector instructions processed by the array control unit. The number of clock periods (160 ns) required for each of these tasks is variable, depending on the number of operands and operations performed. The overlap of this pipeline is controlled using different microcode fragments called templates, which describe the amount of overlap in these tasks.

The arithmetic units were general-purpose and driven by a single broadcast micro-instruction sequence. The control word is over 100 bits wide and provides direct access to the primitive functions of the AUs, which in addition to floating-point operators have a comprehensive set of field manipulation and editing operators. These include special FORTRAN format conversion operators.

Floating-point addition and multiplication both required two clock periods of 160 ns, giving a maximum processing rate for one AU of $3\frac{1}{8}$ Mflop/s. The array of 16 thus gives the peak rate of 50 Mflop/s noted earlier. Floating-point division and square rooting are both implemented using a Newton–Raphson iteration and read only memories (ROMs), which supplied the first approximation.

The 48-bit word, used historically by Burroughs, has 36 bits of significant mantissa giving a precision of 10–11 decimal digits. An 11-bit binary exponent gives a range of $\pm 10^{+307}$. Double-length accumulators and registers permit the hardware implementation of double-precision arithmetic.

The parallel memory in the BSP array section contained 17 memory banks, with a cycle time of 160 ns. 17 is the next highest prime number greater than 16 (the number of processors). This arrangement when coupled with full connection networks provides conflict-free access to all linear vectors, provided that the increment between elements in memory is not a multiple of the prime number, 17. This access technique has been described in §3.3.8.

The connection networks used in the BSP were full crossbar switches and contained error detection and correction logic. These switches also had a cycle time of 160 ns.

Vector instructions processed by the array control unit could have from one to four operations (one to five sets of operands) producing a single vector or scalar result. The use of such high-level instructions enabled more optimisation in the use of the hardware, not only in register usage but in matching templates to maximise the overlap through the pipeline. It can be seen that an overall balance of the elements of the pipeline was achieved for triadic operations, e.g.

$$A = B \text{ op } C \text{ op } D.$$

Here the four memory references of 160 ns and two operations ($+$, $-$ or $*$) at 320 ns can be fully overlapped. However, the input and output routing networks contain some slack, using only three and one of the four cycles respectively.

The BSP was designed as a high-level language processor (FORTRAN), which provided hardware optimisations and transformations of user code at run time. A complicated control processor was necessary to maintain a high loading of the pipeline network:

    (a) fetch operands;
    (b) align operands;
    (c) perform operation;
    (d) align results;
    (e) store results.

From the FORTRAN language viewpoint, the construct most likely to optimise the use of multiple arithmetic units is a nested set of DO loops. In fact BSP vector operations are based on one or two nested DO loops.

However, most DO loops contain some run-time resolution of parameters and because of this the BSP high-level descriptor format for vector operations may also contain run-time parameters.

These 'machine' instructions had the form of memory-to-memory vector operations whose operands were arrays of arbitrary length (volume) having one or two dimensions. A single instruction contained up to four possible different operators and thus up to five different operands. For example the FORTRAN code

```
      DO 10 I=2, N-1
      DO 10 J=2, N-1
10    NEWB(I,J) = (B(I-1,J) + B(I,J-1) + B(I,J+1) + B(I+1,J))/4.0
```

would compile into one BSP high-level instruction. Depending on N, a run-time variable, this instruction would generate a sequence of control signals for each set of 16 elements. These control signals controlled memory, both alignment networks and the arithmetic units. What is more, the control signals were timed so as to optimise the overlap in the array section pipeline.

In addition to the monadic, dyadic, tetradic and pentadic operations provided in the BSP instruction repertoire, there were also a number of special vector instruction forms. The most important of these are summarised in table 3.7.

The format of BSP instructions is given in figure 3.37 and includes loop length fields, instruction form and operations used, and descriptors for each operand and result. Optionally, boolean vectors for operands and results could also be specified. We will use the example in figure 3.37 to illustrate

TABLE 3.7   BSP special vector operation forms.

Vector scalar operations	$A \leftarrow B$ op S
Single precision reduction	$S \leftarrow A_1 + A_2 + A_3 + \ldots + A_n$
Double precision reduction	As above but all in double precision
Sequence	$A_i \leftarrow A_1 + A_2 + \ldots + A_i$
Compress	$A \leftarrow B$ (under boolean control)
Expand	$A \leftarrow B$ (under boolean control)
Merge	$A \leftarrow B$ or C (under boolean control)
Random fetch	$A \leftarrow B(I)$
Random store	$A(I) \leftarrow B$
Dot product	$S \leftarrow A_1 * B_1 + A_2 * B_2 + \ldots + A_n * B_n$
Recurrence (all)	$A_1 \leftarrow C_1, A_i \leftarrow A_{i-1} * B_i + C_i$
Recurrence (last)	Last term in above sequence
Data transfer	Control memory $< - >$ parallel memory

```
DO 10 K=1, 32
DO 10 J=1, 128
DO 10 I=1, 64
10    C(I,J) = C(I,J) + A(I,K)*B(K,J)
```

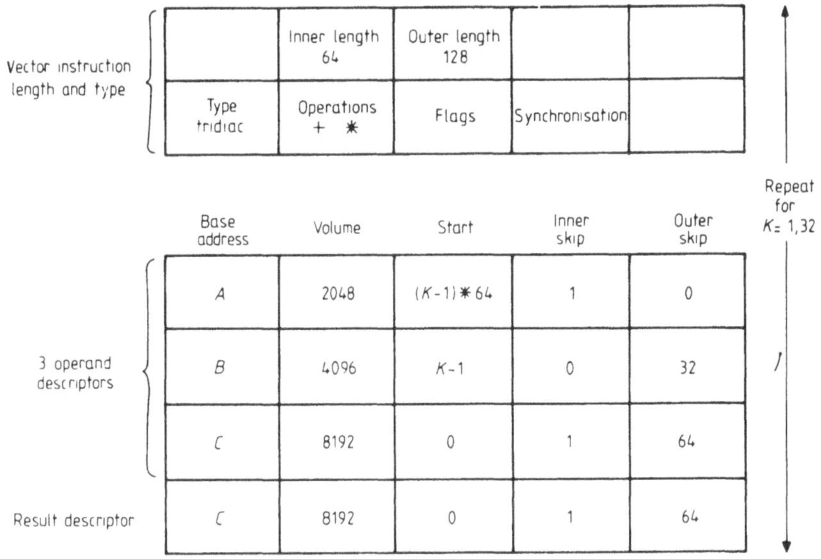

	Inner length 64	Outer length 128		
Type tridiac	Operations + *	Flags	Synchronisation	

Vector instruction length and type

Repeat for K= 1,32

	Base address	Volume	Start	Inner skip	Outer skip
	A	2048	(K-1)*64	1	0
	B	4096	K-1	0	32
	C	8192	0	1	64

3 operand descriptors

Result descriptor

	C	8192	0	1	64

FIGURE 3.37 Format of the BSP high-level machine language and vector descriptors: (*top*) FORTRAN code; (*bottom*) BSP vector form.

the control flow in processing this instruction, which performs matrix multiplication.

The first thing to notice is that the code specifies a triple nested DO loop. Thus this single instruction would be placed within a loop in scalar code which when executed on the scalar processor would generate 32 copies of this instruction with the parameterised start fields resolved. These fields would be planted by the scalar processor while the instruction resided in the 120-bit vector data buffers. When each completed instruction was assembled it would be passed on to the array control unit for further processing and for execution.

The first stage of the control unit pipe assembled the sequence of descriptors provided by the scalar processing unit into a single global description of the operations. It also established any dependencies between successive vector instructions. When this processing was complete the finished package was placed in a queue called the template descriptor memory, where it awaited execution. At this stage the description of the operation is still at the program array level. This was translated into operations on sets of 16 elements only at the final stage of instruction processing, performed by the template control unit, which selected the appropriate micro-instruction sequence for the operation, and then cycled through it while incrementing and decrementing addresses and loop counts.

For any given instruction form there are a number of alternative micro-instruction control sequences held in ROM. These templates and the template control processor select the optimum template at each stage of the instruction execution. These control signals were called templates as they define areas in the array pipeline—time space, which had to be matched as closely as possible in order to maximise the use of the array pipe.

Figure 3.38 illustrates two templates which can be used to execute the instruction of our example (figure 3.37). They define the number of cycles required for each section of the pipe and how these cycles overlap. Because the pipeline formed a ring network and because a single memory was used

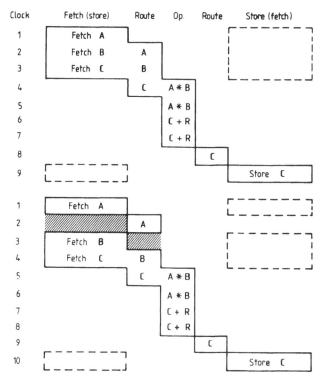

FIGURE 3.38   Two templates for the execution of $C = C + A*B$ on the BSP. The templates show the clock periods required in the BSP pipeline. Fetch and store are performed in the same memory and cannot be overlapped, hence the second template which reserves a store cycle for a previous template.

for both fetching and storing, reservations had to be made which are illustrated by the broken lines. These areas must not be overlapped. The second template in figure 3.38 has a space between fetching A and B; this is to accommodate a reservation from an execution of a previous template.

Returning to our example we can now see how the template control unit selected the appropriate template and incremented the addresses. This is shown in figure 3.39. We assume that the first template starts up the pipe. The control unit will therefore select the first of the two alternative templates shown in figure 3.38, as this minimises the number of clocks required. There is no reservation required when fetching the next set of 16 operands, so the same template is used for a second time. Note that the arithmetic unit is the critical resource here. All successive templates are of the second variety, which allows for the reservation required by the store cycle. It can be seen that after this point all memory and all arithmetic unit cycles are being used.

Figure 3.39 also illustrates a common design constraint when overlapping input and output to memory. It can be seen that although all memory and

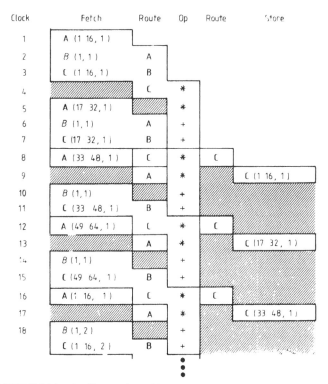

FIGURE 3.39 An illustration of how templates can be overlapped. The operation performed is matrix multiplication (see also figures 3.37 and 3.38).

arithmetic unit cycles are being used, the routing networks are both under-utilised. Although one cycle is used for each memory cycle, routing and memory operations cannot both be overlapped using a single switch. This is because routing is a post-memory operation on fetch and a pre-memory operation on store. Thus both switching networks are required and both will always be under-utilised, leaving the control unit only memory and arithmetic unit cycles to optimise.

What happened between instructions was important to the BSP's performance on short vectors. Let us assume that the example code for matrix multiplication was followed by the following loop:

```
      D0 20 I=1, 200
20    D(I) = E(I)*F(I)
```

The BSP array control unit would determine whether there were any

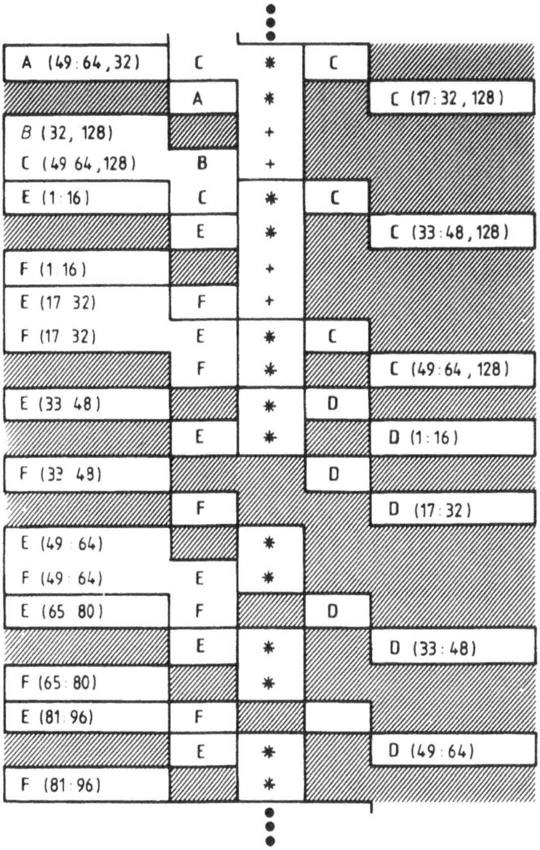

FIGURE 3.40 An illustration of how different operations may be overlapped. Here the triadic operation continued from figure 3.39 is overlapped with a dyadic operation using the templates from figure 3.41.

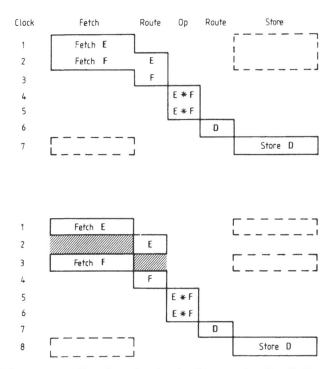

FIGURE 3.41   Templates for the dyadic operation $\mathbf{D} = \mathbf{E} * \mathbf{F}$ on the BSP.

sequential dependencies between the successive instructions and if there are no vector hazards it would overlap the instructions. In our example this is shown in figure 3.40. It can be seen that when this vector instruction is fully operational, it is memory limited and that during the overlap, using the two templates in figure 3.41, all memory cycles have been used. Similarly, had the operation been arithmetic unit dominated, then the optimisation would have utilised all arithmetic unit cycles.

The whole concept of BSP was to provide continuous high performance without interruption from I/O and system. Thus looking at the BSP performance figures without this perspective can be a little misleading. Burroughs had not aimed for massive performance figures which could not be sustained, but instead designed a computer which could maintain 50–100% of its maximum performance continuously. It would do this despite being programmed exclusively in FORTRAN.

The basic operation times for the BSP arithmetic units are given in table 3.8. It should be noted that all overheads associated with these operations (e.g. memory and data alignment) are overlapped to some extent. Therefore these figures represented sustainable performance figures. For example for

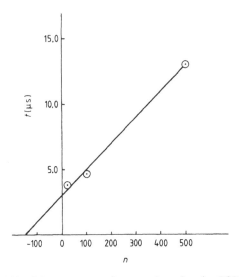

FIGURE 3.42   Measurement of $n_{1/2}$ and r_∞ for the BSP, for a single vector instruction with the pipeline flushed before and after the operation; $n_{1/2} = 150$, $r_\infty = 50$ Mflop/s. (Timings courtesy of Burroughs Corporation.)

TABLE 3.8   Operations on the BSP.

Operation	Time for 16 result (ns)	Dyadic execution rate (Mflop/s)	Triadic execution rate (Mflop/s)
+	320	33	50
−	320	33	50
*	320	33	50
÷	1280	12.5	12.5
$\sqrt{}$	2080	7.7	7.7

+, − and *, the BSP can compute memory-to-memory on triadic operations at a continuous 50 Mflop/s. For dyadic operations, which were memory limited, a continous rate of only 33 Mflop/s was possible.

Although the overheads are overlapped in a pipeline, no or very little start-up time was associated with vector operations. The reason for this was that the BSP was capable of overlapping even dissimilar operations. This is illustrated in figures 3.42 and 3.43, which show real timings for vector operations of different lenghts. In figure 3.42 the pipeline was artificially flushed before and after a single vector operation. It can be seen that an $n_{1/2}$ of 150 is exhibited due to the start-up time associated with the ring network pipeline. However in normal use, succeeding vector operations are overlapped using template descriptors. The effects of this are shown in figure 3.44, where it can be seen that $n_{1/2}$ has been reduced to 25.

3.4.4 The Denelcor HEP

The Denelcor Heterogeneous Element Processor (HEP) was designed by Burton Smith (1978). Although it was discontinued in 1985, due to the financial problems of the company, the architecture (like the BSP) is of sufficient interest to warrant a detailed discussion in this chapter. At the outermost level (figure 3.45) the HEP comprises up to 16 process execution modules (PEMS) connected via a packet-switched network to up to 128 data memory modules (DMMS). The 16 PEMS execute separate programs but all may access data in any of the 128 DMMS, which together form a large shared memory. Each DMM contains up to one million 64 bit words, with an access

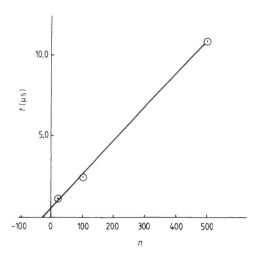

FIGURE 3.43 Measurements of $n_{1/2}$ and r_∞ for the BSP in the steady state, with successive vector operations being overlapped; $n_{1/2} = 25$, $r_\infty = 48$ Mflop/s. (Timings courtesy of Burroughs Corporation.)

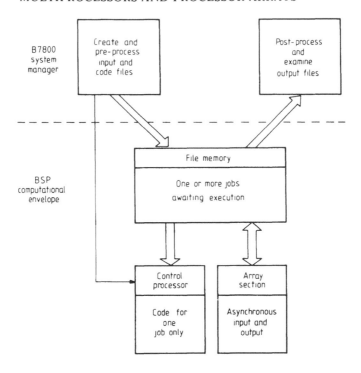

FIGURE 3.44   Control and data flow in a typical BSP program.

time of 50 ns. A PEM accesses memory by sending a 128-bit 'request packet' containing the address of data (i.e. DMM and address within DMM). This traverses the multistage network to the correct DMM, the required data is accessed and entered into the packet, which is returned to the requesting PEM again via the nodes of the switch. The transit time of the packet is 50 ns per node and a typical time between requesting data and it being available in the PEM is 2.4 μs.

At this level of description the HEP can be considered as a shared-memory, multistage-switched MIMD system, as described in the classification in §1.2.6.

In any MIMD system the most important aspects of the design concern the mechanisms for synchronisation and the related protocols for multiple access to shared memory. In the Denelcor HEP both of these mechanisms are provided by a full/empty tag associated with all words in memory. This device provides a handshake protocol on every word of data in the entire

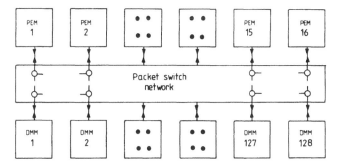

FIGURE 3.45   A diagram showing the Denelcor HEP system, with up to 16 PEMs connected by a packet-switching network to up to 128 DMMs.

system, as in normal use; data may not be read from a location unless this bit signifies 'full' and data may not be written to a location unless the bit signifies 'empty'. Normally, after a read operation the bit is set to 'empty' and after a write operation the bit is set to 'full'. With a little thought it can be seen that this mechanism is sufficient to implement the protocol required for an Occam channel (see §4.4.2). This is sufficient therefore to implement all synchronisation and data protection primitives.

The structure of the individual PEM is particularly interesting because each may process up to 50 user instruction streams, from up to seven user tasks. The multiple instruction streams share a single eight-stage instruction execution pipeline with a great deal of hardware support. Instruction streams or processes are effectively switched on every clock cycle, so the distribution of processor resources is exceedingly fair. The organisation of this pipeline is shown in figure 3.46. Thus a single PEM is itself an example of a pipelined MIMD computer (see figure 1.8). The PEM is controlled by a queue of *process tags* (one for each instruction stream), which rotate around a control loop. The tag contains the *program status* word for the instruction stream or process that it represents. The process tag contains among other things the program counter for that process, which is updated on each pass through the INC PSW box. The instructions themselves are stored in a 1 Mword *program memory*, and local data is held either in 2048 64-bit registers or in 4096 read-only memory locations, which are intended for frequently used constants. Separate pipelined functional units are provided for floating multiply, add and divide, integer (IFU) and create operations (CFU), and references to shared memory (SFU). As the process tags rotate around the control loop, the

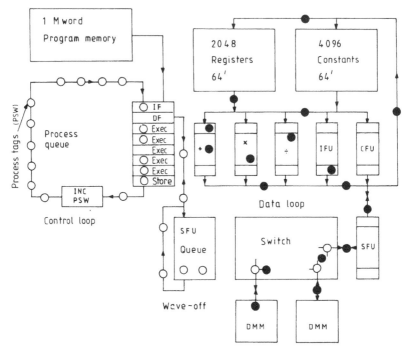

Exec = execute operation

FIGURE 3.46 A diagram illustrating the operation of a single HEP PEM. The diagram shows the control loop on the left and the execution loop on the right. The asynchronous switch is decoupled from the synchronous operation of the functional units. (INC denotes increment, IF denotes instruction fetch and DF denotes data fetch. PSW denotes process status word.)

appropriate data rotates around the data loop. Data leave the registers, pass through the appropriate pipeline, and the result is returned to the register memory under the control of the passage of the process tag through the instruction pipeline. As the process tag passes through the first stage of the instruction pipeline, the instruction is brought from the program memory; during the second stage the data referenced by the instruction are fetched from the register or constant storage to the appropriate functional unit; during stages three to seven the data pass through the unit and the function is completed; and finally in the eighth stage, the result of the operation is stored back in the registers.

The above description applies only to instructions that access data from the registers or constant store. If, however, an instruction refers to data stored in the shared memory (DMMs), then the process tag is removed from the process queue and placed in the SFU queue while data is being retrieved from shared memory. The tag is said to be *waved off* and its place in the process queue can be used by another process or the process queue can be shortened. This mechanism allows the process queue to contain only active processes. In this way each process status word (PSW) in a process tag can always have its program counter updated on each pass around the control queue. Thus all operations are synchronous, with the exception of access to shared memory. This is handled in the SFU queue, which re-inserts the waved-off process into the process queue only when the data packet has arrived from the appropriate DDM.

A new instruction stream (called a process) is initiated with the machine instruction (or FORTRAN statement) CREATE. This uses the CFU to create a new process tag and to insert it into the process queue. It is then said to be an active process. It remains so, until either the process is completed (i.e. a successful QUIT instruction or FORTRAN RETURN statement is executed), or the tag leaves the process queue while waiting for data from the shared memory (wave-off). Unlike the transputer, there is no mechanism for a process to be passively delayed until a time-out.

It is instructive to examine the performance of the instruction pipeline as the number of active processes increases. In the case of a single task (collection of user processes) the process queue may be thought of as a circular queue of up to 50 tags. The minimum length of the process queue is eight and if less than eight processes are active, there will be some empty slots in the queue. These slots can be filled by creating additional processes (until eight are active) without changing the length of the queue, or the timing of the other processes. Thus as more processes are created, the total instruction processing rate increases linearly with the number of active processes, because there are less empty slots in the process queue and hence more instructions being processed in the same time. This continues until there are eight active processes and the pipeline is full. In this condition the processing rate is at a maximum. One instruction leaves the pipeline every 100 ns, giving a maximum rate of 10 Mips per PEM, or 160 Mips for a full system of 16 PEMS. If more than eight active processes are created, the only effect is to increase the length of the process queue. In this way the instruction processing rate for each process decreases, leaving the total instruction processing rate constant at 10 Mips per PEM.

To summarise this mechanism, we would expect the processing rate to rise linearly with number of active processes until the instruction pipeline is full and then to remain constant. The instruction pipeline becomes full when there are eight active processes. Since most active processes will make some use of the shared memory and hence become waved off, then in practice more than eight active processes will be required to maintain a full instruction pipeline. In fact, in FORTRAN programs, about 12 to 14 active processes are required to maintain a full instruction pipeline, which allows from four to six waved-off processes.

The beauty of this system is that even if a particular user is unable to provide sufficient processes to fill the instruction pipeline, the pipeline will be automatically filled with processes from other tasks or jobs. Thus the HEP can effectively multitask at the single-instruction level, in a single processor with no processor overhead. There is of course an overhead for process creation and deletion.

In order to interpret the effective performance of the HEP on applications dominated by floating-point operations, it is necessary to know how many instructions are required per floating-point operation. This variable is called i_3 and modifies the asymptotic performance per PEM as indicated below:

$$r_\infty = 10/i_3 \text{ Mflop/s.}$$

Clearly i_3 must be minimised in order to maximise the useful floating-point performance. Consider the case of a dyadic vector operation in which all variables are stored in shared memory. This may be expressed by the loop:

DO 10 I = 1, N

10   A(I) = B(I)*C(I)

Since the HEP has no vector instructions, this loop must be programmed with scalar instructions. If the loop were coded in assembler it could be coded in six instructions, namely:

(a) fetch B(I) from shared memory to register;
(b) fetch C(I) from shared memory to register;
(c) perform a register-to-register scalar multiply;
(d) store the result to A(I) in shared memory;
(e) increment I;
(f) test and branch to start of loop.

In this case $i_3 = 6$ and $r_\infty = 1.7$ Mflop/s. If, however, all variables were stored in registers, instructions (a), (b) and (d) would be unnecessary, making $i_3 = 3$ and doubling the asymptotic performance to 3.3 Mflop/s. Further optimisations could be made by in-line coding (loop unrolling), giving processing rates of 4.5 and 6.7 Mflop/s for memory and register operations, using four operations per loop. Sorensen (1984) and Dongarra and Sorensen (1985) have shown that a performance of 5 to 6 Mflop/s can be achieved for a variety of common matrix problems by these methods and the intelligent use of the PEM registers as temporary storage of intermediate vector results. This reduces the number of references to shared memory and the value of i_3.

Performance may also be degraded when using many separate processes to solve a single problem, because one must also consider the time required to synchronise the separate instruction streams, as discussed in §1.3.6. To measure the synchronisation overhead on the HEP the $(r_\infty, s_{1/2})$ benchmark has been executed. In this benchmark, the work of a memory-to-memory dyadic vector multiply operation with s floating-point operations is divided among P processes. The time, t, of execution is fitted to

$$t = r_\infty^{-1}(s + s_{1/2})$$

where now both r_∞ and $s_{1/2}$ are functions of P. The measured time includes the time to start P instructions at each synchronisation point, and the time to detect that all streams have finished their work.

A detailed theoretical analysis of the timing of the above benchmark is given in Hockney (1984a), and measured values of r_∞ and $s_{1/2}$ for a variety of different cases are reported in Hockney and Snelling (1984), and more fully in Hockney (1985c). The results show that for a fixed number of processes, t is a linear function of s and therefore fitted well by the above model. On the other hand, figure 3.47 shows the time as a function of the number of processes, for a fixed s. In this case, the time is seen first to decrease to a minimum, as slots in the instruction pipeline are filled. At $P = P_{opt} = 12$ the time is a minimum. If P is increased further, the time gradually increases, due to the greater time required to synchronise a larger number of processes. It is clear from this example that there is no virtue in creating more than 12–14 processes on a single PEM, even if the logic of the program lends itself to a larger number.

Figure 3.48 shows values of r_∞ and $s_{1/2}$ as a function of P for fork/join synchronisation. In this method the processes are created with the FORTRAN CREATE statement and allowed to die as they finish their work. The asymptotic performance is seen to rise linearly to its maximum value and then to remain constant, as expected from the earlier discussion. The maximum value of $r_\infty = 1.7$ Mflop/s is consistent with $i_3 = 6$ as described previously. The theoretical analysis (full curve) predicts that $s_{1/2}$ should rise quadratically with P, and this is observed. The best operating point, shown by the arrow, occurs when the maximum performance has just been reached.

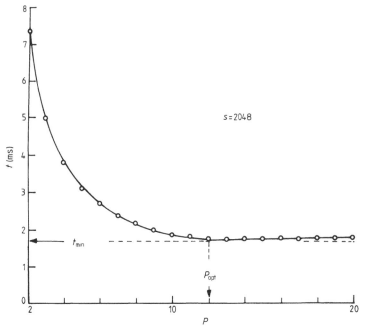

FIGURE 3.47   A diagram showing the optimal number of processes required to drive a single PEM.

Any further increase in P increases the overhead (i.e. the value of $s_{1/2}$) without improving the value of r_∞. At the optimum point we have:

$$\text{FORK/JOIN} \quad r_\infty = 1.7 \text{ Mflop/s} \quad \text{and} \quad s_{1/2} = 828$$

The above method of synchronisation requires the process to be created and destroyed dynamically. A process is created when required by the CREATE statement (FORK) and allowed to die on completion (JOIN). This is obviously an inefficient method of obtaining synchronisation in the program and therefore leads to high values of $s_{1/2}$. The alternative is to create static processes once only, and to achieve synchronisation using other means. The hardware provides the mechanism for synchronisation in its full/empty tags on each location in shared memory. Therefore, using a shared variable, synchronisation can be achieved using a semaphore. A counter initialised to the number of concurrent processes used can be decremented once by each process on completion of its share of the work. All completed processes then wait for the variable to become zero, at which point synchronisation has been achieved.

This is a software implementation of barrier synchronisation, which has substantially less overhead than dynamically creating and destroying

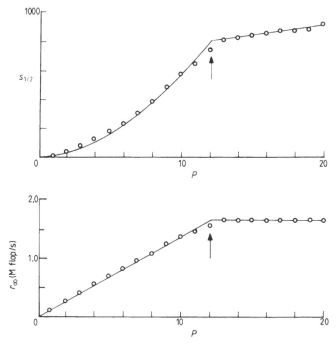

FIGURE 3.48   Diagrams showing $s_{1/2}$ and r_∞ against the number of processes, P, in the process queue.

processes

$$\text{BARRIER}\quad r_\infty = 1.7 \text{ Mflop/s}\quad\text{and}\quad s_{1/2} = 230.$$

In order to test the effect of increasing the amount of arithmetic per memory reference, statement 10 of the test loop (1.5) was replaced with:

IF (A(I).LE.0.0) THEN

C(I) = SIN(A(I)*EXP(B(I)))

ELSE

C(I) = COS(A(I)*EXP(B(I)))

ENDIF

In this case all arithmetic associated with the evaluation of the SIN, COS and EXP functions is performed between temporary variables stored in registers. The effective performance in Mflop/s is increased about three-fold to:

$$\text{FORK/JOIN}\quad r_\infty = 4.8 \text{ Mflop/s}\quad\text{and}\quad s_{1/2} = 710$$
$$\text{BARRIER}\quad r_\infty = 4.8 \text{ Mflop/s}\quad\text{and}\quad s_{1/2} = 190$$

It is clear that in this example the average number of instructions per floating-point operation has been reduced from six to two, compared with the vector dyadic kernel.

3.5   REPLICATION—A FUTURE WITH vlsi

3.5.1   An auspicious start

The advent of lsi and vlsi chips has given a large boost to the research and development of array processor and multiprocessor architectures. A review of multiprocessor design has already been given in Chapter 1 and to some extent the availability of cheap and powerful microprocessors has catalysed this research. However, it is in the area of array processors that the true impact of vlsi is being seen. As outlined in Chapter 6, vlsi chips need to contain great regularity and repetition in order to reduce design time and optimise the utilisation of silicon area. Bit-serial processor arrays are therefore ideal candidates for vlsi (and even wafer-scale integration (wsi)) as they consist of a large number of relatively simple circuits replicated in a very regular manner, usually in a two-dimensional grid.

This is confirmed by the growing body of literature describing projects based on grid-connected arrays. University College, London, for example, has continued to develop designs for the CLIP (Fountain and Goetcherian 1980, Fountain 1983) and ICL are continuing to develope their DAPs, which are described in more detail in §3.5.2. Other projects include the GEC GRID (Robinson and Moore 1982, Arvind *et al* 1983), the LIPP project at Linkoping University, Sweden (Ericsson and Danielson 1983), the NTT APP (Komdo *et al* 1983) in Japan, the connection machine from Thinking Machines Corporation (which is described in §3.5.3) and the RPA project at Southampton University (described in §3.5.4). All of these projects have developed or are developing vlsi chips which contain from 16 to 64 single-bit processors.

Indeed, there is a bit-serial parallel processing chip on the market called the GAP (NCR 1984), which contains 96 single-bit pes. It is a 6×12 array, with nearest-neighbour interconnections, which can be cascaded into larger systems. Its major limitation is the limited on-chip memory, 128 bits per pe and the lack of a pin per pe for connecting external ram.

Another important development, which is set to make a major impact in the field of multiprocessor systems is the INMOS transputer (INMOS 1984). This chip can either be viewed as a state-of-the-art 32-bit microprocessor, as a new generation systems building block, or as a hardware realisation of an

OCCAM process. OCCAM (INMOS 1984) is a parallel processing language based on CSP (Hoare 1986) and is described in some detail in §4.4.2.

What distinguishes the INMOS transputer from its competitors is that it has been designed to exploit vlsi. Like some other recent microprocessors it has a reduced instruction set (risc) architecture, and the silicon real estate thus saved has been used for a large area of on-chip ram, and (of equal importance) a communications system for providing intertransputer communications. The communications system is a direct implementation of the point-to-point asynchronous communications found between OCCAM processes. Parallel OCCAM processes can therefore be directly mapped onto a connected set of transputers. The transputer and related systems are described in more detail in §3.5.5.

Already many projects are underway which hope to exploit the transputer in parallel systems. These include at least two large Alvey projects, a major ESPRIT project originating from Southampton University and many commercially funded projects. It was one of Iann Barron's objectives (he is one of the co-founders of INMOS) that the transputer should sell in volumes comparable to those of memory chips and thus exploit the economies of scale. If this were to happen, it would provide a great deal of processing power (10 Mips) at the cost of a memory chip (about $10). The recently announced T800 floating-point transputer provides 1–2 Mflop/s, and if this chip can be brought down even to the $50 level, then the impact on high-performance computers cannot be underestimated. Already five European and at least two US manufacturers have products on the market, based on multiple transputers, which provide a great deal of computing power at a very realistic cost. As the cost of transputers and hence systems is reduced, parallelism in the form of replicated systems will certainly come of age.

INMOS have been very thorough in their forward planning of transputer products and, to a large extent, systems containing transputers should be very future-proof. This is achieved by making the communications links run at standard speeds (5, 10 and 20 MHz) regardless of processor speed and thus providing a standard and simple interface between generations of products. The first transputer product, the T414, and its successor, the T800, are described in more detail in §3.5.5 and OCCAM is introduced in §4.4.2. The two sections should be read together to obtain a full understanding of the transputer systems concepts. Section 3.5.5 also considers some of the issues involved in building systems in transputers, with illustrations from the transputer projects at Southampton University. It should be noted that the RPA described in §3.5.4 also uses transputers as host and control computers for the array, thus taking full advantage of the process parallelism inherent in OCCAM and the standard interfaces provided by the transputer links.

3.5.2   LSI DAPs and beyond

In 1985 the first of a second generation of DAPs, built from LSI technology, was delivered to a customer site—the Royal Signals and Radar Establishment (RSRE), Malvern, took delivery of the first 32×32 prototype for radar and other signal processing applications. In 1986 the first commercial prototype version of the same machine was delivered to the DAP Support Unit at QMC. This prorotype is interfaced to an ICL Perq single-user system running PNX, ICL's implementation of UNIX for this machine. Since these deliveries, ICL has separated off its commercial development in DAP technology into a start-up company called Active Memory Technology Ltd (AMT). AMT are re-engineering this prototype for commercial exploitation. The first customer deliveries were in the last quarter of 1987. Whereas the prototype mini-DAPs (as they are known) contain a 16-PE VLSI gate array chip, the re-engineered AMT version uses a 64-PE custom VLSI design.

The architecture of the mini-DAP is largely unchanged from the main-frame DAP at the PE level. The major difference is in the addition of a fast I/O path, as found in both GEC GRID (Robinson and Moore 1982, Arvind *et al* 1983) and Goodyear MPP (Batcher 1980) designs. In this scheme, an additional register is connected between PEs to form a parallel shift register running from south to north across the array. The shift register is controlled independently to the normal operation of the rest of the PEs, which are issued with instructions from the master control unit (MCU), see §3.4.2. The operation of this shift or I/O plane is controlled from the fast I/O unit, which also provides buffering for data as well as a facility for reformatting data on the fly. Data can be shifted through this plane at a rate of one 32-bit word into or out of the array every 100 ns, giving a total bandwidth of 40 Mbyte/s. When the plane formed by these registers has been shifted in from the I/O unit, it can be loaded into the DAP memory by stealing a DAP memory cycle. This means that fast I/O and processing in the DAP can proceed concurrently, enabling it to be used for real-time applications such as image and signal processing.

This I/O mechanism signifies a major shift in system philosophy for the DAP. Main-frame DAPs were implemented as an integral component of another system, namely the ICL 2900, and relied on that system to provide all I/O services (see §3.4.2). The mini-DAP, however, is an attached processor, which can be interfaced to a variety of hosts, or indeed used as a stand-alone or embedded system, after applications have been developed. A block diagram of the mini-DAP system is given in figure 3.49, which shows the major data paths. The fast I/O path acts as a 32 serial-to-parallel converter, one for each column of the DAP memory. With the exception of this data path, the array is unchanged from the main-frame DAP. Two-dimensional nearest-

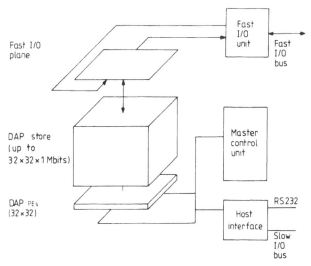

FIGURE 3.49   Diagram showing the structure of the ICL mini-DAP, with the addition of a fast I/O plane and buffer.

FIGURE 3.50   A photograph of the prototype mini-DAP, showing the physical size of the cabinet. (Photograph courtesy of AMT Ltd.)

FIGURE 3.51   A photograph of an array board from the mini-DAP, showing the technology used. The large square packages are LSI gate array chips containing 16 processors. (Photograph courtesy of AMT Ltd.)

neighbour connectivity with optional wrap-around is implemented between processing elements and row and column highways connect PES to bits in the MCU and the HCU (host connection unit). Indeed, the mini-DAP is source-code compatible with its predecessor, in both DAP FORTRAN and APAL (DAP assembler) languages.

The implementation of the prototype mini-DAP is in TTL and CMOS components, with the array of processing elements being implemented in 4×4 subarrays contained in a customised National 6224 gate array. This uses CMOS technology and the chips are packaged in 120 pin-grid array packages. The development of this chip, coupled with the advance in memory technology over the last five to ten years, has brought about a substantial reduction in the number of chip packages in the mini-DAP system. The main-frame DAP used on average five packages per PE, whereas the mini-DAP uses one gate array and eight memory chips (1 Mbyte option) for a 4×4 subarray. This order of magnitude reduction in parts, combined with the smaller array size, allows the machine to be packaged in an office environment pedestal cabinet, dissipating less than 1 kW of power. Figures 3.50 and 3.51, respectively, show the physical size of the mini-DAP and the technology used.

At the next level of packaging, the array is constructed from boards containing 128 pes, in a 16 × 8 subarray. The board also contains the 128 × 8 K of memory required for the 1 Mbyte option or 128 × 16 K required for the 2 Mbyte option. Eight such boards comprise the 32 × 32 array. The remainder of the system requires a further eight boards; two of these are array support boards providing instruction decode, parity memory etc; another two provide the mcu and separate code store; the remainder provide an interface to the outside world, via the hcu and fast I/O unit.

The two major differences in architecture over the main-frame DAP are found in the mcu and additional control units such as the hcu and I/O unit. These facilities make the mini-DAP considerably more powerful than the main-frame DAP. For example, in the mcu itself additional instructions include a multiply instruction, n-place shift and interrupt handling instructions. In the main-frame DAP microcode was stored in the DAP array and fetched and executed in a two-phase cycle, but in the mini-DAP there is a separate code store, which adds no overhead to the instruction cycle. The code store is currently 32K by 36 bits, although addressing capability is there for up to one Mword of code. A block diagram of the new mcu is given in figure 3.52.

Figure 3.53 shows the mini-DAP fast I/O unit, which interfaces to the 32-bit input and output buses of the array I/O plane. It comprises two data buffers of 64 Kbytes each, which can be configured separately or for use together. The buffers can be configured separately or for use together. The buffers can be configured under software control to provide corner turning or reformatting of the data passed to the DAP array. Corner turning is illustrated schematically in figure 3.54.

The prototype mini-DAPs have instruction cycles of 155 ns, which is some 25% faster than the main-frame DAP, although their peak processing rate will be reduced as there are only one quarter of the pes found in the main-frame DAP. Some performance figures for the mini-DAP are given in table 3.9.

With the increase in packing density expected in the re-engineered AMT DAPs, it may be possible to market a similar sized machine comprising 4096 pes, as in the main-frame DAP. This machine is also expected to have a faster clock rate, possibly 100 ns. The combination of these improvements would give a factor of six improvement in performance over those figures above. However, it is unlikely that such architectures will compete in the scientific market where floating-point arithmetic is required. The new T800 could achieve the figures quoted above, with just five to ten chips. The application areas where the bit-serial architectures do best is where their word length flexibility can be exploited; for example, in signal and image processing. It can be seen in table 3.9 that performance for addition is inversely proportional to word length.

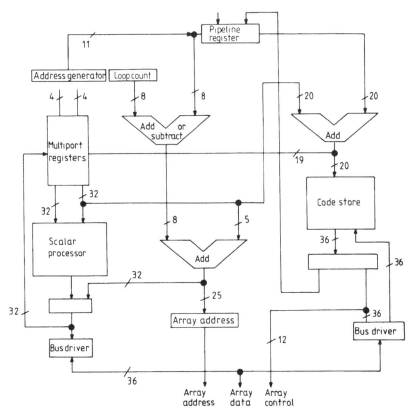

FIGURE 3.52   The control unit of the mini-DAP.

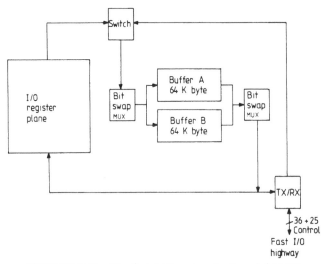

FIGURE 3.53   The fast I/O system in the mini-DAP.

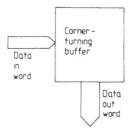

FIGURE 3.54 A schematic illustration of corner turning; the square buffer can be accessed in two orthogonal directions.

TABLE 3.9 Some performance figures for the prototype mini-DAP machine. These figures are for a 32×32 array using a 155 ns clock. All figures are given in millions of operations per second.

Word length	Operation	Performance
8-bit	Add	280
	Multiply	42
	Multiply (constant)	100–200
16-bit	Add	140
	Multiply	10
	Multiply (constant)	30–100
32-bit floating-point	Add	8.6
	Multiply	5.1
	Square	10.9
	Square root	7.4
	Divide	4.1
	Log	4.2

Other applications where this class of architecture excels are where broadcast and reduction operations over data are required. For example, in associative processing, set and database manipulation etc.

3.5.3 The Connection Machine

The Connection Machine has developed from research by Hillis and others at MIT, primarily for use as a parallel artificial intelligence engine. Although claimed by Hillis in his book of the same name (Hillis 1985) to be a 'new type of computing engine', to a large extent the implementation draws heavily on the established bit-serial grid computer developments that have preceded it. What distinguishes the connection machine from its predecessors is the use of a complex switching network which provides programmable connections between any two processors. These connections may be changed dynamically during the execution of programs, as they are based on packet-routing principles. Data is forwarded through the network, which has the topology of a twelve-dimensional hypercube, to an address contained within the packet. Because of the topology of the network, one bit of the binary address corresponds to one of two nodes in each dimension of the hypercube (see §3.3).

The connection machine derives its power and name from this ability to form arbitrary connections between processors; however, there are many compromises in the design and it is not clear that the designers had sufficient experience in the implementation technologies. A twelve-dimensional hypercube requires a considerable amount of wiring and, as is expounded in Chapter 6, this can contribute excessively to cost and performance degradation. However, the principles behind the machine are laudable, in providing a virtual replicated architecture, onto which a user description of the problem to be solved can be transparently mapped.

The Connection Machine is not the only bit-serial development that has considered the problems of information representation and array adaptability. Southampton University's RPA project (§3.5.4) is also an adaptive array architecture which has a communications system capable of creating arbitrary connections between processors and varying these dynamically. The connection machine implements connections in a bit-serial packet network, arranged in a binary cube topology and giving a slow but general connection capability. The RPA, on the other hand, implements connections by circuit switching over a connection network which conforms to the underlying implementation technology, which is planar. However, there is provision for long-range communication, which can be irregular. These long-range connections are implemented at the expense of processing power, by using processing elements as circuit switch elements. This latter trade-off is one which need not be fixed at system design time. For example, in the connection machine, 50% of the array chip is dedicated to packet routing. In the RPA, all of the chip is used for PES, but at compile or run time some of these may be used simply for switching elements.

The reader is encouraged to compare and contrast these two approaches to this communications problem. The direct connection provides an efficient hardware implementation and gives a high bandwidth. The packet switch, on the other hand, is more costly, but gives high network efficiencies over a wide variety of data structures. Moreover, the packet approach provides a system which can be used implicitly and dynamically.

The Connection Machine has been implemented by Thinking Machines Corporation, a company co-founded by Hillis, and is called the CM-1. This prototype consists of a large array of processor/memory cells, which can be connected together using the programmable connection network to form data-dependent patterns called active data structures. The activity of these structures is directed by a host computer, which has a data bus giving it access to the array memory. In this respect the connection machine is similar to the ICL 2900/DAP system and indeed to the transputer/RPA system described in the following section. In all three cases the array memory can be considered as belonging to the host system, so that it may address and modify locations within the array memory. However, for computationally intensive tasks, the host can instruct the array to access and process data stored in this shared memory. In this way processing power is distributed throughout the host's memory sytem, avoiding the von Neumann bottleneck which would otherwise be crippling in a uniprocessor realisation of comparable processing rates.

This active-memory array in the CM-1, DAP and RPA is a very powerful configuration as it provides a very responsive configuration for synchronising activity. Typical applications would require the broadcast of information to the array, keys (for example to activate data), some processing of active data, followed by a reduction of that information for dissemination by the host system using selection (array memory addressing) or reduction (minimum, maximum, sum etc).

The processor/memory array in the CM-1 comprises 65 536 processors each with 4096 bits of memory. This is implemented using a 16-PE custom VLSI chip and 4×4 K static RAM chips. On the chip the PES are connected in a nearest-neighbour grid or 'NEWS' network, although from the pin-out figures given in Hillis (1985) it seems the prototype does not extend this connectivity across chip boundaries. This is illustrated in figure 3.55. The binary hypercube connection provides the interconnect at the chip level, and 16 PES on a single connection machine VLSI chip share the bandwidth of a single packet-router. Thus the machine can be considered as a twelve-dimensional hypercube of nodes comprising a 4×4 grid of single-bit PES at each node. Each chip contains a router, which handles messages for the 16 processors on that chip, which it does by forwarding data packets bit-serially over one of twelve bi-directional wires. This network therefore comprises 4096 routers connected by 24 576 bi-directional wires. This is a very high wiring density when compared to other bit-serial processor arrays.

The processor chip has been implemented in CMOS, is about 1 cm² and

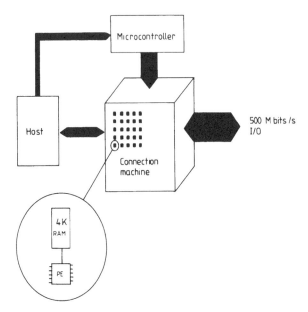

FIGURE 3.55   The connection machine.

contains about 50 000 transistors. It dissipates 1 W at 4 MHz and is packaged in a 68-pin ceramic chip carrier; to provide an isomorphic NEWS network over the entire machine would have required another 32 pins per package. Each processor chip is supported by four 4 k × 4 static memory chips and a subarray of 32 array chips and 128 memory chips is packaged onto a single printed circuit board, giving 512 processor/memory cells in all. These modules are then packaged in backplanes, with two backplanes containing 16 modules housed in a rack. Four racks complete the machine.

This hierarchy in the packaging technology simplifies the wiring for the cube network, with the first five dimensions on a circuit board, the following four within a backplane and the remaining three within racks. Even so, at the top level of the cube, each edge comprises 8192 signal ground pairs wired using flap cables. The whole machine is air-cooled, operates at 4 MHz and dissipates 12 kW of power. Comparing clock speed and array size, one would expect to see a performance which was an order of magnitude greater than the main-frame DAP, but for some applications even this performance has not been achieved.

A single microprogrammed control unit provides control to the array, with this control being synchronised by a single global clock. Instructions from the host (called macro-instructions) and data returning to the host are queued in a first-in, first-out (FIFO) sequence between the host and microcontroller. These buffers even out the flow of micro-instructions and data between the host and microcontroller. Each call to a microroutine, called micro-instruction by Hillis (1985), may require a few or many thousands of clock cycles, depending on the data to be processed. When branching on array-supplied data in the host, the FIFO buffers must of course be flushed.

Each processor cell in the CM-1 typical of many bit-serial array designs and is illustrated in figure 3.56. The basic operation of the cell is a five-address operation with two sources and the destination bits coming from and going to external memory. These are addressed by two 12-bit fields, with the destination sharing an address with one of the source operands. Three further four-bit addresses provide a source and destination flag register for input and output from the ALU; these addresses are provided to the PE chip. A final four-bit address specifies which of the flag registers should be used as a conditional or activity flag. Three clock cycles are required for the sequence comprising this operation. Two eight-bit global control fields completely specify one of the 256 boolean functions possible, with three boolean inputs for each of the outputs. Local on/off control can be provided by any of the 16 processor flags and a sum tree provides a logical OR of a signal from each processor in the array.

One of the major shortcomings of the connection machine design is the under-exploitation of the mass of wire which comprises the hypercube network. It can be argued that wiring is the most costly component in modern VLSI-based systems. The router on the connection machine routes data in cycles, where each cycle proceeds to send a message in each of the dimensions in sequence. This means only 1/12 of the wire is ever being used. It is obvious

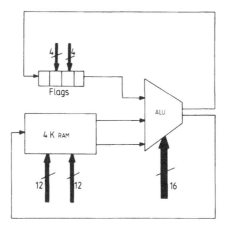

FIGURE 3.56 The connection machine processing element. The ALU can yield two outputs, each of which can be one of the 256 three-input boolean functions.

that it would be costly to provide a router that was capable of handling 12 concurrent transfers, which leads this author to doubt the viability of a twelve-dimensional network in this application. With 1/12 of the bandwidth being wasted, approximately the same long-range bandwidth and greater short-range bandwidth could be achieved with a packet-based NEWS network, with concurrent transmission and reception over the four directions. A software implementation of such a network has been implemented relatively efficiently on the RPA, which is described in §3.5.4.

For further information on the Connection Machine, the interested reader is referred to the book by Hillis (1985) for more detail of the design of the CM-1. In particular, the operation of the packet network is covered in some detail. The performance of the CM-1 is summarised in table 3.10, using Hillis' own measures.

TABLE 3.10   Performance of the 64K prototype CM-1.

Performance characteristic	Value
Size of memory	2.5×10^8 bits
Memory bandwidth	2.0×10^{11} bits/s
Processor bandwidth†	3.8×10^{11} bits/s
Communication bandwidth	
Worst case	3.2×10^7 bits/s
Typical case	1.0×10^9 bits/s
Best case	5.0×10^{10} bits/s

† It should be noted that processor bandwidth measures the number of bits entering and leaving the ALU in a second and not the number of operands per second.

The programming environment for the connection machine is based on LISP, which has a long history at MIT. Connection machine LISP (CMLISP) is an extension to common LISP, which is designed to support the parallel operations of the connection machine. It is a SIMD language, which reflects the control flow of the host computer and and microcontroller while allowing operations to be expressed over parallel data structures. Connection machine LISP is to common LISP what FORTRAN 8X (see §4.3.3) is to FORTRAN 77. The language is fully documented in Hillis and Steele (1985) and is introduced in §4.3.4.

3.5.4   The RPA

(i) Reconfigurable processor arrays

The RPA is an architecture designed for structure processing, defined in

§4.3; it is a processor array which uses many simple processors, but which can adapt its physical structure in a limited way to accommodate a variety of data structures. The RPA is designed to support structure processing over the widest range of data structures, consistent with the adaptable grid network used; however, this is not as restrictive as would be expected, as will be seen later, for it includes structures such as binary trees. The RPA was developed at Southampton University with funding from the UK's Alvey program (Jesshope 1985, Jesshope 1986a, b, c, Jesshope *et al* 1986, Rushton and Jesshope 1986, Jesshope and Stewart 1986, Jesshope 1987a, b). It is similar to the DAP and connection machine in that it is an array of single-bit synchronous processing elements with a common microcontroller. It is not a true SIMD machine, however, as local modification of an instruction is allowed by distributing some of the control word fields across the array. This allows the array to adapt to different situations. Figure 3.57 illustrates this concept for rectangular arrays.

It is well known that a synchronous structure such as a large SIMD machine is limited in the size to which it can be extended, as clock and control information must be distributed to synchronise the system. Any skew in distributing these signals will reduce the clock frequency. There is therefore an optimal size of RPA which will depend on the characteristics of the implementation and

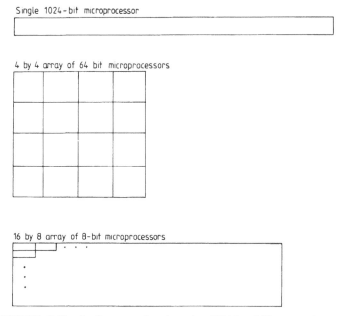

FIGURE 3.57 A diagram showing the RPA's ability to adapt to different rectangular structures.

packaging technologies. The current architecture will support arrays of upto 256 by 256, with 16 processors per VLSI chip. The size of an RPA system, however, is not limited by the size of the optimal synchronous system block, for RPA systems can be further expanded using process parallelism (see §4.4).

The RPA exists within a transputer's memory system as active memory, and processes executed on the RPA can be considered as extending the transputer's instruction set in an OCCAM programming model. Another way of expressing this is that the process parallelism of the transputer and OCCAM is used to bind together many synchronous systems capable of efficiently executing structure parallelism. Therefore, the RPA can either be viewed as an array processor with a transputer host, or as an intelligent memory system to the transputer, which can be replicated like any other transputer system. This replication could exploit the explicit parallelism of OCCAM to describe program structure by mapping communications onto the INMOS links. The technique of algorithmic parallelism (see §4.5.2) could thus be exploited within a multiple RPA system.

The major conceptual problem in any replicated design is one of mapping the problem's description of virtual processors onto a fixed processor structure. In this respect, an important aspect of the RPA design is that it allows the structure of the array to adapt to the problem being implemented. This greatly simplifies the compiler's task of data mapping to obtain conformity between the user's view of the world and the machine's view. It does this at the same time as maintaining a high utilisation of the hardware.

Ideally, in a processor array for structure processing, there should be a single processing element for each individual element of the data structure being processed. Unfortunately, there is no way of knowing *a priori* how large or small a data structure will be required. Indeed, the structure may vary dynamically. Structures larger than the array are well understood, for each processing element can process sequentially as many elements of the data structure as are required. Problems arise, however, in maintaining the topology of the data structure, especially if the structure is complex. The corollary of this problem is the ability to assign more than one PE to a data-structure element, and this is less well understood. In fact, this represents the major departure between the RPA and other bit-serial processor arrays.

To achieve this flexibility, the SIMD execution model is extended by allowing processors to modify the broadcast instruction. In this way the processors may act in a different, but locally preprogrammed, manner. This extended local control provides both flexibility and efficiency in the most common data processing operations and communications. It can also be shown that as well as allowing the rectangular array structure to be varied in size and shape, it also allows efficient implementations of data manipulation operations, including those on irregular data structures. For example, other topologies, such as trees and graphs, may be efficiently implemented, although not with the dynamic flexibility of using a packet-based communications system. However, the mechanisms can also be exploited to provide reasonably efficient software implementations of packet routing over the RPA, as will be shown later.

The RPA approach to virtual structure parallelism allows a single code to be executed on different physical realisations of replicated systems, without any recoding; moreover, this is true at the low level, as microcode routines can adapt to different configurations of array size and structure by parametrisation.

(ii) The RPA implementation

The RPA array is constructed from a very flexible processing element in a two-dimensional, four-connected lattice, with wrap-around at both edges. A single RPA system is illustrated in figure 3.58. The array of RPA pes has

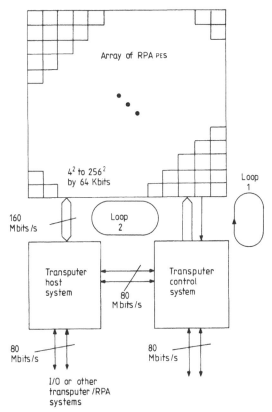

FIGURE 3.58 The RPA computer system, showing the array, the controller and host, with the interaction between the three components.

been designed to act as a pool of virtual processors, connected according to the structure being processed. To achieve this, each PE must be considered as a bit slice of a larger processing unit, which will support a wide range of common microprocessor operations. The composite processing unit supports binary operations, a wide range of shift operations and bit-parallel arithmetic operations. These processing units can be configured, dynamically if required, from any connected path of RPA PEs . The most useful configurations are closed subgraphs of the RPA array, as these support cyclic and double-length shift operations in a single microcycle and also greatly simplify serial arithmetic when applied to words of data stored in processing units. For example, performing 16-bit arithmetic serially using four-bit processing units, the carry bit from one four-bit addition at the most significant bit will be adjacent to the least significant bit, where it will be required on the following cycle. If multi-bit processors are configured as closed loops of PEs, then the configurations will support the following operations:

(1) all bitwise logical operations on two operands (two independently programmed operations can be performed in a single cycle);

(2) single-length cyclic and planar (arithmetic) shifts;

(3) double-length cyclic and planar (arithmetic) shifts;

(4) broadcast of one bit to all other bits within a processor;

(5) ripple-carry arithmetic with fast Manchester carry chain, performed in multiple 'byte' operations, with automatic handling of carry between sub-operations;

(6) multiplication with full double-length result, using multiple ripple-carry adds (5), double-length shifts (3) and broadcast of bit data within 'bytes' (4).

The PE or bit slice has a four-bus structure, as shown in figure 3.59, with

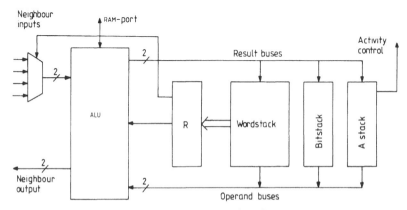

FIGURE 3.59   The RPA processing element (bit-slice).

two operand and two result buses. The ALU provides logical and bit-serial operations, using local operands only and the neighbour-select provides a means of transferring data from one PE to another. Bit-parallel arithmetic is made possible by the use of both ALU and neighbour-select, which together form a fast Manchester carry-chain adder. The key to the flexibility of the array lies in the use of three preset control fields, which are provided in each PE in the array, R in figure 3.59. These control fields determine switch settings and edge effects in arithmetic and shift operations.

Two of these fields, each of two bits, specify the control of the nearest-neighbour switch (left and right shifts) the reconfiguration register also contains a further two-bit field, which codes the significance of the PE in the larger processing unit. The codes used are for most significant bit, least significant bit and any other bit. The remaining code is used for a very powerful feature which allows the inputs from the switch to be connected directly to the outputs. This allows the distributed control of communications such as the row and column highways found in the ICL DAP and GEC GRID. However, it does this in a much more flexible manner, as the combination of direction and significance codes can be used to bus together any connected string of PEs. These bus connections may either be used to distribute data to all PEs so connected, or indeed be used to bypass PEs in order to implement connection structures that would not otherwise be possible, containing long-range connections.

The storage provided in each PE is stack-based. There is a bit stack and an activity stack, for storing single bits of data in a PE. Each comprises eight bits of data and both are identical, with the exception that the top bit of the activity stack can be used to conditionally disable the operation of the PE. Storage is also provided for words of data in the PE. These are arranged as a stack of eight-bit words with parallel-to-serial and serial-to-parallel conversion provided by a pair of shift registers. This structure allows bit-serial (or word-serial) arithmetic to be performed, without the awkward bit reversal encountered using stacks. Finally, a single-bit I/O port allows up to 64K bits of external storage to be connected to each PE.

The internal RAM has dual control mechanisms which allow stack or random access to the bytes of a word with local or global control of either. For example, the stack can be conditionally pushed or popped, depending on the value of the activity stack. The store structure also contains a full n-place shift, which is locally controlled, and comparitor circuits between the two shift registers providing serial–parallel conversion. These facilities provide much enhanced floating-point capability within the array. For a given cycle time and array size, the floating-point performance of the RPA can be a factor of ten above that of the ICL DAP architecture. The price paid for this is a more complex processing element. At the time of writing, a test chip containing a single PE has been fabricated. It uses a 3 μm n-well CMOS process and occupies a little under 4 mm^2. We anticipate going into production with a 16-PE, which will occupy about 0.8 cm^2. Figure 3.60 shows the RPA test chip.

(iii) The RPA control system

Processor arrays, such as the DAP and connection machine, generally use a single controller to control all processors in the array. This gives a very cost effective machine by sharing a complex control mechanism between many processors. The RPA, however, departs from this control structure in two ways.

At one level the RPA can be considered as a MIMD system, with each 'processor' containing an array of processors under SIMD control. This structure (MIMSIMD) has also been proposed by a number of other designers (Lea 1986) and allows the exploitation of all forms of parallelism within applications. The synchronous component can exploit the parallelism found in structure, such as arrays, strings, lists and trees, and the event-synchronised

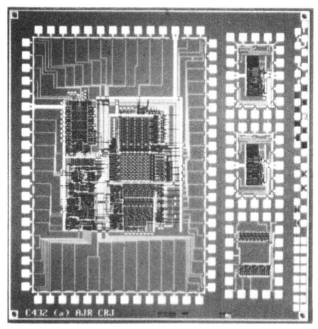

FIGURE 3.60 A microphotograph of the RPA test chip showing one PE and additional test structures. This chip was fabricated at Southampton University.

mimd layer can exploit the irregular or independent parallelism also found in many algorithms.

Although a single RPA/transputer system can be considered as a conventional simd array, each pe may, under locally determined conditions, store or generate some or all of its control word for use in subsequent operations. This adaptive modification of simd control allows each processor some autonomy of action. Most simd computers provide minimal control at the processor level, being usually restricted to a data-dependent on/off switch. This structure maintains the advantage of fine grain synchronisation while allowing a limited adaptability to differing processing requirements in the array.

Figure 3.58 shows the control structure of a single RPA. Two major control loops can be identified. The first is between the array and its microsequencer. This is conventional; the controller supplies microcontrol words and addresses to the array (a wide word data path) and receives condition signals from the array and possibly from the other devices in the system. The second control loop involves data from the array memory, which can be accessed by processes executing on the host system. This loop allows a coarser grain synchronisation that may involve events from other transputer/RPA systems. The use of this loop involves the initiation by the host of some action in the array, which will set data in the array memory; this data will subsequently be retrieved by the host and used to determine subsequent actions.

One of the major design goals of the RPA system has been to map this second control loop into an OCCAM programming model in such a way that the programmer can trade real and virtual concurrency in an application. This will allow for applications code to be ported between different transputer/RPA configurations.

The implementation alternatives in mapping the RPA control structure into OCCAM are determined by the granularity of processes running on the array. Should array processes be complete programs, running concurrently and perhaps communicating with the host? Or should they be considered as indivisible extensions to the host instruction set? It is the communications required between host and array which provide the discrimination between these two alternatives. If we define a basic array process as being an indivisible unit of computation, so far as communication between host and array is concerned, such that the order may be modelled by the following OCCAM fragment

```
    —HOST PERCEPTION OF ARRAY ORDER
    ARRAY! SOMETHING
    ARRAY.PROCESS
    ARRAY?SOMETHING.ELSE
```

then any decision concerning the partitioning of control between host and array controller is in effect a decision about whether constructions of basic array processes may be determined by the array controller, or whether they must be determined by the transputer host.

If sequences of these basic array processes are to be determined by the array controller, allowing complete programs (processes) to run on the array, then for concurrent operation of the host and array, there must be communication between host and array to provide the necessary synchronisation events. This concurrent operation between host and array is illustrated in OCCAM in the example below.

```
PAR
  SEQ - - HOST PROCESS
    ARRAY ! SOMETHING
    HOST.PROCESS.1
    ARRAY ? SOMETHING.ELSE
    HOST.PROCESS.2
  SEQ - - ARRAY PROCESS
    ARRAY? SOMETHING
    ARRAY.PROCESS.1
    ARRAY! SOMETHING.ELSE
    ARRAY.PROCESS.2
```

In this example the host initiates action in the array and proceeds. At some later stage the two processes synchronise by a communication initiated from the array and both then proceed, perhaps modifying their actions based on that communication. In this control model synchronising communications may be initiated by host or array.

If we wish to overlap processing and communication, so that efficiency is not lost in achieving synchronisation, then this model must be modified to allow both host and array to run their respective processes in parallel. This is illustrated in the example below, where the synchronising communications have been hidden within host and array processes.

```
PAR
  PAR - - HOST PROCESS
    HOST.PROCESS.1
    HOST.PROCESS.2
  PAR - - ARRAY PROCESS
    ARRAY.PROCESS.1
    ARRAY.PROCESS.2
```

To implement this model requires an array sequencer that can support multiple, concurrent processes. It would need to be interruptable, so that an array process requiring communication with the host could be suspended, while communications were proceeding, and resumed on completion of the communication. However, the overheads in this process switching are large, and the microcontroller required is complex. Thus in the RPA we have opted for the alternative of indivisible array orders.

This approach excludes synchronisation between processes running on the host and array, except on initiation and termination of the array process. It has a number of advantages. Because there is no need for synchronisation within an array process, there is now no need for array orders to be suspended. Entry microroutines can run to completion using a simpler microsequencer. Some mechanism for supporting concurrent process execution on the array is desirable, however, as there may be communication before and after each microroutine, and it is still desirable to overlap this with processing. Such parallelism may, however, be implemented using a single process queue, containing active processes; an interruptable sequencer would require at least

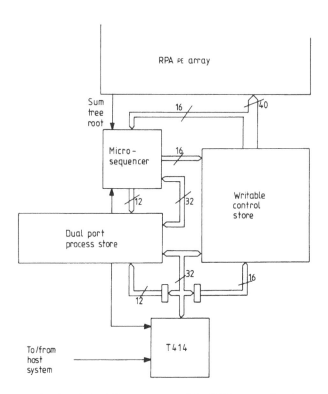

FIGURE 3.61   The structure of the RPA control unit.

two process queues, for active and suspended processes, with support for suspending and activating processes to achieve synchronisation.

Figure 3.61 gives a more detailed view of the transputer control system, which contains a single multiport process queue, through which the transputer and microcontroller communicate. Within this implementation, array microcode routines provide processes which can be considered as indivisible extensions to the host's instruction set, and which map into the OCCAM language. PAR and SEQ constructors, and array and host processes, may now be freely mixed, with communication (hidden in the example below) providing synchronising events and control flow where required:

```
- - HOST/ARRAY PROCESS
SEQ
      .

      .
         .
      PAR
        HOST.PROCESS.1
        SEQ
          ARRAY.PROCESS.1
          ARRAY ? SOMETHING
          PAR
            HOST.PROCESS.2
            IF
              SOMETHING = GOOD
                ARRAY.PROCESS.2
              TRUE
                ARRAY.PROCESS.3
```

Using this system, a sequence of array processes which require no synchronising communications may be buffered in the array controller with control passing from one to the next without recourse to host interaction. They are added to the process queue in the array controller and then executed from this queue in sequence. Likewise, parallel array processes may also be queued for execution on the array.

A considerable volume of low-level software has been implemented in the design of the RPA computer system, using the RPA simulator-based microcode development system (Jesshope and Stewart 1986). This provides a simulator of a 32 × 32 array and controller, implemented on the binary image of the microcode. It uses the graphics hardware of an ICL Perq workstation to implement this. The system provides programming by menu and is illustrated in figure 3.62. It can provide interactive data and

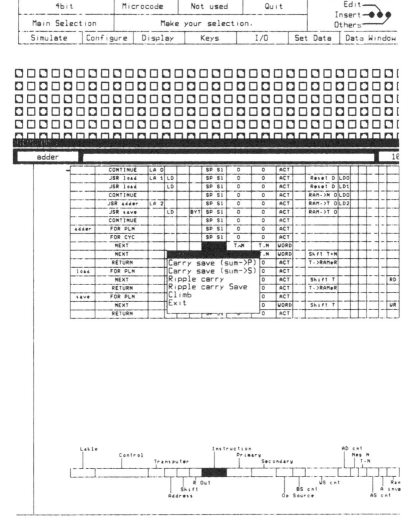

FIGURE 3.62 The RPA microcode development system, showing menu-driven graphical input of microcode.

configuration editing, microcode editing and full array and controller simulation with graphical display of all internal states and buses, if required for debugging.

The RPA as a transputer-based workstation with a 32×32 array will provide between 20 and 100 Mflop/s on 32-bit floating-point numbers, assuming a clock rate of 10 MHz. Because of the serial nature of arithmetic implemented on the array, maximum performance will be dependent on the precision of the data. For example, 32-bit integer addition of 32×32 arrays will be performed at a rate of approximately 200 Mop/s, whereas 8-bit integer addition will be performed at almost four times this rate, which is almost a thousand million operations per second (1 Gop/s). Indeed, boolean operations can be performed at a rate of 20 Gop/s and it is possible to associate on character data at the rate of about 1 Gop/s.

Table 3.11 gives examples of performance estimates provided by running real code on the simulator. The only assumption contained within these figures is a clock period of 100 ns, which is justified by electrical simulation of the test chip.

(iv) Data structure mapping

The RPA array can be viewed as a very flexible array of processing units where, as the array size increases, the number of bits in the processing unit decreases. Figure 3.57 illustrates this concept. It should be noted that rectangular, as well as square, arrays may be directly configured. It is also possible to map multidimensional arrays over the RPA structure, and providing the array bounds are powers of two, any remappings to activated data may be efficiently coded. Figure 3.63 illustrates the shifts required to provide communication within 8-bit processing units. Use is made of the reconfiguration bits to provide boundaries for single shift operations and long-range shifts. The direction registers provide the ability to perform

TABLE 3.11   Performance estimates for the RPA computer system over 1024 operations. (All figures in millions of operations per second.) All floating-point operations are full IEEE specification, with denormalisation and rounding. IBM format floating-point is very much faster.

Operation	Performance
8-bit integer addition	930
16-bit integer addition	465
32-bit integer addition	233
8-bit integer multiplication	128
32-bit integer multiplication	16
32-bit integer multiplication	5
32-bit floating-point addition	18
32-bit floating-point multiplication	6
32-bit floating-point division	6
32-bit maximum	790
string matching	500–1000

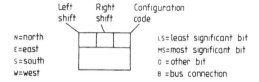

Key: A single RPA processing element.

(a)

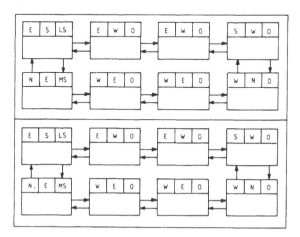

(b)

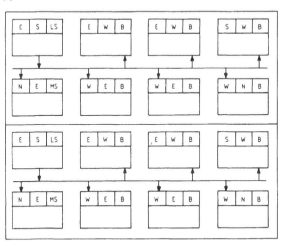

FIGURE 3.63   Illustration of the use of local control fields in the mapping of regular data structures over the RPA; the key gives the significance of the control fields. (a) Two eight-bit microprocessors, configured from closed cycles of RPA PEs (note that ms and ls bits are adjacent, enabling efficient shift operations and carry handling). (b) The same two microprocessors, with connections configured to create a local bus structure, to support broadcast operations from the least significant bit.

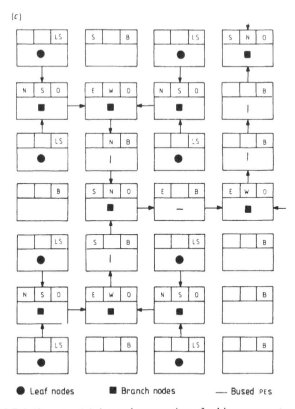

FIGURE 3.63 *cont.* (*c*) A regular mapping of a binary tree structure.

non-laminar shifts. More complex structure mappings and rotations can exploit the theory of musical bits (Flanders 1982).

For nearest neighbour shifts between processing units, it can be shown that although local communication bandwidth is degraded by approximately the square root of the number of processing elements in the processing unit, the global communication bandwidth remains constant for any configuration. Indeed, the communication bandwidth in less regular data structures or in complex data movements can be much improved over the software implementations that would be required in a non-adaptive array. In a software implementation, a great deal of repetition may be required to obtain multiple-shift directions, by alternately masking and shifting. However, in the RPA this can be achieved by configuring the switch network to realise the various shift directions concurrently. Moreover, all stack stores may be conditionally pushed or popped, allowing for irregular data storage, with stacks in different processing units out of step.

Figure 3.63 illustrates the codes required in the local control fields to implement some regular configurations, which together give some idea of the power and flexibility of this adaptive array structure. The three configuration fields are shown in each processing element in figures 3.63(a) to 3.63(c). These are coded as shown in the key and correspond directly to the configuration store state in each RPA PE.

Figure 3.63(a) shows the configuration for two 8-bit processors. Notice that these are closed loops and thus support the full range of 'byte'-oriented operations.

To perform multiplication, it is possible to configure the processors of figure 3.63(a) to broadcast data from one bit to all other bits in a given processor and to do this in each processor simultaneously. This configuration (figure 3.63(b)) gives the facility of a parallel bus structure, in which multiple broadcasts can be performed concurrently.

Figure 3.63(c) shows how the array can be used to map a binary tree of single-bit processors onto the RPA. This is achieved with a 50% utilisation of the PES. Some 25% of the PES are not used at all and another 25% are used in bus configuration and only pass inputs to outputs, providing long-range communication where required.

Less regular structures can be mapped onto the array by implementing a packet-based communications structure over the adaptable nearest-neighbour communications network, using a layer of microcode. Such an implementation has been coded on the RPA and the results of one cycle of the algorithm are illustrated in figure 3.64. We have implemented 32-bit packets, with two absolute address fields for x and y PES. Using this scheme only one data packet may be buffered in on-chip RAM, but simultaneous transmission and reception of packets is possible in the absence of data collision. However, with the global reduction facility implemented over the RPA array, it is possible to detect a potential collision deterministically in very few microcycles and buffer packets out to external RAM.

The total circuit-switched bandwidth of the RPA is two bits in and two bits out of every PE in every microcycle, a total of 4×10^{10} bit/s, approximately equivalent to the CM-1 machine (see §3.5.3), although this machine is 64 times larger. In packet-switched mode, it requires between 100 and 200 microcycles to forward 32 bits of data one PE nearer their destination, giving a total of between 3 and 6×10^9 bit/s over the entire array. Thus, the flexibility of addressed data communication degrades the communications bandwidth by an order of magnitude; however, the static bandwidth is high and this is very much more efficient compared to an implementation on a non-adaptive array. This system allows irregular and dynamic data structures to be implemented.

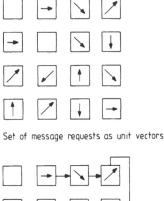

Set of message requests as unit vectors

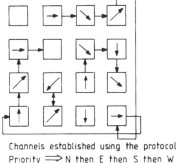

Channels established using the protocol
Priority ⟹ N then E then S then W

FIGURE 3.64   An illustration of the implementation of a packet-switching protocol over the RPA network.

3.5.5   Transputers and related systems

The transputer (INMOS 1985) is shown in figure 3.65; it is a single-chip microprocessor from the British company INMOS Ltd. It is distinguished from other vendors' microprocessors in that it is designed as a building block for parallel processors. To facilitate this it has memory and links to connect one transputer to another, all on a single VLSI chip. It can be viewed therefore as a family of programmable VLSI components that support concurrency.

In the transputer INMOS have taken as many components of a traditional von Neumann computer as possible and implemented them on a single 1 cm² chip, while at the same time providing a high level of support for a concurrent or process-based view of computation. To some extent this is an optimisation of the underlying CMOS technology (see §6.1), for communication between VLSI devices has a very much lower bandwidth than communication between subsystems on-chip. Pins on a VLSI chip are expensive and relatively slow. The support for concurrency includes hardware support for a process queue, with special registers and microcode instructions supporting the creation of parallel tasks; a minimal context for each active process, so that process

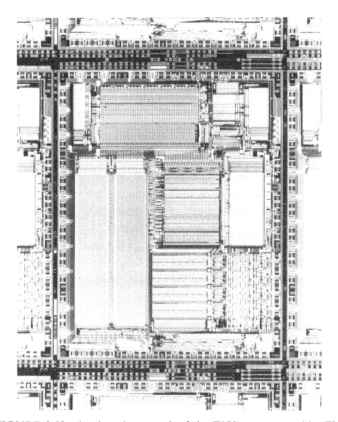

FIGURE 3.65   A microphotograph of the T800 transputer chip. This chip contains 4 Kbytes of ram (the large area to the bottom left); controllers for four INMOS links and the event input (in the relatively random area in the bottom right); a 32-bit data path and rom (in the middle right); and the 64-bit floating-point unit data path flanked by rom (at the top). Each of these blocks occupies approximately 20% of the chip area; the remaining 20% is required for the memory interface and controllers around the periphery of the chip. (Photograph courtesy of INMOS Ltd.)

switching is very rapid; hardware timers; and an external interrupt or event signal.

The processing part of the transputer is a risc or reduced instruction set computer, which provides a very rapid rate of execution (20 MHz) of a small range of instructions and addressing modes. This processor occupies about 25% of the chip area. A further 25–30% is occupied by 2 Kbytes (T414) of on-chip static memory, and the transputer-to-transputer communications links occupy another 25%. The remainder is occupied by the interface circuits and bus logic for the external memory system.

Currently, transputers products available include a 32-bit processor with a multiplexed 32-bit address/data bus (T414), a 16-bit processor with 16-bit address and data buses (T212), and a disc controller chip, with 16-bit address and data buses and a specialised interface for industry standard discs. At the time of writing, INMOS were testing engineering samples of the T800 floating-point version of the transputer which has, on the same chip, a floating-point hardware unit, as well as all of the other T414 components. Indeed, the use of a more dense RAM technology has allowed 4 Kbytes of on-chip memory to be used in the T800 design.

In all products, a range of speeds is available, currently up to 20 MHz internal operation, although INMOS literature for the T800 talks to a 30 MHz part. One unique feature of the transputer, however, is that the external clock rates are maintained at 5 MHz for all parts. This greatly simplifies engineering, as only a relatively low-frequency clock signal need be routed around a system or board. Transputer systems communicate asynchronously and can therefore be combined with each processor using its own clock, provided the crystals are true to $\pm$ 200 PPM.

In transputer parts, all components execute concurrently; each of the four links and the floating-point coprocessor on the T800 can all perform useful work while the processor is executing other instructions. Each link has a DMA channel into the memory system which will reduce, but not significantly so, the memory bandwidth to the processor.

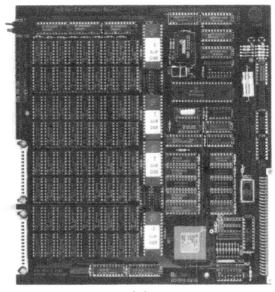

(a)

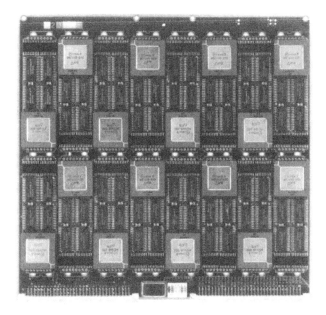

(b)

(c)

FIGURE 3.66 A sequence of transputer board level components, containing (a) 1, (b) 16 and (c) 42 transputers respectively. (Photographs courtesy of INMOS Ltd.)

Figure 3.66 illustrates the power of the T800 transputer chip, by considering some transputer board sub-assemblies. Figure 3.66(a) shows a 2 Mbyte single transputer board, a 1–2 Mflop/s component; figure 3.66(b) illustrates a 16 transputer board, a 16–32 Mflop/s component; and figure 3.66(c) illustrates a 42 transputer board, a 42–84 Mflop/s component! All boards are double-extended eurocard format. Of course one is trading processing power for memory in this sequence, but it can be seen that there is a great incentive to provide computing models which utilise a relatively small amount of RAM, for example pipelines. With the rate at which gate density is increasing in MOS technologies, it would not be surprising to see the demise of the RAM chip, in favour of a processor–memory device, of which the transputer is the harbinger. The achievement of 4 Mflop/s and 64 Kbytes on a single silicon die in 1990 would not be a surprising feat.

(i) The processor architecture

Transputer systems execute the OCCAM programming language (see §4.4.2), in which concurrency may be described between transputers in the system, or indeed within a single transputer. The transputer must therefore support internal concurrency. A process on the transputer is described by six registers (see figure 3.67). The six registers are:

(a) a three-element operand stack (A, B and C in figure 3.67);

(b) a workspace pointer which points to an area of memory holding the processes' local variables;

(c) an instruction pointer which points to the next instruction to be executed by the process; and

(d) an operand register, used for forming literal values or operands.

The stack is used in a conventional manner, with operations referencing the top locations implicitly. Instructions are all of a fixed and compact format, which embody the principles of RISC design, but at the same time allow for the extension of the instruction set within the same instruction format. This design also allows independence of processor word length. The format is given in figure 3.68 and contains a four-bit opcode and a four-bit data value, which can be used as an operand or address.

The instructions encoded into the four-bit opcode include:

load constant;
add constant;
load local;
store local;
load local pointer;
load non-local;

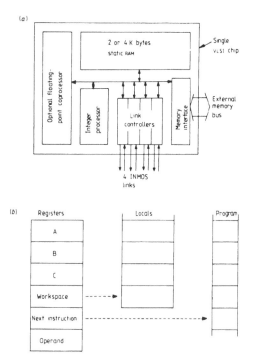

FIGURE 3.67   The INMOS transputer. (a) A block diagram of the chip architecture. (b) The transputer registers.

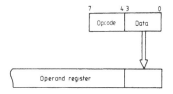

FIGURE 3.68   The transputer instruction format and operand register.

store non-local;
jump;
conditional jump; and
call.

These single instructions can use the four-bit constant contained in the data field, giving a literal value between 0 and 15, or an address relative to the workspace pointer for local references, i.e. an offset of 0 to 15. The non-local references give offsets relative to the top of stack, or A, register. For larger literals, two additional instructions provide the ability to build data values from sequences of these byte instructions. All instructions commence their execution by loading their four-bit data value into the operand register, and the direct functions above terminate by clearing this register. There are, however, additional instructions which load this register and whose action is to shift this result left by four places, therefore increasing its significance by a factor of 16. These instructions do not clear the operand register after execution. Two instructions—*prefix* and *negative prefix*—allow positive and two's complement negative values to be constructed in the operand register, up to the word length of the processor.

This data can be used as an operand to the above direct instructions. Statistics of the analysis of compiler-generated code indicate that the above direct instructions are the most commonly used operations and addressing modes. Moreover, these results also show that the most commonly used literal values are small integer constants. This simple instruction set therefore allows the most commonly used instructions to be executed very rapidly. However, the story does not end there, for even the most ardent RISC advocates would require more than this handful of instructions; 64–128 is typical for other RISC processors.

The operate instruction allows the operand register to be interpreted as an opcode, giving, in a single byte instruction, an additional 16 instructions, which operate on the stack. What is more the prefix instruction can be used to extend the range of functions available. Currently, all instructions can be encoded with a single operand register prefix. The most frequently used indirect operations are encoded into the nibble contained in the first byte. As an example, to multiply the top of stack, with a value of 16 bits significance, it would require the following code sequence:

> prefix—most significant 4 bits
> prefix—next least significant 4 bits
> prefix—last prefix bits
> load constant—contains the last 4 bits
> prefix—indirect function
> operate—decodes operand register as multiply.

This sequence takes four cycles to load the operand and 40 to execute the multiply, not such a large overhead for the simplicity. Indeed, if account were taken of speed improvement in processor cycle time because of the simple decoding, then the benefits of the RISC approach the transputer has taken are well justified. The sequence is also compactly encoded, requiring only six bytes of data (two bus cycles to load the instructions).

(ii) Support for concurrency
The transputer has a high degree of support for concurrency. It has a

microcoded scheduler, which allows any number of processes to compete for the processor's time, subject to memory limitations only. Processes are held on two process queues. One is the active queue, which holds the process being executed and any other active processes waiting to be executed. The inactive queue holds those processes waiting on an input, output or timer interrupt. The scheduler does not need to poll any of these devices so consumes no processor cycles on inactive processes. Figure 3.69 illustrates the linkages involved in a process queue, in this case the active queue. Two registers hold the head and tail of the list, and the executing process has its workspace pointer and program counter loaded into the processor's registers. All other processes hold these values in the workspace.

Process swap times are very small, of the order of microseconds, depending on the instruction being executed. This is because little state has to be saved. The evaluation stack does not need to be saved, and when the current process can no longer proceed, its program counter and workspace pointer are saved in its workspace and the next process is taken from the active list.

Two further micro-instructions provide a means of adding and deleting processes from the active process queue. These are *start process* and *end process*. A start process instruction is required for each component of the PAR constructor in OCCAM (see §4.4.2), and the correct termination of this process is ensured by the use of the end process instruction, which decrements a workspace counter when executed. Only when each component of the PAR construct has terminated will the counter be decremented to zero, signifying correct termination and allowing the process in which the construct was embedded to proceed.

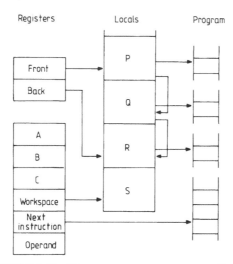

FIGURE 3.69   The transputer process queue linkage.

(*iii*) *The INMOS link*

Communication between processes on the transputer or between processes on different transputers is performed by two instructions *input message* and *output message*. The communication which is supported is a point-to-point, unbuffered message-passing scheme. It therefore requires a handshake between processes, which synchronises them. The same two instructions are used for message passing between processes on the same transputer, as well as between processes running on different transputers. The instructions use the address of the channel to determine which form of communication is required.

Internal channels are represented by a single word in the memory which provides the handshake protocol between communicating processes. The channel holds either a special value 'empty' or the identity of a process. A channel is initialised to empty, and when either a reading or writing process becomes ready, the identity of that process is stored in the channel location; this process would then be descheduled. When the second process becomes ready, it will find the process identification of the first process in the channel location and the message is copied and the waiting process is added to the active process list. The message is defined by a count, a channel location, and a pointer to the message. Figure 3.70 shows the above sequence diagrammatically.

External channels proceed in a similar manner, but the link interface performs the job of copying the message across the link circuit. Each link implements two OCCAM channels in opposite directions over a three-wire TTL level circuit. Communications over these links are controlled by autonomous controllers, which have DMA access to the transputer's memory. Four bi-directional communications and processing can therefore proceed concurrently on the same transputer chip. Each link controller has three registers, holding a pointer to a process workspace, a pointer to a message and a byte count for the message, with which it controls the transfer of the message. The only difference in operation is that on external communication, both processes would need to be descheduled while the autonomous transfer took place.

The operation and performance of the INMOS link is fundamental to the effective exploitation of the transputer. All transputer products will support communications over this asynchronous point-to-point connection, with speeds of 5, 10 and 20 Mbit/s. Although the transfer is autonomous, unless the messages passed over the links are reasonably large, a true overlap will not be possible, as each transfer will use a small but finite amount of processor resource to deschedule the processes and initiate set-up of the link controller's registers. If the link bandwidth is measured as a function of message length, the classic pipeline model is seen, with a start-up without transmission, followed by a constant throughput. The granularity of a message therefore modifies the bandwidth seen. It must also be remembered that the start-up period also degrades processor performance, to such an extent that many concurrent small transfers may saturate the processor.

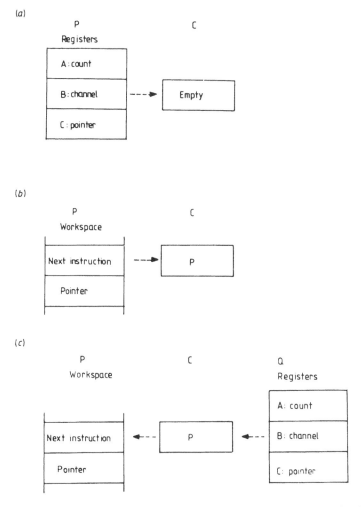

FIGURE 3.70   The operation of internal and external channel communication. (*a*) The first process P finds the channel location empty. (*b*) The process is descheduled and its location stored in the channel. (*c*) The second process Q finds the location of P in the channel, the communication proceeds and process P is rescheduled.

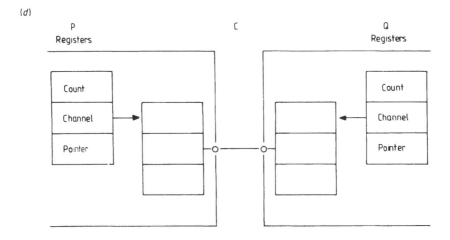

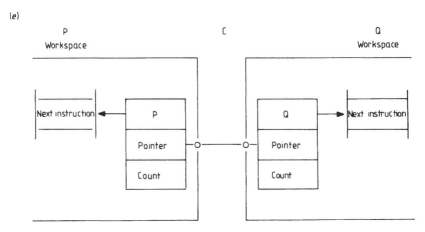

FIGURE 3.70 *cont.* (*d*) The initial state of two yet-to-communicate channels. (*e*) The state during communication; note that both process P and Q are descheduled.

Figure 3.71 shows the time required to transmit a message of a given size on a transputer with 10 MHz links and 12.5 MHz processor clock, as found in the first T414 transputer products. The maximum or asymptotic transfer rate is approximately 0.5 Mbyte/s, which can be achieved in both directions over the link. A half-performance message length of approaching one byte is also observed. Various processor and link rates will modify these observed parameters, with the start-up time being proportional to the processor clock and the asymptotic bandwidth being, to a first order, proportional to the

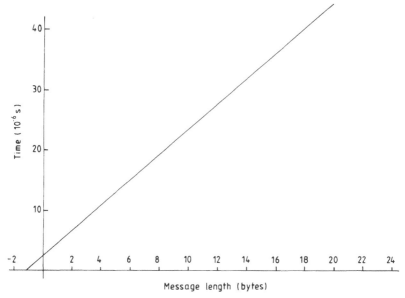

FIGURE 3.71   A graph of time required for communication over an INMOS link against message length for a T414 rev A transputer.

link rate. Thus, on current best T414 silicon (20 MHz processor and 20 MHz links), an asymptotic bandwidth of 1 Mbyte/s should be seen, with a half bandwidth message length of 1 byte.

The link protocol uses an 11-bit data packet, containing a start bit, a bit to distinguish data and acknowledge packets, eight data bits and a stop bit. The acknowledge packet is two bits long and contains a start and stop bit. The protocol allows for a data byte to be acknowledged by the receiving transputer, on receipt of the second (distinguishing) bit. This would allow for continuous transmission of data packets from the transmitting transputer, providing that the signal delay was small compared with the packet transmission time.

This pre-acknowledgement of data packets is not implemented on current T414 transputers, and an acknowledge packet will not be sent until the complete data packet has been received. However, this protocol has been implemented on the T800, giving a maximum theoretical transfer rate of approaching 2 Mbyte/s, per link, per direction. This bandwidth could cause degradation on bi-directional traffic, as data and acknowledge packets must be interleaved.

(iv) Performance
The performance of the transputer is dependent on a number of factors, for

example, the clock speed of the part, which may vary from 12.5 MHz to 20 MHz. Also, if data is coming from off-chip RAM, the access times will depend on the speed of the memory parts used and, for most operations, this is a major factor in the speed of operation. At best, the external memory interface will cycle at three transputer clock cycles (150 ns for the 20 MHz processor). Internal memory, on the other hand, will cycle within a single processor cycle. It is clear from this that on-chip data will provide significant speed gains. The transputer reference manual gives a breakdown of cycles required for the execution of various OCCAM constructs. We have, however, performed a number of benchmarks on a single transputer board, containing a T414 rev A transputer with a 12.5 MHz internal clock. The results of these timings are given in table 3.12 below.

When using OCCAM the programmer is encouraged to express parallelism, even if the code fragment is configured onto a single transputer. This style of programming should be encouraged, but only providing that parallel processes can be created without excessive overhead. INMOS claim that in the transputer this overhead is no larger than a conventional function call in a sequential language. In order to test this we have executed the following code on a single (12.5 MHz) transputer.

```
VAR b,c
b: = ...
c: = ...
PAR [j = 0 FOR m]
  VAR a
  SEQ [i = 0 FOR n]
    a: = b*c
```

TABLE 3.12   A range of performance figures for the INMOS transputer T414 rev A (12.5 MHz). Figures given are in millions of operations per second for operands from internal and external memory respectively.

	Performance	
Operation	Internal memory	External memory
Integer addition	1.78	0.23
Integer multiplication	0.33	0.14
Integer division	0.27	0.13
Floating-point addition	0.03	0.017
Floating-point multiplication	0.03	0.018
Floating-point division	0.03	0.016

This code was executed for various values of m and n, where the product $m*n = 1024$. This was executed in internal and external memory, and the results are summarised in figure 3.72. Using off-chip memory, it can be clearly seen that the overheads of process creation do not become significant until 128 parallel processes are used to perform 1024 integer multiplications; and

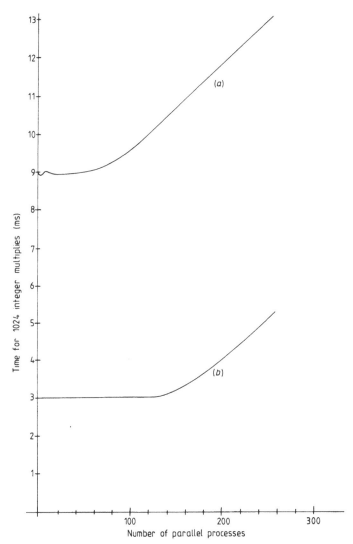

FIGURE 3.72 A graph showing the effect of parallelism on the performance of a single T414 rev A transputer. In this experiment 1024 integer multiplication operations were performed using a number of parallel tasks. (a) Data and program in external memory. (b) Data and program in internal memory.

in the case of on-chip memory, when 256 parallel processes are used to perform the 1024 operations. These correspond to eight and four operations for each process respectively.

The overheads of creating parallel processes are therefore small. These figures also show the advantages of programming transputers so that code and data both use on-chip memory. If the transputer is used in this mode of operation, it is likely that code for the algorithm will be distributed across a number of transputers. This is called algorithmic parallelism, and the technique is illustrated in §4.5.2. In this situation, data will probably be sourced from communications links. We therefore executed code on connected transputers to simulate the effect of various packet sizes on operation times. The OCCAM code executed on the transputer was a sequence of integer multiplications executed from internal memory, with both sets of operands sourced from two externally configured OCCAM channels and the results sent to an externally configured OCCAM channel. All I/O was buffered so that communications and operations could all proceed in parallel. The results are summarised in figure 3.73, where curves are plotted for a given packet length, showing time required against total number of operations performed; the curves are compared with the ideal, which gives time for internally sourced data.

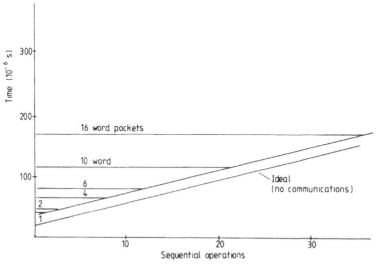

FIGURE 3.73   A graph showing the effect of communications on transputer operations. The operands and results were all sourced from INMOS links. Each line shows the result for a given packet size. The ideal (no communications) is also shown for comparison.

In advance literature for the floating-point transputer, INMOS made claims that the 20 MHz part would execute at a sustained rate in excess of 1.5 Mflop/s. This would make it a very powerful chip. These claims have been verified on engineering samples tested at Southampton University. If the application allows, the transputer could be used with very little support circuitry and one could, for example, imagine of the order of 32 transputer chips on a single IBM format circuit board, using no external memory. This board would add an astonishing 50 Mflop/s to a personal computer system. Table 3.13 gives operation times for IEEE floating-point on the T800, as taken from INMOS' advanced literature.

(v) Building systems with transputers
Transputers can be used in a variety of applications and in a variety of ways, and there are now many transputer-based systems on the market, including racks of transputer development cards from INMOS. However, it does not matter whether code or data may be partitioned over the transputer network (see §4.5); it is the communications issues which determine the mapping of code and the configuration of links.

Communications play an important role in all well designed systems. The communications bottleneck in a conventional processor is typically the processor-to-memory interface, which is usually the off-chip interface. Communications, not surprisingly, can also limit the performance of parallel machines, although here the limit arises in processor-to-processor communications. Moreover, this problem is fundamental, and although it can be moved

TABLE 3.13 Floating-point operation times for the T800 transputer. The -20 is a 20 MHz part and the -30 a 30 MHz part. All times in ns.

	Part			
	T800-30		T800-20	
	Precision			
Operation	Single	Double	Single	Double
Addition	233	233	350	350
Subtraction	233	233	350	350
Multiplication	433	700	650	1050
Division	633	1133	950	1700

up and down the systems' hierarchy, it will not go away. For example, we can trade chip and wiring complexity in a fully switched system against the large diameter of a static network.

In the latter case, when processing a partitioned data structure, one that is shared between the processors in the system, then only in the special case where each partition of the data structure is independent will no communications be required between processors. More generally, data must be shared between processors in the system and in many problems the communications complexity can dominate the complexity of the algorithm. For example, in sorting or computing matrix products each element in the resulting data structure requires information from all other elements in the original structures.

Such problems with global communications properties scale rather unfavourably with the extent of parallelism in the system, unless the connectivity between processors reflects that between partitions of the data structure. It is shown in figure 3.73 above that on local communications the number of operations (integer multiplications) required per word of data received over the INMOS link is two. It is also clear that a significant degradation of the communication bandwidth due to a large-diameter network would require a relatively coarse granularity in the partitioning of the algorithm. The transputer with its four links could be configured directly into the following networks: a two-dimensional grid, a perfect shuffle exchange network, a butterfly network, or other four-connected topologies (see §3.3.4). In all cases, however, if the data partitioning does not match the underlying network, then communication bandwidth will be degraded in relation to the processing rate of the system, in proportion to the diameter of the network. The diameter of the networks above varies from $\log_2 n$ to $n^{1/2}$. On applications that become communications-limited, performance will not scale linearly with processors added to the system.

The technique which avoids this unfavourable scaling is to allow arbitrary permutations to be established between processors, using a crossbar switch or its equivalent. Although the latter has many advantages and allows an arbitrary scaling of parallelism, the costs of such a switch will tend to grow with the square of the number of processors in the system. The costs of the fixed networks, however, vary linearly with the number of processors added to the system. However, because the transputer communicates over a high-bandwidth serial circuit, the costs of fully connecting transputers via their link circuits is not prohibitively expensive, or at least is not so for arrays of up to several thousand transputers, which with the T800 gives multiple gigaflop/s machines.

A transputer system which allows arbitrary networks of transputers to be configured was first proposed by one of the authors and is now the subject of a major ESPRIT project in advanced information processing. One of the motivating factors behind this design was to use switched nodes of transputers to implement program-derived networks, so that the transputers could be configured in algorithmic networks. This effectively creates a more powerful node which has a higher communications performance than a single transputer. Algorithmic parallelism can be exploited, with the placed processes and network configuration forming a static or quasidynamic data flow graph of the algorithm (see §4.5.2 for details).

To understand the arguments behind this, consider the following. A transputer has processing power P and communications bandwidth C, with the ratio of these, C/P, determining how soon a system would become communications-bandwidth-limited as more transputers were added to the system. Now consider taking a small number of transputers, say n, and arranging them as in figure 3.74. It has been shown (Nicole and Lloyd 1985) that this configuration will implement any graph of the n labelled transputers.

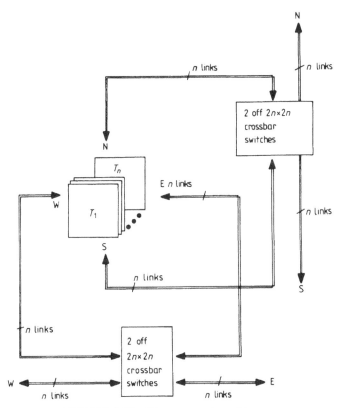

FIGURE 3.74 The supernode architecture.

It does this using two pairs of $2n \times 2n$ complementary crossbar switches. Such a 'supernode' can implement any four-connected data flow graph for a given algorithm. Now consider the communications ratio. The node has n transputers and therefore power nP; it also has communication bandwidth nC, as there is one link to the outside world for each link on a transputer in the node. The ratio of the two is therefore the same as a single transputer, namely, C/P. However, if these nodes are now connected into a static network, the diameter of the network would be smaller, as fewer nodes would be required to achieve the same performance; it would be smaller by a factor of n. This has three effects on communications scaling.

(1) Because the node has more legs, it can be more richly interconnected. For example, up to $4n + 1$ supernodes may be directly connected by one link only, $2n + 1$ by two links, and so on. Because of this, the diameter of the resulting network has been considerably reduced.

(2) If a static network of the same topology is implemented, then its diameter would in any case be reduced, as fewer nodes are required. For example, in a grid network the diameter would be less by a factor of $n^{1/2}$.

(3) Finally, as the fewer nodes can chew on a larger partition of the data structure, there is a surface/volume reduction in the data structure partition which will aid all but global communications requirements.

Of course, having defined a supernode of transputers, one can recursively define larger and larger systems, as shown below, using the notation from §1.2.4;

$$C_{i+1} = nC_i \times 4n^{i-1}S(2n \times 2n \text{ crossbar})$$

$$C_0 = \text{transputer.}$$

Such a description defines a multistage switch, built from $2n \times 2n$ crossbar nodes. For example, the two-level machine being built at Southampton for the ESPRIT collaboration is in fact a set of transputers connected by a three-stage CLOS network. The first level is illustrated in figure 3.74, the second level in 3.75, where here each node is a cluster of n transputers or a supernode.

The supernode can be considered as a supertransputer, for it obeys an OCCAM programming model, with the OCCAM providing transputer orders for each component and instructions to set the switching circuits, which realise the required network of connections. Thus OCCAM is used to define procedure and structure. A supernode would typically comprise some 10–50 transputers and by itself could provide a powerful, stand-alone workstation. Larger machines can be constructed using the supernode as a unit of replication, in regular arrays or in higher-level supernodes.

In the ESPRIT design, each supernode will have up to 36 transputers,

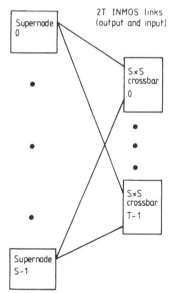

FIGURE 3.75 The second-level switch topology for a multiple-supernode, reconfigurable transputer network.

where most transputers have only 256 Kbytes of fast static RAM (eight chips), packed eight per board using hyper-extended, triple eurocards. At supernode level, one of the transputers, designated the control transputer, will have the switch chips mapped as I/O devices and will be able to configure the local network. This device is also the master on an eight-bit control bus, which is connected to all other transputers in the supernode. This bus provides the ability to set and read signals such as reset, analyse and error on any transputer, and provides a low-bandwidth communications medium between the transputers for control and debug purposes. This bus also provides IF ANY and IF ALL synchronisation between the transputers, so that global events may be signalled to the controller. This is necessary for non-static use of the switch, to provide a time reference when link activity has ceased so that it may be reconfigured. Each first-level supernode will also have RAM and disc servers, both implemented as transputers hooked into the switch.

The controller and bus are linked at the second level by the second-level controller, which sets the outer switches and acts as bus master for a bus on which each first-level controller is a slave. A prototype supernode was completed in the third quarter of 1987, with first samples of the T800 transputer, which was developed by INMOS as part of the same ESPRIT collaboration. The partners in the collaboration are the Apsis SA, Grenoble University, INMOS Ltd, the UK government research establishment RSRE, Southampton University, Telmat SA and Thorn EMI plc.

4 Parallel Languages

4.1 INTRODUCTION

There seem to have been two trends in the development of high-level languages. Those that owe their existence to an application or class of applications, such as FORTRAN, COBOL, and C, and those that have been developed to further the art of computer science, such as ALGOL, LISP, PASCAL and PROLOG. The development of the former to some extent has been stifled by the establishment of standards. Conversely, to some extent, the lack of standards and the desire to invent have led to the proliferation of versions of the latter.

There have also been attempts to produce a definitive language, incorporating the 'best' features of the known art and to bind these into an all-embracing standard. Such an attempt was made by the US Department of Defense (DOD) (1978) and the resulting language, ADA (Tedd *et al* 1984, Burns 1985), has been adopted by both the DOD and the UK Ministry of Defence. Even though validated compilers are now becoming widely available, most implementations seem very inefficient at implementing one of its major assets, its tasking or concurrency. Typical values for task creation and synchronisation range from 4–20 ms and 1–10 ms respectively (Burns 1985, Clapp *et al* 1986, Rhodes 1986). These may be compared to similar figures, but measured in microseconds, for OCCAM implemented on the transputer.

It can be argued that the procedure adopted with ADA was doomed from its inception; it has taken almost a decade from the establishment of the requirements to the widespread use of validated compilers. The resulting compilers are large, because the language is 'complete' and implementations supporting real concurrency on distributed systems are still immature. What is more, during that period we have seen some fundamental changes to the conception and use of computers. In particular, this decade has seen the widespread use of concurrency. Although concurrency is addressed in ADA, there is now far more practical experience of concurrency and new languages have been developed, such as OCCAM (May and Taylor 1984), which treat concurrency in a simpler, more consistent, and more formal manner. OCCAM is not a complete language, but its explicit and minimalist approach makes it an ideal tool to explore techniques for exploiting parallelism.

However, as we will demonstrate in this chapter, the numerous and vastly different applications and underlying models of parallelism will require radically different language structures.

One of the major divisions in language development, which has emerged during the last decade, is that between the imperative and declarative styles of programming. The declarative style of programming has had the most profound effect on computer architecture research during this period. This style of programming does not map well onto the classical von Neumann architecture, with its heavy use of dynamic data structures, which only highlight the deficiencies of the single port into a linear memory (the von Neumann bottleneck). Although this development is not the subject of this chapter, there is an abundance of literature for the interested reader. A good introduction can be found in the books *Functional Programming* (Henderson 1980) and *Distributed Computing* (Chamber *et al* 1984).

Another major development, which aims to promote the concept of re-usable software, is the object-oriented approach to programming. Although it is possible to buy well specified integrated circuits, with an exact specification (data sheet), and to integrate these into a more complex system, the same notion of 'off the shelf' software components has to date been limited to scientific library packages (equivalent say to MSI TTL components). Objective languages give the programmer the ability to construct 'software ICS' of arbitrary complexity; they thus promote the concept of the 'software shop', where a systems engineer may browse and, for example, buy a windowing package and applications package, and be sure (on reading data sheets) that they could be integrated into a single application. For an enthusiastic introduction to these concepts, the reader is referred to the book *Object Oriented Programming* (Cox 1986).

4.1.1 Imperative languages

An imperative language is one in which the program instructs the computer to perform sequences of operations, or if the system allows it, disjoint sequences of instructions operating concurrently. Imperative languages have really evolved from early machine code, by successive abstraction away from the hardware and its limited control structures. This has had two beneficial effects, namely the improvement of programmer productivity and the portability obtained by defining a machine-independent programming environment.

An imperative language, however, even at its highest level of abstraction, will still reflect the algorithmic steps in reaching a solution. In addition to the retention of this notion of sequence, these languages also retain a strong flavour of the linear address space, still found in most machines. This is reflected in the use of arrays, which have direct or addressed access to all elements.

One of the more recent abstractions added to the imperative programming model has been the introduction of concurrency (Harland 1985), where many disjoint sequences of instructions may proceed in parallel. This abstraction or parallelism in instruction streams evolved well before the widespread introduction of hardware concurrency, the requirement being the effective utilisation of the single most expensive resource in the computer, the CPU. With the use of logically separate tasks, CPU cycles could be shared between those tasks, thus avoiding the situation where a single task might have to wait for action on some slow I/O device. By abstracting concurrency, the non-deterministic sharing of the CPU's cycles could be obtained. Any task awaiting action from slow peripherals could be suspended by the system to allow another free task to proceed. Such multiprogramming environments gave rise to the interrupt, a mechanism to enable peripheral devices to attract the attention of the CPU when required.

The paper by Harland (1985) gives a good introduction to concurrency in imperative programming languages. He develops this theme from the early requirements for concurrency, through to more recent abstractions, which include the manipulation of processors and programs, as values within the language.

4.1.2   Declarative languages

The declarative style of programming is based on a more mathematical foundation, with the aim of moving away from descriptions of algorithms towards a rigorous specification of the problem. Hence the use of the name declarative. These languages are based either on the calculus of functions, lambda calculus, or on a subset of predicate logic. Examples of these two mathematical foundations can be found in LISP and PROLOG respectively. A good introduction to these foundations and their application to this style of programming can be found in the book *Logic, Algebra and Databases* (Gray 1984). The natural data structure for these declarative languages is the list, which is an indirect mapping onto a linear address space. This structure, however, is less well suited to implementation on a linear address space than the array and can lead to very inefficient utilisation of what is probably the computer's most critical resource, the processor–memory interface.

Although less efficient than the imperative style of programming, there is a demand for languages based on the formal rigour of mathematics. This is due to the massive complexity of modern software systems and the growing reliance found on such systems in our modern society. The move towards intelligent or expert systems also results in less regular, and hence less imperative, programming styles. Programmer productivity can be greatly improved for many applications by the use of more formal languages (Henderson 1986). Moreover, as these declarative languages are based on mathematics, it is possible to formally verify the software systems created with them (Gries 1976).

A further advantage of the declarative approach, or at least it is claimed to be an advantage by the experts in this field, is that such languages can supply implicit parallelism, as well as implicit sequencing (Clark and Gregory 1984, Shapiro 1984). However, this is not achieved without pitfalls, such as the parallel generation of unproductive work! However, research in this area is very active, being well supported through the UK government's Alvey research program, and major advances in this area may well be expected in the 1990s. ICL, for example, are collaborating on one of the major Alvey projects in this area, which could well lead to designs for efficient parallel-computer-based declarative languages. Whether these computers will be the general-purpose workhorses of the late 1990s, replacing the traditional main-frame, is an open question.

As an introduction to this style of programming, §4.3.4 describes CMLISP or *LISP, a language based on LISP and hence the lambda notation. This language contains extensions to drive the connection machine which involve an explicit expression of parallelism across the list structure. The connection machine has already been described in Chapter 3.

4.1.3 Objective languages

Objective languages are based on two main techniques, encapsulation and inheritance. Objectivity is a more pragmatic foundation than the rigour of logic or functional languages. It does, however, provide a potential solution to the software problem discussed above, which may at the same time be acceptably efficient on conventional computers. What is more, it provides a model of computation which can be implemented on a distributed system.

Of the two techniques, encapsulation is the most straightforward and is often used as a good programming technique (Parnas 1972, Booch 1986), even when not enforced by the language. Encapsulation hides data and gives access to methods or procedures to access that data. The modules in ADA

(Burns 1985) and Modula 2 (Wirth 1981) embody this technique. Access to the data is only provided through the shared procedures. The encapsulation of data and access mechanisms in this way defines the object, and an objective language is one which enforces this regime by building an impenetrable fortress around these objects. This can be done at a high level, such as in OBJECTIVE C and ADA (Cox 1986), or at a very low level in the system, as found in SMALLTALK 80 (Goldberg and Robson 1983).

Having encapsulated a programmer's efforts in the creation of such constrained objects, a mechanism must be provided in order to evolve or enhance that object, without a full-scale assault on its defences. This is the second technique of objective languages—inheritance. Inheritance allows the programmer to create classes of objects, where those classes of objects may involve common access mechanisms or common data formats. For example, given a class of objects 'array', we may wish to use this to implement a subclass of objects, 'string', which are based on array objects but have access mechanisms specific to strings. Conversely, given a set of access mechanisms to an object, we may wish to extend the range of type of that object, but exploit the access mechanisms which already exist. These are examples of inheritance. The mechanism for implementing inheritance is to replace the function or procedure call to an object by a mechanism involving message passing between objects. The message provides a key by which the access mechanism can be selected. This is equivalent to a late or delayed building of an access mechanism to a function call or procedure. Thus only at run time, when a selection is made from a class, is the appropriate access mechanism or data type selected.

The notion of encapsulation can readily be distributed for, as is indicated above, implementations are often based on message passing. The notion of inheritance is not so easily distributed, as it is based on a class tree defining and extending objects. A distributed implementation of a class of objects would, if naively mapped onto a processor tree, become heavily saturated at the root. However, by exploiting the applications parallelism and replicating the class structure (program) where required, an efficient mapping can result. What is more, because of the underlying packet nature of communications, a dynamic load balancing may be readily implemented. Object-oriented languages can then be considered as prime candidates for the efficient exploitation of parallel systems, where 'efficient' implies both programmer and machine utilisation efficiency. The field is new, however, and little work has been published to date in this area. The interested reader is referred to the OOPSLA 1986 proceedings, published as a special issue of the SIGPLAN notices (Meyrowitz 1986).

Although we have briefly reviewed some modern language trends, this chapter (with the exception of CMLISP) follows the trends in the development of imperative languages, and the way they have evolved to deal with various aspects of hardware parallelism. Section 4.3 deals with parallelism which is introduced through structure, as found in the SIMD-like architectures. These languages are ideal for array processors, such as the ICL DAP, and vector processors, such as the CRAY-1. Section 4.4 covers the use of process or task parallelism that is found in MIMD systems. However, the most common exploitation of parallelism is provided by the automatic vectorising compilers used for FORTRAN on most vector computers. This form of parallelism is extracted from loop structures within sequential code. It has the advantage (the only advantage perhaps) that existing sequential code for production applications can benefit from the speed-up obtained from vector instruction sets implemented on pipelined floating-point units. This approach and its potential disadvantages are discussed in detail in §4.2 below.

4.2 IMPLICIT PARALLELISM AND VECTORISATION

4.2.1 Introduction

The high-level language has been developed as a programming tool to express algorithms in a concise and machine-independent form. One of the most common languages, FORTRAN, has its roots in the 1950s and because of this, it reflects the structure of machines from that era: computers which perform sequences of operations on individual items of scalar data. Programming in these languages therefore requires the decomposition of an algorithm into a sequence of steps, each of which performs an operation on a scalar object. The resulting ordering of the calculation is often arbitrary, for example when adding two matrices together it is immaterial in which order the corresponding elements are combined, yet an ordering must be implied in a sequential programming language. This ordering not only adds verbosity to the algorithm but may prevent the algorithm from executing efficiently.

Consider for example the FORTRAN code to add two matrices:

```
      DO 10 J=1, N
      DO 10 I=1, N
      A(I,J) = A(I,J) + B(I,J)
   10 CONTINUE
```

Here the elements of **A** and **B** are accessed in column major order, which is, by definition, the order in which they are stored. On many computers, if this order had been reversed, the program would not have executed as efficiently. In this example, because no ordering is required by the algorithm, it is unwise to encode an ordering in the program. If no ordering is encoded the compiler may choose the most efficient ordering for the target computer. Moreover should the target computer contain parallelism, then some or all of the operations may be performed concurrently, without analysis or ambiguity.

The language APL (Iverson 1962) was the first widely used language which expressed parallelism consistently, although its aim was for concise expression of problems rather than their parallel evaluation. Iverson took mathematical concepts and notations and based a programming language on them. However, the major difference between the mathematical description of an algorithm and a program to execute it is in the description and manipulation of the data structures. APL therefore contains powerful data manipulation facilities.

Chapters 2 and 3 have described new developments in computer architecture. These architectures embed parallelism of one form or another into the execution of instructions in the machine. Computing has arrived therefore (nearly two decades after Iverson) at the situation where there is a need to express the parallelism in an algorithm for parallel execution.

4.2.2   The ideal approach

To illustrate the need described above consider the development of code, from the stated problem, to its execution in machine code on the target machine. The following four stages can be recognised, where the figure in brackets indicates the degree of parallelism to be found at that stage:

(a) choose a suitable algorithm for the problem (P);
(b) express the algorithm in a high-level language (L);
(c) compile the language into machine-readable object code (O);
(d) execute the code on the target machine (M).

The degree of parallelism is the number of independent operations that may be performed simultaneously (see §5.1.2).

In the ideal situation the degree of parallelism should not increase through this development process. We call this the principle of conservation of parallelism, i.e.,

$$P \geqslant L \geqslant O \geqslant M.$$

Programming languages that embody this ideal allow the programmer to express the parallelism of implementations of algorithms in terms of explicit language constructs, without being constrained by the nature of the target hardware. Should this expression of the algorithm give cause for inefficiency, then a transformation must be applied to impose order on the calculation. It takes only a little thought to verify that this principle is desirable; to translate a parallel operation into a sequence of steps requires only an arbitrary ordering to be placed on the elemental operations. To translate a sequential process into a parallel operation, however, requires a more complex analysis. The data flow of the program must be analysed to ensure that any ordering is in fact arbitrary and that there are no sequential dependencies in the proposed parallel operations. Inevitably, this analysis is often prevented by run-time constructs and variables in the code.

4.2.3 The vectoriser

The persistent demand for a new generation of computers to execute existing production codes unchanged has mandated a compromise approach in the exploitation of the performance benefits to be gained from parallelism. This demand has been such that all commercially successful vector processors support some form of vectorising compiler (Wolfe 1986).

How then does this vectorising approach compare with our ideal?

The CRAY-1 provides a good example of the vectorisation approach. Code must be written in sequential FORTRAN (i.e. $L = 1$). However, the natural parallelism of the machine is 64 and therefore $O = M = 64$. In this situation any parallelism in an algorithm is lost when it is expressed in the high-level language and must therefore be regenerated by the compiler. This is shown graphically in figure 4.1: (a) shows the ideal situation and (b) the loss and regeneration of parallelism.

This approach should really be considered as a temporary means of maintaining continuity when moving to a parallel computer. However, quite often the vectorising compiler will provide no parallel structures as an alternative to vectorisation. Thus all new code development must be made in the sequential language, which can only further entrench the sequential attitude towards programming and moreover can lead to obscure programming practices. Vectorisation often requires programmer intervention because many dependencies can only be resolved at run time. Rewriting code to avoid this often results in obscure constructs and bad programming practices, which hide the basic algorithm and make maintenance of the code more difficult.

4.2.4 Machine parallelism

It is unfortunate, but the most common class of language which expresses parallelism does so by the limitation of parallelism to that found on the target computer. Here $L = O = M$ and hopefully $P \geq L$. This is illustrated in figure 4.1(d). Obviously these languages, by their definition, are not portable. They have evolved, in the absence of any parallel processing standard, as the easiest and most efficient language to implement on a given machine. They are efficient, as inefficient constructs are avoided completely. Also, the programmer is made aware of the machine's parallelism and will therefore

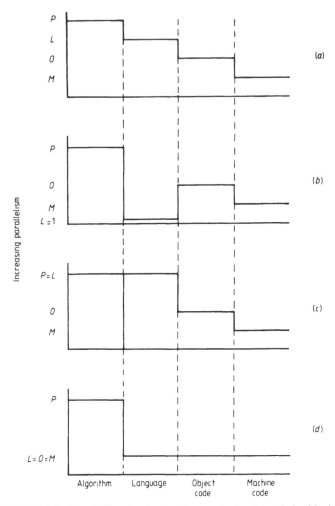

FIGURE 4.1   Parallelism in the development of code: (a) the ideal of conservation of parallelism; (b) vectorising languages; (c) expression of problem parallelism languages; (d) limitation of parallelism languages.

be more likely to avoid a severe mismatch with his problem's data structures. However, as more parallel computers (and languages) become available, the problems of transporting programs between them become very severe (Williams 1979). We will return to this problem of portability later.

4.2.5   Vectorising techniques
Vectorising compilers take code written in a sequential language, usually FORTRAN, and where possible generate parallel or vector machine code

instructions for the target machine. To do this the compiler must match segments of code to code templates which are known to be vectorisable. For example one of the simplest templates would be:

$$DO\langle LABEL \rangle Y = \langle CONST \rangle TO \langle CONST \rangle$$

$$\langle LABEL \rangle \langle ARRAY\ VAR \rangle(Y) = \langle ARRAY\ VAR \rangle(Y) \langle OP \rangle$$
$$\langle ARRAY\ VAR \rangle(Y)$$

This would simply translate into one or more vector operations, depending on target instruction set and difference between the loop-bound constants. More complex templates may involve variable loop bounds, array subscript expressions, conditional statements, subroutine calls and nested constructs.

This process can be considered as an optimisation or transformation, usually performed on the source code or some more compact tokenised form. For example, the compiler for the Texas Instruments Advanced Scientific Computer (ASC NX) performs the optimisation on a directed graph representation of the source code. However, for simplicity, only the source code transformations will be considered in the examples given in this chapter. It is clear that the most likely place in which to find suitable sequences of operations for vectorisation is within repetitive calculations, or DO loops in FORTRAN. Thus the vectorising compiler will analyse DO loops, either the innermost loop, or possibly more. The ASC NX compiler analyses nests of three DO loops and if there are no dependencies can produce one machine instruction to execute the triple loop. The Burroughs Scientific Computer (BSP) vectorising compiler analyses nested DO loops. It will reorder the loops if one of the innermost loops contains a dependency.

In general, the transformation performed on one or more DO loops is a change in the implied sequence or order of execution. In the sequential execution of the DO loop, the order implied is a statement-by-statement ordering, for each given value of the loop index. The order required for parallel execution is one in which each statement is executed for all given index values, before the following statement is executed for any. This transformation can only be performed when there is no feedback in and between any of the statements within the loop. Detecting and analysing these dependencies is the major task in vectorisation.

4.2.6 Barriers to vectorisation

It will be instructive to look at the various constructs, which can inhibit or block the vectorisation process. Some of these constructs will be fundamentally unvectorisable, while others are 'difficult' constructs for the vectoriser to analyse.

(i) Conditional and branch statements

Loops containing IF statements and transfers of control can inhibit the vectorisation process. However, many single- or multiple-statement conditionals can be made parallel (with some attendant loss of efficiency). For example the simple loop below,

```
      DO 10 I=1, N
10    IF (A(I).LT.0.) A(I) = -A(I)
```

reduces to the parallel construction (§4.3.1)

```
      A(A.LT.0.) = -A
```

where A.LT.0 supplies the vector mask which 'controls' the vector operation It is possible to vectorise single- and the more complicated multi-statement conditions, using this masking technique.

(ii) Sequential dependencies

This covers the largest area of possible barriers to vectorisation. Some dependencies are due to the ordering of statements, others are due to the recursive nature of the calculation. The former category can easily be vectorised by statement reordering and the use of temporary storage. Consider the simple example:

```
      DO 10 I=2,N
      A(I-1) = NEW(I)
10    OLD(I) = A(I)
```

This may look contrived, but is typical of the indexing constructs which can inhibit vectorisation. It is not recursive, although simply applying the ordering transformations will produce the wrong results. The implied ordering would put old values of A(I) into OLD(I); the transformed ordering puts new values of A(I) (i.e. NEW($I + 1$)) into OLD(I). The problem then is a simple question of timing and a good compiler would therefore reorder this calculation, or provide temporary storage for the old values of A(I).

The equivalent loops below are both vectorisable:

```
      DO 10 I=2,N
      OLD(I) = A(I)
10    A(I-1) = NEW(I)


      DO 10 I=2,N
      T(I) = A(I)
      A(I-1) = NEW(I)
10    OLD(I) = T(I)
```

The true recursive construct is similar to the above example and can be expressed in one line, or hidden using multiple assignments. Two examples are given below:

```
DO 10 I = 2,N
  T = A(I-1)*B(I)
10  A(I) = T + C(I)


DO 10 I = 2,N
10  A(I) = (A(I-1) + A(I+1))/2
```

These are examples of first-order linear recurrences which cannot be vectorised without resorting to special algorithms (see §5.2).

In general, for a construct matching the following template,

```
DO 10 I = ...
    .
    .
A(I) = A(I + OFFSET)

    .
    .
    .
```

where OFFSET is an integer variable, vectorisation is inhibited. This is because the compiler does not know at compile time that OFFSET has the same sign as the loop increment, or is such that no overlap occurs between left- and right-hand sides of the assignment. Thus the construct may be vectorisable in principle but not in practice, unless runtime checks are compiled.

(iii) Nonlinear and indirect indexing

Certain index expressions can also inhibit vectorisation. Simple linear combinations of loop index variables should cause no problems, unless there are hardware restrictions. However, if index expressions are nonlinear or contain indirect references, then vectorisation is not a trivial process. Examples of index expressions which could possibly inhibit vectorisation are given below:

$$(I*J + K)$$

$$(IV(I))$$

In the first case if either I or J were invariant within the loop being considered, then this expression would be linear and vectorisable, otherwise it would not be.

(iv) Subroutine calls within loops

The final barrier to vectorisation considered here is the inclusion of subroutines or user-defined functional calls within a loop. These are not vectorisable because subprograms are generally compiled separately and the compiler lacks the information required to make its analysis.

4.2.7   The BSP vectoriser

The BSP compiler (Austin 1979) is a good example of the use of vectorising techniques. It was based on a great deal of experience obtained in this field at the University of Illinois. It analyses more than one level of nested loops and can perform certain order transformations on statements and DO loops. In addition to this, the compiler will vectorise loops containing first-order linear recurrences and also certain loops containing IF statements. The following examples of the vectorising capability of the BSP FORTRAN compiler are taken from Austin (1979). The vectorisation transformation is expressed in terms of BSP parallel constructs

```
C          LOOP CONTAINING AN IF STATEMENT

           DO 2 I = 1, N-1
              IF (M.GT.3) A(I) = A(I+1) + C(I)
2          CONTINUE

C          IF LOOP VECTORISED

           WHERE (M.GT.3) A(1:N-1) = A(2:N) + C(1:N-1)

C          MULTI-STATEMENT CONDITIONAL

           DO 3 I = 1, N
              IF (A(I).LT.B(I)) GOTO 31
                 C(I) = E(I) + 2
                 A(I) = 0
              GOTO 3
31               C(I) = F(I)
3          CONTINUE

C          MULTI-STATEMENT LOOP VECTORISED

           LOGICAL L(N)
           L = A(1:N).GE.B(1:N)
           WHERE (L) DO
              C(1:N) = E(1:N) +2
              A(1:N) = 0
           OTHERWISE
              C(1:N) = F(1:N)
           ENDWHERE

C          MULTI-STATEMENT LOOP WITH NON-VECTORISABLE RECURRENCE

           DO 4 I = 2, 100
              A(I) = B(I) + 1
              C(I) = D(I-1) + E(I)
              D(I) = C(I) + A(I)
4             F(I) = D(I) + 2

C          PARTIAL VECTORISATION

           A(2:100) = B(2:100) + 1
           DO I = 2, 100
              C(I) = D(I-1) + E(I)
              D(I) = C(I) + A(I)
           END DO
           F(2:100) = D(2:100) + 2

C          LOOP REQUIRING STATEMENT REORDERING
```

```
      DO 5 I = 1, N
         A(I) = B(I) + C(I)
         C(I) = B(I-1)
5        B(I) = 2*A(I+1)

C     VECTORISED WITH TEMPORARY AND REORDERING

      REAL TEMP(N)
      TEMP(1:N) = A(2:N+1)
      A(1:N) = B(1:N) + C(1:N)
      B(1:N) = 2*TEMP(1:N)
      C(1:N) = B(0:N-1)

C     LOOP REQUIRING REORDERING

      DO 6 I = 1, N
      DO 6 J = 1,N
      DO 6 K = 1, L
         A(I,J+1,K) = B(I,J-1,K)
6        B(I,J,K) = A(I,J,K)

C     VECTORISATION WITH REORDERING

      DO 7 J = 1, N
         A(1:M, J+1, 1:L) = B(1:M, J-1, 1:L)
7        B(1:M, J, 1:L) = A(1:M, J, 1:L)
```

In this last example a recurrence in *J* has been isolated and the innermost and outermost loops in *M* and *L* have been vectorised. Notice that although the BSP hardware can handle first-order linear recurrence, the recurrence vectorisation process will only be invoked if no other vectorisation can be performed on a given set of DO loops. This is because the parallel evaluation of a first-order linear recurrence is not 100% efficient (see §5.2).

4.3   STRUCTURE PARALLELISM

Section 4.2 above gives one solution to the problem of portability, which is to continue to use existing (sequential) languages and to make the system responsible for generating the mapping of the problem or algorithm onto the underlying hardware. However, techniques for automatically generating parallelism are largely limited to optimisations of loop constructs for execution on vector processors. The automatic exploitation of other forms of parallelism, for example replicated systems, is not well understood and results to date have shown poor utilisation of processor and communications resources. One of the major problems in this respect is the lack of a formalism for describing the structure of the problem within a consistent model of computation which will facilitate a mapping of that structure onto the underlying machine's structure. This problem of data mapping is not described well in a language which must use loop structures and indexing to express it. Other more formal approaches treat whole array objects and classes of mappings that can be applied to them (for example, see §3.3 and Flanders (1982)).

Even targeted on a vector computer, the vectorising compiler is not the ideal route to portability. The FORTRAN programmer will of course optimise his sequential code, so that the compiler will recognise as much as possible as being vectorisable. Because of the underlying differences in the vector hardware, optimal loop lengths and access patterns into memory will change from machine to machine. The solution to the portability using vectorisation therefore fails, as the programmer treats FORTRAN as in the past, as an assembler language.

An alternative solution places the onus on the programmer to explicitly declare which parts of his code are to be executed in parallel. In this way a parallel structure can be chosen to express the solution of the problem and not the underlying machine structure. The portability problem now becomes one of implementation, as an efficient mapping of the programmer's expression of parallelism must now be automatically mapped onto the target hardware.

Two techniques are available to explicitly express parallelism: by using a description of the data structure to express parallelism, *structure parallelism*, or by using a description of the program or process structure to express parallelism, *process parallelism*. They can in some circumstances be very similar, as a distributed program's process structure may be designed to exploit the structure of its data. To make a distinction in these circumstances, assume that structure parallelism is defined at the granularity of a single operation and that the operations are carried out as if simultaneously over every element of a data structure. Process parallelism, on the other hand, is defined with a large granularity, with an instruction stream and state associated with each element, or (more likely) with each partition of the data structure.

It is really the underlying computational model that distinguishes these two forms of parallelism, and the manner in which load is balanced across a system. In structure parallelism, one can consider virtual processors to be associated one per data structure element, with activated data being mapped onto the available processors in a manner which balances the load between processors. Thus, load sharing is achieved through data-structure element redistribution—data remapping. For example, if only one row of a distributed matrix was selected for processing, a redistribution of the data elements in that row may be required to maintain a load on all processors. This may also be viewed as a redistribution of virtual processors to actual processors to maintain an even load. In process parallelism, on the other hand, the program partition or process is virtualised and load balancing occurs by the distribution of processes across the processors. Process parallelism is explored in more detail in §4.4: this section is concerned only with the exploitation of structure parallelism.

Structure parallelism is in essence a formal method of expressing the result of the vectorising transformations given in §4.2 above, as the operational semantics of the vector processor are as described above for structure parallelism. The vectorisation approach works for vector computers because pipelined access to a single memory system can mask the data transformation implicit in these semantics, and also because the efficiency of a pipeline is an asymptotic function of vector length (see §1.3).

Structure parallelism has been used in many language extensions, usually to express the SIMD parallelism found in processor arrays, e.g. DAP FORTRAN (ICL 1979a). However, these languages have tended to express the parallelism of the underlying hardware, thus leaving the programmer to define the data transformations required to maintain a high processor load. They can therefore be considered as low-level, machine-specific languages. The emerging standard for FORTRAN, FORTRAN 8X (ANSI 1985), is proposing extensions to allow any array to be manipulated as a parallel data structure. This is a welcome proposal, although somewhat late and incomplete; it will, however, allow the programmer to express the structure and parallelism inherent in an algorithm, rather than in that of the hardware. We explore this emerging language standard later in this section.

Using structure parallelism, the programmer is free to use virtually unbounded parallelism in the expression of an algorithm. Completely general languages of this type allow all data structures to be treated as objects on which operations can be performed in parallel. This departure is best viewed as allocating a virtual processor to each element of the data structure. Then, at a given stage in an algorithm, some of these data-structure elements will be activated, either explicitly by selection from the data structure (for example a row from a matrix), or implicitly by a conditional construct in the language (such as the WHERE statement in FORTRAN 8X). The compiler, or indeed the system at run time, will then allocate activated virtual processors to real processors in the system.

This idea of virtually unbounded parallelism could prove an embarrassment in a process-based approach, where the overhead in creating many instruction streams may be greater than the work involved. In both structure and process parallelism, a transformation from parallel to sequential can be made at compile time for a given target architecture, in order to optimise performance. However, in the case of process parallelism, the transformation is not so straightforward and it would be unlikely that a run-time system could be developed to do this efficiently. In the case of data-structure parallelism, the transformation is simple, merely one of ordering sets of similar operations.

Moreover, any ordering imposed can be such so as to optimise access patterns into memory systems (Jesshope 1984). It is true that this can result in a large requirement for storage, because of the need to perform all operations as if simultaneously. For example, complex array-valued expressions will require array-valued temporaries. However, optimisations such as those used for vectorisation can reduce the requirements to match the parallelism of the target hardware.

4.3.1   Array constructs

In this section we consider a number of constructs which can be used to express the structure of parallelism through the use of arrays. These are based on the FORTRAN language but the syntax does not necessarily reflect any implementation, current or proposed. They attempt to derive a consistent extension to a language, which naturally supports structure parallelism as applied to arrays. These proposals differ in some important issues from these being proposed by the ANSI X3J3 Committee, which has been considering, among other issues, array extensions to the next FORTRAN standard (8X). The ANSI proposals are outlined in detail in §4.3.3.

Because of the higher level of abstraction found in the array construct when compared with sequential code, we would expect to see a contraction in the amount of code required to express a given algorithm. This is analogous to the expressiveness of matrix algebra over that of scalar algebra. Any construct included in the language should therefore reflect this and allow the concise expression of operations between array-valued data objects. APL takes this philosophy to its extreme and results in code which is difficult to read. We do not wish, however, to simply add cumbersome syntax to existing sequential constructs. This opposite extreme can be illustrated by the PAR DO, which has been suggested as an alternative to the existing FORTRAN DO statement, wherein the order of evaluation is modified so that each statement is executed for all values of the DO range, before the succeeding statement is executed. Thus the body of the loop is executed in sequence, but with each statement being executed for the whole range of the DO statement, as if each statement were included in its own loop. This syntax is confusing and verbose, and it is far more natural to imply the DO loop(s) from the structure of the array operands.

(*i*) *Arrays as elemental data objects*
In this approach arrays are treated as distinct data objects of a given rank (dimensionality). Therefore, any reference to the array will imply the whole array. It should be noted that this approach does not preclude the use of an array as a set of sequential objects, for indexing can be used to

reduce the rank of the object by selection. Thus, for example, given a three-dimensional array **A**, it is said to have rank 3, and any reference to the name **A** alone will imply a reference to all elements in parallel. Indexing in any one of the dimensions of **A** may be considered as a selection operation, which reduces the rank of **A** by one. Thus if all three dimensions are indexed, a scalar or data object of rank 0 is selected from **A**. More details of selection mechanisms are given below. It should be noted that this concept already has some precedent in sequential languages; in FORTRAN for example, the reference to an array name in the READ and WRITE statement implies a reference to all elements of that array.

Although this approach is the most general, many language extensions have made a compromise, in that the number of dimensions in which the array can be considered as a parallel object are limited. Any further dimensions must be indexed within the body of the code, thus providing sets of parallel array objects of limited dimensions. This mixed approach has been adopted almost exclusively on processor arrays, where the number of dimensions which may be referenced in parallel corresponds to the size and shape of the hardware. It has the advantage of distinguishing the parallel and sequential access found in processor array memories, but the disadvantage of being machine-dependent and hence non-portable.

Examples of this approach are found in CFD (Stevens 1975), GLYPNIR (Lawrie *et al* 1975) and ACTUS (Perrott 1979), all languages proposed or implemented for the ILLIAC IV. CFD and GLYPNIR manipulate one-dimensional arrays of 64 elements, which map onto the ILLIAC array of processors. ACTUS is more general, but the original implementation developed for the ILLIAC IV allowed only one dimension of a PASCAL array to be referenced in parallel. Another example of this mixed approach is found in DAP FORTRAN (ICL 1979), where either the first or first two dimensions of an array may be referenced in parallel. Again these first two dimensions must correspond to the DAP size. Both DAP FORTRAN and CFD are discussed further in §4.4.

(ii) Selection from array objects

It is assumed that an array is defined as above as a parallel data object and that indexing mechanisms are rank-reducing or selection techniques. Thus slices and subsets of array objects, which may themselves be parallel array objects, may be manipulated within the language. There are many techniques available to perform this selection, some of which can have conflicting meanings.

Selecting reduced rank objects   This mechanism is perhaps the most important, as it provides downward compatibility with the underlying sequential language. At its simplest it is equivalent to indexing in a sequential language. The difference however is that given an array object of rank n say, then all reduced rank array objects are also atomic elements of the language. For example row vectors, column vectors and scalars are all valid atomic objects, which are selections from a two-dimensional array or matrix. Row and column vectors are selected by indexing in one and only one dimension of the array, and a scalar is selected by indexing in both dimensions of the array. Thus in FORTRAN, given a two-dimensional $N \times N$ array or matrix **A**, then

(a) **A** would refer to the whole matrix,
(b) A(I,) would refer to the Ith row of **A** (a vector),
(c) A(,J) would refer to the Jth column of **A** (also a vector),
(d) $A(I,J)$ would refer, as expected, to the I,Jth element of **A** (a scalar).

In some languages the syntax of this selection mechanism has been emphasised by a special symbol, usually an asterisk, which is used in the position where subscripts are elided. This symbol can be thought of as a whole-dimension selector. Thus the examples above become,

(a) **A** or **A**(*,*)
(b) A(I,*)
(c) A(*,J).

However, because of its elegance and simplicity, subscript elision will be used in all the following examples, unless a specific syntax is being described.

Selecting a range of values   It will often be necessary to select a range of values from a given subscript position. Thus instead of reducing the rank of an array object the size of that object is reduced. The range must therefore specify a subset of the whole dimension range. This can be defined, as in loop control, by a pair or triplet of integers or integer expressions. These would specify a start index, an optional step index and a limit index, which would be used as a selector in one of the dimensions of an array. Thus, given the same $N \times N$ matrix used in the example above, $\mathbf{A}(2:N-1, 2:N-1)$ would select the interior points of the matrix and $\mathbf{A}(1:(2)N, 1:3)$ would select the first three elements of all odd rows. In both examples the syntax used is start:(step) limit, where the step, if omitted, is assumed to be unity. These rank- and range-reducing selection mechanisms are illustrated in figure 4.2 for an 8×8 array.

Selection using integer arrays   The simple selection mechanisms described above give as a special case the more familiar sequential indexing. It can also

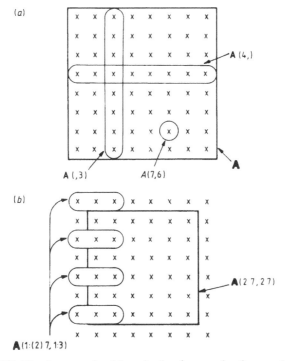

FIGURE 4.2   An example of the selection from an 8 × 8 array: (a) rank-reducing selections; (b) range-reducing selections.

be argued that the same should be true for indexing with arrays of integers and that it should be a special case of the familiar indirect indexing found in sequential languages. However, there are at least two ways in which this indirect indexing may be used, and both cannot be represented by the same syntax.

One interpretation gives a linear mapping or cartesian product of linear mappings over more than one dimension. This is shown below, using familiar sequential constructs, where **IV** and **JV** are the integer vectors defining the mappings:

(a) $\mathbf{A}(I, \mathbf{JV}(J))$
(b) $\mathbf{A}(\mathbf{IV}(I), \mathbf{JV}(J))$

Here (a) gives a linear mapping of the second dimension of **A** and (b) gives a cartesian product of linear mappings in both dimensions of **A**. It can be seen that if this use were applied in parallel, with subscript elision

(a) $\mathbf{A}(, \mathbf{JV})$
(b) $\mathbf{A}(\mathbf{IV}, \mathbf{JV})$

then the resulting arrays are of the same rank as **A**, but have a range in the dimensions selected given by the mapping vector. The mapping produced may be one-to-many; however the values of the elements of the mapping vector must all lie within the range of the dimension in which it is used. Thus if **A** is an $N \times N$ array and **IV** and **JV** are vectors of range M, then all elements of **IV** and **JV** must be less than or equal to N, and (a) would give an $N \times M$ array and (b) an $M \times M$ array. Both are arbitrary mappings of the elements of **A**.

The other interpretation of this construct gives a projection over one or more dimensions of an array. This is a rank-reducing operation and is shown below using similar sequential constructs. Here however **JV** is an integer vector and **JM** an integer array, both of which define a slice of the array which is projected over the remaining dimensions:

(c) $\mathbf{A}(I, \mathbf{JV}(I))$
(d) $\mathbf{B}(I, J, \mathbf{JM}(I, J))$

The important point to notice here is that the index vector **JV** and index array **JM** have the same shape as the array slice they select, and are indexed in the same way. When using subscript elision there is no way to differentiate between the two uses of indirect indexing, as (c) and (d) yield the following parallel constructs

(c) $\mathbf{A}(, \mathbf{JV})$
(d) $\mathbf{B}(,, \mathbf{JM})$

It can be seen that the syntax of the parallel construct given by (c) is identical to that given by (a), but they have completely different semantics. Here if **A** is an $N \times N$ array and **B** an $N \times N \times N$ array, then **JV** must be an N-element vector and **JM** an $N \times N$ array. Both reduce the rank of the arrays they are indexing as if they were scalars; however, the slice obtained is not laminar but projected over the elided dimensions. Again the values of the elements of the index array must be less than or equal to the range of the dimension from which they select. These two techniques are illustrated in figure 4.3, for a 4×4 array.

It will be shown later that the projection variant of this structure, when used in conjunction with another construct, is able to emulate the general mapping technique. The obvious choice of semantics for this construct is therefore as a projection. This convention is adopted in all later examples, unless otherwise specified. This selection technique may be used in conjunction with other indexing techniques in other dimensions in the array, including other index arrays. However all indexed arrays must conform to the slice of the array produced. Thus TABLE $(,, I, J, K)$ is a valid selection from TABLE, even if

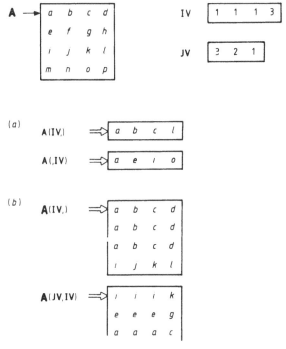

FIGURE 4.3   An example of the selection from an array **A** using integer vectors (**IV** and **JV**): (*a*) the projection selection; (*b*) a general mapping in one and two dimensions of the array.

I, J and K are integer arrays. In general I, J and K must either be integer scalars or integer arrays which conform to the slice produced (i.e. those dimensions in which the whole range is implied). The concept of conformity is discussed in more detail later.

Selection using boolean array objects   Another selection mechanism which can be applied to array objects uses conforming logical arrays or expressions. Selection is made depending on the truth value of the logical selector. Thus the ith element may be selected from a one-dimensional array by indexing it with a one-dimensional boolean array with the value 'true' in the ith element. The arrays must of course conform. Similarly the same boolean array could have been used to select either a row or column vector from a two-dimensional array or matrix. These examples are illustrated in figure 4.4.

All of the examples given above use this form of indexing as a rank-reducing selection mechanism. DAP FORTRAN (see §4.3.2) for example insists on this restriction. However it is possible, although perhaps not desirable, to extend this form of indexing so that it becomes range-reducing. In this case the boolean vector above would have a number of elements set to true, and

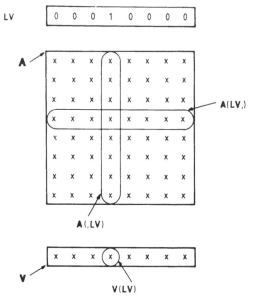

FIGURE 4.4   An example of the selection from an 8 × 8 array using a logical vector **LV**.

the selection would give an object of the same rank, but with only those elements selected by the values 'true'. Similar techniques can be used to selectively update an array object. These are discussed in more detail later.

Shift indexing   Strictly speaking this is not a selection mechanism; however as it is often implemented as an indexing technique it is included in this section for completeness. It is an alignment mechanism and can be used to shift or rotate array objects along a given dimension. A good example of its use is in mesh relaxation techniques (see §5.6.1), where each point on the mesh is updated from some average of its neighbouring points. The simplest case is where a point takes on the average values of its four nearest neighbours in two dimensions. This could be expressed as below, where the syntax is from DAP FORTRAN:

$$\mathbf{A} = (\mathbf{A}(+,) + \mathbf{A}(-,) + \mathbf{A}(,-) + \mathbf{A}(,+))/4.$$

Every element of **A** is updated simultaneously from the four neighbouring elements, selected by shifting the array **A** left (+) and right (−) in each dimension. For the interior values of **A**, this is equivalent to the sequential code below.

```
DO 10 I=2, N-1
DO 10 J=2, N-1
10   ANEW(I,J) = (A(I+1,J) + A(I-1,J) + A(I,J-1) + A(I,J+1))/4
```

```
DO 20 I=2, N-1
DO 20 J=2, N-1
20   A(I,J) = ANEW(I,J)
```

The most natural way of writing this relaxation sequentially would be to replace ANEW(I,J) by **A**(I,J) in loop 10. This, however, is recursive and when evaluated uses both old and new values of **A**, defined by the loop ordering. Numerical analysts will recognise this as the difference between the Gauss–Seidel and Jacobi relaxations. It will be seen in §4.3.3 that FORTRAN 8X proposes that these shifts be implemented using functions, and not employing indexing techniques.

In sequential code the edge effects would normally be considered separately; however in the parallel construct this can be accomplished by assuming some geometry to the array. For example, a dimension can be considered to be cyclic, in which case shifts will be end-around, or to be planar, where shifts will be end-off with appropriate data shifted in at the boundaries. Combinations of these geometries in a two-dimensional array are illustrated in figure 4.5. It can be seen that planar, cylindrical and toroidal topological surfaces are produced.

The syntax given in the example above is obviously limited to shifts of only one position in each direction, as to add an integer to the + and − symbols creates a valid selection from **A**. Shifts of more than one place have been incorporated into parallel languages using a whole-dimension selector

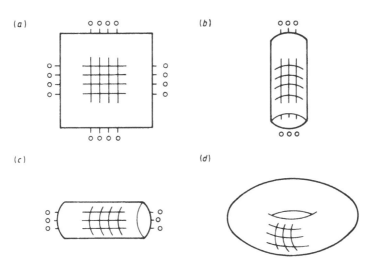

FIGURE 4.5   The topological geometries possible in two dimensions using planar and cyclic geometries for array shifting: (a) planar; (b) cylindrical (axis N–S); (c) cylindrical (axis E–W); (d) toroidal.

symbol, $*$ for example, where $\mathbf{A}(* - N)$ would shift $\mathbf{A}$ in its first dimension N places to the right.

(iii) Array expressions and conformity

Given whole arrays as data objects within a language, it is necessary to introduce some rules concerning their use in expressions. The most fundamental of these is the way in which operations are applied to the individual elements of the array objects. When combining arrays with arithmetic, relational or logical binary operators it is assumed that corresponding elements of the two arrays are combined with the same operator, as if done simultaneously. Similarly unitary operators are applied to all elements within the array object. The results of such operations are also array objects. Having defined operations between arrays in this way, expressions can be built up as in sequential languages, using the accepted rules concerning type, precedence of operators and placement of parentheses. However, the most important considerations are the rules concerning size and shape of array objects as operands within expressions. These are fundamental to the language design.

Unless two objects contain the same number of elements, the definition of the elemental array operation becomes ambiguous. This may be avoided by defining the way in which array operands overlap, together with a value to be given to the undefined elements. For example, array operands may be aligned by their first or last elements, linearly or dimension by dimension. Undefined elements may be given unity or null values, or left undefined etc. Clearly, this situation is not desirable.

The alternative to this is to constrain all operand pairs in an expression to conform. This will be defined here by saying that the two operands must have the same rank and the same range in corresponding dimensions. By limiting expressions in this way, it is possible to have a much tighter control over the occurrence of errors, both at compile and at run time. This is illustrated neatly in the example performing matrix multiplication given at the end of this section. Having introduced this restriction however, it becomes necessary to introduce constructs which will provide the ability to compress, expand or reshape arrays, so that they conform. In this way the onus will be placed firmly on the programmer, to specify exactly how he or she wishes to make arrays conform. The indexing techniques discussed above will provide some of the mechanisms for this; others must be provided either as additional operators or functions within the language.

(iv) Coercion of arrays to obtain conformity

A functional notation will be used to illustrate some of the operations required in order to coerce array objects to conform. The functions given here will be required in some form, in any language that requires array operands to conform.

Rank-reducing functions It has been shown that the rank of an array object may be reduced by indexing or selection. Another way in which it may be reduced is by the repeated use of a binary operator between elements in one or more dimensions of the array. Although this could be described in the language as a sequence of operations, this would not give the opportunity of using parallelism for the reduction. For example the sum of N elements can be performed in $\log_2 N$ steps in parallel (see §5.2.2). Table 4.1 gives a list of the most common reduction operations. Each has two parameters, the array and the dimension along which the reduction is to be performed. Functions should also provide for a reduction over all of the dimensions of the array. In APL similar functions are provided as composite operators where @/gives the reduction using the binary operator @. The syntax of combining operators is considered in a paper by Iverson (1979).

Rank-increasing functions It is often necessary to increase the rank of an object to obtain conformity. The most frequent use of this is in operations between scalars and arrays. This is often termed broadcasting in parallel

TABLE 4.1   Reduction functions for array extensions to FORTRAN. A is integer or real, B is logical. Provision should also be made to extend the action of these functions to all dimensions, perhaps by omitting the second parameter.

Function	Operation	Reduction
$\text{SUM}(A,k)$	$+$	$\sum_{i_k} A(i_1 \ldots, i_k, \ldots, i_l)$
$\text{PROD}(A,k)$	$*$	$\prod_{i_k} A(i_1, \ldots, i_k, \ldots, i_l)$
$\text{AND}(B,k)$	$\wedge$	$\text{and}_{i_k} B(i_1, \ldots, i_k, \ldots, i_l)$
$\text{OR}(B,k)$	$\vee$	$\text{or}_{i_k} B(i_1, \ldots, i_k, \ldots, i_l)$
$\text{MAX}(A,k)$	$\geqslant$	$\max_{i_k}\{A(i_1, \ldots, i_k, \ldots, i_l)\}$
$\text{MIN}(A,k)$	$\leqslant$	$\min_{i_k}\{A(i_1, \ldots, i_k, \ldots, i_l)\}$

computation, because a copy of the scalar is broadcast as the second operand to every element of the array. This action may be regarded more generally as the coercion of a scalar, by repetition, into a conforming array. This particular example is unambiguous and can be implied simply by relaxing the rule which requires conformity between operands, allowing scalars and arrays to be mixed freely within expressions.

This is not true however for the more general case. Operations between vectors and matrices for example can be performed either by repeating the vector as a row, or as a column, to obtain conformity with the matrix. In general there are ^nC_r ways in which an r-dimensional 'square' array may be coerced into an n-dimensional 'square' array, where

$$^nC_r = \frac{n!}{r!(n-r)!}.$$

If arrays have different ranges in each dimension, the number of possibilities may be restricted, or even vanish. However, it is a good idea to introduce this coercion as an explicit operation or function call, so that error checking may be provided. Again the example of matrix multiplication illustrates this. Only one function need be required to perform any coercion by repetition. Thus, for example,

$XPND(\mathbf{A},k,N)$

is an array formed by the repetition of $\mathbf{A}$, N times along its kth dimension. For example if $\mathbf{V}$ is a vector of N elements then

$XPND(\mathbf{V},1,N)$

is an $N \times N$ matrix formed by the repetition of $\mathbf{V}$ as rows of the matrix and

$XPND(\mathbf{V},2,N)$

is formed by the repetition of $\mathbf{V}$ as columns of the matrix.

Any coercion can now be built up by repeated use of this one simple function.

A reshaping construct   Another construct which may be necessary to obtain conformity is the conceptual reshaping of data structures. Operations between arrays of different dimensions but with the same total number of elements are sometimes necessary. A good example is in the fast Fourier transform algorithm (§5.5.2), where it is simpler to treat a linear array of N elements as a variable three-dimensional array. This is illustrated in one of the programming examples given at the end of this section.

As much more emphasis is being placed on the shape of arrays, which must conform in expressions and assignments for example, it is more important to define the shape of array objects passed to subprograms. For example, a parameter and argument would then conform if only the shape or dimensionality of these objects was the same. Thus in FORTRAN, an array parameter to a subprogram should define the principal subarray of the argument passed. In this way a parametrised subset of the range in each dimension may be selected within the subprogram.

Having allowed the DIMENSION statement to define a dynamic range within the subprogram, a statement is required to change the shape of an array with respect to its initial declaration. For the sake of illustration we introduce the MAP statement, which has the format of a dimension statement but which can be considered as an executable statement, defining a remapping of a declared array object. It must be assumed that this mapping is one-to-one in order to give an exact definition.

For example, if we consider the dimension list $n_1, \ldots, n_p$, of the array object

$$\mathbf{A}(n_1, \ldots, n_p)$$

then this list defines the shape and mapping of $\mathbf{A}$ over an ordered set of $\prod_{k=1}^{p} n_k$ data locations (not necessarily contiguous). An element from this set is located by providing an index list, $i_1, \ldots, i_p$, which is used to generate the mapping function f_D, giving the element's position in the set:

$$f_D(i_1, \ldots, i_p) = i_1 + n_1(i_2 - 1) + \ldots + \prod_{k=1}^{p-1} n_k(i_p - 1).$$

The MAP statement would redefine this mapping function over the same ordered set of elements, by providing a MAP list $m_1, \ldots, m_q$ in the MAP statement.

$$\text{MAP } \mathbf{A}(m_1, \ldots, m_q)$$

This redefines the shape of A, so that given a new index list $i_1, \ldots, i_q$, an element is selected from the set by the new mapping function f_M:

$$f_M(i_1, \ldots, i_q) = i_1 + m_1(i_2 - 1) + \ldots + \prod_{k=1}^{q-1} m_k(i_q - 1).$$

It can be seen that the mapping is not completely defined unless

$$\prod_{k=1}^{p} n_k = \prod_{k=1}^{q} m_k.$$

It should be noted that f_D need not define a mapping over a contiguous set of storage locations. This is because the ordered set of elements, over which f_D is defined, may also represent a mapping (reduction in range) of an array passed to a subprogram.

In a serial computer, these address mapping functions would merely be used for address calculation. However, on some parallel machines, remapping of physical data may actually be required in order to provide parallelism.

An example of the use of this MAP statement is given below; it reshapes a linear array into two dimensions.

DIMENSION **A**(N)

.

.

.

MAP **A**($N/2$, 2)

.

.

.

It can be seen that the remapping of **A** is not completely defined if N is odd. The last element of **A** is undefined in the MAP statement, due to the truncation of integer division.

(v) Expression indexing

By the introduction of parallelism into a language, expressions themselves have become array-valued objects, with a given shape. It may therefore be desirable to select from an expression using the same indexing techniques described earlier. This selection is only likely to be required with the use of the functional form and then only to avoid unnecessary assignment to temporary variables. Syntactically the selection can be performed by following the expression with a pair of brackets containing the required selector. Thus the expression is treated exactly as an array object. An example of this is given in the first programming example.

(vi) Array assignment

The value of an array expression must at some stage be assigned to an array variable. The assignment like all other operations is elemental and, using the same arguments as for expression evaluation, the result of the expression must conform to the array variable. Again this conformity can be relaxed for the assignment of a scalar expression to an array variable, as this can be performed unambiguously. In order to obtain conformity at the assignment stage, all of the selection mechanisms described earlier may be used to selectively update an array variable.

Thus for example, a vector expression could be selectively assigned to a row or column of a matrix by:

$$\mathbf{A}(I,) = \langle \text{conforming vector expression} \rangle$$

or

$$\mathbf{A}(,J) = \langle \text{conforming vector expression} \rangle$$

One technique which is particularly powerful is the selective assignment, using a boolean array or expression. This in fact can replace the conditional assignments. The technique is sometimes called masking, as the boolean array may be considered as a mask controlling the assignment. For example the absolute value of **A** may be generated by negating only its negative element as given below:

$$\mathbf{A}(\mathbf{A}.LT.0) = -\mathbf{A}$$

(vii) Programming examples
The constructs above are now used to illustrate how parallel coding can give concise and readily understandable interpretations of an algorithm. Three examples are used; the general one-dimensional mapping problem, matrix multiplication and the fast Fourier transform algorithm.

General mapping   This problem can be neatly illustrated using the selection of a word or phrase from an alphabet. Thus given a one-dimensional array of characters, ALPHABET(27), which contains A through Z and blank, we wish to select a word or phrase, WORD(N), from this alphabet, using an integer array INDEX(N), which points to the corresponding letters in the alphabet.

Remember that the indirect indexing we have allowed is of the projection or rank-reducing type, and not the general mapping required by this example. The technique, therefore, is to expand the alphabet into a two-dimensional object by repetition N times. The word can then be selected by indexing the first dimension, using the N elements of INDEX. This gives the code below. The structures manipulated are illustrated in figure 4.6.

```
      FUNCTION WORD (ALPHABET, INDEX, N)
      CHARACTER ALPHABET(27), WORD(N)
      INTEGER INDEX(N)
C
      WORD = XPND(ALPHABET,2,N)(INDEX,)
      RETURN
      END
```

Matrix multiplication   The second example given is matrix multiplication. It shows the correspondence between the mathematical and programming

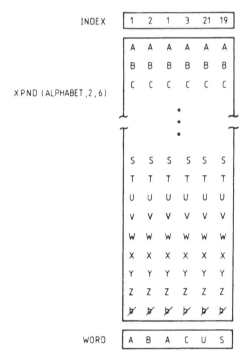

FIGURE 4.6   The data structures manipulated in the general mapping programming example.

notation. Notice also that in this example, all *NLM* multiplications are expressed in parallel without loss of generality in the algorithm.

```
       FUNCTION MATMULT (A,B,N,L,M)
       REAL A(N,M), B(M,L), MATMULT(N,L)
C
C      THE MATHEMATICAL DESCRIPTION OF THE PROBLEM IS GIVEN FOR I1 = 1,...,N
C      AND I3 = 1,...,L, BY
C
C                          M
C                         --
C      MATMULT(I1, I3) =  \ A     *B
C                         / I1,I2  I2,I3
C                         --
C                         I2=1
C
C      THE CODE IS VERY SIMILAR
       MATMULT = SUM(XPND(A,3,L)*XPND(B,1,N),2)
       RETURN
       END
```

The product is formed by first expanding **A** and **B** so they become three-dimensional objects (figure 4.7). In the first **A** is repeated as the first two dimensions of the structure, in the second **B** is repeated as the last two dimensions. It can be seen that both these calls to XPND produce arrays

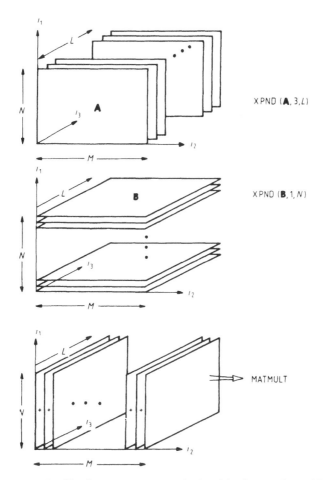

FIGURE 4.7 The data structures manipulated in the matrix multiplication programming example.

which conform (i.e., both have shape (N,M,L)). The product of these arrays is then reduced by summation over the middle dimension (i.e. for all M subspaces of shape (N,L)), giving the required result, an N by L matrix, which is assigned to MATMULT. Notice that in this form, declared with non-square matrices, any error in specifying the appropriate dimension selectors (1, 2, or 3) would produce a compile-time error, as either the multiplication or assignment would not conform.

This expression of the algorithm implies no sequencing whatsoever (this is now left to the compiler), so here all four variants of the algorithm (§5.3.1–§5.3.4) can be extracted from this one code.

(iii) The fast Fourier transform

In §5.5.2 recursive equations are derived which define the fast Fourier transform as a sequence of partial transforms specified over a variable two-dimensional structure (equation (5.88)). The offset between elements combined together in these equations can also be considered as a third dimension of length 2, halving the range of the first dimension. This gives a simple transformation from equation (5.88) to the equations below:

for sequence $l = 0, 1, \ldots, q - 1$

    for all $k = 1, \ldots, 2^l$

        for all $i = 1, \ldots, n/2^{l+1}$

$$\overset{(l+1)}{f}(i, 1, k) = \overset{(l)}{f}(i, 1, k) + \omega_{2^{i+1}}^{k-1} \overset{(l)}{f}(i, 2, k)$$

$$\overset{(l+1)}{f}(i, 2, k) = \overset{(l)}{f}(i, 1, k) - \omega_{2^{i+1}}^{k-1} \overset{(l)}{f}(i, 2, k)$$

    end

  end

repeat

It is now only a trivial transformation to express this in our parallel constructs, using a function RECUR (as in figures 5.13, 5.14 and 5.16). The function performs one pass of the sequence over l.

```
FUNCTION RECUR(F,W,L,N)
COMPLEX F(N), W(N), RECUR(N)
I2L = 2**L
N2L = N/I2L
N2L1 = N2L/2
MAP F(N2L1,2,I2L), W(N2L, I2L), RECUR(N2L1, 2, I2L)
F( ,2, ) = XPND(W(1,),1,N2L1)*F( ,2, )
RECUR( ,1, ) = F( ,1, ) + F( ,2, )
RECUR( ,2, ) = F( ,1, ) - F( ,2, )
RETURN
END
```

This code, when used with a calling sequence assigning RECUR to F, gives the transforms in place. It uses only the temporary storage defined by the array-valued function and returns the transformed values in bit reverse order. It is left as an exercise for the reader to recode the function using the flow diagram in figure 5.12, to produce the results in natural order.

Again it should be noted that this description of the algorithm incorporates all three schemes described in §5.5. Scheme A is obtained if the compiler sequences the last dimension of the arrays and scheme B is obtained if the compiler sequences the first dimension of the arrays. If, however, there is sufficient machine parallelism available, then sequencing neither dimension yields the parallel (PARAFT) scheme (§5.5.4). Thus with a suitable compiler, efficient machine code can be generated for an entire spectrum of computers, with $n_{1/2}$ varying from 1 to N.

4.3.2 DAP FORTRAN—a constrained approach

DAP FORTRAN (ICL 1979a, b) is a language developed as its name suggests for the ICL DAP. This language was the only high-level language supported on the main-frame DAP (see §3.4.2) and has since been released as the programming language for the later mini-DAP (see §3.5.2). This language, although constraining arrays to take on the size and shape of the target array processor, has made major contributions to the further development of the future FORTRAN standards on array facilities. It implements many of the constructs found in the previous section, although on array objects constrained to the size of the target hardware, 64×64 for the main-frame DAP and 32×32 for the mini-DAP. The language has also been the source of many of the array-oriented intrinsic functions proposed in the FORTRAN 8X standard.

(i) Data objects

Arrays are declared in DAP FORTRAN in the usual way, using the DIMENSION or TYPE statement. However vectors are declared with their first dimension elided and matrices with the first two dimensions elided. These are the constrained dimensions, which take on the DAP size of N elements for a vector and $N \times N$ for a matrix. Sets of these objects may also be declared by using other dimensions.

    DIMENSION V( ), VSET(,4)

    REAL   M(,), MSET(,,4)

The examples above therefore define a single vector V and vector set **VSET** of four vectors, followed by a matrix **M** and matrix set **MSET** of four matrices. The unconstrained dimensions are for sequential access only and map onto the DAP store. Details of this mapping are described in §3.4.2. These dimensions are treated exactly as in standard FORTRAN.

As the parallel dimensions are constrained to the DAP array size, objects of like type conform. However when mixing objects of different type, coercion must be applied. DAP FORTRAN allows the coercion of scalars to types vector and matrix, and provides functions to coerce vectors to type matrix. The two functions MATR and MATC give matrices formed by repetition of a vector as rows, or columns, of the matrix respectively.

(ii) Selection mechanisms

Selection mechanisms from constrained dimensions in DAP FORTRAN use integer, logical and integer vector indices, as described in §3.4.2. The vector or indirect indexing can be applied as a projection over all rows or all columns. This technique cannot however be used in the unconstrained dimensions, as it would imply separately indexed memories in each PE, which is not supported in the hardware.

All selection from constrained dimensions is rank-reducing, with the exception of selective updating (see §4.3.1). Thus a selection from a matrix gives either a single vector or a single scalar, and a selection from a vector always gives a scalar. There is no range selection mechanism, and logical selection must yield a unique result. Examples of these selection mechanisms are given below, for the following declarations:

REAL	**M**(,),**V**( )
INTEGER	*I,J*,**VI**( )
LOGICAL	**ML**(,),**VL**( )
M(*I*,)	row *I* of **M**,
M(,*J*)	column *J* of **M**,
M(*I,J*)	element *I,J* of **M**,
V(*I*)	element *I* of **V**,
M(**VL**,)	row of **M**,
M(,**VL**)	column of **M**,
M(**ML**)	element of **M**,
V(**VL**)	element of **V**,
M(**VI**,)	vector containing *M*(**VI**(*I*),*I*) in element *I*,
M(,**VI**)	vector containing *M*(*I*,**VI**(*I*)) in element *I*.

In these examples, the vector and logical matrices must have one and only one element set to .TRUE., and the integer vector **VI** must have all of its elements in the range 1 to *N*. Indexing is generalised in DAP FORTRAN, by allowing suitable expressions in the place of variables in the above examples.

Routing can also be applied as an indexing operation using the + or − symbol in either of the constrained dimensions examples are given below.

V(+)	shifts **V** one place to the left,
V(−)	shifts **V** one place to the right,
M(+ ,)	shifts **M** one place to the north,
M(− ,)	shifts **M** one place to the south,
M(, +)	shifts **M** one place to the west,
M(, −)	shifts **M** one place to the east,
M(+)	shifts **M** one place to the left, treated as a long vector,
M(−)	shifts **M** one place to the right, treated as a long vector.

Shifts of greater than one place are expressed by the use of functions.

(iii) Array assignments

Indexing expressions on the left-hand side of an assignment specify which selection from the variable is to be updated. This selective assignment can be used to make an assignment conform; for example a vector expression can be assigned to any valid vector selection from a matrix. Alternatively, selection may also allow the updating of a range of values, where the expression and left-hand side conform. In this case the selection only indicates which elements are to be assigned and which are to be left unchanged. For example consider the use of logical matrix, or logical vector indexing:

$$\textbf{M}(\textbf{LM}) = \langle \text{matrix expression} \rangle$$

$$\textbf{M}(\textbf{VL},) = \langle \text{matrix expression} \rangle$$

In both of these cases, the restrictions which were placed on right-hand side indexing no longer apply. The logical matrix **LM** or logical vector **LV** can have more than one element true. Thus **LM** selects a set of elements to be updated and **LV** a set of row vectors. In both cases the elements selected for updating are taken from the corresponding elements in the matrix expression, and those not selected are left unchanged.

(iv) Functions

Because expressions in DAP FORTRAN can be matrix- or vector-valued, the function subprogram definition has been extended to include matrix and vector valued functions, as well as the more usual scalar-valued function. The type of the function is declared in the function statement. Thus

REAL MATRIX FUNCTION MATMULT

declares a function MATMULT which returns a real matrix valued result.

In addition to the extension of a user-defined function, the standard or built-in FORTRAN functions have been made polymorphic. They return a value of the same type as their argument, with matrix and vector types being evaluated in parallel. Other built-in functions introduced into DAP FORTRAN include data manipulation functions, shift functions, selection functions and reduction functions. The reduction functions operate over either or both of the constrained dimensions, using arithmetic, relational or logical operators. Some examples are given below.

ALL	.AND. over rows and columns
ANY	.OR. over rows and columns
ANDROWS	.AND. over rows
ORCOLS	.OR. over columns
SUM	+ over rows and columns
MAXP	$\geqslant$ over rows and columns (gives logical position only)

(v) *Examples*

In the general mapping example, the code for DAP FORTRAN looks very similar to that given for the general parallel constructs (§4.3.2). This is because the size of the problem maps onto DAP array or the constrained dimensions in DAP FORTRAN. However, the code is only good for $N \leqslant 64$ (32). The arrays ALPHABET and WORD must also be padded with blanks to length 64 or 32.

```
CHARACTER VECTOR FUNCTION WORD(ALPHABET, INDEX)
CHARACTER ALPHABET( )
INTEGER INDEX( )
WORD = (MATC(ALPHABET))(INDEX,)
RETURN
END
```

Notice here that the syntax requires the expression to be enclosed in parentheses in order to index it.

(vi) *Matrix multiplication example*

Again because of the constrained dimensions, the code given below multiplies matrices less than or equal to the DAP size. The variant of the algorithm used is the outer product, which has N^2 parallelism (§5.3.3).

```
      REAL MATRIX FUNCTION MATMULT(A,B,N)
      REAL A(,), B(,)
      INTEGER K, N
      MATMULT = 0
      DO 10 K = 1, N
   10 MATMULT = MATMULT + MATC(A(,K))*MATR(B(K,))
      RETURN
      END
```

A variant of the general expression of matrix multiplication can also be contrived in DAP FORTRAN, for matrices of size 16×16, 16 being the cube root of 64^2:

```
MATMULT(ROW(I3),AND,COL(J3)) = SUM-4-SQUARE(
                                    MATC-4-SQUARE(A,I1,J1)
                                    *
                                    MATR-4-SQUARE(B,I2,J2)
                               ,I3,J3)
```

where SUM-4-SQUARE, MATC-4-SQUARE and MATR-4-SQUARE are user-defined functions (Jesshope and Craigie 1980), which simulate a 16^3 DAP, using a symmetric low-order interleaving over the 64^2 DAP array (Jesshope 1980c).

4.3.3 FORTRAN 8X—the new FORTRAN standard

Although in DAP FORTRAN above, the parallel data objects in the language map onto the target architecture, other languages proposed at about the same time took an unconstrained approach to introducing structure parallelism into FORTRAN. Two such examples are the BSP FORTRAN compiler, which combined the benefits of both vectorising capability and array constructs (Burroughs 1977e, Austin 1979), and VECTRAN, an experimental language developed by IBM, based on FORTRAN, and described in the papers by Paul and Wilson (1975, 1978).

Clearly, the separate development of many languages targeted on particular array and vector architectures is not an ideal situation. For this reason, the ANSI X3J3 Committee decided to consider array extensions as a feature to be added to the new FORTRAN standard that they were then considering. This decision was first considered in the ANSI committee which produced the FORTRAN 77 standard, but was discarded at that time with a recommendation that it be considered for inclusion in the next standard. Many of these array features are based on the constructs found in languages such as DAP FORTRAN, BSP FORTRAN and VECTRAN. These deliberations began almost a decade ago, but the X3J3 Committee, at the time of writing, has just been voting on a draft specification of the next FORTRAN standard. This is called FORTRAN 8X, and the name reflects the expected data for the full ANSI approval, i.e. by 198X. The sequence (66, 77, 88) is looking increasingly unlikely, as the standard will only be approved after the Committee has considered responses to a public review. Indeed the Committee is having difficulty in agreeing a draft specification for comment, as many delegates believe the language is becoming too large.

The new features of the proposed standard, which are of interest here, are those concerned with arrays and array processing (process or task parallelism is not one of the new additions to the language). The features described here are based on information obtained from the ANSI X3J3 working document (X3J3/S8, version 95) dated June 1985 and should not therefore be considered as fixed. They may be changed, removed or added to prior to acceptance as a proposed standard. Having given this disclaimer, we add that the array features described here have been stable for some time.

The array extensions in the new language follow quite closely the proposals outlined in §4.3.1. There has, however, been a new data type introduced as a consequence of the addition of the array-processing features. This new type is the BIT type, which has two values, '1' and '0' and together with bit operations represents the two-value system of boolean mathematics. This type has been added to support the use of boolean mask variables to enable and disable array operations.

(i) Arrays as elemental data objects

Array objects are notionally rectangular, with an index for each dimension (up to seven) in the structure. The size of each dimension is called the *extent* of that dimension and the number of dimensions the *rank* of the array. A scalar has rank zero. The array size is the total number of elements or the product of extents. An array may have zero size. The *shape* of an array is defined by the rank and the extent in each dimension. Once declared, the rank of an array is constant but the extents need not be constant. Extents may vary for dummy arguments, local procedure arrays and alias arrays, details of which are given below. There is also a facility for changing the rank of an array, while the size remains constant, using the parameter/dummy argument interface on a procedure subprogram.

Two arrays are said to conform in the new language if they have the same shape, with the exception of the scalar, which is conformable with any array. As described in §4.3.1, any operation defined for scalars may be applied to conforming arrays and these are performed as if the scalar operation were applied to each element of the arrays and in such a way that it may be assumed that all operations occur simultaneously.

(ii) Explicit-shape arrays

An explicit-shape array is one in which the array is defined with explicit values for the array dimension bounds, an optional lower bound and an upper bound for each dimension of the array. If variables are used in the definition of these bounds then the arrays must either be dummy arguments or local variables to a procedure, and in the latter case the array is termed automatic. The following are explicit-shape arrays.

$$A(2:10, 5:30), B(10, 10, 10), C(N-1)$$

Here A is a two-dimensional array, with extent from 2 to 10 in the first dimension and extent from 5 to 30 in the second dimension; B is a three-dimensional array with extents of 1 to 10 in each dimension. If N were a variable, then the last array C would have to be a dummy argument or local variable to a procedure.

(iii) Assumed-shape arrays

An assumed-shape array is one in which the shape of the array is inherited from an actual argument passed to a procedure. An assumed-shape array may only appear in a procedure, where it will be declared as a dummy argument to that procedure. It will assume the shape of the actual argument when the procedure is referenced. The dummy argument may also be declared in an ARRAY statement,which confers array attributes to a variable. In this case the dummy argument and all other entities declared within the same ARRAY statement will all assume the shape of the actual argument, which becomes associated with the dummy argument. In this way a number of array objects, not necessarily dummy arguments, will assume the shape of an array parameter.

(iv) Allocate-shape arrays

An allocate-shape array is one in which the type, name and rank are declared, but whose shape or bounds are determined at run time when space is allocated by an ALLOCATE statement. Examples of allocate-shape arrays are given below.

$$A(:,:), \ B(:,:,:), \ C(:)$$

The rank of the array is given by the number of colons, so these arrays have the same rank as the earlier explicit-shape array declarations. The size and shape of such allocate-shape arrays are undefined until the array has been allocated. No reference to it or any of its elements may be made until the array has been allocated.

(v) Array sections

As with the constructs introduced in §4.3.1, an array reference without subscripts implies the use of the whole array. If some computation is inferred, then this will be performed on all array elements, and in any order. There are exceptions to this where the lexical ordering is still required, such as in I/O and data statements, for example. Elements of arrays may still be selected, by supplying a full list of subscripts, in the normal way. The number of subscripts required is given by the rank of the array. Each subscript variable used will reduce the rank of the section selected by one, so that if all subscripts are supplied, the rank of the section selected is zero, which corresponds to a scalar.

Alternatively, an array section may be selected by the use of index ranges in appropriate subscript positions. The section is therefore defined (as is the parent array) by the cartesian product of these index ranges. The subset is selected from the index range by means of a section selector, which takes one of two forms: a triplet defining base with extent and skip indices, giving a monotonic sequence, or a rank-one integer array. Using the triplet (separated by colons)

$$B(1:10,1,1) \quad \text{and} \quad B(1:10:2,1:10,1:10)$$

are both valid sections from the explicit-shape array defined above. The first is a vector of ten elements, defined using subscripts in the last two dimensions, and the second is a three-dimensional array formed using the odd indices from the first dimension. Exactly the same sections could have been defined using a rank-one array in the first subscript, i.e.

$$B(IV,1,1) \quad \text{and} \quad B(IV,1:10,1:10)$$

where in the first case $IV = [1,2,3,4,5,6,7,8,9,10]$ and in the second case $IV = [1,3,5,7,9]$. In general, the number of indices and section selectors must match the rank of the parent array. The vector section selector may contain any sequence of indices which lie within the extent of the dimension in which it is placed.

As noted in §4.3.1, which introduced this mapping variant of vector selection, the list of indices may define a one-to-many mapping from the parent to the array section and this could lead to non-deterministic results if assignment were allowed to such array sections. The 8X definition forbids such assignment and also assignment via a parameter/dummy argument pair, if such implied assignment is non-deterministic.

(vi) ALIAS attribute and IDENTIFY statement

When is an array not an array? The answer is when it has the ALIAS attribute. An array name declared with this attribute is not an array object until it has been associated with an actual object by appearing in an IDENTIFY statement. The IDENTIFY statement has the flavour of a dynamically executed COMMON statement, as it overloads an area of memory with a number of different arrays, which may have different shape. However, unlike the COMMON statement this assignment of a named (ALIASed) array with an area of memory is performed by an executable statement, which may contain variables used to define the mapping of the ALIASed array over the actual array object.

The IDENTIFY statement is probably one of the most controversial new features in the package for handling array processing. It is a means of providing a dynamic address mapping from the alias array into the parent array, although this mapping is constrained to be linear. This is rather like an array section, except that it provides a wider range of subsets of the parent array. A simple example will illustrate the use of this statement.

$$IDENTIFY(DIAG(I) = ARRAY(I, I), I = 1:N)$$

In this example the array DIAG is an array of rank one, which is the ALIAS array; it has been associated with the storage in the parent array ARRAY, in such a way that the elements of DIAG map onto the diagonal of ARRAY. In general, the mapping onto the parent array can be any linear combination of the dummy subscripts of the alias array. There must be a range specifier given for the dummy subscript which defines the range over which the mapping is defined. In one statement, therefore, a linear subscript transformation and dynamic range can be specified.

Another example, given below, redefines the first 100 elements of a rank-one array VECTOR, to create a rank-two alias array ARRAY. In this example, the two dummy subscripts define the extent of the two dimensions of ARRAY, and the expected mapping of this aliasing is defined by the linear combination of these used to subscript VECTOR.

$$IDENTIFY(ARRAY(I,J) = VECTOR(I + 10*(J - 1)),$$

$$I = 1:10, J = 1:10)$$

Any reference or assignment to the alias array will actually modify the elements of the host array, as specified by the subscript mapping. In the example given, therefore, an assignment to DIAG will modify the first N locations of the leading diagonal of ARRAY.

An alias array, once IDENTIFYed can be treated just like any other array object, including being subscripted or sectioned, being passed as an actual argument to a subprogram, or even as a parent object to another IDENTIFY statement. Non-deterministic use of a many-to-one ALIASing is not allowed, as there are similar restrictions to those applied on vector selectors.

It can be seen, therefore, that this is a very powerful feature for selecting subarrays. However, it is likely to prove difficult to implement on many processor arrays, for example the ICL DAP, for it implies the run-time manipulation of a global address space. It is far better suited to the vector supercomputer, where access to memory is sequential and is specified by a constant stride through memory, which simply provides the linear subscript translation. On a processor array it could, for example, be implemented using selective update over the parent structure, or if this were too inefficient it would require the assignment to an alias array, with subsequent mapping and merging with the parent array, as this is a runtime construct it is quite likely that packet-switched communication will be required.

(vii) Array intrinsic functions and procedures
Because the semantics of the new language allows the manipulation of array objects as first-class citizens, it allows their use as parameters in all the existing intrinsic functions in FORTRAN. It does this by making all such functions polymorphic, in much the same way as the operators are overloaded. Thus SIN(A) would produce as a result an object which conforms to its argument. Therefore, if A were a matrix of a given size or shape, the function would return another matrix, of the same size and shape, but whose elements were the values obtained by applying the SIN function to each of the argument's elements.

There are also many new intrinsic functions in the FORTRAN 8X proposal, which have been added to support the array extensions to the language. For example, there are enquiry functions which will provide the range, rank or size of an array. There are also array manipulation functions, such as MERGE, SPREAD, REPLICATE, RESHAPE, PACK and UNPACK. Most of these functions have equivalent operators in the APL language (Iverson 1962).

This section is not intended to give a complete description of the new FORTRAN standard, which in any case is not, at the time of writing, even a draft proposal. It attempts instead to give a flavour of the constructs of interest to users of array and vector parallel computers. The interested reader is referred to the source document (Campbell 1987), or its successor, which may be obtained from the Secretary of the ANSI X3J3 Committee. There is also a paper by the two British delegates on the FORTRAN X3J3 Committee (Reid and Wilson 1985) which gives examples of the use of the FORTRAN 8X features, and the book *Fortran 8X Explained* by Metcalf and Reid (1987).

4.3.4   CMLISP

In contrast to the languages described above, which are based on the array structure and FORTRAN, this section describes a language based on the list data structure and the common LISP language. An introduction to the LISP language is given below (for continuity) however, the interested reader is referred to one of the many introductory books on this language (Winston and Horn 1981, Touretsky 1974).

In LISP, the central objects in the language are lists; even the functions that operate on them and their definitions are lists. Indeed, functions defined in LISP may take other functions as arguments or produce functions as results. These are known as higher-order functions and their use provides a very powerful and expressive programming environment. The reduction operator '\' in APL is a higher-order function; it takes another operator and an array as its arguments and applies the operator between every element of the array. For example, in APL. \ + A would sum the elements of A, and \ * A would form the product of elements of A.

In LISP the list is represented by a sequence of items, separated by space and enclosed by parentheses. For example:

item — is an atom;

(item1 item2) — is a list containing two atoms;

((item1 item2)) — is a list containing one item, which is itself a list containing two atoms;

((item1 item2) item3) — is a list containing two items, one a list and one an atom.

The basic operations defined in LISP provide for the construction and dissolution of list objects, for example:

CAR — returns the first atom of a list;

CDR — returns the tail of the list;

CONS — returns a list constructed from an atom and a list;

LIST — returns a list constructed from atoms.

CMLISP (Hillis 1985) was designed as a programming environment for the connection machine and is based on common LISP, which has a long history at MIT. It is designed to support the parallel operations of the connection machine, which is a SIMD machine. However, because of the expressiveness of the LISP language, it is possible to define MIMD-like operations in CMLISP. However, CMLISP does reflect the control flow of the host computer and microcontroller of the connection machine, while at the same time allowing operations to be expressed over parallel data structures. Connection machine LISP is to common LISP what FORTRAN 8X is to FORTRAN 77. The language is fully documented in Hillis and Steele (1985).

The three artifacts that have been added to the common LISP language to produce CMLISP are the xector, which is an expression of parallelism; the alpha notation, which is a higher-order function expressing parallelism of operation across a xector; and the beta reduction, which is a higher-order function expressing reduction. The beta reduction can be equivalent to the APL \ operator described above.

(i) The xector

The parallel data structure in CMLISP is called the xector, which loosely speaking corresponds to a set of processors or virtual processors and their corresponding values. It is a parallel data object and can be operated on, giving element-by-element results. In this sense it is reminiscent of the FORTRAN 8X array. CMLISP supports parallel operations to create, combine, modify and reduce xectors. Unlike DAP FORTRAN, the CMLISP xector is not hardware-dependent; the xector size or scope is not constrained, but can be of user-defined size. Indeed the xector's size may vary dynamically.

Another difference between FORTRAN 8X and CMLISP arises from the nature of the data structures involved. In FORTRAN the data structure is the array, and a FORTRAN 8X array is, in essence, a parallel set of indices and values. However, because the structure is implicitly rectangular, the indices are not important. Manipulation of these indices is expressed in the language as a set of data movement operators, which shift the array in a given direction and for a given distance. In CMLISP the xector also comprises an index/value pair, where both index and value are LISP objects. Moreover, because the objects are lists, a significant coding will often be placed on the index set. In other words, the xector represents a function between LISP objects. The domain and range are sets of LISP objects and the mapping associates a single object in the range with each object in the domain. Each object in the range is an index and has a corresponding object in the domain, which is its value.

The implementation is such that it is assumed that each element of a xector is stored in a separate processor where the index is the name of that processor, an address stored in the memory of the host machine, and the value is the value stored at that processor. Hillis (1985) introduces a notation for representing xectors, as follows:

$$\{ \text{John} \rightarrow \text{Mary Paul} \rightarrow \text{Joan Chris} \rightarrow \text{Sue} \}.$$

Here the set of symbols John, Paul, Chris is mapped onto the set of symbols Mary, Joan, Sue. This notation reflects the view of a xector as a function. A special case of the xector is the set, where a set of symbols is mapped onto itself;

$$\{ \text{red} \rightarrow \text{red white} \rightarrow \text{white blue} \rightarrow \text{blue} \}$$

which is equivalent to

$$\{ \text{red white blue} \}.$$

Another special case is where the domain of the xector comprises a sequence of integers, starting from zero. This is essentially a rank-one FORTRAN 8X array, as it corresponds to an ordered set of values. The alternative notation suggested by Hillis reflects this:

$$\{0 \to 5 \; 1 \to 2 \; 3 \to 10\}$$

which is equivalent to

[5   2   10].

The last special case is the constant xector, which maps every possible index value onto a constant value. For example, the xector which mapped all values onto the number 100 is represented by:

$$\{\to 100\}.$$

Xectors can be manipulated in LISP just as other LISP objects. For example, a xector can be assigned to a variable using the SETQ function. For instance:

(SETQ Wife—of ' {John → Mary Paul → Joan Chris → Sue}).

This assignment sets the value of the symbol 'Wife—of' to the xector defined above. The ' signifies that the item following is an atom. Having assigned a xector, the symbol can be used in other functions; for example, to reference a value

{XREF Wife—of ' John)

evaluates to Mary. Similarly XSET will change a value for a given index and XMOD will add another index value pair, if they do not already exist. Functions are also provided to convert between xectors and regular LISP objects. For example,

(LIST—TO—VECTOR'(5   2   10))

evaluates to the xector special case

[5   2   10].

(*ii*) *The alpha notation*
The alpha notation in CMLISP is a more formal means of expressing parallel operations than found in the other languages described here, for example FORTRAN 8X. Strictly speaking the alpha notation is the broadcast operator; it creates a constant xector of its argument, i.e. writes the same value in each processor. For example:

$$\alpha 100 \Rightarrow \{\to 100\}$$

$$\alpha(* \ 2 \ 3) \Rightarrow \{ \rightarrow 6 \}.$$

More interestingly

$$\alpha + \Rightarrow \pm \{ \rightarrow + \}$$

which is a xector of $+$ functions. When this object is applied to two xectors, the effect of applying it is to perform an elementwise composition of the values of the xectors. In FORTRAN 8X, this notation is implicit, as the operators have been overloaded, to represent both scalar and parallel operations, depending on context. Thus

$$(\alpha + '\{a \rightarrow 1 \ b \rightarrow 3\}'\{a \rightarrow 2 \ b \rightarrow 4\}) \Rightarrow \{a \rightarrow 3 \ b \rightarrow 7\}$$

and

$$(\alpha \text{CONS}'\{a \rightarrow 1 \ b \rightarrow 3\}'\{a \rightarrow 2 \ b \rightarrow 4\}) \Rightarrow$$
$$\{a \rightarrow (1.2) \ b \rightarrow (3.4)\}.$$

In order to use the algebraic properties of the alpha notation, Hillis introduces its inverse operator, which cancels the effect of the application of alpha. This is useful if alpha is a factor of most items of an expression. For example

$$\alpha(+ ab) \equiv (\alpha + \alpha a \ \alpha b)$$

but if a or b evaluated to a xector, then the use of '$\bullet$' would cancel the application of α. Thus, if a were a xector,

$$\alpha(+ \bullet ab) \equiv (\alpha + a\alpha b).$$

Thus α is a parallel operator and $\bullet$ its inverse, or a way of marking already parallel objects, in an expression to be evaluated in parallel.

(iii) Beta reduction

The beta operator is similar to the APL reduction operator described above. When applied to an expression involving a binary operator and a single xector, it has the effect of modifying that operator to one that performs a reduction operation over that xector by applying the operation between each value of that xector. For example

$$(\beta + '\{a \rightarrow 1 \ b \rightarrow 3\}) \Rightarrow 4.$$

The combined use of alpha and this use of the beta notation in CMLISP provides a very powerful tool for constructing all manner of functions. A number of function definitions which make use of this combination of operators are given below. In these examples alpha provides broadcast and allows all manner of associative operations, and beta gives reduction, a powerful set manipulation operation.

(DEFUN All—Same $(x\,y)(\beta$ AND $(\alpha = x\,y)))$.

This can be read, define a function called 'All—Same', with parameters x and y (xectors), which is defined as the beta reduction using logical AND of the xector result of the elementwise operation of equality between the parameters x and y. Only if all values of the xectors x and y are equal will the function return a value TRUE. A similar function, but which uses the logical operator, OR, is defined below. This would detect an equality between any two values with the same index.

(DEFUN One—Same $(x\,y)(\beta$ OR $(\alpha = x\,y)))$.

The next example illustrates the use of the dot operator. In fact the definition differs only in the use of this symbol, applied to one of the parameters, and the factorisation of the operator α. The function Is—In behaves quite differently, however. It detects whether the item y is in the set of values defined by the xector x. The equality is now defined to be between the item y and the values of x. Alpha and dot are used to provide a consistent xector expression. This is a combination of associative and reduction operators:

(DEFUN Is—In $(x\,y)(\beta$ AND $\alpha\,(= \bullet x\,y)))$.

A final example defines the xector length, by summing a value of one, over every index of the xector parameter. This is achieved by the device of defining a function 'Second—One', which returns its second parameter. The use of this function within an alpha expression, using the dotted xector parameter x as the first parameter to the function, effectively produces a xector of ones, of the same size as x:

(DEFUN Xector—Length $(x)(\beta + \alpha\,($Second—One$\bullet x\,1)))$

(DEFUN Second—One $(x\,y)(y))$

The use of beta as described in the examples above is only a special case of a more generalised function. In its most general form, β takes two xectors as arguments. It returns a third xector, which is created from the values of the first xector and the indices of the second xector. This is illustrated in the example below:

$(\beta'[1\quad 4\quad 9]'[A\quad$B$\quad$C$]) \Rightarrow \{A\rightarrow 1$ B$\rightarrow 4$ C$\rightarrow 9\}$.

In fact this can be viewed as a routing operation, for it sends the values of the first xector to the indices specified by the second xector. For example, in a packet-routing network, the operation performed at each processor would be to emit a packet into the network, whose data was its value from the first xector, and whose address was its value from the second xector. Of course the resulting xector, defined by the arrival of those messages, is undefined if the second xector contains replicated values. In CMLISP this condition is regarded as an error. However, if β is used in conjunction with another operator, this specifies a reduction operation to be performed on any clashes of index (address). So whereas

$$(\beta'[1 \quad 4 \quad 9 \quad 16]'[A \quad B \quad A \quad B]) \Rightarrow \text{error}$$

it is found that

$$(\beta+'[1 \quad 4 \quad 9 \quad 16]'[A \quad B \quad A \quad B]) \Rightarrow \{A \rightarrow 10 \; B \rightarrow 20\}.$$

The special case described above, with only a single xector argument, assumes the second argument to be the constant xector, so that all values of the first xector are reduced by the qualifying operator. The implementation will define which index, or processor number, the reduction is performed to. In a SIMD machine, this is naturally the control sequencer.

4.4   PROCESS PARALLELISM

4.4.1   Introduction

As introduced in the previous section, there are two fundamental techniques available to explicitly express parallelism; structure parallelism and process parallelism. The previous section also described various implementations of the first of these, structure parallelism. However, CMLISP, and even APL, both provide mechanisms which treat operators as the basic values of the parallel data structure. We could, of course, imagine whole programs as being the components of the parallel structure, in which case we have a description of process parallelism. To reiterate then, the distinction is one of granularity, for structure parallelism was defined above as the granularity of a single operation over every element of a data structure. Process parallelism, however, requires distributed sequences of operations.

The underlying computational model and the manner in which load is balanced across a system is also very different between these two approaches. As has been shown in the previous section, in structure parallelism one can consider the processors to be associated one per data structure element, with activated data being mapped onto the available processors in order to balance the load. Thus, load sharing is achieved through data structure element redistribution. In process parallelism, however, the process is virtualised and load balancing occurs by the distribution of processes across the processors. The unit of distribution is the code, and of course its associated data and state. The virtualisation in this case involves maintaining a number of instruction streams on a single processor.

Because each process involves a considerable overhead in terms of establishing an instruction stream, there is a penalty in implementing a high degree of concurrency, using process parallelism. To some extent, the transputer and OCCAM is the first system to reduce this overhead to a manageable level. The penalty paid is a static process structure, established at compile time. Dynamic process creation and deletion, while being more flexible, will necessarily incur a greater overhead for concurrency and will therefore require a coarser granularity for efficiency. This is one of the major problems that is being tackled by researchers in declarative systems—that of controlling the grain size of the processes being created.

Historically, process parallelism has evolved to exploit the non-deterministic sharing of the resources of a single processor and not from the need to exploit concurrent hardware. This sharing allowed the CPU, then the single most expensive resource in a main-frame, to be effectively utilised. For when a program had to wait on a slow input or output device, the 'wasted' CPU cycles could then be used by another program.

Four fundamentals are required for any process-oriented language: a method of initiating and terminating concurrent tasks (for example, the fork and join procedures in UNIX); a means of communicating between concurrent tasks, either implemented as a message-passing system or by shared memory; a means of synchronising tasks, so that they can share a common time reference, however vague; and finally a means of determining choice, or non-determinism. In many languages more than one of these requirements may be merged into a single construct or underlying implementation. For example, the choice of shared memory provides both communication and non-determinism, for it allows information written by one task to be read from another. Moreover, if more than one task has write access to a given location, then a reading task cannot know, without further information, which was the writing task.

Another difference between these two techniques is that whereas the use of structure parallelism results in very concise code, which is easier to understand and debug than sequential code, the use of process parallelism creates more pitfalls for the unwary programmer, such as deadlock, which is where two or more processes are in a state such that each is waiting for an event or communication from one of the other processes. The simplest situation is where two processes wish to communicate with each other, but both are programmed to read from the other before writing; they will both wait forever, or for a system time-out, as neither can write before reading. Breaking the programmed symmetry breaks the deadlock.

It is the fundamental asynchrony, or global non-determinism, which makes programming in this model difficult. Consider that we have n instruction streams, and that it is possible to give a time reference for any instruction in the overall system. Consider what the next state of the system is after a given instruction has executed. In the absence of any synchronisation, there are n possible choices of the next state of the system, following that n^2, etc. It is this exponential growth of trace that makes debugging process-based languages so very difficult. A classical situation is the manner in which a bug will mysteriously vanish when code to trace the system's behaviour is added to a faulty program. This is, of course, because the timing of the system has been altered.

This is one small section in a book covering all aspects of parallelism, and we restrict ourselves here to a discussion of one process-based language, OCCAM The reason this language has been chosen is because it is a complementary language to the transputer. Indeed, the OCCAM language is very intimately related to the transputer hardware, and it is recommended that this section and §3.5.4, describing the transputer, should be read together.

For more details about other process-based languages and a more general discussion on the theory of this approach, the reader is referred to an excellent introduction to this topic (Ben-Ari 1982), and the book *Communicating Sequential Processes* by Hoare (1986).

4.4.2 OCCAM—a minimalist approach
One of the most elegant schemes that has been proposed for process synchronisation is that implemented in communicating sequential processes (CSP) (Hoare 1978, 1986), by unbuffered point-to-point communication. CSP has been chosen by INMOS as the basis for a language for the transputer. That language is called OCCAM (May and Taylor 1984, INMOS 1984, Jones 1985, INMOS 1986, Pountain 1986a) and it is designed to support explicit hardware concurrency. It therefore reflects the concurrency found in the transputer, or to put it another way, transputers are designed to implement the programming language OCCAM very efficiently. Although transputers will also provide efficient implementation of most modern languages, concurrency in transputer systems is only available through OCCAM.

(i) *Processes and process construction*
OCCAM programs can be described as processes which perform actions and terminate. This concept of the process can be viewed at many levels within the program. Indeed, the entire program can be considered as a process which

starts, performs some actions and then terminates. At the lowest level, the primitives of the language are themselves considered as processes and are called primitive processes. Composite processes can be constructed from primitive or other constructed processes by a number of process constructors. The scope of these constructors is indicated in the text of the program by a fixed layout, with indentation of two spaces. For example,

> SEQ
> > A
> > B

is read: perform in sequence first the process A and then the process B. A similar constructor is used to express parallel execution of processes:

> PAR
> > A
> > B

This reads: perform the processes A and B in parallel. Both constructed processes terminate only after all of their constituent processes terminate. However, if process A did not terminate in the SEQ construction, process B would not start; whereas in the PAR construction, if process A did not terminate, process B would still have a chance to execute and terminate.

The parallel and sequential constructors may be freely mixed, with their scope determined by indentation and outdentation. For example:

> PAR
> > SEQ
> > > A
> > > B
> > SEQ
> > > C
> > > D

Here two processes are executed in parallel; one is the sequence of process A followed by B, the other is the sequence of process C followed by process D.

It is shown above how processes, including the language's primitives, can be combined to execute in sequence or in parallel, this latter being the first fundamental described in the introduction. However, what of the rules concerning communication between these processes? Sequentially constructed processes may share memory, for they will be executed on a single processor and reflect the flow of control found in other sequential languages. However, the constituent processes of a PAR constructor are not allowed to share memory, even though they may be executing on the same processor. *All* communication between parallel processes in OCCAM must be via communication links joining those processes. These are called OCCAM channels. They are labelled and provide point-to-point, unbuffered communication between the processes to which they are attached.

All active OCCAM processes are constructed from three primitive processes, namely input, output and assignment. Input and output are the communication primitives and are used in pairs between parallel processes, using a named channel. For example:

    PAR

        chan!a + b

        chan?c

The two parallel primitive processes communicate with each other by writing (!) and reading (?) across the named channel 'chan'. This constructed process is formally equivalent to the assignment process below, assignment being the other primitive process in OCCAM:

    c: = a + b

The only reason to write this action in a parallel way would be to assign from one processor to another. A little more thought will reveal that this transformation from assignment to communication is one which introduces parallelism from a sequential construct.

There are two further primitive processes, but these do no useful work. The SKIP process starts, does nothing and terminates, and is used where the OCCAM syntax requires a process in a constructor, but the algorithm requires no action. The second process that does nothing is the STOP process. As its name implies, this process starts, does nothing, but does not terminate. For example, the process

    SEQ

        A

        B

        STOP

        C

is formally equivalent to

    SEQ

      A

      B

    STOP

as process C will never execute.

(ii) Synchronisation and configuration

As in CSP, the communication channel between parallel processes is unbuffered. Thus, whichever process on a channel becomes ready first will wait for the other process to reach its communication primitive. There is a handshake implicit in the implementation of this communication sequence, for the first process ready must signal the other that it is waiting. For details of this implementation see §3.5.4. The communication therefore gives a global time reference between the two parallel processes. If synchronisation is not important to the execution of a program, for example in a producer–consumer situation, the user can add buffering processes. Ideally, the buffer process should always be ready to accept input or output. It must of course run in parallel with both producer and consumer processes. The communication structure of such a program is shown below. If the buffer provides sufficient storage to even out the flow of traffic between producer and consumer, they can then both proceed in their own time, blissfully unaware of the other's existence.

    PAR

      —producer

      out!...

      —buffer

      PAR

        out?...

        in!...

      —consumer

      in?...

An OCCAM program is said to be configured when OCCAM channels are assigned to the physical links on a transputer. When this is done, the

OCCAM program provides a static process structure, which at some level of its hierarchy is mapped onto the network of transputers. OCCAM programs thus provide code for executing on transputers, as well as information defining the network connecting them. Indeed, it is often useful to consider OCCAM processes as balloons and OCCAM channels as arcs connecting them. Such a process diagram is given in figure 4.8.

The handshake providing synchronisation between transputers is generated by the link hardware during the communication process. If more than one parallel process is executing on a single processor and one process is awaiting

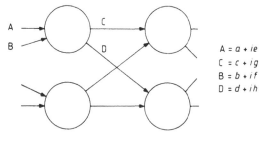

$$A = a + ie$$
$$C = c + ig$$
$$B = b + if$$
$$D = d + ih$$

```
PROC TWIDDLE(CHAN m,n,o,p VAR w,z)
  VAR ptr
  ptr = 4
  WHILE TRUE
    SEQ
      VAR a[6],b[6],c[6],d[6],t[2]
      PAR
        m?a[ptr for 2]
        n?b[ptr for 2]
        SEQ
          PAR
            t[0] = b[ptr − 2]*w − b[ptr − 1]*z
            t[1] = b[ptr − 2]*z + b[ptr − 1]*w
          PAR
            c[ptr − 2] = a[ptr − 2] + t[0]
            c[ptr − 1] = a[ptr − 1] + t[1]
            d[ptr − 2] = d[ptr − 2] − t[0]
            d[ptr − 1] = d[ptr − 1] − t[1]
          o!c[ptr − 4 for 2]
          p!d[ptr − 4 for 2]
      ptr = (ptr + 2)\6
```

FIGURE 4.8   The fast Fourier transform expressed algorithmically as sets of communicating processes, each of which applies complex multiply and add/subtract to the inputs. The communication binding would provide the usual butterfly network.

a synchronising event, then that process is descheduled and other processes may compete for the processor's resources. On receipt of that event, the suspended process is rescheduled. In this way the non-deterministic loading of transputers in a system may be evened out, although the overall static loading must be decided by the programmer. Later implementations of OCCAM may well relax this static process description.

In communication between two processes running on a single transputer, the OCCAM channels are implemented by moving data within the transputer's memory, using the same synchronising protocol as is applied to the link. Moreover, the same transputer orders are used for both situations, allowing precompiled code to have a channel assigned to a link or a memory location. It is thus possible to port an OCCAM program between physically different transputer networks, by exchanging time-shared for real concurrency, but only if a partitioning of the static process network can be found which is isomorphic to the physical transputer network. If such a partitioning exists, then the only required changes to the OCCAM code are the statements which allocate channels to links.

(iii) Choice and non-determinism

Like other sequential languages, OCCAM has other constructors, which allow a choice of action. The first of these is a looping construct. For example:

    WHILE condition

      SEQ

        A

        B

This construct executes its body repeatedly until the controlling condition becomes FALSE, at which point it terminates. In this example the body is a process comprising the sequence of processes A and B.

The second conventional mechanism for choice in OCCAM is a generalised IF constructor, which allows conditional processes. The semantics of the IF statement are similar to the CASE statement from PASCAL, for example. The IF statement can take any number of processes, each of which must have a test or guard associated with it. The IF statement starts, evaluates each test in the order written and executes the first process whose test evaluates to TRUE; it then terminates. An IF statement without any components therefore acts like a STOP process. Each test does not have to partition the test space into disjoint sets, but to ensure termination of the IF statement the space must be covered. For this reason, an IF statement will often finish with a combination of TRUE and SKIP. For example:

```
IF
    a = 1
        A
    a = 2
        B
    TRUE
        SKIP
```

This code provides choices for $a = 1$ and $a = 2$, and covers the test space with the TRUE condition. Because of the semantics of the statement, the SKIP process will only be executed if a is not equal to one and is not equal to two.

The IF statement provides deterministic choice within an OCCAM program; the choice is determined by the value of variables within the scope of the constructor, and the semantics of its execution. The same cannot be said if choice were to be determined by the state of a channel, for the state of channels cannot be determined at any given time because they are asynchronous. Thus, non-deterministic choice is provided by another constructor, the ALT. The ALT constructor provides a number of alternative processes and, like the IF constructor, each process has a choice which determines the process to execute. Unlike the IF constructor, the choices must include an input on a channel; they may also contain a condition (as in the IF constructor). The choices in an ALT constructor are called guards, after Dijkstra (1975). For example:

```
running:= TRUE
VAR x
WHILE running
    ALT
        chan1 ? x and RUNNING
            chan3 ! x
        chan2 ? x and RUNNING
            chan3 ! x
```

onoff ? ANY

running:= NOT running

This program will multiplex inputs received from channels 1 and 2 onto channel 3, until it receives any input on the channel 'onoff'. The ALT process will start, wait for one of the non-deterministic choices to be satisfied (in this case by waiting for an input and evaluating the condition), execute the guarded process, and then terminate.

This program also illustrates a number of other aspects about OCCAM; for example the input to ANY provides a signal, as the data associated with the communication is discarded. Only the synchronisation is relevant. Notice also how the program is terminated. The variable 'running' must be set as a guarded process in this example, otherwise the ALT statement may not terminate (an ALT process behaves like an IF process in this respect). For example, it could be reset after the WHILE test, but before the ALT test, in which case none of the guards would ever be satisfied on that pass of the WHILE process. The program illustrates variable declaration, which can be associated with any process or construct in the language, and its scope is determined by the persistence of that process. Thus the variable x would be allocated off the process stack prior to the execution of the WHILE constructor, and reclaimed after it has terminated. Indeed it is possible, although a little silly, to do the following:

VAR x

$x:=$ VALUE

(*iv*) *Replicators*
With sequential, as well as parallel, constructors, it is often desirable to use replication. OCCAM allows this as an extension of the syntax of the respective constructors. For example:

SEQ $i = 0$ FOR 5

A

B

will perform the sequence of processes A then B, five times. The processes may contain arrays which use i as an index, or may use i as a label in other ways, just as in other sequential loop constructs. The values taken by i are [0 1 2 3 4]. The use of the parallel replicator is more interesting, for with it arrays of parallel processes may be described. For example:

PAR $i = 0$ FOR 32

    PAR $j = 0$ FOR 32

        A

will set up a two-dimensional array of processes A. The replicator variables can now be used to select from arrays of channels, and to describe communication between these replications of this process. It should be noted that the requirement for static networks in current implementations of OCCAM mandate a constant or constant expression for the replicator count.

It is clear that in this text we can only provide an introduction to the OCCAM language. For further reading, the reader may wish to refer to INMOS (1984) and Jones (1985) for an introduction to OCCAM I, to INMOS (1986) and Pountain (1986a) for an introduction to OCCAM II, and to Hoare (1986) and Roscoe and Hoare (1986) for a more theoretical treatment of CSP and OCCAM.

4.5   TECHNIQUES FOR EXPLOITING PARALLELISM

To complete this chapter on parallel languages we summarise a number of techniques which can be applied to the solution of a problem. The choice of technique will depend on many factors, including the languages available, the underlying model of parallelism, and the parameters of its implementation. However, the same algorithm may be implemented in many cases using all three of the techniques outlined below.

For example, using transputer networks the model is one of process parallelism and the most important parameter which must be considered is that of link communication bandwidth, discussed in §3.5.5. This will determine the granularity with which a program or its data may be partitioned for distribution on a network of transputers (assuming, of course, that the program requires communications). For a given algorithm, the granularity and hence efficiency may well be different using the different techniques described below.

4.5.1   The processor farm
This is one of the simplest methods of exploiting parallelism. It can be used in applications where discrete bundles of processing can be 'farmed out' to an available processor. These processes may well be identical, but must be

independent. For example, in experimental high-energy physics vast amounts of data are produced from experiments. This data must be analysed for significant events. Each event, once recognised, may require a large amount of processing to determine whether it is significant or not. The processing of this data may be farmed out to a processor, while other events are being recognised and processed on other processors. The processors can be allocated from a pool of available processors. This is sometimes called event parallelism.

There are many other applications in which this technique may be exploited, for example the processing of independent shapes in ic design mask data, or the calculation of independent pixels of colour in the Mandlebrot set calculation. In all cases, the allocation of units of work of sufficient size to a worker processor would be performed by a supervisor process, whose only task was to distribute work and collate results.

The requirement for efficient operation of this methodology is to have independent units of processing which require no (significant) inter-communication, in terms of their computational requirement, when compared with the overhead of distributing the work and collating the results. There is still a communications requirement, which should ideally be overlapped with the processing activity. The load on the supervisor will also limit the granularity of the unit of computation.

Process structures for implementing processor farms are illustrated in figure 4.9(a). Figure 4.9(b) shows a single-level model, where work is distributed along a linear structure when requested, or even continuously, with results being collected along back channels. Each process, which would be physically distributed, would contain code for receiving and forwarding work (which would depend on the buffering scheme used), and a copy of the program to be executed.

Figure 4.9(c) shows a hierarchical structure for a processor farm, where all but the leaf nodes would contain a supervisor process, receiving and distributing work, and a copy of the program to be executed. The leaf nodes would require only a simple buffering process plus the program.

4.5.2 Algorithmic parallelism

The second technique relates to pipeline structures. In this technique, a program or process is partitioned and distributed across a network of processors, or processing agents. Each partition executes in parallel on its own data, or data that it receives from other partitions of the program. Where there are sequential constraints on the evaluation of the partitions of the program, pipelining may be brought into play to maintain concurrency. Where pipelining is used, there is a requirement for streams of data to be

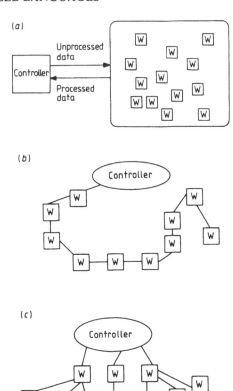

FIGURE 4.9   The concept of the processor farm. (*a*) Model of the processor farm. (*b*) A linear network of farm processes. (*c*) A tree network of farm processes.

processed by each partition, so that any start-up imposed by sequence may be amortised.

The network of processors required for this technique must ideally reflect the data flow of the given algorithm. It should in fact form a static data flow graph of the algorithm. The granularity of the distributed processes could theoretically be at the level of a single operation, as in a fine grain dynamic data flow machine (Gurd and Watson 1980, Watson and Gurd 1982), or even finer if we consider the microprogramming of pipelined architectures. Alternatively, the granularity could be at the program level, as in UNIX piped processes. The granularity of the processes must be determined by the ratio of time required for processing, compared with the time required for communicating any data to a given process.

This is a highly effective way to program the transputer, for example, especially if the code and data required in each transputer is small compared with its on-chip memory, as this gives a system with a high density of processing power and low density of distributed memory. Indeed, if the program partitions are sufficiently small, all data and the program may be held in the on chip memory, 4 Kbytes in the case of the T800.

This technique has often been exploited in special-purpose hardware, for example in digital signal processing. However, given a programmable component, such as the transputer, many applications could exploit common hardware. The problem is that unless the network accurately reflects the data flow graph of the application, the granularity of the distributed processes will become unacceptably large. This is due to the additional latency, larger processor overhead, and the lower bandwidth associated with data forwarding between non-adjacent transputers. An example of a FFT implementation is given in figure 4.8, where a subset of the process network is shown, together with the fragment of OCCAM 1 which executes on each transputer.

4.5.3   Geometric parallelism

The final technique for exploiting replication involves a partitioning of data, rather than code. If a given algorithm can process a large data structure concurrently, then it may be partitioned over a network of processors. In this case, the algorithm normally used to process that data would simply be replicated on every processor in the network. The algorithm would then be applied to the local partition of the data structure, together with any data communicated from other partitions.

It is important, for the reasons outlined in §3.3, that the underlying network reflect the communications patterns required between partitions of the data structure.

5 Parallel Algorithms

5.1 GENERAL PRINCIPLES

In order to obtain the optimum performance from any computer it is necessary to tailor the computer program to suit the architecture of the computer. This has always been the case, even on serial computers, and remains so. It is the reason why carefully written assembler code which takes into account the structure of the computer can still outperform the code produced by the most sophisticated compilers. What has changed with the advent of the parallel computer is the ratio between the performance of a good and a bad computer program. This ratio is not likely to exceed a factor of two or three on a serial computer, whereas factors of ten or more are not uncommon on parallel computers. Quite simply the stakes in the programming game have been substantially raised.

We make no attempt in this chapter to survey parallel algorithms in all major areas of numerical analysis—such a task would require a volume to itself. Instead we have selected a series of related problems that illustrate the kind of considerations that are likely to be important in choosing an algorithm, and show how the relative performance of algorithms can be analysed in a simple way using the concepts introduced in Chapter 1. We start in §5.2 with the simple problem of finding the sum of a set of numbers, and then extend the techniques to the solution of the general first-order recurrence. Matrix multiplication (§5.3) is another simple problem that demonstrates a variety of approaches, each of which is suited to a different type of parallel computer. Tridiagonal systems of equations (§5.4) and transforms (§5.5) both occur very frequently and deserve special consideration. The results obtained in these last two sections are then used in §5.6 to analyse different iterative and direct methods for the solution of partial differential equations. This section lays special emphasis on rapid transform methods that are applicable to certain classes of simple partial differential equations, such as Poisson's equation. Of the many areas that we do not have the space to consider, the most important are probably optimisation, root finding, ordinary differential equations, full and sparse general linear equations, and matrix inversion and eigenvalue determination. The reader is referred to Miranker (1971), Poole and Voight (1974), Sameh (1977), Heller (1978) and the references therein for a discussion of some of these topics. Other perspectives on parallel computation are given by Kulisch and Miranker (1983) and Kung (1980).

The analysis of a parallel algorithm must be performed within the framework of a particular computational timing model. Here we use the $(r_\infty, n_{1/2})$ model of the timing behaviour of the hardware (§1.3.2) to develop the $n_{1/2}$ method of algorithm analysis for SIMD computation (§5.1.6), and the $s_{1/2}$ method of analysis for MIMD computation (§5.1.7). Many other models exist, for example that of Bossavit (1984) for vector computation, and Kuck (1978) and Calahan (1984) for MIMD computation.

Until 1984 there was no journal exclusively devoted to the publication of algorithms for parallel computers. Most publications appeared in the USA in *IEEE Transactions on Computers*, the *Journal* and *Communications of the Association of Computing Machinery* or the *Journal of Computational Physics* (Academic Press). In Europe the principal source is *Computer Physics Communications* (North-Holland), which contains the proceedings of the conferences on *Vector and Parallel Processors in Computational Science* (VAPP 1982, 1985) held in Chester in 1981 and Oxford in 1984. Other recent European conferences with published proceedings containing parallel algorithms were *Conpar 81* (Händler 1981) and *Parallel Computing 83, 85* (Feilmeier, Joubert and Schendel 1984, 1986). The principle conference in the USA is the *International Conference on Parallel Processing* which takes place yearly and is published by the IEEE Computer Society Press. In 1984 two new journals appeared devoted entirely to the problems of parallel computation: these are *Parallel Computing* (North-Holland) and the *Journal of Parallel and Distributed Computing* (Academic Press). The journal *Computers and Artificial Intelligence*, published by the Slovak Academy of Sciences since 1982 with papers in English or Russian, also covers parallel algorithms. Another principal source for parallel algorithms are edited collections of papers by the leading practitioners, often resulting from specialist one-off conferences. Examples of these are: *High Speed Computers and Algorithm Organization* (Kuck *et al* 1977); *Parallel Computations* (Rodrigue 1982); *Parallel Processing Systems* (Evans 1982); *High-speed Computation* (Kowalik 1984); *Algorithms, Software and Hardware of Parallel Computers* (Miklosko and Kotov 1984); *Distributed Computing* (Chambers *et al* 1984); *Distributed Computing Systems Programme* (Duce 1984); *Introduction to Numerical*

Methods for Parallel Computers (Schendel 1984); *Parallel MIMD Computation* (Kowalik 1985); *Vector and Parallel Processors for Scientific Computation* (Sguazzero 1986); and *Scientific Computing on Vector Computers* (Schönauer 1987). It seems, from the explosive growth of publications since 1984, that parallel computation has established itself as an independent field of study.

5.1.1 Algorithm performance

We will define the performance of a computer program or algorithm to be inversely proportional to the CPU time† consumed during the execution of the program. In other words a high-performance program reaches its goal in the least time. Thus in this context the objective is to minimise the CPU time that is used to obtain the solution to the problem. This is not the only definition of performance that could have been given—we might have asked for minimum cost on a particular computer installation or for the least use of memory. However, the improvement in performance by our definition is usually desirable and is the easiest to quantify. Any reference to cost is subject to the vagaries of individual charging algorithms, and the availability of very large fast computer memories (256 Mword seems to be becoming the norm on parallel computers) means that the use of memory is not likely to be an important constraint.

It is interesting to note that on the first generation of computers, for example the EDSAC which had a total fast memory of only 512 words (Wilkes *et al* 1951), the minimum use of memory *was* the most important design criterion and special algorithms were devised to satisfy it, for example Gill's variation of the Runge–Kutta method for the solution of ordinary differential equations (Gill 1951). It is also important to realise that on parallel computers, the goal of minimum execution time is *not necessarily* synonymous with performing the minimum number of arithmetic operations in the way that it is on a serial computer. This will become evident from the examples given in this chapter, and arises because the gains in speed resulting from increasing the amount of parallel execution in an algorithm may outweigh the cost of introducing extra arithmetic operations. Thus although it is sensible to measure the hardware performance of a computer in terms of the number of floating-point operations executed per second, such a measure alone is insufficient to assess the performance of an algorithm on a parallel computer (see §§5.1.6 and 5.1.7).

The performance of a computer program depends both on the suitability of the numerical procedure—known as the algorithm—that is used to solve

---

† Central processing unit time—the time during which the computational units of a computer (arithmetic and logic) are used by a program. It does not include time for input and output.

the problem and on the skill with which the algorithm is implemented on the computer by the programmer or compiler during the operation of coding. In this chapter we examine the algorithms that are suitable for the solution of a range of common problems on parallel computers. If the parallelism in the algorithm matches the parallelism of the computer it is almost certain that a high-performance code can be written by an experienced programmer. However, it is outside the scope of this book to consider the details of programming for any particular computer, and the reader is referred to the programming manuals for his particular computer. Most parallel computers provide a vectorising compiler from a high-level language, usually FORTRAN with or without array processing extensions (see Chapter 4), and it should be possible to realise most of the potential performance in such a language In most cases it will be necessary to code the key parts of the program in assembler language—thereby controlling all the architectural features of the machine—if the ultimate performance is to be obtained (for example supervector performance on the CRAY X-MP, see Chapter 2).

5.1.2 Parallelism

At any stage within an algorithm, the parallelism of the algorithm is defined as the number of arithmetic operations that are independent and can therefore be performed in parallel, that is to say concurrently or simultaneously. On a pipelined computer the data for the operations would be defined as vectors and the operation would be performed nearly simultaneously as one vector instruction. The parallelism is then the same as the vector length. On a processor array the data for each operation are allocated to different processing elements of the array and the operations on all elements are performed at the same time in response to the interpretation of one instruction in the master control unit. The parallelism is then the number of data elements being operated upon in parallel in this way. The parallelism may remain constant during the different stages of an algorithm (as in the case of matrix multiplication, §5.3) or it may vary from stage to stage (as in the case of SERICR, the serial form of cyclic reduction, §5.4.3).

The architecture of parallel computers is often such as to achieve the best performance when operations take place on vectors with certain lengths (that is to say, certain numbers of elements). We shall refer to this as the natural hardware parallelism of the computer. The 64×64 ICL DAP, for example, provides three types of storage and modes of arithmetic for vectors of, respectively, length one (horizontal storage and scalar mode), length 64 (horizontal storage and vector mode) and length 4096 (vertical storage and matrix mode). Although these modes are achieved through software, they are chosen to match the hardware dimensions of the DAP array and thus constitute three levels of natural parallelism, each with its own level of performance (see §3.4.2). On pipelined computers without vector registers, such as the CYBER 205, the average performance (equations (1.9) and figure 1.16) increases monotonically as the vector length increases, and one can only say the natural hardware parallelism is as long as possible (up to maximum vector length allowed by the hardware of the machine, namely $64K - 1$). On pipelined computers with vector registers, such as the CRAY X-MP, the performance is best for vector lengths that are multiples of the number of elements held in a vector register. In the case of the CRAY X-MP with vector registers holding 64 elements of a vector, the natural parallelism is 64 and multiples thereof.

The objective of a good programmer/numerical analyst is to find a method of solution that makes the best match between the parallelism of the algorithm and the natural parallelism of the computer.

5.1.3   The paracomputer and efficiency

In order to assess the performance of processor arrays, Schwartz (1980) has introduced the concept of the *paracomputer*. This computer is an infinite SIMD array of processing elements, each of which may access a common memory in parallel for any piece of data. Any algorithm would have its maximum performance on such a computer because the normal causes of inefficiency are eliminated. The paracomputer suffers no routing delays or memory conflicts (§5.1.4 and §5.1.5) and always has sufficient processing elements. Although the paracomputer can never be built, it is a useful concept for assessing the performance of algorithms and actual processor arrays. For example, one can define the para-efficiency as

$$\varepsilon_p = \frac{\text{time to execute on the paracomputer}}{\text{time to execute on an actual computer}} \tag{5.1}$$

where we assume that the processing elements have the same hardware performance on the para- and actual computers.

One may also use the ratio of the execution times on the paracomputer for two algorithms as a relative measure of their performance. However, this measure may not be a good indication of the relative performance on actual computers, because it ignores the time required to transfer data between distant processing elements of the array (the routing delays). Grosch (1979) has used the concept of the paracomputer to compare the performance of different algorithms for the solution of Poisson's equations on processor arrays with different interconnection patterns between the processing elements. He compares the common nearest-neighbour connections, as available in the

ICL DAP (see §3.4.2), with and without long-range routing provided by the perfect shuffle interconnections as proposed by Stone (1971).

Although devised for the comparison of processor arrays, the concept of the paracomputer can be related to the study of pipelined processors. The para-computer corresponds to a pipelined processor with zero start-up time, no memory-bank conflicts and a half-performance length $n_{1/2}$ of infinity. Remembering that $n_{1/2} = s + l - 1$ (see equation (1.6b)), where $s\tau$ is the set-up time and l the number of subfunctions that are overlapped, one can see that the paracomputer is approached as l, the extent of parallel operation in the pipeline, becomes large. At the other extreme one can define a perfect serial processor as a pipeline processor with only one subfunction (i.e. the arithmetic units are not segmented) and which also has no set-up time or memory conflicts. The perfect serial processor therefore corresponds to a pipelined processor with $n_{1/2} = 0$. Actual pipelined designs, with a finite number of subfunctions and therefore a finite value of $n_{1/2}$, then lie between the perfect serial processor and the para-computer, depending on their half-performance length. Characterised in this way, we obtain the spectrum of computers described in §1.3.4 and shown in figure 1.15.

5.1.4 Routing delays

We have already mentioned that the time to route data between different processing elements on a processor array can have an important effect on the para-efficiency of algorithms on such computers. These routing delays may be relatively unimportant, if the time for a typical arithmetic operation is very much longer than the time to pass data between a pair of processing elements—as is the case for floating-point arithmetic on the ICL DAP. However, if the arithmetic operation time is comparable to the routing time, the latter plays an important role in determining the performance of an algorithm and cannot be ignored. This occurs on the ICL DAP when short words and integer arithmetic are used, as is likely to be the case in picture processing. For example Jesshope (1980a) has shown that, in an implementation of the number theoretic transform on the ICL DAP, routing delays can account for over half of the execution time of the algorithm. Other results on data routing are given in Jesshope (1980b, c).

5.1.5 Memory-bank conflicts

Routing delays represent the problem of bringing together in one processor data from different parts of the memory. A similar problem on a typical pipelined vector computer, such as the CRAY X-MP, is that of the memory conflicts that may occur when data is brought from the banked memory of such a machine to one of its pipelined arithmetic units. Although the memory of such machines is often described as a large multimillion-word random-access memory, this description is deceptive. It would be quite impossible to provide a separate parallel connection between each of the several million words and the pipelined arithmetic units. In practice such memories are divided into a relatively small number of independent banks. On a CRAY X-MP with a 2 Mword memory, for example, there are 16 such banks of 128 Kwords, each of which may be servicing simultaneously a different memory request. The banks are numbered 0 to 15 and ordered cyclically in numerical sequence with bank 0 following bank 15. A contiguous vector is one in which successive elements in a vector are stored in successively numbered, and therefore different, banks of memory. Successive elements may therefore be accessed on successive clock cycles of the machine. This is the maximum rate of delivery of data (described as a bandwidth of one word per clock period) and matches the maximum rate at which data can be accepted by a pipelined arithmetic unit. Banking delays are said to occur if the actual rate of transfer is less than this maximum. This happens if a memory request is made to a memory bank that is still busy servicing a previous request. This is known as a memory-bank conflict.

On the CRAY X-MP a memory bank is busy for four clock periods when servicing a memory request and a second request to the same bank cannot be accepted during this time. However, requests to different banks that are not busy may be made on successive clock cycles. It is obvious that, on a 16-bank machine every fourth element of a contiguous vector can be accessed at the maximum rate because the requests to any bank occur at intervals of four clock periods, just as the bank becomes free from its previous request. If, however, requests were made on successive clock periods to every eighth element of the vector, they would arrive at each bank every two clock periods, while the bank was still busy with the previous request. Therefore every eighth element of a vector, although successive values are stored in different memory banks, may only be accessed at half the maximum rate. Every sixteenth element of a vector is stored in the same memory bank and successive values can be accessed at best every four clock periods, that is to say at a quarter of the maximum rate.

It is evident, therefore, that in selecting an algorithm for a machine with a banked memory, care must be taken to avoid memory-bank conflicts. Unfortunately many good numerical algorithms are based on recursive halving or doubling, and involve successive reference to data separated in memory location by a power of two (e.g. the fast Fourier transform, see §5.5.2). It is also common practice to perform matrix manipulation with matrices in which the number of rows and columns is a power of two. Since the number of memory banks is usually chosen also to be a power of two, memory-bank conflicts can arise very easily. They are one reason for not achieving supervector performance on the CRAY X-MP and other vector computers.

Such conflicts can be minimised by considering carefully the layout of a matrix in store. Let us consider a 16×16 matrix stored column by column across a memory split into 16 banks. This is the normal pattern of storage provided by a FORTRAN compiler. Successive elements of any column, or of a forward or backward diagonal may be accessed at the maximum rate because they are stored in different memory banks. However, all elements of the same row are stored in the same bank and cannot be accessed in succession for row operations without memory conflict. This problem may be solved in software by adding a dummy row and column to the matrix. It is then stored as if it were a 17×17 matrix. Successive elements of any column (storage interval of one bank), any row (interval 17 banks) or forward diagonal (interval 18 banks) can be accessed without memory conflict. Only access to the backward diagonal (storage interval 16 banks) leads to memory conflict, but this pattern of access is rare in matrix manipulation.

Memory conflicts can be minimised in the common instances mentioned above, if the number of memory banks is chosen as a prime number. The Burroughs Scientific Computer adopts this hardware approach and has 17 memory banks. In the above example of a 16×16 matrix, all rows, columns and diagonals can be accessed without conflict. An important feature of a prime number of memory banks (other than 2) is that references to elements of a vector separated by a power of two, as occur repeatedly in successive doubling algorithms, cannot be references to the same memory bank.

5.1.6   The $n_{1/2}$ method of vector (SIMD) algorithm analysis

The advent of the new generation of vector computers presents the numerical analyst with new problems. After questions of numerical convergence and accuracy have been satisfactorily answered, there remains the question of the selection of the best algorithm to solve a particular problem on a particular computer. If we consider 'best' to be synonymous with the least execution time, then it is necessary to take into account the timing characteristics of the computer hardware and its associated software. We present the $n_{1/2}$ method of algorithm analysis as a rational way of introducing these characteristics into the method of choice (Hockney 1982, 1983, 1984b). The parameter $n_{1/2}$ has also been used for performance evaluation by Arnold (1983), Gannon and Van Rosendale (1984), Neves (1984), Strakos (1985, 1987) and Schönauer (1987).

In order to time an algorithm on the serial computer it is only necessary to know its total amount of arithmetic—often called the work—because the computer time is directly proportional to this quantity. In this case, optimisation is straightforward because minimising the time is the same as minimising the total amount of arithmetic, which is the condition that has been traditionally used ever since the first serial von Neumann computers were built. On a vector computer, however, the situation is much more complex because some of the work is collected into a smaller number of vector operations, in each of which many arithmetic operations are executed in parallel by either pipelining or the simultaneous use of many arithmetic units. There is a timing overhead associated with the initiation of each of these vector operations. This overhead is proportional to the $n_{1/2}$ of the computer and must be included in the timing comparison. In fact, vector computers often differ more in the extent of this overhead than they do in their asymptotic performance. In estimating the timing of an algorithm on a vector computer it is therefore necessary to know, and include the effect of, both the amount of arithmetic *and* the number of vector operations. The $n_{1/2}$ method of algorithm analysis provides a methodology for including both these quantities in the timing analysis, and thereby solving the much more difficult problem of time minimisation on a vector computer. The traditional condition of minimising the arithmetic is, quite simply, incorrect, and in particular we will find that the minimum arithmetic algorithm does not necessarily execute in the minimum time.

(i) *Vectorisation*
The first stage, however, in converting an existing program to run on a vector computer is the reorganisation of the code so that as many as possible of the DO loops are replaced by vector instructions during the process of compilation using a *vectorising compiler*. This process of *vectorisation* may be all that is done, and leaves a program comprised of two parts: a scalar part to be executed by the scalar unit of the computer, and a vector part to be executed by the vector unit. It is quite usual for the r_∞ of the scalar unit to be ten times slower than the r_∞ of the vector unit, so that the execution time of the algorithm may depend primarily on the size of the scalar part of the code, and rather little on the efficiency with which the vector part of the code is organised (see §1.3.5 and figure 1.19). However, there are many algorithms associated particularly with the solution of partial differential equations, in which all the floating-point arithmetic can be performed by vector instructions, and we give some examples in this chapter. For these vector algorithms the best organisation of the vectors (i.e. the choice of the elements composing them and their length) is of critical importance. The

$n_{1/2}$ method of vector algorithm analysis is presented as a technique for rationally optimising vector algorithms, or the vector parts of more complex codes that also have unavoidable scalar parts.

A survey of vectorisation techniques with reference to the CRAY-1, CYBER 205, HITACHI S9 with IAP, ICL DAP and the Denelcor HEP has been made in a monograph by Gentzsch (1984). Other publications on the topic are Wang (1980), Swarztrauber (1982), Arnold (1983), Bossavit (1984) and Neves (1984). The best known automatic vectoriser is probably that developed at the University of Illinois by Kuck (1981) and co-workers, called the *parafrase* system. Vectorisation techniques are also discussed by Yasamura *et al* (1984) and Schönauer (1987), and the performances of three automatic vectorisers, including parafrase, are compared by Arnold (1982).

(ii) Algorithm timing

The simplest timing assumption to make is that a vector algorithm is a sequence of vector instructions, and that the time to execute a vector instruction is linearly dependent on the length of the vector involved, and characterised by the two parameters $(r_\infty, n_{1/2})$ and the timing formula (1.4a), namely

$$t = r_\infty^{-1}(n + n_{1/2}). \tag{5.2}$$

The time T for the execution of the algorithm can then be written

$$T = \sum_{i=1}^{i=q} r_\infty^{-1}(n_i + n_{1/2}) \tag{5.3}$$

where q is the number of vector operations making up the algorithm and n_i is the vector length of the ith vector operation. If the parameters r_∞ and $n_{1/2}$ are approximately the same for all the operations, or suitable average values are taken, then r_∞ and $n_{1/2}$ may be taken out of the summation, and equation (5.3) may be written

$$T = r_\infty^{-1}(s + n_{1/2}q) \tag{5.4a}$$

where

$$s = \sum_{i=1}^{i=q} n_i$$

is the total amount of arithmetic in the algorithm.

Alternatively, the timing formula (5.4a) for the execution of an algorithm can be written in terms of the average vector length, $\bar{n}$, of the algorithm:

$$T = r_\infty^{-1}q(\bar{n} + n_{1/2}) \tag{5.4b}$$

or as

$$T = r_\infty^{-1} s[1 + (n_{1/2}/\bar{n})] \qquad (5.4c)$$

where $\bar{n} = s/q$. The expression in square brackets in equation (5.4c) is the factor by which a traditional serial complexity analysis, $r_\infty^{-1} s$, is in error when applied in a vector environment. It also shows that it is not the absolute value of $n_{1/2}$ which is important, but its ratio to the average length of a vector operation in the algorithm: i.e. $n_{1/2}/\bar{n}$.

(iii) Serial and parallel complexity

Equation (5.4a) is interesting because it reveals the link between traditional serial numerical analysis and recent work on parallel algorithms. Expressed differently, it shows the link between the serial and parallel complexity analysis of algorithms. A serial computer is one with a small or zero value of $n_{1/2}$, hence only the *first* term of equation (5.4a) is important, and the minimum execution time is obtained by minimising s, the total amount of arithmetic. However, parallel algorithm analysis is based on the use of the paracomputer, which in our formalism is obtained by taking $n_{1/2} = \infty$ (see §5.1.3). In this case only the *second* term of equation (5.4a) is important, and the minimum execution time is obtained by minimising the number of vector operations q, *regardless* of the amount of arithmetic involved. These two contrasting views of algorithm optimisation have led to a dichotomy in the numerical analysis/algorithm community between those living in the 'serial' and 'infinitely parallel' worlds, because quite different algorithms are frequently recommended for the solution of the same problem. We believe that the $n_{1/2}$ method of algorithm analysis provides a simple bridge between the two views.

Both the above views of algorithm optimisation are unrealistic extremes, because actual vector computers usually have finite, non-zero values of $n_{1/2}$ (for example, the CRAY-1 has a value approximately equal to 20, and the CYBER 205 a value approximately equal to 100). The question then arises: 'What should one minimise on a computer with a finite value of $n_{1/2}$?' Equation (5.4a) shows that the answer is that one should minimise neither s nor q, but should minimise the quantity $(s + n_{1/2}q)$. The $n_{1/2}$ method of analysis is a formalism for doing just that. The half-performance length $n_{1/2}$ is thus seen to be the parameter that linearly interpolates between the extreme views of serial and infinitely parallel computation. It is the way through which the finite parallelism of a computer system can be introduced into the analysis. The main point of the method is that one does not have to decide, unrealistically, whether one's computer is serial or parallel; one can actually include numerically the amount of parallelism by including the correct value of $n_{1/2}$.

(iv) Algorithmic phase diagrams

When algorithms are compared on the same computer, we take the ratio of two timing expressions of the form of equation (5.4). The algorithm is fully described by giving the operations counts s and q (or $\bar{n}$ and q, or $\bar{n}$ and s) and the computer described by the two parameters r_∞ and $n_{1/2}$. In taking the ratio of two timing expressions to determine the best algorithm the value of r_∞ cancels out, since it affects the timing of both algorithms by the same factor. Thus, within the approximations of this analysis, $n_{1/2}$ is the only property of the computer that affects the choice of algorithm.

When comparing algorithms, *equal performance lines* (EPL) play a key role. If $T^{(a)}$ and $T^{(b)}$ are the execution times for algorithms (a) and (b) respectively, then the performance of (a) equals or exceeds that of (b), $P^{(a)} \geqslant P^{(b)}$, if $T^{(b)} \geqslant T^{(a)}$, from which one obtains

$$n_{1/2} \geqslant \frac{s^{(a)} - s^{(b)}}{q^{(b)} - q^{(a)}}.$$ (5.5)

The inequality in (5.5) determines the regions of a parameter plane in which each algorithm has the better performance, and the equality gives the formula for the EPL between algorithms (a) and (b).

Because of the nature of the characterisation, $n_{1/2}$ always appears linearly in equations for equal performance lines. It may therefore be set explicitly on the left-hand side of such equations. It is a property of the computer, as seen through the compilers and assemblers that are used. The right-hand side, however, depends only on the operations' counts of the two algorithms, and is likely to be a complicated nonlinear function of the size of the problem n. This might be the order of the matrices in a matrix problem, or the number of mesh points along a side in a finite difference approximation to a partial differential equation problem.

The manner of presenting the results of an algorithm comparison is important. Formulae such as (5.5) contain the result, but cannot be used without extensive computation, and in complicated cases tables of numbers soon become unmanageable. Clearly, a graphical presentation is to be preferred, which will allow the choice of algorithm to be made directly from the information that defines the problem. Such a presentation can be made by plotting equal performance lines between pairs of algorithms on the $(n_{1/2},n)$ parameter plane. These lines divide the plane into regions in which each of the competing algorithms has the best performance. We call such a presentation a 'phase diagram' in analogy with such diagrams in physical chemistry. In the case of the chemical phase diagram, the values of parameters describing the conditions (e.g. temperature and pressure) determine a point in a parameter plane that is divided into regions in which the different states of matter have the lowest energy. One could say that nature then chooses this state from all others as the best for the material. In the case of the *algorithmic phase diagram*, parameters describing the computer and problem size determine a point in a parameter plane that is divided into regions in which each algorithm has the least execution time. This algorithm is then chosen as the best.

It is helpful to adopt certain standards in the presentation of such algorithmic phase diagrams, in order to make comparisons between them easier. It is good practice to make the x axis equal to or proportional to $n_{1/2}$, and the y axis equal to or proportional to the problem size n. In this way algorithms suitable for the more serial computers (small $n_{1/2}$) appear to the left of the diagram, and those suitable for the more parallel computers (large $n_{1/2}$) to the right. Similarly, algorithms suitable for small problems are shown at the bottom of the diagram, and those suitable for large problems at the top. A logarithmic scale is usually desirable for both axes, and in the simplest case of the $(n_{1/2}, n)$ plane the horizontal axis specifies the computer and the vertical axis the problem size. Some examples of algorithmic phase diagrams are given in figures 5.3, 5.9, 5.10, 5.26, 5.27 and 5.28 (Hockney 1982, 1983). A simple example of their preparation and interpretation is given in §5.2.3. This compares two algorithms. A more complicated example comparing four algorithms is given in Hockney (1987a). Figure 5.28 compares two algorithms, giving the best value of an optimisation parameter to use in any part of the phase plane.

We have seen in equation (5.4c) that the ratio $n_{1/2}/\bar{n}$ is more important in the timing than $n_{1/2}$ itself. Similarly, we find that algorithmic phase diagrams are usually more compactly drawn if the x axis is equal to the ratio of $n_{1/2}$ to problem size: $n_{1/2}/n$. This ratio, rather than $n_{1/2}$ itself, determines whether one is computing in a serial environment (small values) or a parallel environment (large values). The ratio also has the advantage of being independent of the units used to measure n.

(v) Serial and parallel vector algorithms
In the design of vector algorithms it is useful to distinguish two extreme formulations of the same basic method, that is to say two different organisations of the same algebraic/numerical relationships which define the numerical method. These are as follows:

(1) The *serial variant* of the vector algorithm, which is obtained by minimising the total amount of arithmetic, s—this formulation of the vector

algorithm is so called because it would run in the minimum time if it were executed on a serial computer for which $n_{1/2} = 0$;

(2) the *parallel variant* of the vector algorithm, which is obtained by minimising the number of vector operations q (or equivalently maximising the average vector length $\bar{n}$)—this formulation would execute in minimum time on an infinitely parallel computer for which $n_{1/2} = \infty$, or on an array of processors with a processor for every element of the longest vector.

The analyses that we give below limit consideration to these two possible formulations. In more complex situations, formulations may be desirable which are hybrids of the above extreme cases. For example, if the vector length in a problem is much larger than the number of processors in an array of processors, then the serial variant can often be used to reduce the vector lengths to less than the array size, after which the parallel variant is used.

(vi) Scalar and vector units
The $n_{1/2}$ method of analysis, expressed in algorithmic phase diagrams, can also be used to compare the performance of a scalar algorithm in the scalar unit of a computer with the performance of an equivalent vector algorithm in the vector unit of the same computer. Suppose the scalar algorithm has $s^{(s)}$ floating-point operations, then the time for the scalar algorithm executing in the scalar unit is

$$T_s = s^{(s)}/r_{\infty s} \tag{5.6a}$$

where $r_{\infty s}$ is the performance of the scalar unit. If the vector unit is characterised by the parameters $(r_{\infty v}, n_{1/2})$, then the time to execute the vector algorithm will be

$$T_v = (s^{(v)} + n_{1/2}q^{(v)})/r_{\infty v} \tag{5.6b}$$

where the superscript (v) indicates the counts of the operations for the vector algorithm. The formula for the equal performance line between the two alternatives is given by the condition $T_s = T_v$, whence

$$n_{1/2} = (R_\infty s^{(s)} - s^{(v)})/q^{(v)} \tag{5.7}$$

where $R_\infty = (r_{\infty v}/r_{\infty s})$ is the ratio of the asymptotic performance of the vector and scalar units. An example of such a comparison is given in figure 5.3(*a*) for the problem of summing a set of numbers.

(vii) Variation of r_∞, $n_{1/2}$
For simplicity, the derivation of the $n_{1/2}$ method of algorithm analysis given above is based on the assumption that the two parameters $(r_\infty, n_{1/2})$ may be

regarded as constant, and may therefore be taken out of the summations in equation (5.3). Manifestly, this assumption is not always true, because in Chapter 2 we have derived different values for the parameters for the different cases of, for example, register-to-register or memory-to-memory operations, contiguous or non-contiguous vectors, dyadic or triadic operations, etc. If the variation in $(r_\infty, n_{1/2})$ is large for these different cases, the summation in equation (5.3) can be split into several sums, one for each type of operation, and different values of the parameter used in each summation. Alternatively, the correct average values $(\bar{r}_\infty, \bar{n}_{1/2})$ can be used for the parameters, which are given by

$$\bar{r}_\infty = \left(\sum_k \frac{(s_k/s)}{r_{\infty,k}} \right)^{-1} \qquad \bar{n}_{1/2} = \bar{r}_\infty \sum_k \left(\frac{n_{1/2,k}}{r_{\infty,k}} \right) \left(\frac{q_k}{q} \right) \qquad (5.8a)$$

where $(r_{\infty,k}, n_{1/2,k})$ are the parameters for the kth type of arithmetic operation, and s_k and q_k are the operations' counts for the kth type of arithmetic operation. The time for the algorithm is then given by

$$T = \bar{r}_\infty^{-1}(s + \bar{n}_{1/2}q) \qquad (5.8b)$$

where $s = \sum_k s_k$ and $q = \sum_k q_k$ are the total operations' counts, as before.

The calculation of the above average values is similar to calculating the average performance of a computer, using different mixes of instructions (for example, the Gibson mix and Whetstone mix). The weights used in the above expressions (5.8), (s_k/s) and (q_k/q), are the fraction of the arithmetic which is of type k and the fraction of the vector operations which are of type k respectively. The important fact is that these ratios will be relatively independent of the problem size n, and the analysis can proceed as before using $\bar{n}_{1/2}$ instead of $n_{1/2}$. In most cases, however, it will be adequate to treat r_∞ and $n_{1/2}$ as constant and interpret the algorithmic phase diagrams for the range of parameter values that arises.

5.1.7   The $s_{1/2}$ method of MIMD algorithm analysis

The extension of the above methodology to MIMD computation with multiple instruction streams requires a slightly different computational model, and we use the concept of the work segment introduced in §1.3.6 (Hockney 1984a, 1985a, 1985c, Hockney and Snelling 1984). A work segment is a section of program in which the work can be divided between several logically independent (i.e. not communicating) instruction streams. These instruction streams may be implemented as separate *processes* in, for example, a single PEM of the Denelcor HEP, or be executed in separate *processors* in a multimicroprocessor architecture or a multi-PEM HEP. From the point of view of the computational model, however, all these cases are treated with the same analysis, simply as separate instruction streams. The essence of the work segment is that the computation is synchronised before and after each segment. That is to say all the work in a segment must be completed before the next segment can begin. The time t_i to execute a work segment is therefore the sum of the time to execute the longest of the instruction streams plus the time to synchronise the multiple instruction streams: that is to say, from equation (1.16)

$$t_i = r_\infty^{-1}(s_i/E_i + s_{1/2}) \tag{5.9}$$

where s_i is the number of floating-point operations between pairs of numbers in the ith work segment, subsequently called the amount of work in the segment, or the grain of the segment; E_i is the efficiency, E_p, of process utilisation in the ith work segment (the subscript p is now dropped); r_∞ is the asymptotic (or maximum possible) performance in Mflop/s as before; and $s_{1/2}$ is the synchronisation overhead measured in equivalent floating-point operations.

In equation (5.9) the value of both r_∞ and $s_{1/2}$ will depend on the number of instruction streams or processors used. For example, if there are p processors the r_∞ in equation (5.9) is p times the asymptotic performance of one processor. The value of $s_{1/2}$ also depends on the type of synchronisation and the efficiency of the software tools provided for synchronisation. Measured values of $(r_\infty, s_{1/2})$ for a variety of different cases are given for the CRAY X-MP in §2.2.6 (Hockney 1985a), for the Denelcor HEP in Chapter 3 (Hockney 1984a, 1985c, Hockney and Snelling 1984), and for the FPS-5000 in Curington and Hockney (1986).

Having established the timing for a work segment, we are now in a position to consider a MIMD program. In any MIMD program there will be a critical path, the time of execution of which is the time of execution of the whole program. In some cases the critical path is obvious (there may only be one path), and in other cases it may be very difficult to determine or even be data-dependent and therefore unknown until run time. However, to proceed further with a timing analysis we must assume that the critical path is known. The time T for a MIMD algorithm is calculated by summing equation (5.9) for each work segment along the critical path, giving

$$T = r_\infty^{-1}[(s/\bar{E}) + s_{1/2}q] \tag{5.10}$$

where q is the number of work segments along the critical path of the algorithm, $s = \sum_{i=1}^{q} s_i$ is the total amount of work along the critical path and $\bar{E} = s/(\sum_{i=1}^{q} s_i/E_i)$ is the average efficiency of process utilisation along the critical path of the algorithm.

For algorithms that fit into the above computational model, we see that the programming environment (computer hardware, system software and compiler facilities) is described by the parameter pair $(r_\infty, s_{1/2})$ and the algorithm itself is described by the triplet $(s, q, \bar{E})$. As in the case of SIMD computing, the value of r_∞ cancels when comparing the performance of two MIMD algorithms on the same computer. That leaves the value of $s_{1/2}$ as the computer parameter that determines the choice of the best MIMD algorithm.

Comparing equations (5.9) and (5.10) with (5.2) and (5.4a), one can see the analogy between SIMD and MIMD computation. The grain of a segment, s, or more correctly $s/\bar{E}_p$, is analogous to vector length, n; and the overhead of synchronisation, measured by $s_{1/2}$, is analogous to the half-performance length, $n_{1/2}$. Similarly, MIMD algorithms can be compared using algorithmic phase diagrams.

It is shown in §1.3.6, and it is obvious from the above analogy, that the grain size of a work segment must exceed $s_{1/2}$ if the average processing rate of the segment is to exceed 50% of the maximum (we assume $E_p \approx 1$).

Because the cost of communication between instruction streams is often high in MIMD systems, values of $s_{1/2}$ tend to be quite large, often several hundred or thousand floating-point operations. It is important, therefore, to divide a MIMD algorithm into large independent blocks of code which are given to different instruction streams, thereby obtaining very large values of s. This can often be best achieved by the parallel execution of different cases of the outermost DO loop, and is therefore called parallelisation at the outermost level of the program. This philosophy of programming is in sharp contrast to SIMD programming, in which it is the innermost DO loops of a program that are parallelised by being replaced with vector instructions. Unfortunately, parallelisation at the outermost level requires overall knowledge of the structure of the program, and is much more difficult to automate than the vectorisation of single or double DO loops, which can be done by a local examination of the program near the loops. Consequently, the parallelisation of a program for MIMD computation is likely to require much more programmer intervention than the vectorisation of a program for SIMD computation, and might even necessitate the complete restructuring of the whole program.

5.2   RECURRENCES

A recurrence is a sequence of evaluations in which the value of the latest term in the sequence depends on one or more of the previously computed terms. Such recurrences pervade numerical analysis: in the solution of linear equations by Gaussian elimination; in all matrix manipulations that require the inner product of vectors; in all iterative methods, because a better approximation to a solution is calculated from previous approximations; in all solutions of ordinary differential equations in time because values at one time depend on those found at earlier times; and in marching methods for the solution of differential equations in space.

The evaluation of a recurrence presents a special problem for a parallel computer because the definition itself is given in terms of sequential evaluation, and it would appear that only one term could be evaluated at a time, giving no scope for parallel evaluation. Needless to say the problem can be rephrased (at the expense of introducing extra arithmetic operations) so as to allow parallel evaluation and we shall explain this first in terms of the simple problem of evaluating the partial sums of a sequence of numbers (§5.2.1 and §5.2.2). In the final section (§5.2.3) we consider the introduction of parallelism into—that is to say the vectorisation of—the general linear first-order recurrence. Parallel algorithms for solving recurrences are also described by Kogge and Stone (1973) and Ladner and Fisher (1980).

5.2.1   Sequential sum

The general linear first-order recurrence can be expressed as the evaluation of the sequence x_j from the recurrence relation

$$x_j = a_j x_{j-1} + d_j, \qquad \text{for } j = 1, 2, \ldots, n, \tag{5.11a}$$

given the values of $x_0, a_1, \ldots, a_n$, and $d_1, \ldots, d_n$. There is no loss of generality, and considerable convenience, if we assume $x_0 = a_1 = 0$. This can be done if we redefine d_1 as $d_1 + a_1 x_0$, and this will now be assumed.

As a special case of the above $(a_j = 1)$ we consider first the evaluation of the partial sums x_j defined by

$$x_j = \sum_{k=1}^{j} d_k, \qquad \text{for } j = 1, 2, \ldots, n, \tag{5.11b}$$

where x_j is the sum of the first j numbers in the sequence $d_1, \ldots, d_n$.

The partial sums may be evaluated simply from the recurrence

$$x_j = x_{j-1} + d_j, \qquad \text{for } j = 1, 2, \ldots, n. \tag{5.11c}$$

This sequential sum method of evaluation may be realised with $(n-1)$ additions by the obvious FORTRAN code:

```
      X(1) = D(1)
      DO 1 J = 2, N
1        X(J) = X(J-1) + D(J)
```
(5.12a)

The relationship between the data storage pattern, arithmetic operations and time may be shown in a routing diagram. Such a diagram is shown for

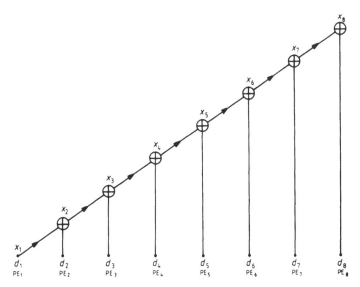

FIGURE 5.1   The routing for the sequential sum method for forming
all the partial sums of eight numbers, $x_j = \sum_{k=1}^{j} d_k$, $j = 1, \ldots, 8$. The
vertical axis is time, the horizontal axis is storage location or processing
element number. Routing of data across the store is indicated by an
arrow.

this algorithm in figure 5.1 for the case $n = 8$. The sequence of evaluations
takes place from the bottom moving upwards; operations that can be
evaluated in parallel are shown on the same horizontal level. It is clear that
at each time level only one operation can be performed (parallelism = 1),
and we say that the sequential sum algorithm has

$$n - 1 \text{ additions with parallelism 1.} \tag{5.12b}$$

and the operations' counts are

$$s = n - 1 \qquad q = n - 1 \qquad \bar{n} = 1. \tag{5.12c}$$

In order to quantify the routing, we suppose that the horizontal axis in
figure 5.1 indicates the relative location of the data in the memory of a serial
or pipelined computer, or in the case of a processor array the processing
element number. Variables in the same horizontal position do not necessarily
overwrite each other. If overwriting is not required, such variables may be
stored in different locations in the same PE of a processor array, or in a
different sequence of locations in a serial or pipelined computer.

A unit parallel routing operation is defined as a shift of all elements of an
array in parallel to a set of neighbouring PEs. In the simplest case of

nearest-neighbour connectivity in a one-dimensional processor array, this can be to shift all elements one PE to the right or one PE to the left. In a two-dimensional processor array such as the ICL DAP, a unit parallel routing can shift all elements of a two-dimensional data matrix, which is mapped over the processor array, to their nearest-neighbour PES to the north, south, east or west. In a processor array, storage after a parallel operation may be suppressed according to the state of an activity bit in each PE. Thus the number of elements actually shifted after a parallel routing operation is under program control. We refer to the number of elements actually shifted (i.e. the usefulness of the parallel shift operation) as the parallelism of the routing operation.

Inspection of figure 5.1 shows that the sequential sum algorithm requires a routing of one unit to the right at each time level. Only one number (the value of the sum accumulated so far) is involved in the shift, hence the parallelism is 1. This is confirmed by the fact that there is only one slanting arrow, indicating a shift of one number, at each time level. The complete algorithm therefore requires

$$n - 1 \text{ routing operations with parallelism 1.} \qquad (5.12d)$$

5.2.2 Cascade sum

The alternative parallel approach to the evaluation of partial sums is most easily understood from the routing diagram for the algorithm. This is shown in figure 5.2(a) for the case $n = 8$, and the algorithm is called the cascade partial sum method. A set of n accumulator registers are first loaded with the data to be summed. At level one, a copy of the accumulators is shifted one place to the right and added to the unshifted accumulators in order to form the sum of the data in adjacent pairs. At the next level, the process is repeated but with a shift of two places to the right, thereby producing sums of groups of four numbers. As the shifts are made, zeros are brought in from the left as required. In general, at the lth time level a shift of 2^l places is made and at level $l = \log_2 n$ the accumulators contain the required partial sums.

The cascade partial sum method may be expressed in an obvious vector FORTRAN-like language by:

```
        X = D
        DO 1 L = 1, LOG2N                              (5.13a)
    1       X = X + SHIFTR(X,2**(L-1))
```

where X and D are vectors of n elements, $+$ is a parallel addition over the n elements, and SHIFTR(X, L) is a vector function that places a copy of X shifted L storage locations to the right into the temporary vector SHIFTR. The elements of the vector X are not disturbed. The cascade partial sum

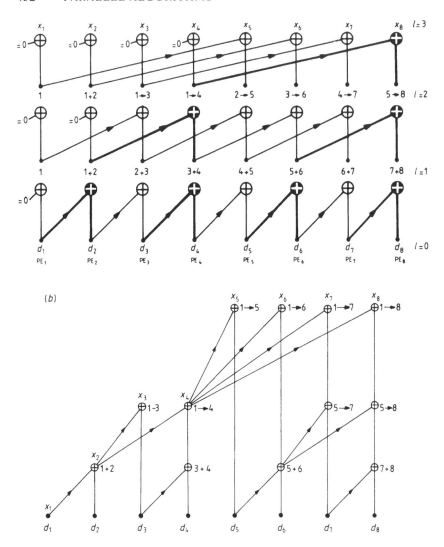

FIGURE 5.2   (a) The routing diagram for the parallel cascade sum method of forming partial sums. Zeros are brought in from the left as the vector of accumulators is shifted to the right. If only the total sum x_8 is required, then only the operations shown as full circles and bold lines need be performed. (b) Martin Oates' method of computing all partial sums.

method therefore requires

$$\log_2 n \text{ additions with parallelism } n \tag{5.13b}$$

giving operations' counts of

$$s = n\log_2 n \qquad q = \log_2 n \qquad \bar{n} = n. \tag{5.13c}$$

If we assume nearest-neighbour connectivity, 2^{l-1} unit routing operations are required at level l, giving a total of $1 + 2 + +4 + \ldots + n/2$ or

$$n - 1 \text{ routing operations with parallelism } n. \tag{5.13d}$$

The formalism for the cascade method described above, which keeps the vector length at its maximum value of n, is appropriate for use on processor arrays, because the redundant additions by zero do not consume extra time. On a vector computer, however, less time is required if the redundant operations are omitted, and the vector length is reduced. In this case the method requires

$$1 \text{ addition with parallelism } n - 2^{l-1}, \qquad \text{for } l = 1, 2, \ldots, \log_2 n \tag{5.13e}$$

giving the operations' counts

$$s = n\log_2 n - n + 1 \qquad q = \log_2 n \qquad \bar{n} \simeq n[1 - (\log_2 n)^{-1}]. \tag{5.13f}$$

In the algorithm analyses made in §5.2.3 we will assume that the vector length is reduced in this way.

Every operation in the cascade partial sum algorithm has approximately a parallelism of n (actually varying from $n - 1$ to $n/2$ if the vector length is reduced). The total number of arithmetic operations has, however, increased from $n - 1$ for the sequential sum method to approximately $n\log_2 n$. The cost of increasing the parallelism from 1 to n has therefore been a substantial increase in the amount of arithmetic. For $n = 1024$, for example, the cascade partial sum method requires ten times the number of additions. On a serial computer that can only perform one addition at a time, it is obvious that the cascade partial sum method will never be worthwhile. On a processor array or a pipelined processor we must consider whether the increased performance that these processors obtain by performing n additions in parallel is enough to outweigh the increase in the total amount of arithmetic (see §5.2.3).

The cascade partial sum algorithm simplifies considerably in the important special case in which only the single result of the *total sum* is required (in our case x_8). In figure 5.2 we have emphasised as bold lines and full circles those routings and arithmetic operations that contribute to the calculation of x_8, and note that these are only a small fraction of the operations that are required for the calculation of all the partial sums. The calculations contributing to the total sum form a binary tree (with the appearance of a cascade), in which the number of operations and therefore the parallelism is halved at each level l of the calculation. Assuming, for simplicity, that n is a power of two, the number of operations for the cascade total sum method is

1 addition with parallelism $n2^{-l}$,   for $l = 1, 2, \ldots, \log_2 n$

$$(5.14a)$$

The number of scalar additions in the algorithm is therefore

$$\frac{n}{2} + \frac{n}{4} + \ldots + 2 + 1 = n - 1, \tag{5.14b}$$

and we note that the cascade total sum algorithm has the same number of scalar operations as the sequential sum method, even though the reorganisation of the calculation into a binary tree has introduced the possibility of parallel operation. This total sum method is the algorithm normally referred to as the cascade sum method.

It has been pointed out by Martin Oates (private communication) that all partial sums can also be computed in a parallel fashion with approximately half the number of redundant arithmetic operations that are present in the original cascade partial sum method shown in figure 5.2(a). Martin Oates' variation is shown in figure 5.2(b). The partial sums are built up hierarchically. First, adjacent pairs are added, then pairs of these are combined to form all the partial sums of groups of adjacent four numbers, then these are combined in pairs to form sums of eight, and so on. At each level $n/2$ additions are performed in parallel, and there are $\log_2 n$ levels as before, giving operations' counts of

$$s = (n/2)\log_2 n \qquad q = \log_2 n \qquad \bar{n} = n/2 \tag{5.14c}$$

compared with approximately $s = n\log_2 n$ additions in the original method.

There is, however, a programming problem with Oates' method, in that the routing or indexing at each level is not as regular as that for the original method. If a vector computer, such as a CRAY X-MP, is being used, although the operations at each level are independent they cannot all be performed with a single vector operation. They would have to be implemented as a sequence of 'scalar + vector' operations, thus not realising the full potential for parallel execution. For example, at the last stage of the calculation in figure 5.2(b), the scalar x_4 is added to the vector made up of the current values of the last four elements of the d array. At the previous stage, two such operations must be executed in sequence, on vectors of length two, and so on.

If, however, a processor array such as the ICL DAP is being used, both shifting and masking are required to arrange that the correct components of the vectors are added. Clearly, the routing is much simpler in the original method. In any particular case, it would require a detailed analysis of many alternatives to decide whether the reduction in redundant arithmetic made up for the extra complexity of the program. Oates' method turns out to be a special case of a class of algorithms for the so called 'parallel prefix calculation' that have been described by Ladner and Fisher (1980). In the following section we will analyse the performance of the original method, and leave it as an exercise for the reader to calculate the effect of using the Oates' variation.

5.2.3 Relative performance

The performance of the cascade and sequential sum methods will now be compared using the $n_{1/2}$ method for analysis (§5.1.6). This model of computation assumes that there is an infinite memory, and that there are no memory bank conflicts. Although possibly violated in practice, the above assumptions give a good first basis for the choice of algorithm. A programmer responsible for evaluating sums on a real computer must, of course, give careful consideration to the properties of its memory.

We consider first the problem of finding only the total sum, and compare the sequential method with the cascade total sum method. Both these have the same amount of arithmetic, $(n-1)$ floating-point operations. However the cascade has fewer vector instructions ($\log_2 n$) than the sequential ($n-1$). It is obvious, therefore, that if both algorithms are executed in the vector unit of a computer, then the cascade method will always be the best, because these are fewer vector start-ups. However, one may ask the question whether it might be better to execute the sequential method in the scalar unit, and thereby avoid suffering the vector start-up overhead altogether. The scalar unit has a slower asymptotic rate, and therefore there will be a break-even problem size, above which it will be better to use the cascade method in the vector unit and gain the benefit of its higher asymptotic performance, and below which it is better to avoid the vector start-up overhead by using the scalar unit, even though its asymptotic performance is lower. This comparison can be quantified by using equation (5.7), with $s^{(s)} = (n-1)$, $s^{(v)} = (n-1)$, $q^{(v)} = \log_2 n$, whence the equation for the equal performance line is

$$n_{1/2} = [R_\infty(n-1) - (n-1)]/\log_2 n \qquad (5.15a)$$

or

$$n_{1/2}/n = (R_\infty - 1)(1 - n^{-1})/\log_2 n. \qquad (5.15b)$$

The algorithmic phase diagram corresponding to equation (5.15b) is given in figure 5.3(a) for a ratio of vector-to-scalar asymptotic speeds of $R_\infty = 2, 5, 10, 50, 100$. The diagram is interpreted as follows. Lines of constant $n_{1/2}$ lie at $45°$ to the axis, and the line for $n_{1/2} = 20$ corresponding to the CRAY-1 is shown. This computer has $R_\infty \approx 10$, and the intercept between

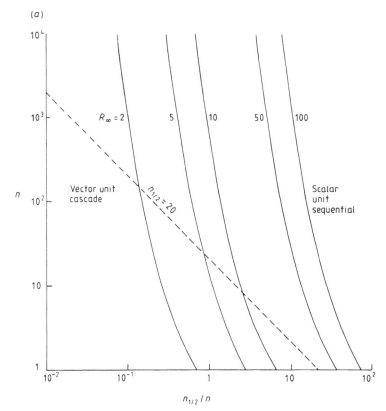

FIGURE 5.3   (a) Comparison of the sequential sum method in the scalar unit, and the cascade total sum method in the vector unit. R_∞ is the ratio of vector to scalar speeds. The broken line is the line of constant $n_{1/2} = 20$.

the line of constant $n_{1/2}$ and the line of constant R_∞ gives the break-even problem size. Roughly speaking, the conclusion is that the summation of up to about eight numbers should be done in the scalar unit (or by scalar instructions), whereas larger problems should be done in the vector unit. It is obvious, and borne out by the diagram, that as the rate of vector-to-scalar speed decreases, the scalar unit should be used for longer vector problems.

In considering the calculation of all partial sums, we have three alternatives to consider; the use of the sequential and cascade partial sum methods in the vector unit, and the sequential method in the scalar unit. We need not consider the use of the cascade partial sum method in the scalar unit but it will always perform worse than the sequential method, since it has more arithmetic. There are therefore three equal performance lines to compute for each of the three possible pairs of algorithms.

(b)

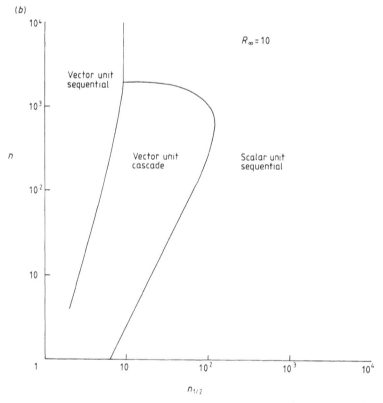

FIGURE 5.3 *cont.* (*b*) Comparison of three algorithms for the calculation of all partial sums. The sequential and cascade method executed in the vector unit are compared with the sequential method executed in the scalar unit. The ratio of vector to scalar speed is $R_\infty = 10$.

In order to compute the sequential and cascade partial sum methods in the vector unit we need only calculate the total arithmetic, s, and the number of vector operations, q, for each method. For the *sequential sum*

$$s = n - 1 \qquad q = n - 1 \qquad \bar{n} = 1. \tag{5.16a}$$

For the *cascade partial sum*

$$s = \sum_{l=0}^{\log_2 n - 1} (n - 2^l) = n\log_2 n - (n - 1)$$

$$q = \log_2 n \tag{5.16b}$$

$$\bar{n} = [n\log_2 n - (n - 1)]/\log_2 n \approx n[1 - (1/\log_2 n)].$$

Given the above operations' counts, the formula for the equal performance line between the two methods can be immediately written down from equation (5.5):

$$n_{1/2} = [n\log_2 n - 2(n-1)]/(n - \log_2 n - 1). \tag{5.16c}$$

The comparison of the sequential method in the scalar unit with the above two methods in the vector unit can be immediately written down using the operations' counts (5.16a,b). The equal performance lines obtained by substituting in equation (5.7) are: with the *sequential* method

$$n_{1/2} = R_\infty - 1 \tag{5.17a}$$

and with the *cascade partial sum* method

$$n_{1/2} = [(R_\infty + 1)(n-1)/\log_2 n] - 1. \tag{5.17b}$$

The algorithmic phase diagram is obtained by drawing the three equal performance lines (5.16c), (5.17a) and (5.17b) on the $(n_{1/2}, n)$ plane. This is shown in figure 5.3(b).

The phase diagram shows us that for any particular computer (described by a fixed value for $n_{1/2}$, i.e. a vertical line in figure 5.3(b)) none of the methods is always the best. We cannot say, for example, that because we are using a parallel computer the cascade algorithm which is designed for parallel computation is always the best. The choice depends in a complex way on the relation between the problem size n and the $n_{1/2}$ and R_∞ of the computer, which can only be adequately expressed by an algorithmic phase diagram.

For $n_{1/2} < R_\infty - 1(R_\infty = 10$ in figure 5.3(b)), the choice of algorithm lies between the sequential or cascade partial sum method, both executed in the vector unit. In this case, for any particular computer there is always a problem size above which the sequential sum method is best, and this is given by the position of the equal performance line. For example, if one is using a computer with $n_{1/2} = 8$, then the sequential method is best for summing more than about 1000 numbers. In this large problem case, the vector lengths in the calculation have become so long that the $n_{1/2}$ of the computer is negligible (i.e. $n/n_{1/2}$ is large). We are then in the regime of serial computation (see equations (1.9c) and (5.4c)), and it is not surprising that the algorithm designed for serial sequential computation is best. Put another way, one might say that the problem is so large (as measured by n) that the parallelism of the computer (as measured by $n_{1/2}$) is negligible, and the computer looks and behaves to problems of this size as though it were a serial computer even though it has finite parallelism, and serial variations of algorithms which minimise s are the best.

If $n_{1/2} > R_\infty - 1$, the choice is between the cascade partial sum method executed in the vector unit, and the sequential method executed in the scalar unit. For example, if $n_{1/2} = 20$, problem sizes less than $n \approx 10$ and greater than $n \approx 2000$ should be solved in the scalar unit. In the former case the vectors are not long enough to make up for the vector start-up overhead, and in the latter case too much extra arithmetic is introduced to make the cascade method worthwhile. In the intermediate region $10 \leqslant n \leqslant 2000$ the use of the vector unit with its higher asymptotic performance is worthwhile. However, for $n_{1/2} \geqslant 120$ (in the case of $R_\infty = 10$), the vector start-up overhead is too large ever to be compensated by the higher vector performance, and the sequential sum method in the scalar unit is always the best. Figure 5.3(b) can be drawn for other values of R_∞, as was done in figure 5.3(a). The boundary line between the use of the scalar and the vector unit will move further to the right as R_∞ increases and to the left as it decreases. The equal performance line between the two algorithms executed in the vector unit, however, will remain unchanged.

If the partial sum problem is being solved on the paracomputer, or a finite processor array with more processing elements than there are numbers to be summed, then we take $n_{1/2} = \infty$. In this case the time is proportional to the number of vector operations, q. Consequently, the cascade method is always the best, because it only has $\log_2 n$ vector operations compared with $(n-1)$ for the sequential method. If an algorithm is being chosen for a processor array, it is also necessary to consider carefully the influence of the necessary routing operations on the time of execution. The number of routing operations is the same for the cascade and sequential methods (both $n-1$ routings, see equations (5.12d) and (5.13d)). Therefore, the inclusion of the time for routing will not affect the choice of algorithm. However, depending on the relative time for a routing and an arithmetic operation, and the algorithm under consideration, the time for routing may be an important component of (or even dominate) the execution time. If routing is γ times faster than arithmetic, that is to say

$$\gamma = \frac{\text{time for one parallel arithmetic operation}}{\text{time for one parallel routing operation}} \quad (5.18)$$

then the ratio of time spent on routing to the time spent on arithmetic is, for the cascade sum method,

$$t_R/t_A = \gamma^{-1}[(n-1)/\log_2 n] \quad (5.19a)$$

from which we can conclude that routing dominates arithmetic if

$$\begin{aligned} n &> \quad 64 \quad \text{if} \quad \gamma \simeq 10 \\ n &> 1024 \quad \text{if} \quad \gamma \simeq 100. \end{aligned} \quad (5.19b)$$

The above shows that routing will always dominate arithmetic for vectors longer than n_3, where

$$n_3 = 1 + \gamma \log_2 n_3. \tag{5.19c}$$

Since $\gamma = 10$ is a typical value, this will happen for any but the most trivial problems. This shows that there is no point in improving the arithmetic speed of a processor array unless a corresponding reduction in routing time can be made.

One way of eliminating the time spent on routing is to provide some long-range connections between the processors. In the above introductory analysis we have assumed for simplicity a one-dimensional array with nearest-neighbour connections. Most large processor arrays are, however, configured as two- or multi-dimensional arrays. The ICL DAP, for example, is configured as a two-dimensional array of 64×64 processors. If these are treated as a single vector of 4096 elements written row by row across the array, then a single routing operation to the nearest-neighbour connections between adjacent rows results in a shift of 64 places. The more general case of data routing in k-dimensional arrays is discussed in §5.5.5 and in Jesshope (1980a, b, c). Other interconnection patterns, such as the perfect shuffle, offer other long-range connections (see §3.3.4 and §3.3.5). The reader is invited to consider the effect of such connections on the comparisons made above.

5.2.4   Cyclic reduction

The general linear first-order recurrence (equation (5.11a)) can be evaluated sequentially from the definition of the recurrence by the following FORTRAN code:

```
      X(1)  =  A(1)*X(0)  +  D(1)
   DO 1  J  =  2, N
1     X(J)  =  A(J)*X(J-1)  +  D(J)
```
(5.20)

This requires

$2n$ arithmetic operations with parallelism 1,   (5.21a)

and

n routings with parallelism 1.   (5.21b)

The routing diagram for the sequential algorithm is given in figure 5.4. For simplicity, we shall not count separately the different types of arithmetic operation although they may have different execution times. The ratio of a multiplication time to an addition time rarely exceeds two and is often quite close to unity. In particular, on a pipelined computer both the addition and

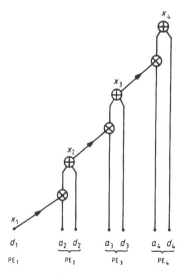

FIGURE 5.4   The routing diagram for the sequential evaluation of the
general first-order recurrence. In this case $n = 4$. Variables linked by a
brace are stored in the same PE. One PE is used to evaluate each term of
the recurrence.

multiplication pipes, when full, deliver one result every clock period. For
$n > n_{1/2}$ the average time for an addition or multiplication operation is very
nearly the same.

The equivalent parallel algorithm to the cascade sum method is known as
cyclic reduction, and has a wide application in numerical analysis, particularly
when one is trying to introduce parallelism into a problem. For example we
will use it again to solve tridiagonal systems of algebraic equations in a
parallel fashion (see §5.4.3). The original recurrence (equation (5.11a)) relates
neighbouring terms in the sequence, namely x_j to x_{j-1}. The basic idea of
cyclic reduction is to combine adjacent terms of the recurrence together in
such a way as to obtain a relation between every other term in the sequence,
that is to say to relate x_j to x_{j-2}. It is found that this relation is also a linear
first-order recurrence—although the coefficients are different and the relation
is between alternate terms. Consequently the process can be repeated (in
a cyclic fashion) to obtain recurrences relating every fourth term, every eighth
term, and so on. When the recurrence relates every n terms (i.e. after $\log_2 n$
levels of reduction), the value at each point in the sequence is related only
to values outside the range which are known or zero, hence the solution has
been found. When the method is used on serial computers, the number of
recurrence equations that are used is halved at each successive level, hence
the term 'cyclic reduction'. On a parallel computer we are interested in

keeping the parallelism high and will not, in fact, reduce the number of equations that are used at each level. Therefore the name of the algorithm is somewhat misleading.

We will now work out the algebra of the cyclic reduction method. Let us write the original recurrence relation for two successive terms as:

$$x_j = a_j x_{j-1} + d_j, \tag{5.22a}$$

and

$$x_{j-1} = a_{j-1} x_{j-2} + d_{j-1}. \tag{5.22b}$$

Substituting equation (5.22b) into equation (5.22a) we obtain

$$x_j = a_j a_{j-1} x_{j-2} + a_j d_{j-1} + d_j \tag{5.23a}$$

$$= a_j^{(1)} x_{j-2} + d_j^{(1)}, \tag{5.23b}$$

where equation (5.23b) is a linear first-order recurrence between alternate terms of the sequence with a new set of coefficients given by

$$a_j^{(1)} = a_j a_{j-1}, \qquad d_j^{(1)} = a_j d_{j-1} + d_j. \tag{5.23c}$$

The repeated application of the above process may be summarised by the reduced equations for level l. Superscripts denote the level number.

$$x_j = a_j^{(l)} x_{j-2^l} + d_j^{(l)}, \qquad \begin{cases} l = 0, 1, \ldots, \log_2 n \\ j = 1, 2, \ldots, n \end{cases} \tag{5.24a}$$

where

$$a_j^{(l)} = a_j^{(l-1)} a_{j-2^{l-1}}^{(l-1)}, \tag{5.24b}$$

$$d_j^{(l)} = a_j^{(l-1)} d_{j-2^{l-1}}^{(l-1)} + d_j^{(l-1)}, \tag{5.24c}$$

and initially

$$a_j^{(0)} = a_j, \qquad d_j^{(0)} = d_j. \tag{5.24d}$$

If the subscript of any a_j, d_j or x_j is outside the defined range $1 \leqslant j \leqslant n$, the correct result is obtained by taking its value as zero. When $l = \log_2 n$ all references to $x_{j-2^l} = x_{j-n}$ in equation (5.24a) are outside the defined range, hence the solution to the recurrence is

$$x_j = d_j^{(\log_2 n)}. \tag{5.24e}$$

The method is therefore to generate successively the coefficients $a_j^{(l)}$ and $d_j^{(l)}$ defined by equation (5.24b, c) until $d_j^{(\log_2 n)}$ is found. This is the solution to the recurrence.

At first sight it might appear as if equation (5.24c) could not be evaluated in parallel. After all, the equation for $d_j^{(l)}$ has superficially the same appearance as the original sequential recurrence with x replaced by d. The fundamental difference between equation (5.24c) and the original recurrence (5.22a) is that the values of $d_{j-2^{l-1}}^{(l-1)}$ and $d_j^{(l-1)}$ on the right-hand side of equation (5.24c) are all known values that were computed at the previous level $(l—1)$. These $d_j^{(l-1)}$ are distinct variables from the $d_j^{(l)}$ on the left-hand side. The latter $\{d_j^{(l)}; j = 1, \dots, n\}$ may therefore be evaluated by a single operation or vector instruction. These relationships are clarified by the routing diagrams for the evaluation of $a_j^{(l)}$ (figure 5.5) and $d_j^{(l)}$ (figure 5.6).

In these diagrams we show only the values of $a_j^{(l)}$ and $d_j^{(l)}$ that are actually used, and the arithmetic operations that are necessary. We note that only about half the values of $a_j^{(l)}$ are required. In particular none is required at $l = 3$. The amount of parallelism varies from about n at the start to approximately $n/2$ at the final level. On pipelined computers which have a variable vector length, it would increase performance to reduce the vector length to the correct value at each level. On processor arrays with $n \leqslant N$ or the paracomputer, in which the execution time is not affected by the vector length, the parallelism can be kept equal to n at each level by loading all $a_j = 0$ for $-n/2 \leqslant j \leqslant 1$, and $d_j = 0$ for $-n \leqslant j \leqslant 0$ or otherwise contriving that out-of-range values are picked up as zeros.

It is obviously quite complex to evaluate the performance of the cyclic reduction algorithm, taking into account the reduction in parallelism at each level. We may however get a lower bound by assuming that the parallelism remains as n at each level, and that all the $a_j^{(l)}$ are calculated. The average parallelism is $[(n-1) + (n-2) + \dots + (n-2^r) + \dots + n/2]/\log_2 n = n[1 - (1 - n^{-1})/\log_2 n]$, hence, asymptotically for large n, we have

$3\log_2 n$ arithmetic operations with parallelism n

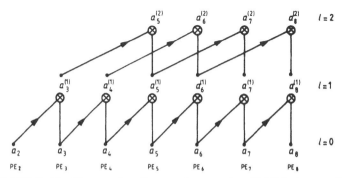

FIGURE 5.5 The routing diagram for equation (5.24b), the parallel calculation of the coefficients $a_j^{(l)}$ in the cyclic reduction algorithm applied to the general linear first-order recurrence for the case $n = 8$.

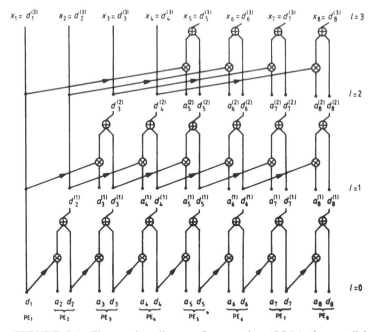

FIGURE 5.6   The routing diagram for equation (5.24c), the parallel calculation of the coefficients $d_j^{(l)}$ in the cyclic reduction algorithm applied to the general linear first-order recurrence. The values of $a_j^{(l)}$ are supplied from the calculation shown in figure 5.5, which proceeds in parallel with this figure.

and   (5.25)

$$2(n-1) \text{ routing operations with parallelism } n.$$

We leave it as an exercise for the reader to carry out a similar analysis to that performed earlier for the sequential and cascade sum methods, and to take into account fully the reduction in parallelism at each level. The above general first-order recurrence becomes the cascade sum for the special case $a_j = 1$ for $2 \leqslant j \leqslant n$, which makes all the multiplications in figure 5.6 and the whole of figure 5.5 redundant.

The cyclic reduction algorithm can be implemented in a vector form of FORTRAN by the code:

```
      X = D
   DO 1 L = 1, LOG2N
      X = A*SHIFTR(X,2**(L-1)) + X
 1    A = A*SHIFTR(A,2**(L-1))
```
(5.26a)
(5.26b)

where **X**, **D** and **A** have been declared as vectors. In the implementation of the above code, we note that the main memory address of the vector **X** and of the vector SHIFTR, which are both required for the evaluation of statement (5.26a), are separated by powers-of-two memory banks. The same is true for the vector **A** in statement (5.26b). Memory-bank conflicts (see §5.1.5) are therefore likely to be a serious impediment to the rapid evaluation of the cyclic reduction algorithm for serial and pipelined computers with the number of memory banks equal to a power of two. The Burroughs BSP, because of its choice of a prime number of banks (17), does not suffer from this problem (se §3.3.8).

5.3   MATRIX MULTIPLICATION

Matrix multiplication is the simplest example of matrix manipulation and illustrates rather well the different ways in which a simple algorithm should be restructured to suit the architecture of the computer on which it is to be executed. Mathematically, the elements $C_{i,j}$ of the product matrix are related to the elements $A_{i,j}$ and $B_{i,j}$ of the matrices being multiplied, by the equation

$$C_{i,j} = \sum_{k=1}^{n} A_{i,k} B_{k,j}, \qquad 1 \leqslant i,j \leqslant n, \tag{5.27}$$

where the first subscript is the row number and the second subscript is the column number.

5.3.1   Inner-product method
Invariably on serial computers, matrices have been multiplied using a nest of three DO loops, using FORTRAN code that is a direct translation of the above definition, namely:

```
DO 1 I = 1, N
DO 1 J = 1, N
```
(5.28a)

```
      DO 1 K = 1, N
1   C(I,J) = C(I,J) + A(I,K)*B(K,J)
```
(5.28b)

where we assume that all elements $C(I,J)$ of the matrix are set to zero before entering the code. The assignment statement in the code (5.28b) forms the *inner product* of the ith row of **A** and the jth column of **B**. It is a special case of the sequential evaluation of the sum of a set of numbers that was discussed in §5.2 (let $d_k = A_{i,k} B_{k,j}$ in equation (5.11b), then $x_n = C_{i,j}$). We therefore have the option of evaluating it sequentially as in the code (5.28) or by using the cascade sum method. The considerations are the same as those given in §5.2. Some computers provide an 'inner or dot product' instruction (e.g. the

CYBER 205, see table 2.4), and the use of this must be considered in any comparisons.

There is, however, more parallelism inherent in the evaluation of the matrix product than is present in the problem of evaluating a single sum. This is because a matrix multiplication involves the evaluation of n^2 inner products and these may be performed n at a time (the *middle-product* method) or n^2 at a time (the *outer-product* method).

5.3.2   Middle-product method

The middle-product method is obtained by interchanging the order of the DO loops in the code (5.28). If we bring the loop over rows, I, to the innermost position, we have code that computes the inner product over all elements of a column of **C** in parallel:

$$
\begin{aligned}
&\text{DO 1 J = 1, N}\\
&\text{DO 1 K = 1, N}
\end{aligned}
\tag{5.29a}
$$

$$
\begin{aligned}
&\text{DO 1 I = 1, N}\\
&\text{1   C(I,J) = C(I,J) + A(I,K)*B(K,J)}
\end{aligned}
\tag{5.29b}
$$

Every term in the loop over I can be evaluated in parallel, so that the loop (5.29b) can be replaced by a vector expression. In an obvious notation, the code can be written:

$$
\begin{aligned}
&\text{DO 1 J = 1, N}\\
&\text{DO 1 K = 1, N}
\end{aligned}
\tag{5.30a}
$$

$$
\text{1   C(,J) = C(,J) + A(,K)*B(K,J)}
\tag{5.30b}
$$

where $C(\quad,J)$ and $A(\quad,K)$ are vectors composed of the Jth and Kth columns of **C** and **A**. The addition, $+$, is a parallel addition of n elements, and the multiplication, $*$, is the multiplication of the scalar $B(K,J)$ by the vector $A(\quad,K)$. The parallelism of the middle-product code is therefore n, compared with 1 for the original inner-product method. This has been obtained by the simple process of interchanging the order of the DO loops. Note that we could have moved the J loop to the middle, thus computing all the inner products of a row in parallel. However, elements of the column of a matrix are usually stored in adjacent memory locations (FORTRAN columnar storage), consequently memory-bank conflicts are reduced if vector operations take place on column vectors. Thus the code of (5.30) is usually to be preferred.

The middle-product method, when programmed in assembler, is found to be the best code on the CRAY-1 computer. Supervector performance of 138 Mflop/s is observed. This implies that, on average, almost two arithmetic operations are being performed per clock period (one operation per clock is equivalent to 80 Mflop/s). This is possible because the multiplication and addition operations in statement (5.30b) can be chained to act as a single pipelined composite operation delivering one element of the result vector $C(\ ,J)$ per clock period.

It is interesting to note that the middle-product method is found to have superior performance to the inner-product method, even on computers such as the CDC 7600 that do not have explicit vector instructions, and are not usually classified as parallel computers. However the CDC 7600 does have pipelined arithmetic units and their performance is improved if requests for arithmetic are received in a regular fashion, as in the serial code (5.29b) for vector operations. This idea has been exploited in a set of carefully optimised assembler routines, called STACKLIB (see §2.3.5), that has been written at the Lawrence Livermore Laboratory. These routines perform various dyadic and triadic vector operations, such as the 'vector + scalar × vector' statement in code (5.30b), and obtain performance improvements of about a factor of two over code written for serial evaluation, such as (5.28b).

5.3.3 Outer-product method

The outer-product method is obtained by moving the loop over K in the code (5.28) to the outside, as follows:

$$\text{DO 1 K = 1, N} \tag{5.31a}$$

$$
\begin{array}{l}
\text{DO 1 I = 1, N} \\
\text{DO 1 J = 1, N} \\
1 \quad C(I,J) = C(I,J) + A(I,K)*B(K,J)
\end{array} \tag{5.31b}
$$

The code (5.31b) can be replaced by a single array-like statement in which one term of the inner product is evaluated in parallel for all n^2 elements of **C**. Using the notation introduced in Chapter 4, **C** stands for the whole array and we may write:

$$
\begin{array}{l}
\text{DO 1 K = 1, N} \\
1 \quad C = C + A(\ ,K)*B(K,\ )
\end{array} \tag{5.32}
$$

where the multiplication operation is an element-by-element multiplication of an $n \times n$ matrix made by duplicating the Kth column of **A**, and an $n \times n$ matrix made by duplication of the kth row of **B**. The addition operation is an element-by-element addition of $n \times n$ elements. This is expressed in DAP FORTRAN by:

$$
\begin{array}{l}
\text{DO 1 K = 1, N} \\
1 \quad C = C + \text{MATC}(A(\ ,K))*\text{MATR}(B(K,\ ))
\end{array} \tag{5.33}
$$

where the function subprogram MATC(X) forms a matrix all of whose columns are the vector X. MATR(X) similarly forms a matrix all of whose rows are the vector X. Two functions are required because otherwise the coercion of a vector into a matrix is ambiguous (see §4.3.1 (iii) and (vi)).

The outer product is clearly highly suitable for a processor array that has the same dimensions as the matrices (e.g. multiplying 64×64 matrices on a 64×64 ICL DAP). In this case the parallelism of the hardware is exactly matched to the parallelism of the algorithm. Because the parallelism has been increased from n to n^2, compared with the middle-product method, the outer-product method is also likely to be superior on pipelined computers with a large value of $n_{1/2}$. For the case of 64×64 matrices the vector length is increased from 64 for the middle product to 4096 for the outer product. The ratio of the performance of the outer-product method, P_o, to the performance of the middle-product method, P_m, is in the ratio of the time to perform n vector operations of length n to the time to perform one vector operation of length n^2:

$$\frac{P_o}{P_m} = \frac{n(n + n_{1/2})}{n^2 + n_{1/2}}$$

$$= \frac{1 + n_{1/2}/n}{1 + n_{1/2}/n^2} \tag{5.34a}$$

$$\approx \begin{cases} 2 & \text{for } n_{1/2} = n > 1, \\ n_{1/2}/n & \text{for } n^2 > n_{1/2} > n. \end{cases}$$

For the case of $n = 64$ we have

$$\frac{P_o}{P_m} = \begin{cases} 1.12 & \text{for } n_{1/2} = 8, \\ 2.5 & \text{for } n_{1/2} = 100, \\ 13 & \text{for } n_{1/2} = 1000. \end{cases} \tag{5.34b}$$

It is clear that there is little to choose between the two methods from this point of view for computers such as the CRAY-1 with small values of $n_{1/2}$. Other considerations, such as the ability to work in the vector registers and to use chaining, favour the middle product. However there are obvious advantages to the outer product in pipelined computers in the last case in which $n^2 > n_{1/2} > n$.

5.3.4   Using n^3-parallelism

An interesting combination of the techniques that have been discussed for introducing parallelism has been used by Jesshope and Craigie (1980) in considering the multiplication of matrices on processor arrays that do not match the size of the matrices (also in Jesshope and Hockney 1979). For example, how should one best compute the product of two 16×16 matrices on a 64×64 ICL DAP? Obviously it would be very wasteful to adopt the outer-product method and only use one sixteenth of the available processors.

Jesshope and Craigie note that there are n^3 multiplications in the product of two $n \times n$ matrices (n^2 inner products, each with n multiplications; see equation (5.27)), and that all these products can be evaluated simultaneously with a parallelism of n^3. The summation of the n terms of all the n^2 inner products can be achieved in $\log_2 n$ steps, also with parallelism n^3, using the cascade sum method. The equivalent FORTRAN code is:

```
        DO 1 I = 1, N
        DO 1 J = 1, N
        DO 1 K = 1, N                                        (5.35a)

1       C(I,J,K) = A(I,J,K)*B(I,J,K)

        DO 2 L = 1, LOG2N
             K1 = 2**(L-1)
        DO 2 K = 1, N - K1
        DO 2 J = 1, N                                        (5.35b)
        DO 2 I = 1, N

2   C(I,J,K) = C(I,J,K) + C(I,J,K+K1)
```

In the above, the loop (5.35a) performs all the n^3 multiplications and the loop (5.35b) evaluates the cascade sum. After the execution of the four nested loops (5.35b) the $C(I,J)$ element of the product is found in location $C(I,J,1)$. The above code can be expressed succinctly, using the parallel constructs outlined in §4.3.1 (see pp 400–1) by the following single statement:

```
C=SUM(XPND(A,3,N)*XPND(B,1,N),2)                             (5.35c)
```

For the case of 16×16 matrices on a 64×64 ICL DAP, we have $n = 16$ and $n^3 = 4096$, hence the parallelism of the algorithm exactly matches the parallelism of the hardware. The three indices of the matrices in the code (5.35) are mapped onto the two indices of the processor array in some regular fashion. In order to reduce the routing during the evaluation of the cascade sum, it is important to store all the values with the same first two subscripts (I and J) near each other. Such mapping is

$$C(I,J,K) \text{ stored in } \text{PE}(4I + K/4, \ 4J + \text{MOD}(K,4)), \tag{5.36}$$

where the subscripts on the PE (,) array designate the row and column number of the PE and are to be evaluated as FORTRAN statements. The result of this mapping is that the 16 values with the same first two subscripts are stored in a compact 4×4 array of processing elements, and routing during the summation is kept to a minimum. The long-range routing that is required during the expansion of **A** and **B** can be effectively performed on the ICL DAP using the broadcast facility (see §3.4.2). The SUM function in statement (5.35c) can also be optimised using bit-level algorithms.

5.4   TRIDIAGONAL SYSTEMS

Triadiagonal systems form a very important class of linear algebraic equations. They occur repeatedly as finite-difference approximations to differential equations with second derivatives: for example, simple harmonic motion and the Helmholtz, Laplace, Poisson and diffusion equations. Consequently efficient methods for the solution of such equations lie at the heart of many important numerical algorithms for the solution of the above equations: for example line-iterative methods such as ADI and SLOR, and the rapid FACR and Buneman direct methods for the solution of certain classes of these equations. Some of these methods will be discussed later in §5.6. Unfortunately many of the techniques, for example Gaussian elimination, that have proved most effective on serial computers are sequential algorithms and therefore unsuitable for use on parallel computers. In this section we discuss the problem of introducing parallelism into these methods and alternatively the adoption of new algorithms, such as cyclic reduction, that are inherently more parallel. In developing these techniques we can use much of the experience gained during the study of vectorising recurrences in §5.2.

The solution of tridiagonal systems of equations has been quite widely discussed in the literature. Methods for vector computers are given by Stone (1973, 1975), Lambiotte and Voight (1975), Swarztrauber (1979a,b), Kershaw (1982), Kascic (1984b), Gentzsch (1984) and Schönauer (1987). On the other hand, methods suitable for parallel implementation on MIMD computer systems are given by Evans (1980), Kowalik and Kumar (1985) and Wang (1985).

5.4.1   Gaussian elimination

The general tridiagonal set of linear algebraic equations may be written as:

$$
\begin{pmatrix}
b_1 & c_1 & 0 & \cdots & & 0 \\
a_2 & b_2 & c_2 & & & \\
0 & & & & & \\
\vdots & & & & & 0 \\
& & & a_{n-1} & b_{n-1} & c_{n-1} \\
0 & \cdots & 0 & & a_n & b_n
\end{pmatrix}
\begin{pmatrix}
x_1 \\ x_2 \\ \cdot \\ \cdot \\ \cdot \\ \cdot \\ x_n
\end{pmatrix}
=
\begin{pmatrix}
k_1 \\ k_2 \\ \cdot \\ \cdot \\ \cdot \\ \cdot \\ k_n
\end{pmatrix}
\tag{5.37}
$$

or in matrix–vector notation:

$$\mathbf{A}\mathbf{x} = \mathbf{k}. \tag{5.38}$$

The Gaussian elimination algorithm may be stated as follows:

(i) *Forward elimination*

$$w_1 = c_1/b_1,$$

$$w_i = c_i/(b_i - a_i w_{i-1}), \qquad i = 2, 3, \ldots, n-1, \tag{5.39a}$$

$$e_i = w_i/c_i \tag{5.39b}$$

and

$$g_1 = k_1/b_1,$$

$$g_i = (k_i - a_i g_{i-1})/(b_i - a_i w_{i-1}), \qquad i = 2, 3, \ldots, n, \tag{5.39c}$$

or

$$g_i = (k_i - a_i g_{i-1})e_i, \tag{5.39d}$$

or

$$g_i = (k_i - a_i g_{i-1})w_i/c_i. \tag{5.39e}$$

(ii) *Back substitution*

$$x_n = g_n,$$

$$x_i = g_i - w_i x_{i+1}, \qquad i = (n-1), (n-2), \ldots, 1. \tag{5.39f}$$

In the forward elimination stage two auxiliary vectors are computed, $\mathbf{w}$ and $\mathbf{e}$, which are functions only of the coefficients in the matrix $\mathbf{A}$. These vectors are the coefficients in the triangular decomposition of $\mathbf{A}$ into the product of a lower triangular matrix $\mathbf{L}$ and an upper triangular matrix $\mathbf{U}$:

$$\mathbf{A} = \mathbf{L}\mathbf{U} \tag{5.40}$$

where

$$\mathbf{L} = \begin{pmatrix} e_1^{-1} & 0 & \cdots & \cdots & 0 \\ a_2 & e_2^{-1} & & & \\ 0 & & \ddots & & \\ & & & \ddots & 0 \\ 0 & \cdots & 0 & a_n & e_n^{-1} \end{pmatrix} \qquad \mathbf{U} = \begin{pmatrix} 1 & w_1 & 0 & \cdots & 0 \\ 0 & 1 & w_2 & & \\ & & 1 & \ddots & 0 \\ & & & \ddots & w_{n-1} \\ 0 & \cdots & & 0 & 1 \end{pmatrix}$$

Using this decomposition the solution of equation (5.37) for a particular right-hand side, k, can be expressed in two stages:

$$\mathbf{L}\mathbf{g} = \mathbf{k}, \tag{5.41a}$$

$$\mathbf{U}\mathbf{x} = \mathbf{g}. \tag{5.41b}$$

Hence $\mathbf{Ax} = \mathbf{LUx} = \mathbf{Lg} = \mathbf{k}$ as required. The forward elimination stage (5.39d) is the expression in component form of equation (5.41a) and the back substitution stage, equation (5.39f), is the expression of equation (5.41b). During this process $\mathbf{g}$ may overwrite $\mathbf{k}$ and the solution $\mathbf{x}$ may overwrite $\mathbf{g}$, hence no extra storage is implied by the appearance of the intermediate vector $\mathbf{g}$.

If the auxiliary vectors are not precalculated we evaluate equations (5.39a) and (5.39c) together, followed by equation (5.39f). The denominator $(b_i - a_i w_{i-1})$ is only evaluated once, and a total of $8n$ scalar arithmetic operations are required. However, if several equations are to be solved with the same left-hand-side matrix $\mathbf{A}$ but different right-hand-side vectors $\mathbf{k}$, the number of operations can be reduced to $5n$ if the two auxiliary vectors $\mathbf{w}$ and $\mathbf{e}$ are precalculated and stored. In this case we use equations (5.39a) and (5.39b) in a preliminary calculation, and equations (5.39d) and (5.39f) to obtain the solution for each right-hand side. If only the one vector, $\mathbf{w}$, is precalculated, equation (5.39b) is omitted and equations (5.39e) and (5.39f) are used. The number of scalar arithmetic operations increases to $6n$ but, of course, storage is saved. If the tridiagonal system arises from the finite difference of a second derivative on a regular mesh, then $a_i = c_i = 1$. In this case $e_i = w_i$ and the algorithm simplifies considerably:

$$w_1 = c_1/b_1,$$
$$w_i = (b_i - w_{i-1})^{-1}, \qquad i = 2, 3, \ldots, n-1; \tag{5.42a}$$

$$g_1 = k_1/b_1,$$
$$g_i = (k_i - g_{i-1})w_i, \qquad i = 2, 3, \ldots, n; \tag{5.42b}$$

$$x_n = g_n,$$
$$x_i = g_i - w_i x_{i+1}, \qquad i = n-1, n-2, \ldots, 1. \tag{5.42c}$$

The scalar operations are now reduced to $6n$ without precalculation and $4n$ with precalculation of the single vector $\mathbf{w}$.

The three loops (5.39a, c, f) or (5.42) of the Gaussian elimination algorithm are all sequential recurrences that must be evaluated one term at a time. Hence the parallelism of the algorithm is 1. This, together with the fact that vector elements are references with unit increment and that the number of arithmetic operations is minimised, makes the algorithm ideally suited to serial computers. Equally, the absence of any parallelism prevents the algorithm from taking any advantage of the parallel hardware features on a computer. Hence the algorithm is most unsuitable for solving a single tridiagonal system of equations on a parallel computer.

However, if one is faced with solving a set of m independent tridiagonal systems (say $m = 64$), as frequently occurs in the solution of PDEs (see §5.6), then Gaussian elimination will be the best algorithm to use on a parallel computer. In this case all m systems would be solved in parallel by changing all variables in the algorithm into vectors of length m running across the tridiagonal systems. For example the variable w_i would become the vector $\{w_{i,k}; k = 1, \ldots, m\}$ where $w_{i,k}$ is the value of w_i in the kth tridiagonal system. When this is done, all the scalar operations in equations (5.39) become vector operations with vector length m. All statements vectorise fully, and the maximum vector performance is obtained. We thus obtain both the minimum number of arithmetic operations (by the choice of the Gaussian procedure) *and* the maximum parallelism. We note that normally these two aims cannot be satisfied simultaneously (see e.g. §5.2 on recurrences). The main disadvantage of this method is that the total storage required increases m-fold compared with solving the tridiagonal systems one at a time. The choice between these two alternatives is discussed further in §5.4.4.

5.4.2 Recursive doubling

Several methods have been proposed to introduce parallelism into the sequential Gaussian algorithm, notably the recursive doubling algorithm of Stone (1973). We will describe a variation of his improved algorithm (Stone 1975). The performance of several of these algorithms on the CDC STAR 100 pipelined computer is given by Lambiotte and Voight (1975). Another comparison, between algorithms on the CDC 7600 and CDC STAR 100, is given by Madsen and Rodrigue (1976).

An examination of the Gaussian algorithm given in equations (5.39a, d, f) or in equations (5.42) shows that the algorithm consists of three recurrences. The recurrences for g_i and x_i are both linear and first-order; they may therefore be evaluated by cyclic reduction using the method for the general linear first-order recurrence given in §5.2.4. No further discussion is required. The recurrence for w_i, although also a two-term recurrence, is nonlinear since it relates w_i to $(b_i - a_i w_{i-1})^{-1}$. Cyclic reduction cannot therefore be directly applied to introduce parallelism into this recurrence. It may however be applied after some transformations.

The recurrence for w_i is:

$$w_1 = c_1/b_1,$$

$$w_i = c_i/(b_i - a_i w_{i-1}), \qquad i = 2, 3, \ldots, n - 1. \tag{5.43a}$$

If we now let

$$w_i = y_i/y_{i+1} \tag{5.43b}$$

and rearrange, we have

$$a_i y_{i-1} + b_i y_i + c_i y_{i+1} = 0, \qquad i = 2, 3, \ldots, n - 1, \tag{5.44a}$$

with

$$y_1 = 1, \qquad y_2 = -c_1/b_1. \tag{5.44b}$$

Equation (5.44a), not surprisingly, is the homogenous form of the original equations. It is a linear second-order (or three-term) recurrence. Hence the problem of finding w_i is the same as the problem of solving the original equations. In §5.4.3 we will show how to use cyclic reduction directly to solve such recurrences. The recursive doubling algorithm, although it uses cyclic reduction, proceeds somewhat differently, as we now show.

For the sake of generality we will solve the second-order recurrence:

$$a_i y_{i-1} + b_i y_i + c_i y_{i+1} = k_i. \tag{5.45a}$$

Equation (5.45a) can be expressed as follows:

$$\begin{pmatrix} y_i \\ y_{i+1} \end{pmatrix} = \begin{pmatrix} 0 & 0 \\ -a_i/b_i & -c_i/b_i \end{pmatrix} \begin{pmatrix} y_{i-1} \\ y_i \end{pmatrix} + \begin{pmatrix} 0 \\ k_i/c_i \end{pmatrix} \tag{5.45b}$$

or as

$$\mathbf{v}_i = \mathbf{Q}_i \mathbf{v}_{i-1} + \mathbf{h}_i, \qquad i = 2, 3, \ldots, n - 1, \tag{5.45c}$$

where

$$\mathbf{v}_i = \begin{pmatrix} y_i \\ y_{i+1} \end{pmatrix} \qquad \mathbf{Q}_i = \begin{pmatrix} 0 & 1 \\ -a_i/b_i & -c_i/b_i \end{pmatrix} \qquad \mathbf{h}_i = \begin{pmatrix} 0 \\ k_i/c_i \end{pmatrix} \tag{5.45d}$$

with

$$\mathbf{v}_1 = \begin{pmatrix} 1 \\ -c_1/b_1 \end{pmatrix}$$

Equation (5.45c) is a linear first-order recurrence for the vector $\mathbf{v}_i$ of the general form of equation (5.11a) except that the multiplying factor is now a matrix. The recurrence can be solved using the cyclic reduction procedure described for the scalar recurrence in §5.2.4, with appropriate interpretation in terms of vectors and matrices. Having found $\mathbf{v}_i$ (taking $\mathbf{h}_i = \mathbf{0}$), the values of y_i are known from its components, and w_i is calculated from equation (5.43b). This completes the recursive doubling algorithm for the parallel evaluation of the LU decomposition of a tridiagonal system by the Gaussian elimination recurrences.

It might be thought that one could solve the original equations by this method, because they are the same as equation (5.45a). However, this approach is thwarted because one does not have a starting value for $\mathbf{v}_1$. This is because such a tridiagonal system arises from a second-order PDE with boundary conditions at the two ends (i.e. one may consider y_0 and y_{n+1} to be known). However, the recurrence can only be evaluated progressively if starting conditions are given for two adjacent values—say y_0 and y_1, giving a starting value for $\mathbf{v}_1$. If such starting values are available, the method above does provide a method of parallel solution for the general linear second-order recurrence. This would occur if the equations arose from a second-order initial-value problem, rather than the boundary-value problem presently under discussion.

In order to estimate the number of operations we will make the approximation that the parallelism remains at n throughout. Then we have, neglecting constant terms:

$$18\log_2 n \text{ operations to obtain } \mathbf{v}_i$$

$$3\log_2 n \text{ operations to obtain } g_i$$

$$3\log_2 n \text{ operations to obtain } x_i$$

$$\overline{\text{total} \quad 24\log_2 n} \text{ operations with parallelism } n. \qquad (5.46a)$$

The time to execute the algorithm on a computer with a half-performance length of $n_{1/2}$ is therefore proportional to

$$t_{\text{RD}} \approx 24(n_{1/2} + n)\log_2 n. \qquad (5.46b)$$

5.4.3   Cyclic reduction

Cyclic reduction was first used to solve tridiagonal equations by Hockney (1965) in collaboration with Golub. The method was implemented on a serial computer, the IBM 7090, and was chosen in preference to Gaussian elimination because cyclic reduction deals with periodic boundary conditions in a much neater way, eliminating the need for the calculation of auxiliary vectors. The equations solved were those arising from the finite-difference approximation to Poisson's equation and had the same coefficients at each mesh point. These equations are discussed in detail in §5.6. For convenience the number of mesh points and therefore the number of equations was taken to be a power of two. Neither of these restrictions is necessary to the method as has been shown by Swarztrauber (1974) and Sweet (1974, 1977). For simplicity here, however, we will assume that $n = n' - 1$, where $n' = 2^q$ and q is an integer, and solve the general coefficient problem defined in equation (5.37).

Writing three adjacent equations we have for $i = 2, 4, \ldots, n' - 2$:

$$a_{i-1}x_{i-2} + b_{i-1}x_{i-1} + c_{i-1}x_i \qquad\qquad = k_{i-1}$$

$$a_i x_{i-1} + \quad b_i x_i + \quad c_i x_{i+1} \qquad\qquad = k_i \qquad (5.47)$$

$$a_{i+1}x_i + b_{i+1}x_{i+1} + c_{i+1}x_{i+2} = k_{i+1}$$

The special end equations are included correctly if we set $x_0 = x_{n'} = 0$. If the first of these equations is multiplied by $\alpha_i = -a_i/b_{i-1}$, and the last by $\gamma_i = -c_i/b_{i+1}$, and the three equations added, all reference to the variables x_{i-1} and x_{i+1} is eliminated. One then obtains

$$a_i^{(1)}x_{i-2} + b_i^{(1)}x_i + c_i^{(1)}x_{i+2} = k_i^{(1)}, \qquad (5.48a)$$

where

$$a_i^{(1)} = \alpha_i a_{i-1},$$

$$c_i^{(1)} = \gamma_i c_{i+1},$$

$$b_i^{(1)} = b_i + \alpha_i c_{i-1} + \gamma_i a_{i+1}, \qquad (5.48b)$$

$$k_i^{(1)} = k_i + \alpha_i k_{i-1} + \gamma_i k_{i+1}.$$

The equations (5.48) relate every second variable and, if written for $i = 2, 4, \ldots, n' - 2$, are a tridiagonal set of equations of the same form as the original equations (5.47) but with different coefficients $(a^{(1)}, b^{(1)}, c^{(1)})$. The number of equations has been roughly halved. Clearly the process can be repeated recursively until, after $\log_2(n') - 1$ levels of reduction, only the central equation for $i = n'/2$ remains. This equation is

$$a_{n'/2}^{(r)}x_0 + b_{n'/2}^{(r)}x_{n'/2} + c_{n'/2}^{(r)}x_{n'} = k_{n'/2}^{(r)} \qquad (5.49a)$$

where the superscript $r = \log_2(n') - 1$ indicates the level of reduction. Since $x_0 = x_{n'} = 0$, the solution for the central equation is obtained by division

$$x_{n'/2} = k_{n'/2}^{(k)}/b_{n'/2}^{(r)}. \qquad (5.49b)$$

The remaining unknowns can now be found from a filling-in procedure. Since we know x_0, $x_{n'/2}$ and $x_{n'}$ the unknowns midway between these can be found from the equations at level $r - 1$ using

$$x_i = (k_i^{(r-1)} - a_i^{(r-1)}x_{i-n'/4} - c_i^{(r-1)}x_{i+n'/4})/b_i^{(r-1)},$$

for $i = n'/4$ and $3n'/4$. The filling-in procedure is repeated until, finally, all the odd unknowns are found using the original equations.

The cyclic reduction procedure therefore involves the recursive calculation of new coefficients and right-hand sides, for levels $l = 1, 2, \ldots, q - 1$, from

$$a_i^{(l)} = \alpha_i a_{i-2^{(l-1)}}^{(l-1)},$$

$$c_i^{(l)} = \gamma_i c_{i+2^{(l-1)}}^{(l-1)},$$

$$b_i^{(l)} = b_i^{(l-1)} + \alpha_i c_{i-2^{(l-1)}}^{(l-1)} + \gamma_i a_{i+2^{(l-1)}}^{(l-1)}, \qquad (5.50)$$

$$k_i^{(l)} = k_i^{(l-1)} + \alpha_i k_{i-2^{(l-1)}}^{(l-1)} + \gamma_i k_{i+2^{(l-1)}}^{(l-1)},$$

where

$$\alpha_i = -a_i^{(l-1)}/b_{i-2^{(l-1)}}^{(l-1)}$$

$$\gamma_i = -c_i^{(l-1)}/b_{i+2^{(l-1)}}^{(l-1)}$$

and

$$i = 2^l \text{ step } 2^l \text{ until } n' - 2^l,$$

with the initial values $a_i^{(0)} = a_i$, $b_i^{(0)} = b_i$ and $c_i^{(0)} = c_i$, followed by the recursive filling-in of the solution, for $l = q, q - 1, \ldots, 2, 1$ from

$$x_i = (k_i^{(l-1)} - a_i^{(l-1)}x_{i-2^{(l-1)}} - c_i^{(l-1)}x_{i+2^{(l-1)}})/b_i^{(l-1)}, \qquad (5.51)$$

where

$$i = 2^{(l-1)} \text{ step } 2^l \text{ until } n' - 2^{(l-1)}$$

and $x_0 = x_{n'} = 0$ when they occur. The routing diagram for this algorithm is shown in figure 5.7 for the case $n' = 8$. For convenience we define the vector $\mathbf{p}_i(a_i, b_i, c_i, k_i)$ to indicate all the values calculated by equations (5.50). The number of operations involved in the evaluation of equations (5.50) is

$$12 \text{ with parallelism } n'2^{-l} - 1 \qquad \text{for } l = 1, 2, \ldots, \log_2(n') - 1.$$
$$(5.52)$$

The time for this part of the calculation on a computer with a half-performance

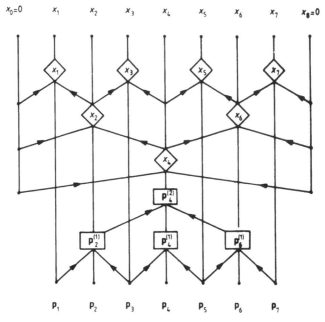

FIGURE 5.7   Routing diagram for the serial cyclic reduction algorithm (SERICR) for $n' = 8$. Rectangular boxes indicate the evaluation of equation (5.50) and the diamonds the evaluation of equation (5.51). The variables calculated are written inside the boxes with the notation $\mathbf{p}_i = (a_i, b_i, c_i, k_i)$.

length of $n_{1/2}$ is proportional to (i.e. omitting the common factor r_∞^{-1})

$$t_{5.50} = 12 \sum_{l=1}^{\log_2(n')-1} (n_{1/2} + n'2^{-l} - 1)$$

$$= 12[n' + n_{1/2}\log_2(n') - n_{1/2} - \log_2(n') - 1]. \tag{5.53a}$$

In these comparisons we will only keep terms of order n' and $\log_2 n'$, hence for $n_{1/2} > 1$ and $\log_2 n' > 1$ approximately we have

$$t_{5.50} \simeq 12(n' + n_{1/2}\log_2 n'). \tag{5.53b}$$

The evaluation of equation (5.51) requires

5 operations with parallelism $n2^{-l}$ for $l = \log_2 n', \ldots, 2, 1$.

Hence

$$t_{5.51} = 5 \sum_{l=1}^{\log_2 n'} (n_{1/2} + n2^{-l}), \tag{5.54a}$$

$$= 5[n' + n_{1/2}\log_2(n') - 1], \tag{5.54b}$$

$$\simeq 5(n' + n_{1/2}\log_2 n'). \tag{5.54c}$$

The manner of performing cyclic reduction that has just been described has the least number of scalar arithmetic operations and is therefore the best choice for a serial computer. We will therefore call the algorithm the serial variant of cyclic reduction with the acronym SERICR. The total time required for its execution is therefore proportional to:

$$t_{SERICR} = t_{5.50} + t_{5.51} \simeq 17(n' + n_{1/2}\log_2 n'). \tag{5.55}$$

On processor arrays it is generally desirable to keep the parallelism as high as possible, and an alternative method of carrying out cyclic reduction keeps the parallelism at n throughout the reduction phase. At the last level of reduction the solution for all variables is found in parallel—instead of only the central value in SERICR—and the filling-in phase is not required. Because this method is most suitable for the paracomputer we call it the parallel variant of cyclic reduction with the acronym PARACR.

The routing diagram for the parallel variant is shown in figure 5.8. At each level the reduction equations (5.50) are applied in parallel to all n equations. A difficult appears to arise when data is required that is not in the defined range $1 \leqslant i \leqslant n$. However, it may be seen that the correct results are obtained

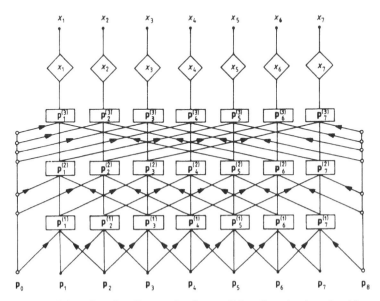

FIGURE 5.8   Routing diagram for the parallel cyclic reduction algorithm (PARACR) for $n' = 8$. The special vector $\mathbf{p}_0 = (0, 1, 0, 0)$.

if one takes

$$a_i^{(l)} = c_i^{(l)} = k_i^{(l)} = 0, \\ b_i^{(l)} = 1, \quad\quad\quad \text{for } i \leqslant 0 \text{ and } i \geqslant n + 1. \quad\quad (5.56a)$$

or

$$\mathbf{p}_i^{(l)} = (0, 1, 0, 0),$$

The above special values, when inserted into the defining equation (5.48a), lead to the equation

$$x_i = 0, \quad\quad \text{for } i \leqslant 0 \text{ and } i \geqslant n + 1 \quad\quad (5.56b)$$

which gives the correct boundary values. We may therefore either consider the solution of the original finite set of equations, or alternatively an infinite set extended with the coefficients (5.56a). This only amounts to adding equations such as (5.56b) outside the range of the original problem. Either point of view is equally valid, but the latter view is more appropriate for the parallel variant of cyclic reduction because it defines the required values of $\mathbf{p}_i$ and x_i outside the originally defined problem. After calculating $\mathbf{p}_i^{(q)}$ (actually only the values of b_i and k_i are required), the value of x_i is obtained from equation (5.51):

$$x_i = k_i^{(q)} / b_i^{(q)}. \quad\quad (5.57)$$

The terms in x on the right-hand side of equation (5.51) do not occur because, at this level of reduction, they refer to values outside the range $1 \leqslant i \leqslant n$ and by equation (5.56b) are zero.

The number of operations in the PARACR algorithm is clearly

$$12[\log_2(n') - 1] \text{ with parallelism } n \quad\quad (5.58a)$$

and the time of execution is proportional to

$$t_{\text{PARACR}} = 12(n_{1/2} + n)[\log_2(n') - 1] \\ \simeq 12(n_{1/2} + n)\log_2 n'. \quad\quad (5.58b)$$

An advantage of the cyclic reduction algorithm is that, under certain conditions, the reduction process may be stopped before completion without loss of accuracy. This is possible if the tridiagonal system is sufficiently diagonally dominant. Let us define the diagonal dominance of the original system of equations (5.47) as δ, the minimum over all the equations of the ratios $|b_i|/|a_i|$ and $|b_i|/|c_i|$. We can then consider the solution of the simpler set of constant coefficient equations:

$$ax_{i-1} + bx_i + ax_{i+1} = k_i, \qu\quad\quad (5.59)$$

with $|b/a| = \delta$, which is equally or less diagonally dominant than the original. If this simpler set of equations can be solved to a certain approximation, then the original equations will be solved more accurately. The cyclic reduction recurrence (5.50) for this case is

$$a^{(0)} = a,$$

and then

$$a^{(l)} = -(a^{(l-1)})^2/b^{(l-1)}, \qquad (5.60a)$$

$$\left.\begin{matrix}\\ \\ \\ \\ \\ \\ \end{matrix}\right\} \quad l = 1, 2, \ldots, \log_2(n') - 1,$$

$$b^{(l)} = b^{(l-1)} - 2(a^{(l-1)})^2/b^{(l-1)}, \qquad (5.60b)$$

where the subscript i is dropped because the coefficients are the same for all equations, and we use the fact that $c^{(l)} = a^{(l)}$. Dividing equation (5.60b) by equation (5.60a) we obtain the recurrence relation for the diagonal dominance:

$$\delta^{(0)} = \delta,$$

$$\delta^{(l)} = |b^{(l)}|/|a^{(l)}|,$$

$$= |(\delta^{(l-1)})^2 - 2|, \qquad l = 1, 2, \ldots, \log_2(n') - 1. \qquad (5.61a)$$

Hence if the initial diagonal dominance $\delta > 2$, the diagonal dominance will grow quadratically at least as fast as equation (5.61a), and

$$\delta^{(l)} \approx \delta^{2^l}, \qquad \text{for } \delta \gg 2. \qquad (5.61b)$$

The finite difference approximation of the Helmholtz equation,

$$\frac{d^2\phi}{dx^2} - \beta\phi = s(x), \qquad (5.62a)$$

for a mesh separation of h, is

$$\phi_{i-1} - (2 + \beta h^2)\phi_i + \phi_{i+1} = s_i h^2. \qquad (5.62b)$$

Hence

$$\delta = 2 + \beta h^2, \qquad (5.62c)$$

and for $\beta > 0$ the difference equations satisfy the condition for the increase in diagonal dominance with depth of reduction. The harmonic equations that are obtained in the solution of partial differential equations are an important set of equations falling in this category (see §5.6.2).

If at any stage in the reduction the inverse of the diagonal dominance is less than the required accuracy (or the rounding error of the computer being used), and the solution x is known to be of order unity, then the reduction may be stopped. The solution equation (5.51) is

$$x_i = k_i^{(l-1)}/b^{(l-1)} \pm (x_{i-2^{(l-1)}} + x_{i+2^{(l-1)}})/\delta^{(l-1)}, \tag{5.63a}$$

and, by postulation, the terms in x on the right-hand side may be neglected compared with the left-hand side. The solution at this level can then be found by simple division:

$$x_i = k_i^{(l-1)}/b^{(l-1)}, \tag{5.63b}$$

and in PARACR the solution is known everywhere or in SERICR the filling-in may start.

For large δ (say > 3), the level, $\hat{l}$, at which reduction may stop is obtained approximately using equation (5.61b):

$$\delta^{(\hat{l})} \simeq \delta^{2^{\hat{l}}} = \varepsilon^{-1}, \tag{5.64a}$$

where ε is the permitted relative error in the solution, or

$$\hat{l} = \log_2[(\log_2\varepsilon^{-1})/(\log_2\delta)],$$
$$= \log_2(\log_2\varepsilon^{-1}) - \log_2(\log_2\delta). \tag{5.64b}$$

The first term shows how a greater demand for accuracy (larger ε^{-1}) necessitates more levels of reduction and the second term shows how the number of levels is reduced as the diagonal dominance increases. If δ is close to two, the full recurrence (5.61a) must be evaluated. In any case, the practical approach is to measure the diagonal dominance, $\delta^{(l)}$, at each level of reduction from the values of $a^{(l)}$, $b^{(l)}$ and $c^{(l)}$, and to stop the reduction if equation (5.64a) is satisfied.

Of course there is no saving in the algorithm unless $\hat{l} < \log_2(n') - 1$, the maximum number of levels for complete reduction. This leads to the result that truncated reduction can produce savings when

$$n' > \hat{n}, \tag{5.65a}$$

where

$$\hat{n} = 2^{\hat{l}} = (\log_2\varepsilon^{-1})/(\log_2\delta), \qquad \text{for } n' \gg 1. \tag{5.65b}$$

If we take the example $\varepsilon = 2^{-20} \approx 10^{-6}$ (32-bit single-precision on the IBM 360) and $\delta = 4$ (as occurs in the harmonic equations of §5.6.2), we obtain

$$\hat{l} = 3.32, \qquad \hat{n} = 10. \tag{5.65c}$$

Hence it is clear that truncated reduction can produce worthwhile savings even for small numbers of equations, provided they are sufficiently diagonally dominant. A simple test for the condition (5.64a) is likely to be worth including in any subroutine for cyclic reduction. The savings in execution time can be substantial if hundreds or thousands of equations are involved.

5.4.4   Choosing the algorithm

We will now consider the issues involved in choosing the best algorithm for the solution of m tridiagonal systems each of which is a set of n equations in n unknowns. The alternatives available and the conclusions are fairly typical of those arising in the choice of the best algorithm for any problem on a parallel computer. We again compare the performance of algorithms on a computer with a finite half-performance length using the $n_{1/2}$ method of algorithm analysis (see §5.1.6). We assume, as usual, an infinite memory with no bank conflicts. The results are taken from Hockney (1982).

On a serial computer we have only one consideration, namely the minimisation of the number of arithmetic operations. On a parallel computer there are many more operations available and the selection of the best method becomes more complex. For the above problem of m tridiagonal systems, one may take the best sequential algorithm and apply it to the m systems, in parallel, or take the best parallel algorithm for solving a single system and apply it sequentially or in parallel to the m systems. In the latter case one has the further complication that the best parallel algorithm will depend on the amount of hardware parallelism available on the computer (i.e. the value of $n_{1/2}$). On a serial computer the first option is unavailable, and all computers have the same value of $n_{1/2}$ ($= 0$), hence the alternatives telescope.

We start by considering the best algorithm for solving a single tridiagonal system of n equations. Using equations (5.46b), (5.58b) and (5.55), and ignoring the unimportant difference between n' and n, we find the operations' counts and average vector length $\bar{n}$ of the competing algorithms to be as follows.

(1) For recursive doubling (RD) the operations' counts are

$$s = 24n\log_2 n \qquad q = 24\log_2 n.$$

Hence using

$$T_{RD} \propto s + n_{1/2}q \qquad P_{RD} = T^{-1} \qquad \bar{n} = s/q$$

we obtain

$$P_{RD} \propto [24(n + n_{1/2})\log_2 n]^{-1} \qquad \bar{n}_{RD} = n. \qquad (5.66a)$$

(2) For cyclic reduction with vector length kept at n and no filling-in stage (termed PARACR),

$$s = 12n\log_2 n \qquad q = 12\log_2 n$$

hence

$$P_{\text{PARACR}} \propto [12(n + n_{1/2})\log_2 n]^{-1} \qquad \bar{n}_{\text{PARACR}} = n. \tag{5.66b}$$

(3) For cyclic reduction with vector length halving at each level of reduction and including a filling-stage (termed SERICR),

$$s = 17n \qquad q = 17\log_2 n$$

hence

$$P_{\text{SERICR}} \propto [17(n + n_{1/2}\log_2 n)]^{-1} \qquad \bar{n}_{\text{SERICR}} = n/\log_2 n. \tag{5.66c}$$

In the above we also ignore possible savings from truncated reduction which should certainly be taken into account if the systems are strongly diagonally dominant.

The recursive doubling algorithm has a poorer performance than either variant of cyclic reduction and will not be further considered. It would, of course, be used if it were required to find the LU decomposition of the equations. For the stated problem of actually solving the equations, cyclic reduction is always better. The choice is therefore between the parallel and serial variants of cyclic reduction (PARACR and SERICR respectively).

Given the operations' counts s and q, the algorithmic phase diagram can now be drawn by using equation (5.5). We find that the parallel variant has the better or equal performance ($P_{\text{PARACR}} \geq P_{\text{SERICR}}$) when

$$n_{1/2}/n \geq 2.4(1 - 1.42/\log_2 n). \tag{5.67a}$$

The equality gives the boundary line of equal performance. This is plotted in figure 5.9 together with the regions of the diagram in which each algorithm is superior. For large $n_{1/2}$ we have:

$$n_{1/2}/n \geq 2.4. \tag{5.67b}$$

This shows that for the paracomputer ($n_{1/2} = \infty$) the parallel variant, PARACR, is the best algorithm for all orders n of the equations (hence the name of the variant).

For finite values of $n_{1/2}$ there is always some $n(\approx 0.42n_{1/2})$ for values greater than which the serial variant is superior. This result is analogous to that found in §5.2.3 for the solution of recurrences. The value of $n_{1/2}$ measures the amount of hardware parallelism. If the vector length is much greater than this for a particular problem, then for this problem the computer will act like a serial computer—that is to say the parallelism of the computer is too small to have an influence on the performance. In this circumstance the criterion of performance on a serial computer will be relevant: namely that

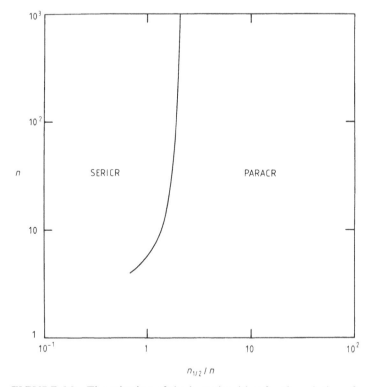

FIGURE 5.9   The selection of the best algorithm for the solution of a
single tridiagonal system of n equations on a computer with a half-
performance length of $n_{1/2}$: SERICR, cyclic reduction with reduction of
vector length; PARACR, cyclic reduction without reduction of vector
length. (From Hockney (1982), courtesy of North-Holland.)

the best algorithm is that with the least number of scalar arithmetic
operations, i.e. the serial variant of cyclic reduction (SERICR).

If, alternatively, we are presented with the problem of solving a set of m
tridiagonal systems each of n equations, we have the choice of either
applying SERICR or PARACR in parallel or sequentially to the m systems,
or using the best serial method (the Gaussian elimination recurrence described
in the last paragraph of §5.4.1) to all systems in parallel. For computers
with a large natural parallelism $\geqslant mn$, for example large processor arrays
(the ICL DAP) or pipelined computers with a large $n_{1/2}$ (the CYBER 205),
the best algorithm is likely to be the one with the most parallelism. For such
cases we compare Gaussian elimination applied in parallel to all systems
(MULTGE) with parallelism m, with $SERICR_{par}$ and $PARACR_{par}$ in which
the named cyclic reduction algorithm is applied in parallel to all systems.

The operations' counts and average vector lengths for these alternatives are as follows.

(1) For MULTGE $s = 8nm$, $q = 8n$, hence

$$P_{\text{MULTGE}} \propto [8n(m + n_{1/2})]^{-1} \qquad \bar{n}_{\text{MULTGE}} = m. \qquad (5.68a)$$

(2) For PARACR$_{\text{par}}$ $s = 12mn \log_2 n$, $q = 12 \log_2 n$, hence

$$P_{\text{PARACR}_{\text{par}}} \propto [12(mn + n_{1/2}) \log_2 n]^{-1} \qquad \bar{n}_{\text{PARACR}_{\text{par}}} = mn. \qquad (5.68b)$$

(3) For SERICR$_{\text{par}}$ $s = 17mn$, $q = 17 \log_2 n$, hence

$$P_{\text{SERICR}_{\text{par}}} \propto [17(mn + n_{1/2} \log_2 n)]^{-1} \qquad \bar{n}_{\text{SERICR}_{\text{par}}} = mn/\log_2 n. \qquad (5.68c)$$

Using equation (5.5) we obtain the following relationships, in which the inequality determines which algorithm has the better performance, and the equality gives the equation for the equal performance line on the $(n_{1/2}/m, n)$ parameter plane.

$P_{\text{PARACR}_{\text{par}}} \geqslant P_{\text{MULTGE}}$ when

$$n_{1/2}/m \geqslant (1.5 \log_2 n - 1)/[1 - (1.5 \log_2 n/n)] \qquad (5.69a)$$

$$\geqslant 1.5 \log_2 n \qquad \text{as } n \to \infty.$$

$P_{\text{SERICR}_{\text{par}}} \geqslant P_{\text{MULTGE}}$ when

$$n_{1/2}/m \geqslant 1.1/[1 - (2.1 \log_2 n/n)] \qquad (5.69b)$$

$$\geqslant 1.1 \qquad \text{as } n \to \infty.$$

$P_{\text{PARACR}_{\text{par}}} \geqslant P_{\text{SERICR}_{\text{par}}}$ when

$$n_{1/2}/m \geqslant 2.4n[1 - (1.42/\log_2 n)] \qquad (5.69c)$$

$$\geqslant 2.4n \qquad \text{as } n \to \infty.$$

The lines defined by equations (5.69) are shown in figure 5.10 and divide the $(n_{1/2}/m, n)$ parameter plane into regions in which each of the three methods has the best performance. This is rather like a chemical phase diagram, and there is even a 'triple point' at $n \approx 7$, $n_{1/2}/m \approx 7.7$ where the three algorithms have the same performance. We find that for $n_{1/2} = \infty$ (the paracomputer), PARACR$_{\text{par}}$ is, as expected, the best algorithm for all n. However for finite $n_{1/2}$ and $n_{1/2} > 10m$ there is always some n, greater than which SERICR$_{\text{par}}$ is favoured. If the number of systems m is greater than $n_{1/2}$, we find that the application of the best serial algorithm in parallel to the m systems is always the best. In the region $1 < n_{1/2}/m < 10$ all three algorithms may be favoured in the complex way displayed by the diagram.

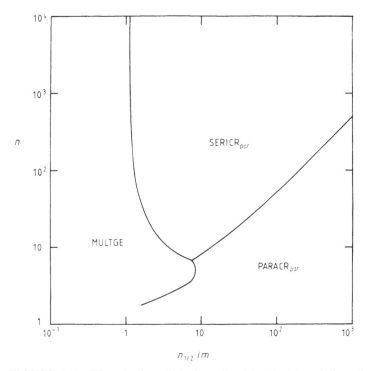

FIGURE 5.10 The selection of the best algorithm for the solution of
m tridiagonal systems of n equations for computers with a large natural
parallelism $\geqslant mn$. We compare the Gaussian elimination (MULTGE),
and the serial (SERICR$_{par}$) and parallel (PARACR$_{par}$) versions of cyclic
reduction when applied in parallel to all m systems. (From Hockney
(1982), courtesy of North-Holland.)

For computers with a limited parallelism approximately equal to n or m,
there is insufficient parallelism to apply SERICR or PARACR in parallel.
This would apply, typically, to computers such as the CRAY X-MP (which has
a natural parallelism of 64) when applied to the solution of 64 tridiagonal
systems of length 64. In this case we must compare SERICR$_{seq}$ and
PARACR$_{seq}$ (in which the named cyclic reduction algorithm is applied
sequentially to the m systems) with the best serial method applied in parallel
(MULTGE). The performance and average vector lengths for these cases
are

$$P_{\text{PARACR}_{seq}} \propto [12m(n+n_{1/2})\log_2 n]^{-1} \qquad \bar{n}_{\text{PARACR}_{seq}} = n, \quad (5.70a)$$

$$P_{\text{SERICR}_{seq}} \propto [17m(n+n_{1/2}\log_2 n)]^{-1} \qquad \bar{n}_{\text{SERICR}_{seq}} = n/\log_2 n. \quad (5.70b)$$

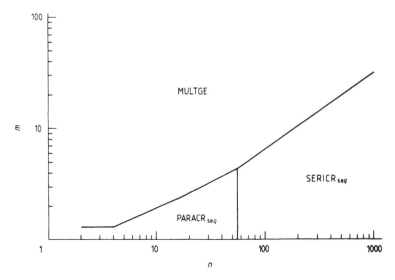

FIGURE 5.11   The selection of the best algorithm for the solution of m tridiagonal systems of n equations for a computer with limited parallelism of approximately m or n (here $n_{1/2} = 100$). We compare Gaussian elimination applied in parallel to all m systems (MULTGE), with the serial ($\text{SERICR}_{\text{seq}}$) and parallel ($\text{PARACR}_{\text{seq}}$) versions of cyclic reduction applied sequentially to the m systems.

Figure 5.11 shows the comparison between these two algorithms and MULTGE for the case $n_{1/2} = 100$. The vertical line separating the $\text{PARACR}_{\text{seq}}$ and $\text{SERICR}_{\text{seq}}$ methods is obtained from the equality in equation (5.67a) or figure 5.9, and is true for all m. Multiple Gaussian elimination will have a performance better than or equal to the parallel variant of cyclic reduction applied sequentially when

$$P_{\text{MULTGE}} \geqslant P_{\text{PARACR}_{\text{seq}}}$$

or when

$$m \geqslant n_{1/2}[(1.5\log_2(n) - 1) + (1.5n_{1/2}\log_2 n)/n]^{-1},$$
$$\geqslant 0.67n/\log_2 n \qquad \text{as } n_{1/2} \to \infty. \tag{5.70c}$$

Similarly we find that MULTGE is superior to $\text{SERICR}_{\text{seq}}$ when

$$P_{\text{MULTGE}} \geqslant P_{\text{SERICR}_{\text{seq}}}$$

or when

$$m \geqslant n_{1/2}[1.125 + 2.125(n_{1/2}\log_2 n)/n]^{-1}. \tag{5.70d}$$

We find, broadly speaking, that the multiple application of the sequential

Gaussian algorithm is the best when the number of systems exceeds one tenth of the number of equations in each system. This will be the case for most applications arising from the solution of partial differential equations (see §5.6). There is a relatively small part of the diagram favouring the parallel variant of cyclic reduction, and this becomes smaller as $n_{1/2}$ decreases. Indeed for $n_{1/2} \leqslant 10$, PARACR$_{seq}$ is never the best algorithm.

5.5 TRANSFORMS

Mathematical transforms play an important role in mathematical and numerical analysis. Amongst these the finite Fourier transform (see Bracewell 1965) and the related number theoretic transform are the most important because especially fast algorithms exist for their evaluation. The fast Fourier transform (FFT) algorithm (§5.5.1) obtains all components of the transform of n data values in $O(n \log_2 n)$ real scalar arithmetic operations, compared with $O(n^2)$ such operations for other transforms. The number theoretic transform (NTT) (see §5.5.6) has not only the logarithmic operation count of the FFT, but also the advantage of using only integer additions and shifts; there are no multiplication operations. The NTT therefore is particularly well suited for use on bit-oriented architectures such as the ICL DAP or Goodyear MPP (see Chapter 3).

Some of the applications of the Fourier transform are:

(a) The Fourier transformation of data as a function of time (time-series analysis) in order to determine the frequency spectrum. The Fourier transform is often the best way of computing the auto- and cross-correlations of such time series, and smoothing of the data is often best achieved in frequency space (for example the reduction or elimination of high noise frequencies). For a full discussion see Blackman and Tukey (1959).

(b) The Fourier transformation of data as a function of space in order to obtain the wavenumber spectrum. In many physical situations a wave analysis leads to the clearest understanding of a phenomenon (for example the realisation that certain wavelengths are unstable). Any linear system (or small disturbances of a nonlinear system) can be completely described by giving its dispersion equation, that is to say the frequency of oscillation as a function of the wavenumber of the disturbance. The measurement of this relation clearly requires a Fourier analysis in both space and time, for a large number of wavenumbers and frequencies.

(c) Spatial correlations between particles in, for example, a liquid are described by the pair (and higher) correlation functions, that is to say the probability that two particles are separated by a distance, as a function of that distance. Physical measurements, however, by x-ray and neutron diffraction determine the structure factor, which is the Fourier transform of the spatial correlation. Many transformations from wavenumber to coordinate space are required in order to interpret the experiments.

(d) Particle simulations by the method of molecular dynamics (see Hockney and Eastwood 1988, Chapter 12), trace the orbits of a large collection of atoms in a liquid. The interpretation of the dynamic properties of the liquid and its comparison with theory is best undertaken via the dynamic structure factor. This is determined by a Fourier analysis of the particle coordinates in space and time.

(e) The solution of linear partial differential equations is frequently undertaken in wavenumber space (so called Galerkin or spectral methods, see Orszag 1971). Nonlinear terms are included by transforming to coordinate space, carrying out the nonlinear interaction, and then transforming back to wavenumber space. Clearly such methods are only possible if rapid methods for evaluating the transforms are available and used.

(f) As a special case of (e) it is found that some of the fastest methods for solving the discrete form of Poisson's equation in simple geometries are based on a partial Fourier transform of the data using the FFT (e.g. the FACR algorithm of Hockney (1965) that is discussed in §5.6.2). In this circumstance the requirement is to perform many Fourier transforms in parallel, for example to transform the data on each of 64 lines of a 64×64 mesh in parallel. In a meteorological application using a semi-implicit timestepping procedure thousands of real Fourier transforms are required per timestep (Temperton 1979b).

(g) The solution of many field problems may be expressed as the *convolution* of a source distribution with an influence function that describes the influence on the solution of a point source. Using the convolution theorem, one has equivalently that the Fourier transform of the solution is the *product*, wavenumber by wavenumber, of the Fourier transform of the source and the Fourier transform of the influence function. The solution may be found by: (i) taking the transform of the source; (ii) multiplying it by the transform of the influence function, which may be precalculated and stored; and finally (iii) taking the inverse transform to obtain the solution. Such a method is only practical with the advent of the FFT, and has been extensively used in the calculation of the gravitational potential due to an isolated system of stars, that is to say a galaxy (Hockney 1970, Brownrigg 1975, Eastwood and Brownrigg 1979) and in the calculation of electrostatic interactions in ionic crystals with the P^3M algorithm (Eastwood 1976, Amini and Hockney 1979, Hockney and Eastwood 1988).

This list of applications is by no means exhaustive. Other applications involving digital filtering, calculation of covariances and spectral densities, averaging and smoothing and the Laplace transform are discussed by Cooley *et al* (1967).

5.5.1 Fast Fourier transform

The fast Fourier transform is a collective term for a group of methods for the rapid evaluation of the finite Fourier transform (or Fourier analysis):

$$\bar{f}^k = \frac{1}{n} \sum_{j=0}^{n-1} \exp(-2\pi i j k/n) f_j, \qquad 0 \leqslant k \leqslant n-1, \tag{5.71a}$$

where f_j are the set of n complex data to be transformed and $\bar{f}^k$ are the set of n complex harmonic amplitudes resulting from the transformation. A direct evaluation of the definition (5.71a) would require n complex multiplications and n complex additions per harmonic, or a total of $8n^2$ real arithmetic operations for the evaluation of all n harmonics. The value of the FFT algorithm is that it reduces this operation count to approximately $5n \log_2 n$ real arithmetic operations. The ratio of the performance of the FFT to the direct evaluation on a serial computer is therefore

$$P_{FFT}/P_{DE} = 8n^2/5n \log_2 n) \simeq n/\log_2 n. \tag{5.71b}$$

This ratio varies from about 18 ($n = 128$) to about 102 ($n = 1024$) and to about 5×10^4 ($n \approx 10^6$). It is not surprising therefore that the publication of the FFT by Cooley and Tukey (1965) has led to a major revolution in numerical methods. Although Cooley, Lewis and Welch (1967) cite earlier works containing the idea for $n \log_2 n$ methods (e.g. Runge and König 1924, Stumpff 1939, Danielson and Lanczos 1942, Thomas 1963) these were not generally known. Before 1965 Fourier transformation was thought to be a costly n^2 process that could only be conducted sparingly and was usually best avoided; after 1965 it became a relatively cheap process, orders of magnitude faster than had been previously thought.

The transform (5.71a) has the inverse (or Fourier synthesis):

$$f_j = \sum_{k=0}^{n-1} \exp(2\pi i j k/n) \bar{f}^k, \qquad 0 \leqslant j \leqslant n-1, \tag{5.72a}$$

which may be proved using the orthogonality relation:

$$\sum_{k=0}^{n-1} \exp[2\pi i k(j-j')/n] = n\delta_{jj'}, \tag{5.72b}$$

where $\delta_{jj'}$ is the Kronecker delta:

$$\delta_{jj'} = \begin{cases} 0 & j \neq j', \\ 1 & j = j'. \end{cases} \tag{5.72c}$$

Ignoring the factor of n^{-1} in equation (5.71a), which can be included outside the routine for Fourier transformation, both the direct and inverse transform can be expressed as

$$y_k = \sum_{j=0}^{n-1} \omega_n^{jk} x_j, \qquad 0 \leqslant j \leqslant n-1, \tag{5.73a}$$

where

$$\omega_n = \begin{cases} \exp(-2\pi i/n) & \text{in analysis,} \\ \exp(+2\pi i/n) & \text{in synthesis.} \end{cases} \tag{5.73b} \tag{5.73c}$$

The association of the negative sign with Fourier analysis in equation (5.73b) and the positive sign with Fourier synthesis is arbitrary. We have followed the conventions of transmission theory and Bracewell (1965). Clearly the choice is immaterial and can be accommodated by the appropriate choice of ω_n. The algorithm is the same in both cases. Being the nth root of unity, ω_n has the important property that

$$\omega_n^n = 1, \tag{5.74a}$$

and thus

$$\omega_n^{ns+t} = (\omega_n^n)^s \omega_n^t = \omega_n^t,$$

where s and t are integers. The definitions (5.73b, c) also show that any common factor may be removed from or inserted into both the subscript and superscript of ω_n^m without altering its value. Thus, for example,

$$\omega_n^{n/2} = \omega_1^{1/2} = \omega_2 = -1. \tag{5.74b}$$

Frequently the function to be transformed is real, say g_j, and the transforms are expressed in terms of sines and cosines. Thus, assuming n is even,

$$g_j = \tfrac{1}{2}[a_0 + a_{n/2}(-1)^j] \\ + \sum_{k=1}^{n/2-1} [a_k \cos(2\pi jk/n) + b_k \sin(2\pi jk/n)], \tag{5.75a}$$

where

$$a_k = \frac{2}{n} \sum_{j=0}^{n-1} g_j \cos(2\pi jk/n) \tag{5.75b}$$

and

$$b_k = \frac{2}{n} \sum_{j=0}^{n-1} g_j \sin(2\pi jk/n). \tag{5.75c}$$

The coefficients of a_k and b_k of the real transform are simply related to the real and imaginary parts of the complex transform (5.71a) of g, by:

$$a_k = 2\,\text{Re}(\bar{g}^k), \qquad b_k = -2\,\text{Im}(\bar{g}^k). \tag{5.75d}$$

A real transform can therefore be obtained by performing a complex transform of length n on data of which the real part is g_j and the imaginary part is zero. The real and imaginary parts of the first $n/2$ harmonics of the complex transform yield the coefficients of the real transform by equation (5.75d). The second $n/2$ harmonics are redundant, being the complex conjugates of the first $n/2$ harmonics by:

$$\bar{g}^{n-k} = (\bar{g}^k)^* = \text{Re}(\bar{g}^k) - i\,\text{Im}(\bar{g}^k), \qquad 0 \leqslant k \leqslant n/2. \tag{5.76}$$

The above method of computing the real transform is clearly wasteful as half the input data is zero (all the imaginary parts) and half the results contain no new information and are discarded. The number of scalar real arithmetic operations is $5n\log_2 n$.

A more economic method of performing a real transform is to treat the even values of the real input data as the real parts of a function to be transformed, and the odd values of the real input data as the imaginary parts, thus

$$f_j = g_{2j} + ig_{2j+1}, \qquad 0 \leqslant j \leqslant n/2 - 1. \tag{5.77}$$

The complex transform of f_j is now of length $n/2$ and takes $2\frac{1}{2}n\log_2 n$ scalar real operations. The coefficients of the real transform are then found as follows:

(a) Calculate for $k = 1, 2, \ldots, n/4 - 1$ the transforms of the $n/2$ even and $n/2$ odd values:

$$\overline{\text{evens}}\,k = \frac{2}{n}\sum_{j=0}^{n/2-1} g_{2j}\omega_{n/2}^{jk} = \tfrac{1}{2}[\bar{f}^k + (\bar{f}^{n/2-k})^*], \tag{5.78a}$$

$$\overline{\text{odds}}\,k = \frac{2}{n}\sum_{j=0}^{n/2-1} g_{2j+1}\omega_{n/2}^{jk} = \frac{1}{2i}[\bar{f}^k - (\bar{f}^{n/2-k})^*], \tag{5.78b}$$

where the asterisk denotes the complex conjugate.

(b) Calculate for $k = 1, 2, \ldots, n/4 - 1$ the intermediate transform C_k of length $n/2$, defined by

$$C_k = \overline{\text{evens}}\,k + \omega_n^k\,\overline{\text{odds}}\,k \tag{5.79a}$$

$$C_{n/2-k} = \overline{\text{evens}}\,k - \omega_n^k\,\overline{\text{odds}}\,k \tag{5.79b}$$

$$C_{n/4} = (\bar{f}^{n/4})^*. \tag{5.79c}$$

(c) Finally obtain the coefficients of the real transform from:

$$a_0 = \text{Re}(\bar{f}^0) + \text{Im}(\bar{f}^0), \tag{5.80a}$$

$$a_{n/2} = \text{Re}(\bar{f}^0) - \text{Im}(\bar{f}^0), \tag{5.80b}$$

$$\left.\begin{array}{l} a_k = \text{Re}(C_k), \\ b_k = -\text{Im}(C_k), \end{array}\right\} \quad k = 1,\ldots,n/2 - 1. \tag{5.80c}$$

The number of real scalar arithmetic operations is $2\frac{1}{2}n\log_2 n + 3\frac{1}{2}n$, where the second term arises from the post-manipulation of $\bar{f}_k$ in equations (5.78)–(5.80).

The calculation of the Fourier series (5.75a) from the coefficients (5.75b,c) can be performed by reversing the above procedure as follows:

(a) Set

$$C_0 = a_0, \qquad C_{n/2} = a_{n/2} \tag{5.81a}$$

and

$$C_k = a_k - ib_k, \qquad k = 1, 2, \ldots, n/2 - 1. \tag{5.81b}$$

(b) Evaluate

$$\left.\begin{array}{l} \overline{\text{evens}}\,k = \tfrac{1}{2}(C_k + C^*_{n/2-k}), \\[2ex] \overline{\text{odds}}\,k = \tfrac{1}{2}(C_k - C^*_{n/2-k})\omega_n^{-k}, \end{array}\right\} \quad k = 1, 2, \ldots, n/4 - 1. \tag{5.82a}\;\tag{5.82b}$$

(c)

$$\left.\begin{array}{l} \bar{f}^k = \overline{\text{evens}}\,k + i\,\overline{\text{odds}}\,k, \\[2ex] \bar{f}^{n/2-k} = (\overline{\text{evens}}\,k - i\,\overline{\text{odds}}\,k)^*, \end{array}\right\} \quad k = 1, 2, \ldots, n/4 - 1 \tag{5.83a}\;\tag{5.83b}$$

and also

$$\bar{f}^0 = \tfrac{1}{2}(C_0 + C_{n/2}) + \tfrac{1}{2}i(C_0 - C_{n/2}) \tag{5.83c}$$

$$\bar{f}^{n/4} = C^*_{n/4}. \tag{5.83d}$$

We then calculate the complex Fourier synthesis of the $n/2$ complex values $\bar{f}^k$ and the resulting $n/2$ complex values contain the required synthesised values in successive real and imaginary parts, thus

$$f_j = \sum_{k=0}^{n/2-1} \exp\left(\frac{2\pi ijk}{n/2}\right)\bar{f}^k = g_{2j} + ig_{2j+1}, \qquad j = 0, 1, \ldots, n/2 - 1. \tag{5.83e}$$

The above procedure involved a pre-manipulation of the n real data into the appropriate $n/2$ complex values which are then transformed using the complex FFT. As in analysis, the number of real scalar arithmetic operations is $2\frac{1}{2}n\log_2 n + 3\frac{1}{2}n$.

The calculation of the real Fourier transform is discussed by Cooley *et al* (1967) and by Bergland (1968). A different approach, involving the folding of data before multiplying by sines and cosines, is described by Hockney (1970). The latter method is the amalgamation of hand computation methods as given, for example, by Runge (1903, 1905) and Whittaker and Robinson (1944) and gives in addition to the periodic transform described above, the finite sine and cosine transformations. The latter together with the calculation of the Laplace transform are discussed by Cooley *et al* (1970). Other important publications on the FFT are Gentleman and Sande (1966), Singleton (1967, 1969), Uhrich (1969) and the book on the subject by Brigham (1974). In this chapter we aim only to bring out the main features of the FFT algorithm and discuss some of the considerations affecting its implementation on parallel computers. For this purpose we will limit the discussion to the binary (or radix-2) case for which $n = 2^q$ (where q is integral) although the algorithm can be applied efficiently to any n that is a product of small primes preferably repeated many times. Such mixed radix algorithms are described by Singleton (1969) and Temperton (1977, 1983a, 1983c). A form of the binary algorithm particularly well suited to parallel processing has been given by Pease (1968), and the performance of several of the above algorithms has been compared on the CRAY-1 by Temperton (1979b), and on the CYBER 205 by Temperton (1984) and Kascic (1984b). Vectorisation of the fast Fourier transform is discussed by Korn and Lambiotte (1979), Wang (1980) and Swarztrauber (1982, 1984). The implementation of fast radix-2 algorithms on processor arrays is considered by Jesshope (1980a).

The above algorithms, which are all variations and extensions of the algorithm published by Cooley and Tukey (1965), may be described as the *conventional fast Fourier transform*. Their efficiency relies on the factors of n being repeated many times. Curiously, there is another form of the fast Fourier transform, called the *prime factor algorithm* (PFA), in which n is split into non-repeating factors which are all mutually prime to each other. These methods were first introduced by Good (1958, 1971) and Thomas (1963), and subsequently developed by Kolba and Parks (1977), Winograd (1978), and Johnson and Burrus (1983). The method has found most application in signal processing (Burrus 1977, Burrus and Eschenbacher 1981), and is fully described in the books by McClellan and Rader (1979), and Nussbaumer (1982). The practicalities of implementing the PFA on current vector computers have been discussed by Temperton (1983b, 1985, 1988).

One aspect of the above methods is the manipulation of the algebraic formulae for the FFT in such a way as to minimise the number of multiplications. The theory of this technique has been developed by Winograd (1978, 1980), and the method has been used extensively for the computation of the FFT and convolutions by Cooley (1982) and Auslander and Cooley (1986). The numerical stability of the resulting algorithms is discussed by Auslander *et al* (1984).

5.5.2   Deriving the FFT

The key to the derivation of the fast Fourier transform algorithm is the appropriate definition of intermediate partial transforms. Here we adopt the definition used by Roberts (1977) although the notation has been slightly changed. The partial transforms of the n data, $f_0, f_1, \ldots, f_{n-1}$ at level l of the transform are defined by†

$$\overset{(l)}{f}{}_i^k = \sum_{j=0}^{2^l - 1} \omega_{2^l}^{jk} f_{jn2^{-l} + i} \tag{5.84a}$$

where

$$j, k = 0, 1, \ldots, 2^l - 1 \tag{5.84b}$$

and i is an integer subscript with values

$$i = 0, 1, \ldots, n2^{-l} - 1. \tag{5.84c}$$

The above defines a collection of $n2^{-l}$ transforms. Each transform is distinguished by its identification number i which is the index of the first datum from which the transform is calculated. The remaining data for the transform are separated from the first by the interval $n2^{-l}$. The length of each transform is 2^l. Thus $\overset{(l)}{f}{}_i^k$ is the kth harmonic of the ith transform at level l.† When $l = 0$ we have $j = k = 0$ and

$$\overset{(0)}{f}{}_i^0 = f_i, \qquad i = 0, 1, \ldots, n - 1. \tag{5.85a}$$

Hence the level-zero transforms are the initial data. At level $l = q$, where $q = \log_2 n$, we have $i = 0$ and

$$\overset{(q)}{f}{}_0^k = \sum_{j=0}^{n-1} \omega_n^{jk} f_j = n\tilde{f}^k. \tag{5.85b}$$

Thus at level $\log_2 n$ there is only one partial transform which is proportional to the required complete transform over all the original n data.

† The notation $\overset{(l)}{\overline{f}}{}_i^k$ will be used in the text only.

It is also helpful to consider the partial transforms $^{(l)}\bar{f}_i^k$ as the elements of a two-dimensional matrix of 2^l rows and $n2^{-l}$ columns, formed by writing the harmonics of each partial transform as a column vector and placing these column vectors side by side. Then we start at level zero with a single row of data, and finish at level $\log_2 n$ with a single column of results. In going from one level to the next the number of columns is halved and the number of rows is doubled. This view should not be taken to imply that the partial transforms are necessarily stored in the computer as a variable two-dimensional array. One-dimensional indexing is invariably used in FORTRAN and we will see that storage may be saved if the harmonic amplitudes are calculated in a shuffled—rather than their natural—order. Two-dimensional indexing can be used if the necessary features are available in the parallel constructs of the language and efficiently implemented, as is shown in §4.3.1(iii).

The arithmetic of the algorithm is obtained by deriving a recurrence for the partial transforms at level $l+1$ in terms of those at level l:

$$\overset{(l+1)}{f}{}_i^k = \sum_{j=0}^{2^{l+1}-1} \omega_{2^{l+1}}^{jk} f_{jn2^{-(l+1)}+i}. \tag{5.86a}$$

Dividing the summation into two parts by setting $j = 2s + t$ with $s = 0, 1, \ldots, 2^l$ and $t = 0, 1$ one obtains:

$$\overset{(l+1)}{f}{}_i^k = \sum_{s=0}^{2^l-1} \sum_{t=0}^{1} \omega_{2^{l+1}}^{(2s+t)k} f_{(2s+t)n2^{-(l+1)}+i}. \tag{5.86b}$$

Remembering the definition of $\omega_{2^{l+1}}$ as the 2^{l+1}th root of unity we have:

$$\omega_{2^{l+1}}^{(2s+t)k} = \omega_{2^{l+1}}^{tk} \omega_{2^{l+1}}^{2sk} = \omega_{2^{l+1}}^{tk} \omega_{2^l}^{sk}, \tag{5.86c}$$

where in the last step we have removed the common factor 2 from the subscript and superscript of the second factor. Substituting equation (5.86c) into (5.86b) and writing separately the terms for $t = 0$ and 1 one obtains

$$\overset{(l+1)}{f}{}_i^k = \sum_{s=0}^{2^l-1} \omega_{2^l}^{sk} f_{sn2^{-l}+i} + \omega_{2^{l+1}}^{k} \sum_{s=0}^{2^l-1} \omega_{2^l}^{sk} f_{sn2^{-l}+i+n2^{-(l+1)}}.$$

Hence

$$\overset{(l+1)}{f}{}_i^k = \overset{(l)}{f}{}_i^k + \omega_{2^{l+1}}^{k} \overset{(l)}{f}{}_{i+n2^{-(l+1)}}^k. \tag{5.87a}$$

Replacing k by $k + 2^l$ gives:

$$\overset{(l+1)}{f}{}_i^{k+2^l} = \overset{(l)}{f}{}_i^k - \omega_{2^{l+1}}^{k} \overset{(l)}{f}{}_{i+n2^{-(l+1)}}^k. \tag{5.87b}$$

since $^{(l)}\bar{f}_i^{k+2^l} = {}^{(l)}\bar{f}_i^k$, from equation (5.84a), and

$$\omega_{2^i+1}^{k+2^l} = \omega_{2^i+1}^{2^l}\omega_{2^i+1}^k = \omega_2\omega_{2^i+1}^k = -\omega_{2^i+1}^k. \tag{5.87c}$$

In the last step of equation (5.87c) we have removed the common factor 2^l from the subscript and superscript of the first factor, and used the fact that $\omega_2 = -1$.

To summarise, the Cooley–Tukey (1965) formulation of the FFT is given by:

$$\left.\begin{array}{l} \overset{(l+1)}{f}{}_i^k = \overset{(l)}{f}{}_i^k + \omega_{2^i+1}^k \overset{(l)}{f}{}_{i+n2^{-(l+1)}}^k \\[2ex] \overset{(l+1)}{f}{}_i^{k+2^l} = \overset{(l)}{f}{}_i^k - \omega_{2^i+1}^k \overset{(l)}{f}{}_{i+n2^{-(l+1)}}^k \end{array}\right\} \quad \text{for } l = 0, 1, \dots, q-1, \qquad \begin{array}{l}(5.88a)\\[2ex](5.88b)\end{array}$$

where $k = 0, 1, \dots, 2^l - 1$ and $i = 0, 1, \dots, n2^{-(l+1)} - 1$ with starting condition $^{(0)}\bar{f}_i^0 = f_i$. The result of the transform is found from:

$$\bar{f}^k = \frac{1}{n}\overset{(q)}{f}{}_0^k. \tag{5.88c}$$

If $^{(l+1)}\bar{f}_i^k$ overwrites $^{(l)}\bar{f}_i^k$ and $^{(l+1)}\bar{f}_i^{k+2^l}$ overwrites $^{(l)}\bar{f}_{i+n2^{-(l+1)}}^k$ then the transform can be performed in-place without the need for any auxiliary storage. However the final harmonics are then obtained in reverse binary order. That is to say, if $k = k_{q-1}2^{q-1} + k_{q-2}2^{q-2} + \dots + k_1 2 + k_0$, where k_p is the pth digit in the binary representation of k, then this harmonic will be found in location $^{(q)}\bar{f}^{k'}$, where $k' = k_0 2^{q-1} + k_1 2^{q-2} + \dots + k_{q-2}2 + k_{q-2}$. If the harmonic analysis is to be followed by a synthesis step which reverses the above steps, then there is no need to sort the harmonics into natural order. This is the case during the solution of a field problem by the convolution method. If the harmonics are required as output, however, a sorting step must be inserted after the basic algorithm. Such a sorting is often referred to as bit-reversal, because the harmonics occur in bit-reversed order. Alternatively, if overwriting is eliminated by placing the result of the recurrence in a second array, sorting may take place at each level as is shown in the data flow diagram in figure 5.12.

The fast Fourier transform of equations (5.88) may be reversed by solving for level l in terms of level $l+1$ and the inverse transform obtained as follows:

$$\left.\begin{array}{l} \overset{(l)}{f}{}_i^k = \tfrac{1}{2}\left(\overset{(l+1)}{f}{}_i^k + \overset{(l+1)}{f}{}_i^{k+2^l}\right) \\[2ex] \overset{(l)}{f}{}_{i+n2^{-(l+1)}}^k = \tfrac{1}{2}\omega_{2^i+1}^{-k}\left(\overset{(l+1)}{f}{}_i^k - \overset{(l+1)}{f}{}_i^{k+2^l}\right) \end{array}\right\} \quad \text{for } l = q-1, q-2, \dots, 0, \qquad \begin{array}{l}(5.89a)\\[2ex](5.89b)\end{array}$$

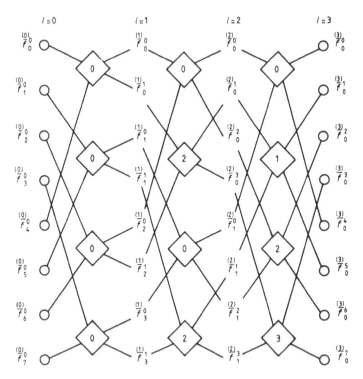

FIGURE 5.12 Data flow diagram for the Cooley–Tukey formulation of the fast Fourier transform. The diamonds evaluate the recurrence (5.88). The number in the diamond is the power of the nth root of unity that is used as a phase factor. (After Temperton 1977.)

where $k = 0, 1, \ldots, 2^l - 1$ and $i = 0, 1, \ldots, n2^{-(l+1)} - 1$ with the starting condition,

$$\overset{(q)}{f}{}^k_0 = n\tilde{f}^k. \tag{5.89c}$$

The synthesised results are found from:

$$f_i = \overset{(0)}{f}{}^0_i, \qquad i = 0, 1, \ldots, n - 1. \tag{5.89d}$$

An examination of equations (5.89) shows that the final result is unchanged if the factor n is dropped from equation (5.89c) and the factors of $\frac{1}{2}$ are dropped from equations (5.89a) and (5.89b). This amounts to inserting a factor 2^{-l} on the right-hand side in the definition of the partial transform (5.48a).

Since an inverse transform can be converted to a forward transform by replacing ω by ω^{-1}, interchanging the roles of f and $\bar{f}$, and multiplying by n^{-1} at a convenient place, equations (5.89) also provide an alternative formulation for the forward transform:

$$\overset{(l)}{\bar{f}}{}^k_i = \overset{(l+1)}{\bar{f}}{}^k_i + \overset{(l+1)}{\bar{f}}{}^{k+2^l}_i \qquad\qquad (5.90a)$$

$$\left.\begin{array}{l} \\ \\ \end{array}\right\} \text{ for } l = q-1, q-2, \ldots, 0,$$

$$\overset{(l)}{\bar{f}}{}^k_{i+n2^{-(l+1)}} = \omega^k_{2^{l+1}}(\overset{(l+1)}{\bar{f}}{}^k_i - \overset{(l+1)}{\bar{f}}{}^{k+2^l}_i) \qquad\qquad (5.90b)$$

where $k = 0, 1, \ldots, 2^l - 1$ and $i = 0, 1, \ldots, n2^{-(l+1)} - 1$ with the starting values

$$\overset{(q)}{\bar{f}}{}^k_0 = f_k \qquad\qquad (5.90c)$$

and the results found

$$\bar{f}^i = n^{-1}\overset{(0)}{\bar{f}}{}^0_i. \qquad\qquad (5.90d)$$

This is the Gentleman–Sande (1966) formulation of the fast Fourier transform.

The FFT may be based on either the recurrence (5.88) or (5.90). Implementations also differ in the storage pattern chosen for the intermediate partial transforms. For example Uhrich (1969) uses equations (5.88) and Pease (1968) uses equations (5.90).

5.5.3 Vectorisation

The techniques for the implementation of the FFT on vector pipelined computers can be illustrated by considering the FORTRAN code for the Cooley–Tukey algorithm of equations (5.88) that is given in figure 5.13. This program comprises a control routine (top) which calls the subroutine RECUR (bottom) to evaluate the recurrence (5.88). RECUR (F, G, W, L, N) performs the recurrence once for $L = l$ on the input data in the complex vector **F** and puts the output in the complex vector **G**. The vector **W** contains powers of the nth root of unity and $N = n$ is the total number of complex variables. At successive levels the input and output alternate between the arrays **F** and **G**, and calls to RECUR are therefore made in pairs in the control routine. Since **F** and **G** are different vectors the output never overwrites the input. This permits the harmonics to be kept in natural order and allows most compilers to vectorise the statements in the DO 11 loop. We assume that **W** has been previously loaded with powers of the nth root of units such that

$$\omega^k_n \text{ is stored in } \mathbf{W}(k+1) \qquad\qquad (5.91a)$$

```
C      MAIN CONTROL PROGRAM

       N = 2**LOG2N
DO 10 L1 = 1, LOG2N, 2
       CALL RECUR (F, G, W, L1-1, N)
       IF (L1.EQ.LOG2N) STOP
10     CALL RECUR (G, F, W, L1, N)

STOP


       SUBROUTINE RECUR (F, G, W, L, N)
       COMPLEX F(N), G(N), W(N), V(N), WK

       I2L  = 2**L
       I2L1 = 2*I2L
       N2L1 = N/I2L1

       INC1 = I2L
       IF (L.GE.2) INC1 = INC1+1
       INC2 = I2L1
       IF (L+1.GE.2) INC2 = INC2+1
       INC3 = N2L1*INC1-INC1
       INC4 = I2L-INC2

DO 20 K1 =1, I2L
       WK = W(N2L1*K1-N2L1+1)

DO 20 I1 = 1, N2L1
       V(I1) = WK*F(INC1*I1+K1+INC3)
       G(INC2*I1+K1-INC2) = F(INC1*I1+K1-INC1) + V(I1)
20     G(INC2*I1+K1+INC4) = F(INC1*I1+K1-INC1) - V(I1)

RETURN
END
```

FIGURE 5.13   The control program (top) and the subroutine RECUR for scheme A (bottom) with vectorisation of the DO I1 loop.

and the partial transforms are arranged such that

$$\overset{(l)}{F}_i^k \text{ is stored in } \mathbf{F} \text{ or } \mathbf{G} \ (2^l i + k + 1). \tag{5.91b}$$

The other principal FORTRAN variables are $L1 = l + 1, K1 = k + 1, I1 = i + 1$, $I2L = 2^l$, $N2L1 = n2^{-(l+1)}$, $I2L1 = 2^{l+1}$, $N2L2 = n2^{-(l+2)}$, $LOG2N = \log_2 n$. The statements in the DO 20 loop are the implementation of the Cooley–Tukey recurrence (5.88). They are written with all the indexing explicitly in terms of the loop control variables in order to display the opportunity for vectorisation. As written, most compilers would replace the DO 20 I1 loop by vector instructions. Since $\mathbf{W}$ is not a function of I1, it is replaced by the scalar WK in the inner loop which may be implemented by instructions of the form

$$\text{vector} = \text{vector} + \text{scalar} * \text{vector}$$

The vector length is $n2^{-(l+1)}$ and decreases at each call to the subroutine RECUR, taking on the values $n/2, n/4, \ldots, 4, 2, 1$. The storage interval between successive elements of the vector $\mathbf{F}$ that are referenced in the DO I1 loop is 2^l and takes on the values $1, 2, 4, 8, 16, \ldots, n/2$. Consequently, in the later stages of the algorithm, there will be memory-bank conflicts on computers with a power-of-two number of memory banks. On the CRAY-1 for example, there will be conflicts for $l \geqslant 2$ or intervals that are multiples of four complex or eight real numbers.

The above memory-bank conflicts can be eliminated at the expense of a small amount of extra storage, by adding unity to the storage intervals INC1 and INC2 for $l \geqslant 2$. This increases the storage requirement by $\frac{1}{4}$, but the storage intervals between successive elements of the vector $\mathbf{F}$ become $1, 2, 5, 9, 17, \ldots, n/2 + 1$ at levels $l = 0, 1, 2, 3, 4, \ldots, \log_2(n) - 1$. This modification of the storage pattern takes place in the IF statements just prior to the DO 20 loop.

There are ten scalar real arithmetic operations in the DO 20 loop, hence on a computer with a half-performance length of $n_{1/2}$, the time to execute this algorithm (which we call scheme A) is proportional to:

$$t_A = \sum_{l=0}^{\log_2(n)-1} 10(n_{1/2} + n2^{-(l+1)})2^l, \tag{5.92a}$$

$$= 10n_{1/2} \sum_{l=0}^{\log_2(n)-1} 2^l + 5n \sum_{l=0}^{\log_2(n)-1} 1, \tag{5.92b}$$

$$= 10n_{1/2}(n-1) + 5n\log_2 n. \tag{5.92c}$$

The last term in equation (5.92c) is the normal serial operation count for the FFT algorithm (take $n_{1/2} = 0$), and the first term shows the effect of hardware parallelism through the value of $n_{1/2}$.

It is obvious that the DO K1 and DO I1 loops of the above code can be interchanged without altering the effect of the statements in the DO 20 loop. If this is done, as is shown in figure 5.14, the DO K1 loop becomes the innermost and is replaced by vector instructions. The vector length is 2^l, or $1, 2, 4, \ldots, n/2$ in successive passes, and the storage interval is $n2^{-(l+1)}$, or $n/2, n/4, \ldots, 4, 2, 1$. Memory-bank conflicts are a serious problem in the early stages of this algorithm but can be avoided in the manner described above. The variable $\mathbf{W}$ is now a vector in the inner loop and the multiplication is a vector * vector instruction. The time to execute this alternative algorithm (scheme B) is proportional to:

$$t_B = \sum_{l=0}^{\log_2(n)-1} 10(n_{1/2} + 2^l)n2^{-(l+1)}, \tag{5.93a}$$

$$= 10n_{1/2} \sum_{l=0}^{\log_2(n)-1} n2^{-(l+1)} + 5n \sum_{l=0}^{\log_2(n)-1} 1. \tag{5.93b}$$

```
SUBROUTINE RECUR (F, G, W, L, N)
COMPLEX F(N), G(N), V(N), W(N)

      I2L  =  2**L
      I2L1 =  2*I2L
      N2L1 = N/I2L1

      INC1 = I2L
      IF (L.GE.2) INC1 = INC1+1
      INC2 = I2L1
      IF (L+1.GE.2) INC2 = INC2+1
      INC3 = N2L1*INC1-INC1
      INC4 = I2L-INC2

  DO 20 I1 = 1, N2L1

  DO 20 K1 = 1, I2L
      V(K1) = W(N2L1*K1-N2L1+1)*F(INC1*I1+K1+INC3)
      G(INC2*I1+K1-INC2) = F(INC1*I1+K1-INC1) + V(K1)
20    G(INC2*I1+K1+INC4) = F(INC1*I1+K1-INC1) - V(K1)

  RETURN
  END
```

FIGURE 5.14   Subroutine RECUR for scheme B with vectorisation of the DO K1 loop.

Setting $l' = \log_2(n) - 1 - l$, and reversing the order of the first summation one obtains:

$$t_B = 10n_{1/2} \sum_{l=0}^{\log_2(n)-1} 2^{l'} + 5n\log_2 n \qquad (5.93c)$$

$$= 10n_{1/2}(n-1) + 5n\log_2 n \qquad (5.93d)$$

$$= t_A. \qquad (5.93e)$$

We therefore find that, if taken to completion, the two alternative schemes execute in the same time. However they have different characteristics. Scheme A starts with long vectors which reduce in length as the algorithm proceeds, whereas scheme B starts with short vectors which increase in length as the algorithms proceed. The performance of any parallel computer improves as the vector length increases, hence a combined algorithm suggests itself. Perform the first p levels of the FFT using scheme A and the last $q - p$ levels using scheme B. This combined algorithm executes in a time proportional to

$$t_{A+B} = 10n_{1/2}\left(\sum_{l=0}^{p-1} 2^l + \sum_{l=p}^{\log_2(n)-1} n2^{-(l+1)}\right) + 5n\log_2 n, \qquad (5.94a)$$

$$= 10n_{1/2}(2^p + n2^{-p} - 2) + 5n\log_2 n. \qquad (5.94b)$$

By differentiating equation (5.49b) with respect to p, we find the condition for the minimum execution time

$$\frac{dt_{A+B}}{dp} = 10n_{1/2}\log_e 2(2^p - n2^{-p}) = 0. \qquad (5.95a)$$

Hence

$$2^p = n/2^p, \qquad 2^{2p} = n, \tag{5.95b}$$

and

$$p = \tfrac{1}{2}\log_2 n. \tag{5.95c}$$

With this optimal selection of p, the minimum vector length is $2^p = \sqrt{n}$ (or $\sqrt{\tfrac{1}{2}n}$ if n is not a multiple of four) and the execution time (assuming n is a multiple of four) is given by

$$t_{A+B} = 20n_{1/2}(\sqrt{n} - 1) + 5n\log_2 n. \tag{5.96}$$

This type of combined algorithm was developed independently by Roberts (1977) and Temperton (1979b). Temperton's version is based on the Gentleman–Sande recurrence (5.90) and forms the basis of the subroutine CFFT2 written in CAL assembler by Petersen (1978) for the CRAY X-MP computer scientific subroutine library.

Two further simplifications are usually incorporated in any efficient algorithm. These concern the value of the multipliers $\omega_{2^{l+1}}^k = \omega_n^{kn/2^{l+1}}$ that occur in both the Cooley–Tukey and the Gentleman–Sande recurrences. When $k = 0$ the multiplier is unity and an unnecessary multiplication by this value can be avoided by writing a separate loop for this case. In the calculation of the first level of partial transforms, when $l = 0$ in the recurrences (5.88), this occurs for every evaluation of equations (5.88). At other levels, it occurs at the first use of the equations. In the calculation of the second level of partial transforms, when $l = 1$ in equations (5.88), $k = 0, 1$ and the multipliers are 1 and i. Since multiplication by i merely interchanges the real and imaginary parts and changes the sign of the real part, no multiplications are required in the calculation of the second-level transformations either. Again a separate loop is justified.

Some arithmetic may be saved if the number of elements n being transformed contains factors of four (Singleton 1969). It is then advantageous to combine two applications of the recurrence (5.88) into a single recurrence involving four input values $^{(l-1)}\bar{f}_i^k$ and four output values $^{(l+1)}\bar{f}_i^k$. If we suppose n is a power of four and therefore $\log_2 n$ is even, the transform can be performed by evaluating the recurrence

for $l = 1, 3, 5, \ldots, \log_2(n) - 1$

for $k = 0, 1, \ldots, \tfrac{1}{2}2^l - 1; \qquad i = 0, 1, \ldots, n2^{-(l+1)}$

$$a = \omega_{2^{l+1}}^k \, {}^{(l-1)}\bar{f}_{i+\frac{1}{2}n2^{-l}}^k, \qquad b = \omega_{2^{l+1}}^{2k} \, {}^{(l-1)}\bar{f}_{i+n2^{-l}}^k, \qquad c = \omega_{2^{l+1}}^{3k} \, {}^{(l-1)}\bar{f}_{i+\frac{3}{2}n2^{-l}}^k$$

$$\tag{5.97a}$$

$$d = \overset{(l-1)}{f}{}^k_i + b, \qquad g = a + c \qquad\qquad (5.97b)$$

$$e = \overset{(l-1)}{f}{}^k_i - b, \qquad h = a - c \qquad\qquad (5.97c)$$

$$\overset{(l+1)}{f}{}^k_i = d + g, \qquad \overset{(l+1)}{f}{}^{k+\frac{1}{2}2^l}_i = e + ih \qquad\qquad (5.97d)$$

$$\overset{(l+1)}{f}{}^{k+2^l}_i = d - g, \qquad \overset{(l+1)}{f}{}^{k+\frac{3}{2}2^l}_i = e - ih. \qquad\qquad (5.97e)$$

The data flow diagram for the power-of-four transform ($n = 4^2$) is shown in figure 5.15. The evaluation of the recurrence (5.97) requires three complex multiplications in order to calculate a, b and c, and eight complex additions occur in equations (5.97b, c, d, e). That is 34 real operations for $\frac{1}{2}\log_2 n$ values of l and $n/4$ combinations of k and i. This is a total of $4.25 n \log_2 n$ real operations compared with $5n \log_2 n$ for the power-of-two transform, or a saving of 15%. When $l = 1$, $k = 0$, all the multipliers are unity and as in the case of the power-of-two transform, special code is justified to avoid unnecessary multiplications. The power-of-two and power-of-four transforms can easily be combined to give an efficient transform for any power of two, that first removes all factors of four from n and then, if necessary, removes a final factor of two.

5.5.4 Parallel implementation

On processor arrays there is a high premium in performance to be obtained if the parallelism of the algorithm can be made to match the hardware parallelism of the array—that is to say the number of processors in the array. The algorithms discussed in §5.5.3 have parallelism varying from $n/2$ to $\sqrt{n}$ and are therefore not very suitable for implementation on processor arrays which have a fixed number of processors and therefore fixed parallelism. If, however, the DO K1 loop and the DO I1 loop of figure 5.13 or figure 5.14 are combined into a single loop, all the arithmetic can be performed with a fixed parallelism of n. If this parallelism should match the array (e.g. transforming 4096 values on a 64×64 ICL DAP), then the ideal has been achieved.

We shall call the above algorithm PARAFT because it is the most suitable for execution on the paracomputer. The details of the subroutine RECUR are given in figure 5.16. The DO 20 loop contains no arithmetic and consists solely of data movements that are necessary to prepare for the arithmetic

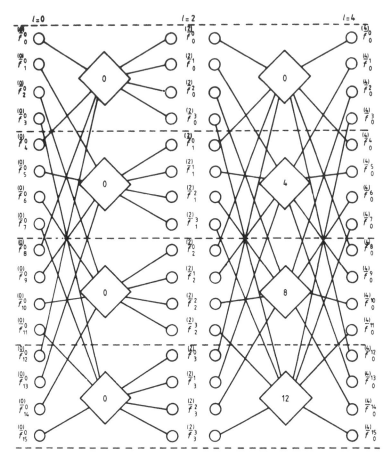

FIGURE 5.15 The data flow diagram for the power-of-four transform
for $n = 16$. The diamonds evaluate the recurrence (5.97). The number in
the diamond is the power of the nth root of unity used in the evaluation
of the constant a.

that is performed in the DO 30 loop. The complex arrays U and E are copies
of the data array F with a sign change corresponding to the negative sign
in equation (5.88b). The complex array V contains copies of the multipliers
$\omega_{2^{l+1}}^{k}$ in the correct place so that all n multiplications of equations (5.88) can
be performed in parallel in the first statement of the DO 30 loop. The second
statement of the DO 30 loop then performs all the n additions of equations
(5.88) in parallel.

As described in the previous section, there are no genuine multiplications
in the first two calls to RECUR, when $l = 0$ and 1. Special code for these
cases and the use of a power-of-four transform when possible would be
incorporated in an efficient program. The routing inherent in the DO 20

```
SUBROUTINE RECUR (F, G, W, L, N)
COMPLEX F(N), G(N), V(N), W(N), U(N), E(N)

      I2L   = 2**L
      I2L1  = 2*I2L
      N2L1  = N/I2L1

      INC1  = I2L
      INC2  = I2L1
      INC3  = N2L1*INC1-INC1
      INC4  = I2L - INC2

   DO 20 K1 = 1, I2L
   DO 20 I1 = 1, N2L1

      U(INC2*I1+K1-INC2) = F(INC1*I1+K1-INC1)
      U(INC2*I1+K1+INC4) = F(INC1*I1+K1-INC1)
      E(INC2*I1+K1-INC2) = F(INC1*I1+K1+INC3)
      E(INC2*I1+K1+INC4) = -F(INC1*I1+K1+INC3)
      V(INC2*I1+K1-INC2) = W(N2L1*K1-N2L1+1)
20    V(INC2*I1+K1+INC4) = W(N2L1*K1-N2L1+1)

   DO 30 J = 1, N
      E(J) = V(J)*E(J)
30    G(J) = U(J)+E(J)

   RETURN
   END
```

FIGURE 5.16   Subroutine RECUR for PARAFT with vector length N in the DO 30 loop.

loop must be considered when assessing the time of execution of the algorithm. In processor arrays where arithmetic is slow compared with routing, such as the ICL DAP, the DO 20 loop will not consume a major part of the execution time. For example, data routing accounts for 10–20% of the overall time when performing 1024 complex transforms on a 32×32 ICL DAP (Flanders *et al* 1977). On arrays of more powerful processors the DO 20 routing loop will assume relatively more importance. Several processor arrays have special routing connections and circuitry designed to perform efficiently the routing necessary in the fast Fourier transform. Examples are the Goodyear STARAN and the Burroughs BSP (see Chapter 3).

Examination of figure 5.16 shows that the PARAFT algorithm requires one complex multiplication and one complex addition to be performed at each level with parallelism n. This is equivalent to eight real operations with parallelism n at $\log_2 n$ levels or, assuming routing is unimportant, an execution time proportional to

$$t_{\text{PARAFT}} = 8(n_{1/2} + n)\log_2 n. \tag{5.98a}$$

By comparing equation (5.98a) with the time for the A + B scheme in equation (5.96) we find that PARAFT has a higher performance than A + B if

$$n_{1/2} > 3n\log_2 n / [20(\sqrt{n} - 1) - 8\log_2 n]. \tag{5.98b}$$

The regions of the $(n, n_{1/2}^{-1})$ plane favoured by the two algorithms on this

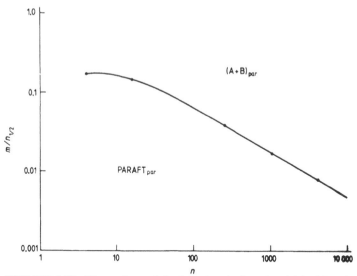

FIGURE 5.17 The regions of the $(n, m/n_{1/2})$ plane in which either the $(A + B)$ scheme or the PARAFT algorithm has the higher performance when applied in parallel to m fast Fourier transforms each of length n. When $m = 1$ the graph applies to the selection of the best algorithm for a single transform.

basis are shown in figure 5.17 (with $m = 1$). The above considerations apply accurately to pipelined computers and to processor arrays in an average sense (see §1.3.3). Other alternatives exist, and Jesshope (1980a) has shown that a combination of the two methods is advantageous on processor arrays when the transform length exceeds the array size. In this case the length of the transform is successively halved using the $A + B$ scheme, until the transfer length is less than the array size. At this stage the PARAFT algorithm is used to complete the transform.

If faced with the problem of calculating m independent transforms each of length n, the approach will depend on the amount of parallelism desired. We first consider the methods which have the maximum parallelism and compare the performance of the $A + B$ scheme and PARAFT when applied in parallel across the m transforms. These methods are suitable for machines with large natural parallelism, for example the CYBER 205 and the ICL DAP. The execution times will be proportional to

$$t_{(A + B)_{par}} = 20n_{1/2}(\sqrt{n} - 1) + 5mn \log_2 n, \tag{5.99a}$$

$$t_{PARAFT_{par}} = 8(n_{1/2} + mn) \log_2 n. \tag{5.99b}$$

Compared with equation (5.4b) these have average vector lengths of

$$\bar{n}_{(A+B)_{par}} = \frac{mn \log_2 n}{4(\sqrt{n}-1)}, \tag{5.99c}$$

$$\bar{n}_{PARAFT_{par}} = mn. \tag{5.99d}$$

Equations (5.99) show that the $(A+B)_{par}$ algorithm has the better performance when

$$\frac{m}{n_{1/2}} > \frac{20(\sqrt{n}-1) - 8 \log_2 n}{3n \log_2 n}. \tag{5.99e}$$

The regions of the $(n, m/n_{1/2})$ plane favouring the two methods are shown in figure 5.17. We see that the $(A+B)_{par}$ scheme is favoured whenever either the number of transformations or the length of the transforms becomes large.

If, alternatively, there is advantage to restricting the parallelism of the algorithm to approximately n or m, then we can ask whether it is advantageous to repeat the best algorithm for a single transform m times, or to take the best serial algorithm and perform it in parallel on the m systems (the MULTFT algorithm). The best serial algorithm is either scheme A or scheme B with $n_{1/2} = 0$ and requires $5n \log_2 n$ real scalar arithmetic operations. The MULTFT algorithm has average vector length $\bar{n}_{MULTFT} = m$, and will execute in a time proportional to:

$$t_{MULTFT} = 5n(n_{1/2} + m) \log_2 n. \tag{5.100a}$$

When applied sequentially m times, on the m different systems to be transformed, the other algorithms will execute in a time proportional to:

$$t_{(A+B)_{seq}} = m[20n_{1/2}(\sqrt{n}-1) + 5n \log_2 n], \tag{5.100b}$$

$$t_{PARAFT_{seq}} = 8m(n_{1/2} + n) \log_2 n, \tag{5.100c}$$

and with average vector lengths

$$\bar{n}_{(A+B)_{seq}} = \frac{n \log_2 n}{4(\sqrt{n}-1)}, \qquad \bar{n}_{PARAFT_{seq}} = n. \tag{5.100d}$$

From these timings we conclude that the MULTFT algorithm will have a superior performance to the $(A+B)_{seq}$ scheme if

$$m > \frac{n}{4} \frac{\log_2 n}{(\sqrt{n}-1)} \tag{5.101a}$$

and will have a superior performance to the $PARAFT_{seq}$ algorithm if

$$\frac{m}{n_{1/2}} > \frac{0.625n/n_{1/2}}{1 + 0.375n/n_{1/2}}. \tag{5.101b}$$

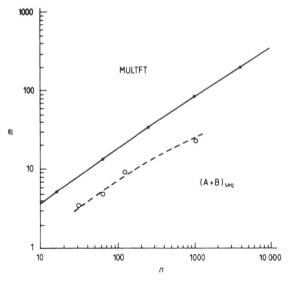

FIGURE 5.18   The selection of the best algorithm for calculating m fast Fourier transforms each of length n, on a computer with natural parallelism approximately m or n, for which the A + B scheme is the best method for calculating a single transform (see figure 5.17). The best serial FFT is applied in parallel to the transforms, MULTFT, or the A + B scheme is applied sequentially, $(A + B)_{seq}$. In this case a single curve applies for all values of $n_{1/2}$, and divides the regions of the (n, m) plane in which either the MULTFT or $(A + B)_{seq}$ scheme has the better performance. The open circles are interpolated from the measured results of Temperton (1979b) on the CRAY-1.

The full curve in figure 5.18 is a plot of equation (5.101a) and shows the region of the (n, m) plane in which either the MULTFT or the $(A + B)_{seq}$ scheme has the superior performance. We note that equation (5.101a) is independent of $n_{1/2}$ and a single curve applies for all computers. We conclude, roughly speaking, that if the number of systems to be transformed exceeds one tenth of the length of the systems, then it is advantageous to use the best serial algorithm on all systems in parallel, rather than the best parallel algorithm on each system in sequence. Temperton (1979b) compared the actual performance of two FORTRAN codes on the CRAY-1 for algorithms comparable to MULTFT and the $(A + B)_{seq}$ scheme. His measurements are shown as the open circles and the broken line. The trend of Temperton's measurements agrees with the predictions of the simple theory given above, and would be in absolute agreement if the predicted speed of MULTFT relative to the $(A + B)_{seq}$ scheme were increased by a factor between two and three. Such relative behaviour might be expected because the MULTFT algorithm has much simpler indexing than the $(A + B)_{seq}$ scheme (the increment of the innermost vectorised loop over the m systems is always unity) and memory is therefore accessed in the most favourable fashion.

Figure 5.19 is a similar comparison between MULTFT and the sequential application of the PARAFT algorithm. It is for use when we conclude from figure 5.17 that PARAFT is the best algorithm for performing a single transform. The following limiting values may be useful:

$$m = \begin{cases} 0.625n, & n \ll n_{1/2} \qquad\qquad (5.102a) \\ 0.454n_{1/2}, & n = n_{1/2} \qquad\qquad (5.102b) \end{cases}$$

and for large n a limiting value of m is reached

$$m = 1.67n_{1/2}, \qquad n \gg n_{1/2}. \qquad\qquad (5.102c)$$

5.5.5 Routing considerations
The above performance comparisons have been made on the assumption that routing delays (see §5.1.4) do not make an important contribution to

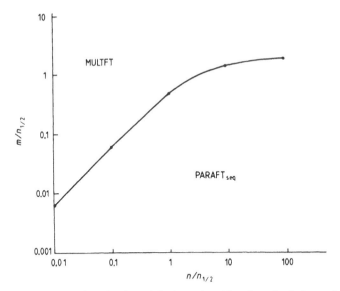

FIGURE 5.19   The selection of the best algorithm for calculating m fast Fourier transforms each of length n, on a computer with a natural parallelism of approximately n or m, and for which the PARAFT algorithm is the best method of calculating a single transform (see figure 5.17). PARAFT is applied sequentially to the m transforms, PARAFT$_{seq}$. The curve shows the boundary in the $(n/n_{1/2}, m/n_{1/2})$ plane between regions in which either the MULTFT or PARAFT$_{seq}$ method has the better performance.

the execution time. Such delays only affect the performance of processor arrays on which the PARAFT algorithm is most likely to be the best from the point of view of arithmetic operations (see figure 5.17). We will therefore consider the routing problem for this algorithm. The routing delay will obviously depend on the connectivity pattern between the processors of the array. We will therefore consider a general class of such processor connections, which includes most of the processor arrays that have been manufactured. The treatment here follows that of Jesshope (1980a) who gives a comprehensive analysis of the implementation of fast radix-2 transforms on processor arrays. Further results on routing and transpositions in processor arrays are given by Jesshope (1980b, c). In addition to the above, Nassimi and Sahni (1980) consider the implementation of bit-reversal and the perfect shuffle.

Let us consider a processor array with P processors, arranged in a cartesian k-dimensional array with Q processors in each coordinate direction, then

$$P = Q^k \tag{5.103a}$$

and, in accordance with normal practice, we take Q to be a power of 2

$$Q = 2^q, \qquad q = 1, 2, \ldots. \tag{5.103b}$$

Each processor has access to data in its own memory and that of its immediate neighbours in each coordinate direction. Note, in particular, that diagonal connections are not present. For the case $q \geq 2$ this connectivity pattern requires $2k$ connections to each processor and kP data paths for the whole array, if we assume that a data path can be used for transferring data in either direction. For the case $q = 1$ the above counts are halved. At the edges of the array the processors are connected in a periodic sense in each dimension. If l represents the coordinate direction and i_l the coordinate in the lth direction, periodicity means that all coordinates are interpreted mod Q†, hence

$$0 \leq i_l \leq Q - 1, \qquad l = 1, 2, \ldots, k, \tag{5.104a}$$

and nearest-neighbour connectivity means that processors are connected if their coordinates differ by ± 1 in one, and only one, coordinate direction.

We wish to use the above k-dimensional array as a store for a one-dimensional array $\{f_i, i = 1, \ldots, n\}$ of data to be transformed. To avoid complication we shall assume that the number of processors equals the number of data. Numbering the processors sequentially along the 1st, 2nd to the kth dimension we have the correspondence

$$i = i_1 + i_2 Q + i_3 Q^2 + \ldots + i_k Q^{k-1}, \tag{5.104b}$$

† mod is short for modulo.

and nearest-neighbour connectivity means that processor i is connected to:

$$(i \pm 1) \bmod Q, (i \pm Q) \bmod Q^2, \dots, (i \pm Q^{k-1}) \bmod Q^k. \qquad (5.104c)$$

Computers that fall into the above class of connectivity are:

(a) ILLIAC IV $P = 64$, $k = 2$, $Q = 8$, $q = 3$, and processor i connects with $(i \pm 1, i \pm 8) \bmod 64$. Note that although the internal connections are two-dimensional, the edge connections treat the array as a periodic one-dimensional structure. For this reason the machine is not strictly speaking in the class;

(b) ICL DAP $P = 4096$, $k = 2$, $Q = 64$, $q = 6$, and processor i connects with $(i \pm 1) \bmod 64$, $(i \pm 64) \bmod 4096$;

(c) HYPERCUBE (binary) $P = 16$, $k = 4$, $Q = 2$, $q = 1$, and processor i connects with $(i \pm 1) \bmod 2$, $(i \pm 2) \bmod 4$, $(i \pm 4) \bmod 8$, $(i \pm 8) \bmod 16$.

We see that the problem of providing kP data paths limits the dimensionality of practical arrays to two when the number of processors is large. The highest level of connectivity is given by the Hypercube (Millard 1975) which takes $Q = 2$, 3 or 4 and $k = 4$. Computers that do *not* fall into the above class are those that provide diagonal connections within the network, for example CLIP at University College, London (Duff 1978, 1980a, 1980b) that provides connections to the nearest eight processors in a two-dimensional array, whereas our definition only allows connections to the nearest four processors.

In the PARAFT algorithm the routing takes place entirely in the DO 20 loop of figure 5.16 in which a partial reordering takes place at every stage In the first and third statements, values of **F** separated by $\text{N2L1} = n/2^{l+1}$ are shifted to lie under each other. A similar shift and negation takes place in the second and fourth statement. If we take the case of the Cooley–Tukey algorithm without reordering, the PARAFT algorithm can be simplified and rephrased as follows:

For $l = 0, 1, \dots, \log_2(n) - 1$ do $\qquad$ (5.105a)

(a) Disable even groups of $n/2^{l+1}$ processors and multiply by a prestored array of phase factors.

(b) Route all data $n/2^{l+1}$ nodes through the linear connectivity network modulo $2n/2^{l+1}$ and store in a temporary workspace. If no cyclic connections exist in the network modulo $2n/2^{l+1}$, then this step must be performed in two masked steps, routing the odd and even groups of $n/2^{l+1}$ elements in opposite directions.

(c) Disable odd groups of $n/2^{l+1}$ processors and change the sign of the work space.

(d) Add the work space to the data array.

End

In the above formulation the numbers of unit routing operations are

$$n/2, n/4, n/8, \ldots, 4, 2, 1 \tag{5.105b}$$

as l takes on the value 0 to $\log_2(n) - 1$. If the P processors are connected in a linear array (i.e. $k = 1$) then the total number of routing operations is

$$R^{(1)} = \frac{n}{2} + 2\left(\frac{n}{4} + \frac{n}{8} \ldots, +2 + 1\right), \tag{5.106a}$$

$$= \frac{n}{2} + 2\left(\frac{n}{2} - 1\right) = \frac{3n}{2} - 2. \tag{5.106b}$$

The first term in equation (5.106a) takes into account that, for $l = 0$, the periodicity of the array allows the required routing to be achieved by a single routing to the left or right by $n/2$ places. The multiplier two in the second term takes into account that in step (b) above, the odd and even groups are routed in separate operations.

If the array is multidimensional and, as we are assuming, the data fills the processor array exactly, then the first relative movement of data by $n/2$ vector elements can be achieved by a routing of $Q/2$ along the kth dimension, the second movement of $n/4$ by a routing of $Q/4$ in the kth dimension. The successive routings by half the previous value can be continued in the kth dimension until the routing is unity. To route by half this amount we must now turn to motion in the $(k-1)$th dimension by an amount $Q/2$ which again can be continued until the routing is unity. To route by half this amount we must now turn to motion in the $(k-2)$th dimension and so on until all the dimensions have been used up. This is illustrated for a 16×16 array of processors in figure 5.20. Obviously the above process requires routings equal to equation (5.106a), but with Q replacing n, for each of the k dimensions. Accordingly the total number of routings of complex numbers for k-dimensional connectivity is

$$R^{(k)} = k(3Q/2 - 2). \tag{5.107a}$$

The minimum number of such complex routings occurs if $Q = 2$, that is to say for a binary hypercube, in which case

$$R^{(k)} = k = \log_2 P = \log_2 n. \tag{5.107b}$$

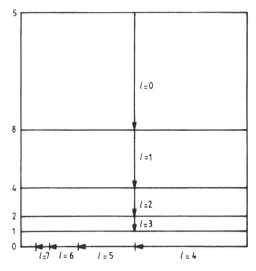

FIGURE 5.20 The magnitude and direction of the routings required for the evaluation of a transform of $2^8 = 256$ values on a 16×16 processor array. $l = 0, 1, \ldots, 7$ is the loop control variable in statement (5.105a). The level of the partial transform calculated after the shift is $l + 1$.

The numbers of complex routing operations that are required by the other computers are, using equation (5.107a):

(a) 'ILLIAC IV'†

$$P = n = 64, \; Q = 8, \; k = 2,$$

$$R^{(2)} = 20 = 3.3 \log_2 n.$$

(5.108a)

(b) ICL DAP

$$P = n = 4096, \; Q = 64, \; k = 2,$$

$$R^{(2)} = 188 = 15.7 \log_2 n,$$

(5.108b)

or since $k = (\log_2 n)/q$ for the special case of $n = P$

$$R^{(k)} = \frac{3Q/2 - 2}{q} \log_2 n.$$

(5.108c)

The number of real parallel arithmetic operations after the above parallel complex routings is $8 \log_2 n$ [see equation (5.98)]. Therefore, if we assume

† The quotes indicate a hypothetical ILLIAC IV-like machine with two-dimensional periodic edge connectivity.

that a parallel routing operation is γ times faster than a parallel arithmetic operation, then again for $n = P$

$$\frac{t_R}{t_A} = \frac{\text{time spent in routings}}{\text{time spent on arithmetic}} = \frac{2(3Q/2 - 2)}{8\gamma q}, \tag{5.109a}$$

where the leading factor of two takes into account that each complex routing in equation (5.108c) is a movement of two real numbers. For the computers under consideration, we have approximately:

(a) 'ILLIAC IV'

$$\gamma = 2, \qquad t_R/t_A = 42\%, \tag{5.109b}$$

(b) ICL DAP

$$\gamma = 20, \qquad t_R/t_A = 19\%, \tag{5.109c}$$

(c) HYPERCUBE

$$\gamma = 10, \qquad t_R/t_A = 2.5\%. \tag{5.109d}$$

It appears therefore that routing delays, although significant, do not dominate the calculation of complex Fourier transforms on practical designs of processor arrays, and may justifiably be ignored in a first estimate of the performance of fast transform algorithms. However, we see in §5.5.6 that the reverse is true for the calculation of number theoretic transforms on certain types of processor arrays.

In general, routing will dominate over arithmetic when $t_R/t_A \geqslant 1$ or, recalling that $q = \log_2 Q$, when

$$3Q/2 - 2 \geqslant 4\gamma \log_2 Q, \tag{5.110a}$$

or for large Q when

$$Q/\log_2 Q \geqslant 2.67\gamma. \tag{5.110b}$$

Then to the nearest power of two we have:

$$Q \geqslant \begin{cases} 32 & \text{if } \gamma = 2 \\ 64 & \text{if } \gamma = 4 \\ 256 & \text{if } \gamma = 10 \\ 512 & \text{if } \gamma = 20 \end{cases} \tag{5.110c}$$

This result, which is shown in figure 5.21, is independent of the dimensionality of the processor array, and states that it is unlikely to be worthwhile, because of routing delays, to build arrays with a linear dimension greater than a few hundred, depending on the value of γ. Indeed the more complex the processors

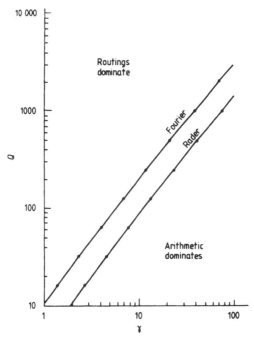

FIGURE 5.21   Linear dimension Q of a processor array above which routing delays exceed the arithmetic time in the FFT, plotted as a function of γ, the ratio between the time for an arithmetic operation to the time for a routing operation. Curves for the calculation of the Fourier transform in complex arithmetic and for the Rader transform in integer modulo arithmetic are shown.

and hence the smaller the value of γ, the smaller the arrays must be kept to prevent routing from dominating the execution time. Small values of γ will also occur in bit-organised computers such as the ICL DAP, when they are used for short-word-length integer arithmetic (say eight-bit words when $\gamma \approx 2$) as might arise in picture processing (see §5.5.6).

The above result is important because the advent of very large-scale integration technology, with 10^4 or more logic elements per chip, makes possible the manufacture of very large arrays. Such large arrays are attractive from the cost point of view, and may apparently give very high performance. However, the above result shows that, when one considers actual algorithms, one must be aware of the limitations when the linear dimensions become large. Although we have specifically discussed the routing problem for the fast Fourier transform algorithm, very similar routings occur in many parallel algorithms — e.g. the parallel algorithm for the evaluation of a first-order recurrence given in §5.2.2. The result must therefore be considered to be fairly general.

In order to concentrate on essentials, the above discussion of routing has assumed that one is performing a single one-dimensional transform of length equal to the number of available processors. For a discussion of routing delays for multiple and multi-dimensional transforms which do not necessarily match the size of the processor array, the reader is referred to the original papers, particularly Jesshope (1980a, b).

5.5.6   Number theoretic transforms

Number theoretic transforms, in particular the Fermat number transform, have important applications in picture processing, and in the solution of partial differential equations (see §5.6.2) as was first considered by Eastwood and Jesshope (1977). These transforms have also been advantageously employed in digital signal processing by Pollard (1971), Rader (1972) and Agarwal and Burrus (1974, 1975). They play an analogous role to the Fourier transform when arithmetic is conducted modulo $F_t = 2^{2^t} + 1$, the tth Fermat number, and functional values are confined to the integers $0, 1, \ldots, F_t - 1$ (technically a ring of integers). In this case one may define the number theoretic transforms as

$$\hat{x}_k = \sum_{j=0}^{n-1} \alpha^{-kj} x_j \pmod{F_t}, \tag{5.111a}$$

which has the inverse

$$x_j = n^{-1} \sum_{k=0}^{n-1} \alpha^{kj} \hat{x}_k \pmod{F_t}, \tag{5.111b}$$

where α is the nth root of unity (modulo F_t). Thus

$$\alpha^n = 1. \tag{5.111c}$$

Comparing equations (5.111) with the definition of the finite Fourier transform (5.73) and (5.74), we see that α has the same properties in the ring of integer values as does ω in the field of complex numbers. Consequently all the fast algorithms that were developed for the Fourier transform in §5.5.2 to §5.5.5 can be applied equally well to the calculation of the number theoretic transform, provided that the arithmetic is conducted modulo F_t. Note however that it is usual to associate the scale factor n^{-1} with the inverse transform (5.111b). This is the opposite convention to that used for the Fourier transform, in which the factor was associated with the forward transform (5.71a).

The integer functions α^{kj} and α^{-kj} are orthogonal, like the functions $\omega^{\pm kj}$,

(a)

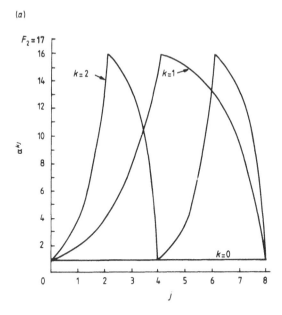

(b)

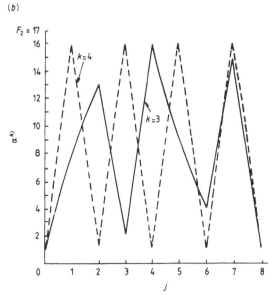

FIGURE 5.22 The discrete orthogonal functions α^{kj} used in the number theoretic Rader transform ($\alpha = 2$) for $t = 2$. Functional values are in the range $0, 1, \ldots, F_2 - 1 = 16$, and are represented by five bits. The transform length is $n = 8$ ($n^{-1} = 15$), and $k = 0, 1, \ldots, n - 1 = 7$. Functions for $k \geqslant 4$ are the mirror image about the vertical line $j = 4$ of the function that is drawn for $8 - k$ (i.e. $\alpha^{kj} = \alpha^{(n-k)(n-j)}$).

as may easily be proved:

$$\sum_{j=0}^{n-1} \alpha^{kj} \alpha^{-k'j} = \sum_{j=0}^{n-1} \alpha^{(k-k')j}. \tag{5.112a}$$

Summing the geometric series and using the fact that $\alpha^n = 1$, we obtain

$$\sum_{j=0}^{n-1} \alpha^{kj} \alpha^{-k'j} = \frac{\alpha^{(k-k')n} - 1}{\alpha^{k-k'} - 1} = 0, \qquad \text{if } k \neq k'. \tag{5.112b}$$

Returning to the definition (5.112a) we obtain directly

$$\sum_{j=0}^{n-1} \alpha^{kj} \alpha^{-k'j} = \sum_{j=0}^{n-1} (\alpha^0)^2 = \sum_{j=0}^{n-1} 1 = n, \qquad \text{if } k = k'. \tag{5.112c}$$

This property is analogous to the orthogonality relation (5.72b). The functions α^{kj} are plotted for the case $t = 2$ in figure 5.22.

The most useful types of Fermat number transforms arise when α is a power of two, when they are called the Rader (1972) transform (RT). In this case all the multiplications by powers of α in equation (5.111) may be achieved by shifting, and especially rapid computer codes can be written. There is also a relation between the number of binary digits (bits) that are used in the number representation and the length n of the transform. Integers in the range $0, 1, \ldots, F_t - 1 = 2^b$, where $b = 2^t$, require $b + 1$ bits for their representation, because the number $2^b = -1 \bmod F_t$ (a one followed by b zeros) is included in the range. All other numbers in the range may be represented by b bits.

If $\alpha = 2$ the length of the transform is, in general,

$$n = 2b = 2^{t+1}, \tag{5.113a}$$

and its reciprocal, that is required in equation (5.111b), is given by

$$n^{-1} = F_t - 2^{b-(t+1)}. \tag{5.113b}$$

If the binary digits are numbered from 0 to b, starting at the least significant digit, n^{-1} has ones in the zeroth bit and in bits $b - 1$ to $b - 1 - t$ inclusive. To prove equation (5.113b) we calculate

$$nn^{-1} = 2^{t+1}(F_t - 2^{b-(t+1)}) = 2^{t+1}F_t - 2^b,$$

and, since $2^b = F_t - 1$,

$$nn^{-1} = 2^{t+1}F_t - (F_t - 1)$$

$$= (2^{t+1} - 1)F_t + 1 = 1 \bmod F_t. \tag{5.113c}$$

Some particular cases of interest are:

$$\alpha = 2, t = 3, b = 8, \text{bits} = 9, n = 16, n^{-1} = 2^8 + 1 - 2^4, \qquad (5.114\text{a})$$

$$\alpha = 2, t = 4, b = 16, \text{bits} = 17, n = 32, n^{-1} = 2^{16} + 1 - 2^{11}, \quad (5.114\text{b})$$

$$\alpha = 2, t = 5, b = 32, \text{bits} = 33, n = 64, n^{-1} = 2^{32} + 1 - 2^{26}, \quad (5.114\text{c})$$

$$\alpha = 2, t = 6, b = 64, \text{bits} = 65, n = 128, n^{-1} = 2^{64} + 1 - 2^{57}. \quad (5.114\text{d})$$

The number theoretic transform can be performed on real variables if they are first scaled to the allowed integer range, and then discretised to the nearest integer value. In this case the discretisation error is about 2^{-b}, and for higher accuracy one must use larger b. If the length of the transform is then too long, it may be halved by squaring α; then $\alpha = 4$, $n = b$ and $n^{-1} = F_t - 2^{b-t}$. For example,

$$\alpha = 4, t = 6, b = 64, \text{bits} = 65, n = 64, n^{-1} = 2^{64} + 1 - 2^{58}, \quad (5.115\text{a})$$

and by repeating the process

$$\alpha = 16, t = 6, b = 64, \text{bits} = 65, n = 32, n^{-1} = 2^{64} + 1 - 2^{59}. \quad (5.115\text{b})$$

In this way it should be possible to choose a precision b, transform length n and root of unity α, suitable for most circumstances.

The usefulness of the Fermat number transform depends critically on the efficiency with which the necessary arithmetic modulo F_t can be performed. Many computers perform two's complement arithmetic which in b bits is arithmetic modulo $2^b = F_t - 1$. To convert this to arithmetic modulo F_t, we must subtract one from the sum of two numbers whenever a carry is detected from the highest-order bit, except in the case when bits 0 to $b - 1$ are all zero. This latter case is a valid result and represents 2^b or $-1 \bmod F_t$. If the computer only has b bits in its integer number representation some other method, such as an auxiliary logical variable associated with each number, must be used to represent this number. Similarly we must add one to the two's complement negative of a number, or to the difference of two numbers if the result is negative. Multiplication by an arbitrary number modulo F_t is performed by subtracting the high-order b bits of the $2b$-bit product from the lower-order b bits. Note that $F_t = 2^b + 1$ contains only two bits, a one in the least-significant-bit position of the low-order b bits and a one in the least-significant-bit position of the high-order b bits. Thus the above operation is the same as normalising the product to the allowed integer range by subtracting F_t until the high-order bits are all zero. Again a special case must be made for the occurrence of 2^b and for multiplying by 2^b. Multiplying by powers of α is performed by shifting followed by normalisation as described above.

It is clear that arithmetic modulo F_t requires several tests and special cases, and it is unlikely that the number theoretic transforms will have any speed advantages over the Fourier transform on computers that provide hardware floating-point arithmetic. On bit-serial and bit-addressable computers such as the ICL DAP or the Goodyear STARAN, which do not provide hardware floating-point, the situation is radically different, for in these cases special arithmetic routines can be microcoded at the bit level for modulo arithmetic. In particular the shifts apparently involved when multiplying by α can be avoided by addressing the appropriate bits in the memory and any multiplication by α^{kj} takes no longer than the single subtraction that is necessary for normalisation. The comparison between modulo F_t and floating-point arithmetic on the ICL DAP is shown in table 5.1. It is clear that modulo arithmetic is about 10 times faster than floating-point on this machine.

If the Rader transform is computed by the same technique as was used in the PARAFT method for the Fourier transform (see §5.5.4), there will be one parallel addition operation and one parallel multiplication operation (by a power of two) for each of the $\log_2 n$ levels of the algorithm. The total number of modulo F_t arithmetic operations, each with parallelism n, for the k-dimensional array of $n = Q^k = 2^{qk}$ processors of §5.5.5, is therefore

$$A = 2\log_2 n = 2kq. \tag{5.116a}$$

The routing requirements are identical to those in the PARAFT algorithm, therefore

$$R = k(3Q/2 - 2), \tag{5.116b}$$

TABLE 5.1   Comparison of modulo $F_5 = 2^{32} + 1$ and real floating-point arithmetic (IBM 32-bit format) on the ICL DAP. Time in μs for 4096 results on a 64×64 array of processors.

Operation	Modulo F_5†	Real floating-point‡
+	36	150
−	30	150
×	24§	250
Route	10	10

† From Eastwood and Jesshope (1977).
‡ From Reddaway (1979).
§ Multiplication by powers of 2.

and the ratio of the time spent in routing to that spent on arithmetic is, for the Rader transform,

$$t_R/t_A = \frac{3Q/2 - 2}{2\gamma q},$$ (5.116c)

where γ is the ratio of the time for a modulo F_t arithmetic operation to the time for a unit routing operation. Referring to table 5.1, we see that $\gamma = 3$ for 33-bit working with $t = 5$ on the 64×64 ICL DAP ($k = 2$, $Q = 64$, $q = 6$), leading to

$$R = 188, \qquad A = 24, \qquad R/A = 7.8,$$ (5.117a)

and

$$t_R/t_A = 261\%.$$ (5.117b)

Hence routings dominate the arithmetic time for the calculation of Rader transforms on the ICL DAP and similar machines.

The time spent on routing will equal or exceed the time spent on arithmetic when

$$3Q/2 - 2 \geqslant 2\gamma q,$$ (5.118a)

or for large Q when

$$Q/\log_2 Q \geqslant 1.33\gamma.$$ (5.118b)

The curve for equality and the regions of the $Q - \gamma$ plane in which routing or arithmetic dominate are shown in figure 5.21. It can be seen that the linear dimensions of processor arrays should be kept as small as possible if the Rader transform is to be efficiently calculated. Since γ is likely to be less than four for modulo arithmetic, linear dimensions greater than 32 will lead to algorithms that are dominated by the time spent in rearranging the data in store, rather than performing useful calculations on the data.

The above analysis is based on a transform equal to the size of the processor array. However in solving a three-dimensional partial differential equation, the transform is likely to be much larger than the array size, in which case the overheads due to routing decrease dramatically. For example, the 261% overhead given in equation (5.117b) for a 64×64 transform becomes only 25% for a $64 \times 64 \times 64$ transform on a 64×64 ICL DAP (Jesshope 1980a).

Notwithstanding the fact that routing may dominate arithmetic, the overall performance of the Rader transform far exceeds that of the Fourier transform on machines such as the ICL DAP. If we compare the amount of arithmetic needed to transform n integer values in the Rader transform with that needed to transform n real values (i.e. $n/2$ complex values) in the Fourier transform, we find $2 \log_2 n$ parallel operations compared with $4 \log_2 n$. Furthermore we have seen in table 5.1 that each modulo arithmetic operation in the Rader transform is about 10 times faster than the corresponding floating-point operation in the Fourier transform. On this basis a maximum speed advantage of about 20 can be anticipated for the Rader transform. Actual values will be less than this maximum because of the overhead due to routing delays which is the same for both transforms. For example, in a three-dimensional transform of $32 \times 32 \times 16$ values on a 32×32 ICL DAP using the same fast radix-2 algorithm, a Rader transform took 50 ms and a Fourier transform 700 ms (Eastwood and Jesshope 1977). This is a speed ratio of 14 in favour of the Rader transform and is consistent with the above estimates.

5.6   PARTIAL DIFFERENTIAL EQUATIONS

In considering the solution of partial differential equations (PDES) on parallel computers, we can draw upon the results obtained earlier in this chapter for the solution of tridiagonal systems (see §5.4) and the calculation of transforms (see §5.5). Almost all of the sound numerical techniques for the solution of PDES, which have been used successfully since the 1950s on serial computers, involve the repeated solution of independent tridiagonal systems or the repeated transformation of independent sets of data. Such methods are well suited to efficient implementation on parallel computers because the independent systems or independent transforms may be computed in parallel. Thus the problem is not to insert parallelism into a known sequential method, as has been the case in previous sections of this chapter, but rather to match the natural parallelism of the computer to the existing parallelism of the algorithm. For example, the algorithm for the solution of a three-dimensional problem on an $n \times n \times n$ mesh can be expressed in terms of operations with a parallelism of either n, n^2 or n^3. The selection of the most suitable technique for a particular computer involves recognising which of these three levels of parallelism matches most closely the natural parallelism of the computer.

We describe first (§5.6.1) the common iterative (or relaxation methods) in two dimensions because these can be applied to the solution of the most general linear PDE with arbitrary coefficients. We make the important point that the method of simultaneously adjusting all points (Jacobi method), which has the maximum parallelism of n^2, cannot be used because of its impossibly slow convergence. This is a warning that in one's quest for methods with the maximum parallelism, one should not ignore the long established results of numerical analysis on convergence as given, for example, by Varga (1962) or Forsythe and Wasow (1960). The method of successive over-relaxation with odd/even ordering and Chebyshev acceleration is recommended because of its vastly superior convergence properties, even though the parallelism is halved. This method may be applied by points (SOR) or by lines (SLOR), and it is interesting that we find the former method most favourable on machines such as the ICL DAP and the latter on machines such as the CRAY X-MP. The more usual algorithm of SOR with typewriter ordering is at first sight a sequential process. However, Hunt (1979) has shown how it may be applied in parallel by operating on alternate diagonal lines in the manner of a pipeline. The parallelism approaches $n^2/2$ if the number of iterations is much greater than n. The parallel implementation of the SOR method has also been discussed by Adams and Ortega (1982), and Evans (1984a, b). The application of multigrid iterative methods on parallel computers has been investigated by Grosch (1979). Another good iterative method that we discuss is the alternating direction implicit method (ADI). Parallel marching methods are considered by Vajtersic (1984), and vector implementations by Schönauer (1987).

The solutions of simple PDES with constant coefficients (for example Poisson's equation $\nabla^2 \phi = \rho$) in simple regions (such as the square or rectangle) with simple boundary conditions (given value, slope or periodicity) have important applications in many areas of physics and engineering. Especially fast direct (i.e. non-iterative) methods, based on the FFT, are available for the solution of this class of problem and these are described in §5.6.2. The speed of these methods makes possible the time-dependent simulation of, for example, stars in galaxies, electrons in semiconductor devices, and atoms in solids and liquids (see Hockney and Eastwood 1988). Again the algorithms, which were first developed on serial computers, are inherently parallel and may be efficiently implemented on parallel computers without alteration. Such methods have been extended to the biharmonic equation by Vajtersic (1982) and implemented on the EGPA computer (§1.1.8).

The first two sections on iterative and direct methods are written in terms of two-dimensional problems, both for simplicity of presentation and because such problems are well within the capabilities of current, c1988, computers and are therefore most frequently posed. The extra speed associated with the advent of the parallel computer, however, for the first time makes the solution of three-dimensional problems with reasonable resolution (say 64^3 mesh) a practical proposition. We therefore consider in the last section (§5.6.3) some of the alternative strategies that are possible in the solution of such three-dimensional problems.

5.6.1 Iterative methods: SOR, SLOR, ADI
The most general linear second-order PDE in two dimensions may be expressed as

$$A(x, y)\frac{\partial^2 \phi}{\partial x^2} + B(x, y)\frac{\partial \phi}{\partial x} + C(x, y)\frac{\partial^2 \phi}{\partial y^2} + D(x, y)\frac{\partial \phi}{\partial y}$$

$$+ E(x, y)\phi = \rho(x, y), \tag{5.119a}$$

where the coefficients A, B, C, D, E are arbitrary functions of position. This equation encompasses the principal equations of mathematical physics and engineering (the Helmholtz, Poisson, Laplace, Schrödinger and diffusion equations) in the common coordinate systems (cartesian (x, y), polar (r, θ), cylindrical (r, z), axisymmetric spherical (r, θ) and spherical surface (θ, ϕ)). If the above equation is differenced on an $n \times n$ mesh of points using standard procedures (see e.g. Forsythe and Wasow 1960), one obtains a set of algebraic equations, each of which relates the values of the variables on five neighbouring mesh points:

$$a_{p,q}\phi_{p,q-1} + b_{p,q}\phi_{p,q+1} + c_{p,q}\phi_{p-1,q} + d_{p,q}\phi_{p+1,q} + e_{p,q}\phi_{p,q} = f_{p,q},$$

$$\tag{5.119b}$$

where the integer subscripts $p, q = 1, \ldots, n$, label the mesh points in the x and y directions respectively. The coefficients a, b, c, d and e vary from mesh point to mesh point and are related to the functions A, B, C, D, E and the separations between the mesh points in a complicated way through the particular difference approximation that may be used. The right-hand-side variable $f_{p,q}$ is a linear combination of the values of $\rho(x, y)$ at the mesh points near (p, q). In the simplest case it is the value of $\rho(x, y)$ at the mesh point (p, q).

Iterative procedures are defined by starting with a guess for the values of $\phi_{p,q}$ at all the mesh points, and using the difference equation (5.119b) as a basis for calculating improved values. The process is repeated and, if successful, the values of ϕ converge to the solution of equation (5.119b) at all mesh points. In the simplest procedure the values of ϕ at all mesh points are simultaneously adjusted to the values they would have by equation (5.119b) if all the neighbouring values of ϕ are assumed to be correct, namely ϕ at each mesh point is replaced by a new value:

$$\phi_{p,q}^* = \frac{f_{p,q} - a_{p,q}\phi_{p,q-1} - b_{p,q}\phi_{p,q+1} - c_{p,q}\phi_{p-1,q} - d_{p,q}\phi_{p+1,q}}{e_{p,q}}.$$

$$\tag{5.120}$$

Since the replacement is to take place simultaneously, all values of ϕ on the right-hand side are 'old' values from the last iteration and the starred values on the left-hand side are the 'new' and, hopefully, improved values.

The above method of simultaneous displacements was first considered by Jacobi (1845), and is often called the Jacobi method. It is ideally suited for implementation on parallel computers. Simultaneous adjustment means that equation (5.120) can be evaluated at all mesh points in parallel with the maximum possible parallelism of n^2. Typically $n = 32$ to 256, so that the parallelism varies from about 1000 to about 64 000. This leads to satisfactorily long vectors for efficient manipulation on pipelined computers. On the other hand for processor arrays such as the ICL DAP, the mesh can probably be chosen to fit the PE array exactly or be a simple multiple of it, thus allowing the computer to be used with all its processors active (for example, the solution of a 64 × 64 mesh problem on a 64 × 64 PE ICL DAP). Furthermore all the variables referred to in the adjustment equation (5.120) are nearest neighbours in the PE array and so may be referenced without the need for routing operations. In fact the ICL DAP was obviously designed with this kind of application in mind.

The fact that a single iteration of the simultaneous replacement method can be efficiently implemented on parallel computers does not necessarily mean that it is a good method to use, because we must also consider the number of iterations that are required to obtain satisfactory convergence. The convergence rate cannot be found simply for the general equation (5.119b) but analytic results are well known for the simpler case of Poisson's equation in the square with zero values (Dirichlet conditions) on the boundary. This corresponds to taking $a_{p,q} = b_{p,q} = c_{p,q} = d_{p,q} = 1$ and $e_{p,q} = -4$ for all mesh points, and is often called the model problem. We emphasise that iterative methods should never be used actually to solve the model problem, because the direct transform methods of §5.6.2 are at least 10 times faster. The model problem is introduced here only to indicate the rate of convergence that might be expected when iterative methods are used to solve the general difference equation (5.119b), on which fast transform methods cannot be used.

It may be shown (see e.g. Varga 1962) that, asymptotically after a large number of iterations, the factor by which the norm† of the error vector is reduced at every iteration of the Jacobi method is given by:

$$\lambda_J = \cos(\pi/n) \simeq 1 - \tfrac{1}{2}\pi^2/n^2. \tag{5.121a}$$

The above convergence factor, λ_J, can be used to calculate t_J^*, the number of iterations required to reduce the error by a factor 10^{-p}:

$$t_J^* = \frac{-p}{\log_{10}\lambda_J} \simeq \frac{2}{\pi^2 \log_{10}e}\, pn^2 \simeq \frac{pn^2}{2}. \tag{5.121b}$$

---

† The sum over all mesh points of the square of the difference between the approximate and exact solution of the equations.

Thus a modest error reduction of 10^{-3} on a typical 128×128 mesh would require about 24 000 iterations. Such slow convergence makes the Jacobi method useless for practical computation, even though it is highly suitable for implementation on parallel computers.

The most commonly used iterative method on serial computers is the method of successive over-relaxation by points or SOR. In this method a weighted average of the 'old' and starred values is used as the 'new' value according to

$$\phi_{p,q}^{new} = \omega\phi_{p,q}^* + (1 - \omega)\phi_{p,q}^{old}, \tag{5.122a}$$

where ω is a constant relaxation factor, normally in the range $1 \leqslant \omega \leqslant 2$, that is chosen to improve the rate of convergence. For the model problem, it may be shown that the best rate of convergence is obtained with:

$$\omega = \omega_b = \frac{2}{1 + (1 - \rho^2)^{1/2}}, \tag{5.122b}$$

when $\lambda = \omega_b - 1$ and where ρ is the convergence factor of the corresponding Jacobi iteration. Hence $\rho = \cos(\pi/n)$ for the model problem. Normally the mesh points are processed sequentially point-by-point and line-by-line, as in reading the words of a book or typing a page of text. The improved convergence relies on the fact that new values replace old values as soon as they are computed; hence the values on the right-hand side of equation (5.120), that are used to calculate ϕ^*, are a mixture of old and new values, and equation (5.120) cannot be computed for all points in parallel as it can in the Jacobi method. It would appear therefore that the SOR method, although it has superior convergence properties, is essentially a sequential method and unsuitable for implementation on parallel computers. However, Hunt (1979) has shown how the principle of macroscopic pipelining can be used to implement SOR with a maximum parallelism of $n^2/2$.

Fortunately the improved convergence of the SOR method can be obtained for other patterns of sweeping the mesh points. One of the best procedures is odd/even ordering with Chebyshev acceleration. In this method the mesh points are divided into two groups according to whether $p + q$ is odd or even. This is shown in figure 5.23(a), where the even points are shown as circles and the odd points as crosses. The method proceeds in half iterations, during each of which only half the points are adjusted (alternately the odd and even set of points) according to equations (5.120) and (5.122). In addition the value of ω changes at each half iteration according to:

$$\omega^{(0)} = 1$$

$$\omega^{(1/2)} = 1/(1 - \tfrac{1}{2}\rho^2) \tag{5.123a}$$

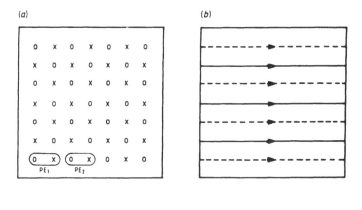

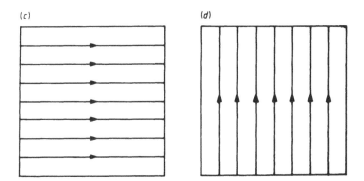

FIGURE 5.23   The pattern of related data in various iterative methods: (a) point successive over-relaxation (SOR); (b) successive line over-relaxation (SLOR); (c) and (d) the alternating direction implicit method (ADI). The multiple Fourier transform direct method (MFT) relates data in the same way as ADI. The arrowed lines link data that are related in either a Fourier transform or a tridiagonal system of equations.

$$\omega^{(t+1/2)} = 1/(1 - \tfrac{1}{4}\rho^2\omega^{(t)}), \qquad t = \tfrac{1}{2}, 1, \ldots, \infty,$$

where the superscript t designates the iteration number. It may be shown that ω tends, for large t, to the limit $\omega^{(\infty)} = \omega_b$, the constant relaxation factor that is used throughout the traditional SOR procedure of equation (5.122b). The asymptotic convergence factor is therefore the same for both methods of formulating SOR and is given by:

$$\lambda_{SOR} = \omega_b - 1 \simeq 1 - 2\pi/n. \tag{5.123b}$$

The number of iterations for a 10^{-p} error reduction is therefore

$$t^*_{SOR} \simeq \frac{np}{2\pi \log_{10} e} \simeq \frac{np}{3}. \tag{5.123c}$$

Using equation (5.123c) we conclude that a 10^{-3} error reduction on a 128×128 mesh would require about 128 iterations, compared with 24 000 for the Jacobi method. The importance of using a well convergent iterative method is apparent. It may also be shown (see Varga 1962) that the maximum norm of the error vector is bound to decrease in the Chebyshev accelerated SOR method with variable ω that has just been described, whereas it may and frequently does increase in the early stages of the traditional SOR with constant ω and typewriter ordering (see Hockney 1970, figures 7 and 9). Since the Chebyshev SOR method requires no extra arithmetic yet has more favourable initial error decay properties, there seems to be no sound reason for not always using it.

The Chebyshev SOR method has additional advantages when one considers its implementation on parallel computers. We note from equation (5.120) that the starred value at an odd mesh point depends only on old values at the neighbouring even points that were computed during the last half iteration. Thus all the odd points can be adjusted in parallel with parallelism $n^2/2$ during one half iteration, and similarly all the even points in parallel during the next half iteration. The time for one complete iteration on a computer with a half-performance length of $n_{1/2}$ is therefore proportional to

$$t_1 = 24(n_{1/2} + n^2/2), \tag{5.124a}$$

since equations (5.120) and (5.122a) require 12 arithmetic operations per mesh point (note: $1 - \omega$ is precalculated and stored as a single constant). The SOR method can alternatively be implemented with a parallelism of $n/2$ by defining a vector to be the $n/2$ odd (or even) values from a line of the mesh, and progressing sequentially through the n lines. The time for a complete iteration is then proportional to

$$t_2 = 24n(n_{1/2} + n/2). \tag{5.124b}$$

An alternative strategy of Chebyshev over-relaxation is illustrated in figure 5.23(b). The points are divided into odd and even sets by lines, as indicated by the broken and full lines. The adjustment equation (5.120) is written for all points of a line, assuming all points in the line above and the line below are correct. The starred values are then computed a line at a time for $p = 1, 2, \ldots, n$ from:

$$c_{p,q}\phi^*_{p-1,q} + e_{p,q}\phi^*_{p,q} + d_{p,q}\phi^*_{p+1,q} = f_{p,q} - a_{p,q}\phi_{p,q-1} - b_{p,q}\phi_{p,q+1}. \tag{5.125}$$

Equation (5.125) is a tridiagonal system for all the starred values along a line, with the right-hand side depending on known values from the line above and below. The Chebyshev accelerated successive line over-relaxation (SLOR) proceeds line-wise using equations (5.122) and (5.123a), only now ρ the convergence factor of the Jacobi method performed line-wise is, for the model problem,

$$\rho = \cos(\pi/n)/[2 - \cos(\pi/n)] \simeq 1 - \pi^2/n^2. \tag{5.126a}$$

Equation (5.122b) still applies with this revised value of ρ, leading to

$$\lambda_{\text{SLOR}} = \omega_b - 1 \simeq 1 - \sqrt{2}(2\pi/n), \tag{5.126b}$$

and

$$t^*_{\text{SLOR}} = t^*_{\text{SOR}}/\sqrt{2} \simeq np/4. \tag{5.126c}$$

Thus, asymptotically, the number of iterations required for a given error reduction in the SLOR method is $2^{-1/2} = 0.7$ of the number of iterations required by SOR.

The SLOR method therefore requires the solution of $n/2$ tridiagonal systems of length n each half iteration. If the systems are solved in parallel with the best serial algorithm (the MULTGE procedure of §5.4.4 with $m = n/2$) we have a parallelism of $n/2$ and a computer time for a complete iteration proportional to

$$t_3 = 24n(n_{1/2} + n/2). \tag{5.127a}$$

In arriving at this estimate we have taken four arithmetic operations per point to form the right-hand side of equation (5.125) in addition to the eight operations per point to solve the general tridiagonal system. A factor of two is introduced because there are two half iterations in the complete iteration. Alternatively the systems can be solved with parallelism $n^2/2$ by solving the systems in parallel with the parallel $\text{PARACR}_{\text{par}}$ algorithm (see §5.4.4). The computer time per complete iteration is now proportional to

$$t_4 = 2(n_{1/2} + n^2/2)(4 + 12\log_2 n), \tag{5.127b}$$

where the four in the last factor is for calculating the right-hand sides and the logarithmic term is for solving the tridiagonal system with the PARACR algorithm [see equation (5.66b)].

We have seen that both the SOR and SLOR methods can be implemented with a parallelism of either $n/2$ or $n^2/2$. Which of these implementations is the best will depend primarily on whether $n/2$ or $n^2/2$ more closely matches the natural hardware parallelism of the computer. We will consider the two extreme cases of the ICL DAP with natural parallelism of 64×64, and the CRAY X-MP with a natural parallelism of 64. Practical finite-difference meshes with useful resolution are likely to have between 32 and 256 mesh points along each dimension. Thus it is natural to match the parallelism of the CRAY X-MP with a side of the mesh and use algorithms which have a parallelism of n or $n/2$. On the other hand it is more natural to match the parallelism of the ICL DAP with two dimensions of the problem and use algorithms which have a parallelism of n^2 or $n^2/2$. This choice is made even more compelling when one considers that the ICL DAP is wired as a two-dimensional array of processors.

Having chosen the level of parallelism that is best suited to the computer, one may ask whether Chebyshev SOR or Chebyshev SLOR is the best algorithm to use. On the CRAY X-MP using $n/2$ parallelism, SOR will be the best algorithm if

$$t_2 < t_3/\sqrt{2}, \tag{5.128a}$$

where the factor $\sqrt{2}$ takes into account the better convergence of the SLOR method. Since $t_2 = t_3$ [see equations (5.124b) and (5.127a)], the above condition (5.128a) can never arise, and we conclude that the SLOR will always be the best on the CRAY X-MP. Put another way, the time per iteration is the same when computed with a parallelism of $n/2$, and SLOR therefore wins because of its faster convergence.

For the ICL DAP using $n^2/2$ parallelism, we find that SOR will be the best algorithm if

$$t_1 < t_4/\sqrt{2}, \tag{5.128b}$$

or

$$24 < \sqrt{2}(4 + 12\log_2 n). \tag{5.128c}$$

This condition is satisfied for $n \geqslant 2$, that is to say for all useful meshes. We thus find the SOR method the best implementation on the ICL DAP. This is because the parallel solution of tridiagonal equations by cyclic reduction that is required if we use the SLOR method introduces extra arithmetic that is not needed in the simpler SOR scheme. The better convergence of SLOR is not enough to make up for this.

The last iterative method to be considered is the alternating direction implicit method or ADI. This is illustrated in figure 5.23(c and d). When it is applied to the solution of the difference equation,

$$(\mathbf{L}_x + \mathbf{L}_y)\phi = \rho, \tag{5.129a}$$

where $\mathbf{L}_x$ and $\mathbf{L}_y$ are matrices representing the finite-difference operators in the x and y directions, the solution is obtained iteratively from a guess $\phi^{(0)}$

by solving repeatedly

$$(I + r_n L_x)\phi^{(n+1/2)} = (I - r_n L_y)\phi^{(n)} + r_n q, \qquad (5.129b)$$

$$(I + r_n L_y)\phi^{(n+1)} = (I - r_n L_x)\phi^{(n+1/2)} + r_n q, \qquad (5.129c)$$

where $n = 0, 1, \ldots$, is the iteration number and the parameter r_n, which changes every double sweep, is adjusted to improve the convergence of the approximations $\phi^{(n)}$ to the solution ϕ. If we are concerned with the general linear second-order PDE (equation (5.119a)) L_x and L_y are tridiagonal matrices. Equation (5.129b) involves computing a right-hand side at a cost of seven operations per mesh point, followed by the solution of n independent tridiagonal systems (one for each horizontal line of the mesh) each of length n (see figure 5.23(c)). Clearly these systems may be solved in parallel either with a parallelism of n or n^2, as in the case of the SLOR method. The iteration is completed by performing a similar process, only now the tridiagonal systems are solved along every vertical line of the mesh (see figure 5.23(d)).

If a parallelism of n is used (e.g. on the CRAY X-MP) we use the MULTGE method and the computer time for a complete iteration is proportional to

$$t_5 = 30n(n_{1/2} + n). \qquad (5.130a)$$

On the other hand using a parallelism of n^2 (e.g. on the ICL DAP) and the PARACR$_{par}$ method the computer time is proportional to:

$$t_6 = 2(n_{1/2} + n^2)(7 + 12\log_2 n). \qquad (5.130b)$$

The fact that data is first referenced by horizontal lines and then by vertical lines may complicate the implementation of ADI on some computers. If the mesh is stored so that adjacent elements in vertical lines are adjacent in the computer memory (FORTRAN columnar storage) then the solution of equation (5.129b) will present no problems because the increment between vector elements is unity (remember we run the vector *across* the systems being solved). However, in solving equation (5.129c) the increment between vector elements will be equal to the number of variables in a column. If n is a power of two, memory-bank conflicts can be a problem even in computers such as the CRAY X-MP, that allow increments other than unity. This may be overcome by storing the mesh as though it had a column length one greater than its actual length. On the other hand, on computers such as the CDC CYBER 205 which only permit vectors with an increment of unity, the second step of the ADI can only be performed after a rotation of the whole mesh in the store. The cost of this manipulation may make ADI an unattractive method on such machines, compared with the SLOR method that only references data in the horizontal direction. Experience with ADI and SLOR is varied and there is no *a priori* way of determining which will be the most effective method on a particular problem. It may be necessary to try both methods.

5.6.2   Direct methods: MFT, FACR (*l*)

Direct methods are those which obtain the solution in a finite number of steps or arithmetic operations. Since no iteration is involved, their effectiveness does not depend on the quality of the initial guess for the solution or on one's ability to judge correctly when to stop the iteration. On the other hand direct methods for the solution of the general linear finite-difference equation (5.119b) can only be applied to relatively coarse meshes with up to about a thousand points which do not give a great deal of scope for effective parallel computation. We will therefore discuss the special transform techniques that are available for the solution of the Poisson equation in the square or rectangle with simple boundary conditions:

$$\phi_{p,q-1} + \phi_{p,q+1} + \phi_{p-1,q} + \phi_{p+1,q} - 4\phi_{p,q} = f_{p,q},$$

$$p, q = 0, 1, \ldots, n-1, \tag{5.131}$$

in which the coefficients are constant and given by: $a_{p,q} = b_{p,q} = c_{p,q} = d_{p,q} = 1$ and $e_{p,q} = -4$. These techniques, often called rapid elliptic solvers or RES algorithms, are an order of magnitude faster than the iterative methods described in §5.6.1 and can be applied to meshes containing 10 000 or more points. RES methods are inherently highly parallel and therefore especially well suited to exploiting the architecture of parallel computers. Rapid elliptic solvers have been reviewed by Swarztrauber (1977) and by Hockney (1980), and the performance of different algorithms compared by Hockney (1970, 1978) and Temperton (1979a) for serial computers, and Grosch (1979) and Temperton (1979b, 1980) for parallel computers.

Rapid elliptic solvers are defined as those methods with an operation count of order $n \log_2 n$ or better [some e.g. FACR(*l*) are of order $n \log_2(\log_2 n)$], which immediately suggests the involvement of fast transform algorithms (see §5.5). The simplest method is obtained by taking the double Fourier transform of equation (5.131):

$$[2\cos(2\pi k/n) + 2\cos(2\pi l/n) - 4]\bar{\phi}^{k,l} = \bar{f}^{k,l}, \tag{5.132a}$$

where

$$\bar{\phi}^{k,l} = \frac{1}{n^2} \sum_{p,q=0}^{n-1} \phi_{p,q} \exp[-2\pi i(kp + lq)/n], \tag{5.132b}$$

and

$$\phi_{p,q} = \sum_{k,l=0}^{n-1} \bar{\phi}^{p,q} \exp[+2\pi i(kp + lq)/n], \tag{5.132c}$$

and similarly for $\bar{f}^{k,l}$ and $f_{p,q}$. Equation (5.132a) permits the calculation of the Fourier harmonic amplitudes of the solution, $\bar{\phi}^{k,l}$, by dividing the harmonics of the right-hand side, $\bar{f}^{k,l}$, by the known numerical factors in the square brackets. The following method of multiple Fourier transform (MFT) therefore suggests itself:

(a) Fourier analyse $f_{p,q}$ using FFT, $f_{p,q} \rightarrow \bar{f}^{k,l}$;

(b) Parallel divide by $[\ldots]$ of equation (5.132a), $\bar{f}^{k,l} \rightarrow \bar{\phi}^{k,l}$;   (5.133)

(c) Fourier synthesis $\bar{\phi}^{k,l}$ using FFT, $\bar{\phi}^{k,l} \rightarrow \phi_{p,q}$.

Both the double Fourier analysis and the synthesis can conveniently be performed by first transforming all lines of data in the x direction as in figure 5.23(c) and then transforming the resulting data by lines in the y direction as in figures 5.23(d). This may clearly be seen by re-expressing equation (5.132b) equivalently as

$$\bar{\phi}^{k,l} = \frac{1}{n} \sum_{q=0}^{n-1} \left(\frac{1}{n} \sum_{p=0}^{n-1} \phi_{p,q} \exp(-2\pi i k p/n) \right) \exp(-2\pi i l q/n), \quad (5.134)$$

where the inner summation is the transform in x and the outer summation the transform in y. We have assumed doubly periodic boundary conditions in the above explanation. However, Dirichlet (given value) conditions can be achieved if the finite sine transform is used and Neumann (given gradient) conditions if the finite cosine transform is used. These refinements do not affect the issues involved in parallel implementation.

The line transforms in equation (5.134) are all independent and may be performed in parallel with a parallelism of n or n^2. In the first option we perform n real transforms in parallel using the best serial algorithm, that is to say the MULTFT method of equation (5.100a) with $m = n$. The total time for the algorithm is therefore:

$$t_{\text{MULTFT}} = 4 \times \tfrac{1}{2} \times 5n(n_{1/2} + n)\log_2 n, \quad (5.135a)$$

$$= 10n(n_{1/2} + n)\log_2 n. \quad (5.135b)$$

The factor of four in equation (5.135a) comes from the need to transform all points four times — an analysis and synthesis in both x and y directions — and the factor $\tfrac{1}{2}$ arises because we are performing real rather than complex transforms. Alternatively if we perform n transforms in parallel using the parallel PARAFT$_{\text{par}}$ algorithm then we have a parallelism of n^2 and

$$t_{\text{PARAFT}_{\text{par}}} = 4 \times \tfrac{1}{2} \times 8(n_{1/2} + n^2)\log_2 n, \quad (5.136a)$$

$$= 16(n_{1/2} + n^2)\log_2 n. \quad (5.136b)$$

The choice between n or n^2 parallelism will follow the same rationale as used in the choice of iterative methods. We would expect to use n parallelism on computers such as the CRAY X-MP and n^2 parallelism on computers such as the ICL DAP. Given, however, that both levels of parallelism are equally suited to the hardware of the computers, as would be the case for the CDC CYBER 205, we may still ask which algorithm is the best. The PARAFT$_{\text{par}}$ algorithm will have the higher performance if:

$$t_{\text{MULTFT}} > t_{\text{PARAFT}_{\text{par}}},\qquad\qquad(5.137a)$$

whence

$$\frac{n_{1/2}}{n} > \frac{6n}{10n - 16}.\qquad\qquad(5.137b)$$

The regions of the $(n_{1/2}/n, n)$ plane which favour the use of the two algorithms, using equation (5.137b), are shown in figure 5.24. We see that, for vectors of any reasonable length (say $n > 10$), the PARAFT$_{\text{par}}$ method is favoured if $(n_{1/2}/n) \gtrsim 0.5$. That is to say, not surprisingly, that if the computer 'looks' parallel when measured on the scale of the vector $(n_{1/2} \gtrsim n/2)$, then an algorithm that is designed to maximise the extent of parallel operation is favoured.

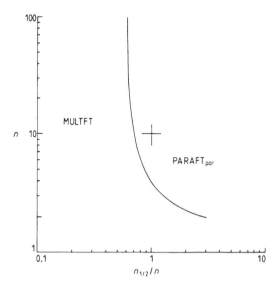

FIGURE 5.24   Regions of the $(n_{1/2}/n, n)$ plane favoured for the implementation of the MFT method by either the MULTFT algorithm or the PARAFT$_{\text{par}}$ algorithm. From equation (5.137b).

The multiple Fourier transform is also used in the solution of related problems that are posed as convolutions. For example, the potential at a mesh point may be expressed as a sum of contributions from sources at all the other mesh points, multiplied by a known influence or Green's function G which is a function only of the separation between the source and potential mesh point. Thus

$$\phi_{p,q} = \sum_{s=0}^{n-1} \sum_{t=0}^{n-1} G_{p-s,q-t} f_{s,t}. \tag{5.138}$$

This manner of expressing a potential problem is appropriate when the sources form an isolated cluster and the only boundary condition is that the potential decays to infinity correctly. An example of such a problem is the calculation of the potential of a cluster of galaxies or stars (see Hockney 1970), in which case G is the coulombic r^{-1} interaction, where r is the separation between the source and potential points. The above expression is also used in the P^3M method (Hockney and Eastwood 1988) for the calculation of the potential in large systems of interacting ions.

Equation (5.138) is equivalent to the statement that the potential ϕ is the convolution of the source distribution f with the Green's function G. The convolution theorem (see Bracewell 1965) states that the Fourier transform of a convolution is proportional to the product of the Fourier transforms of the items convolved. Hence taking the transform of equation (5.138) we obtain:

$$\bar{\phi}^{k,l} = n^2 \bar{G}^{k,l} \bar{f}^{k,l}, \qquad k, l = 0, 1, \dots, n-1. \tag{5.139a}$$

The method of solution proceeds:

$$f_{p,q} \xrightarrow{\text{FFT}} \bar{f}^{k,l} \xrightarrow{\times \bar{G}^{k,l}} \bar{\phi}^{k,l} \xrightarrow{\text{FFT}} \phi_{p,q}. \tag{5.139b}$$

This is the same as the procedure (5.133) with multiplication by $\bar{G}$ replacing division by [ ] of equation (5.132a), and the MULTFT or PARAFT algorithms may be used, as described above.

The transform method outlined in the previous paragraph relies only on the existence of the convolution theorem for converting the multiple sum (5.138) to the simple parallel multiplication in equation (5.139a). It may easily be shown from equation (5.111) that the Fermat number theoretic transform (NTT) also has a convolution theorem. The method described by equation (5.139b) may therefore be used with the FFT replaced by an NTT, and the floating-point arithmetic replaced by the integer modulo arithmetic of §5.5.6. The details of this method are given by Eastwood and Jesshope (1977), who recommend it for the solution of certain three-dimensional

PDES on computers such as the ICL DAP. James and Parkinson (1980) have assessed the effectiveness of the method in the solution of the Poisson equation for a gravitational problem on a $65 \times 65 \times 65$ mesh.

One of the earliest and most successful rapid elliptic solvers is the method of Fourier analysis and cyclic reduction or FACR algorithm (Hockney 1965, 1970, Temperton 1980). This was devised to minimise the amount of arithmetic on serial computers by reducing the amount of Fourier transformation, and has an asymptotic operation count equal to $4.5n^2 \log_2(\log_2 n)$, when optimally applied (see Hockney and Eastwood 1988). The FACR algorithm also proves to be highly effective on parallel computers.

The discrete Poisson equation (5.131) may be rewritten line-wise as follows:

$$\phi_{q-1} + \mathbf{A}\phi_q + \phi_{q+1} = f_q, \qquad q = 0, 1, \ldots, n-1, \tag{5.140}$$

where the elements of the column vectors ϕ_q and f_q are, respectively, the values of the potential and the right-hand side along the qth horizontal line of the mesh. The matrix $\mathbf{A}$ is an $n \times n$ tridiagonal matrix with a diagonal of -4 and immediate upper and lower diagonals of unity. It is a matrix that represents the differencing of the differential equation in the x direction.

If we multiply every even-line equation like equation (5.140) by $-\mathbf{A}$, and add the equation for the odd line above and the odd line below, we obtain a set of equations relating the even lines only, namely:

$$\phi_{q-2} + \mathbf{A}^{(1)}\phi_q + \phi_{q+2} = f_q^{(1)}, \tag{5.141a}$$

where

$$\mathbf{A}^{(1)} = 2\mathbf{I} - \mathbf{A}^2 = (\sqrt{2}\mathbf{I} + \mathbf{A})(\sqrt{2}\mathbf{I} - \mathbf{A}), \tag{5.141b}$$

and

$$f_q^{(1)} = f_{q-1} - \mathbf{A}f_q + f_{q+1}. \tag{5.141c}$$

This constitutes one level of cyclic reduction, line-wise, of the original equations. The resulting equations (5.141a) are half in number (even lines only) and have the same form as the original equations (5.140). The process of cyclic reduction may therefore be repeated recursively, yielding reduced sets of equations for every fourth, eighth, sixteenth line etc at levels $r = 2, 3, 4$ etc. At each level there is a new central matrix $\mathbf{A}^{(r)}$ that can be expressed as a product of 2^r tridiagonal matrices:

$$\mathbf{A}^{(r)} = -\prod_{k=1}^{2^r} (\mathbf{A} - \beta_k \mathbf{I}), \qquad \beta_k = 2\cos\left(\frac{(2k-1)\pi}{2^{r+1}}\right). \tag{5.142}$$

The factorisation for $r = 1$ is shown directly in the second equality of equation (5.141b).

If cyclic reduction is stopped at level $r = l$, the resulting equations are

$$\phi_{q-2^l} + \mathbf{A}^{(l)}\phi_q + \phi_{q+2^l} = f_q^{(l)}, \qquad q = (0, 1, \ldots) \times 2^l. \qquad (5.143a)$$

These equations are solved by Fourier analysis in the x direction, along the vectors, leading to independent harmonic equations for each harmonic k:

$$\bar{\phi}_{q-2^l}^k + \lambda_k \bar{\phi}_q^k + \bar{\phi}_{q+2^l}^k = \bar{f}_q^k, \qquad k = 0, 1, \ldots, n-1, \qquad (5.143b)$$

where λ_k is a constant for each harmonic, depending on the boundary conditions in x and the roots β_k of equation (5.142). Equations (5.143b) are n tridiagonal systems of length $n2^{-l}$ and are easily solved. The procedure for solving equation (5.143a) is therefore

$$f_q \xrightarrow[\text{analysis}]{\text{FFT}} \bar{f}_q^k \xrightarrow[\text{equations (5.143b)}]{\text{solve}} \bar{\phi}_q^k \xrightarrow[\text{synthesis}]{\text{FFT}} \phi_q. \qquad (5.143c)$$

Having found the solution of every 2^lth line from equation (5.143c), the intermediate lines can be filled in successively from the intermediate level equations

$$\mathbf{A}^{(r)}\phi_q = f_q^{(r)} - \phi_{q-2^r} - \phi_{q+2^r}, \qquad \text{for } r = l-1, l-2, \ldots, 0. \ (5.144)$$

It will be found, when applied to the $n2^{-(r+1)}$ unknown lines, that the values of ϕ on the right-hand side of equation (5.144) are values that have been found at the previous deeper level. Because of the factorisation (5.142), the solution of equation (5.144) requires the successive solution of 2^r tridiagonal systems.

The above algorithm is referred to as FACR(l) where the argument gives the number of levels of cyclic reduction that are performed before the equations are solved by Fourier analysis. It was shown numerically by Hockney (1970) and proved analytically by Swarztrauber (1977) that there is an optimum value $l = l^* \approx \log_2(\log_2 n)$ which leads to the minimum total number of arithmetic operations. As more cyclic reduction is performed (increasing l) less Fourier transformation takes place (there are fewer lines to do it on); however, it is necessary to solve more tridiagonal systems in using equation (5.144) to fill in the intermediate lines. The exact position of the optimum l^* therefore depends on the relative efficiencies of the computer codes that are used for the FFT and the solution of tridiagonal systems. A better FFT code will lead to lower l^* and a better tridiagonal solver to higher l^*. The best strategy is probably to write a computer code for general l and measure the optimum value. This has been done by Temperton (1980) with his code PSOLVE on the IBM 360/195 (optimum $l^* = 2$), and by Temperton (1979b) on the CRAY-1 (optimum $l^* = 2$ in scalar mode, $l^* = 1$ in vector mode). It is evident that the introduction of parallelism into the implementation of FACR(l) drives the optimum l^* to a smaller value. We shall see that this empirical result agrees well with the simple theoretical estimates of performance that we shall derive below.

It will be evident from the above description that if cyclic reduction is taken to $l \approx \log_2 n$ level, there remains only a single equation to be solved for the central line. This is of the form of equation (5.144) where the values of ϕ on the right-hand side are, depending on the boundary conditions, either known boundary values or the same as the line on the left-hand side. The equation may therefore be solved without Fourier analysis, as may those for all intermediate lines. This method of complete line cyclic reduction (CLCR), which can therefore be described as a FACR($\log_2 n$) procedure, was devised by Buneman (1969), who showed how, at the expense of extra arithmetic, the procedure could be made numerically stable. The theoretic analysis of the numerical stability of line cyclic reduction was developed afterwards by Buzbee *et al* (1970).

The FACR(l) method may be programmed using either the unstable form of cyclic reduction described above or the stabilised procedure of Buneman, and this has a marginal effect on the optimum value of l^*. In practice it is found that unstabilised reduction may be used for the small values of $l = 1, 2$ that are used by FACR(l), but that the stabilised reduction must be used if cyclic reduction is taken to completion with l typically in the range 5 to 7. Because the CLCR method works with a value of l which is far from the optimum and requires extra arithmetic for stabilisation, we will not consider it further here. Notwithstanding these facts, that were both known and published in 1970, the development of the CLCR method dominated the numerical analysis literature during the 1970s (see e.g. the review by Swarztrauber 1977). The relative performance of complete reduction of CLCR and the partial reduction of FACR(l) was put to the test in a competition between about 20 different Poisson-solvers held at Karlsruhe in 1977 (Schumann 1978). This was won by Temperton's program PSOLVE, a stabilised FACR(3) algorithm, which was 1.8 times faster than the best program using complete cyclic reduction. It is also interesting that the measured rounding error in the whole calculation was actually less for an 'unstabilised' FACR(1) procedure POT1, than for the stabilised program. The above results favouring the FACR(l) method were all obtained on serial computers. We point out that the arguments against the use of complete reduction are stronger when we consider implementation on parallel computers, as we now see.

The steps in the FACR(l) algorithm are now summarised and the time of execution estimated on a parallel computer with a half-performance length of $n_{1/2}$. The patterns of related data in the different steps are illustrated in

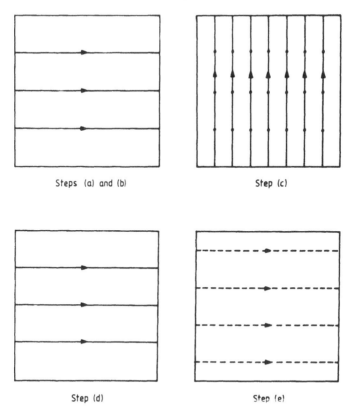

FIGURE 5.25   The patterns of related data in the different steps of the FACR(1) algorithm. The arrows link data that are related in either a Fourier transformation or a tridiagonal system of equations.

figure 5.25 for the case of a FACR(1) algorithm, in which there is one level of initial cyclic reduction.

The FACR(l) algorithm consists of five stages, and can be implemented either to minimise the total amount of arithmetic, s (serial variant), or to minimise the number of vector operations, q (parallel variant), as was described in §5.1.6. The serial variant is called SERIFACR and the parallel variant PARAFACR. We will now describe both methods, and then compare the two, using the $n_{1/2}$ method of algorithm analysis. The objective of the analysis is first to find which variant is best on a particular computer, and secondly to choose the optimum value of l. This will be done by drawing the appropriate algorithmic phase diagrams (Hockney 1983).

(i) The SERIFACR algorithm
In this version of the algorithm the vector length is proportional to n, the

number of points along one side of the mesh. The five stages of the FACR(l) algorithm are:

(a) *Modify* RHS—for $r = 1, 2, \ldots, l$ modify the right-hand side on $n2^{-r}$ lines in parallel using the generalisation of equation (5.141c), namely:

$$f_q^{(r)} = f_{q - 2^{r-1}} - \mathbf{A}^{(r-1)} f_q + f_{q + 2^{r-1}}, \tag{5.145a}$$

at the cost of $(3 \times 2^{r-1} + 2)n$ arithmetic operations per line:

$$t_a = \sum_{r=1}^{l} (n_{1/2} + n2^{-r})(3 \times 2^{r-1} + 2)n; \tag{5.145b}$$

(b) FFT *analysis*—on $n2^{-l}$ lines in parallel using the MULTFT algorithm. Each transform is real and of length n:

$$t_b = (n_{1/2} + n2^{-l})2\tfrac{1}{2}n \log_2 n; \tag{5.145c}$$

(c) *Solve harmonic equations*—n tridiagonal equations, each of length $n2^{-l}$, solved in parallel using the MULTGE algorithm:

$$t_c = 5(n_{1/2} + n)n2^{-l}. \tag{5.145d}$$

The leading coefficient is taken as five, rather than eight, because the equations (5.143b) have two coefficients that are unity ($a_i = c_i = 1$).

(d) FFT *synthesis*—on $n2^{-l}$ lines in parallel using the MULTFT algorithm. Each transform is real and of length n:

$$t_d = (n_{1/2} + n2^{-l})2\tfrac{1}{2}n \log_2 n. \tag{5.145e}$$

(e) *Filling-in*—of the intermediate lines by solving equations (5.144), involving $(2 + 5 \times 2^r)n$ operations per line on $n2^{-(r+1)}$ lines, for $r = l - 1$, $l - 2, \ldots, 0$

$$t_e = \sum_{r=0}^{l-1} (n_{1/2} + n2^{-(r+1)})(5 \times 2^r + 2)n, \tag{5.145f}$$

$$= \sum_{r=1}^{l} (n_{1/2} + n2^{-r})(5 \times 2^{r-1} + 2)n. \tag{5.145g}$$

The total time for the FACR(l) algorithm is therefore:

$$t_{\text{SERIFACR}} = \sum_{r=1}^{l} (n_{1/2} + n2^{-r})(8 \times 2^{r-1} + 4)n + 5(n_{1/2} + n2^{-l})n \log_2 n$$

$$+ 5(n_{1/2} + n)n2^{-l}. \tag{5.146a}$$

Evaluating the sums we obtain the time per point:

$$n^{-2}t_{SERIFACR} = [4l + 4 + (1 + 5\log_2 n)2^{-l}]$$

$$+ \frac{n_{1/2}}{n}[4l + 8 \times 2^l - 8 + 5 \times 2^{-l} + 5\log_2 n], \quad (5.146b)$$

where the first square bracket is the normal serial computer operation count†
and the second square bracket takes into account the effect of implementation
on a parallel computer. We note that the first bracket has a minimium for
positive $l \simeq \log_2(\log_2 n)$ but that the second bracket increases monotonically
with l. Thus, for parallel computers which have $n_{1/2} > 0$, the minimum in
$t_{SERIFACR}$ moves to smaller l, as asserted earlier.

The equal performance line between the algorithm with l levels of reduction
and that with $l + 1$ is easily found to be given by

$$n_{1/2}/n = [(1 + 5\log_2 n)2^{-(l+1)} - 4]/(4 + 8 \times 2^l - 5 \times 2^{-(l+1)}).$$

$$(5.147)$$

The form of equation (5.147) suggests that a suitable parameter plane for
the analysis of SERIFACR is the $(n_{1/2}/n, n)$ phase plane, and this is shown
in figure 5.26. The equal performance lines given by equation (5.147) divide
the plane into regions in which $l = 0, 1, 2, 3$ are the optimum choices. Lines
of constant value of $n_{1/2}$ in this plane lie at 45° to the axes, and the lines for
$n_{1/2} = 20, 100, 2048$ are shown broken in figure 5.26. These lines are considered
typical for the behaviour, respectively, of the CRAY-1, the CYBER 205, and
the average performance of the ICL DAP. For practical mesh sizes (say
$n < 500$) we would expect to use $l = 1$ or 2 on the CRAY-1, $l = 0$ or 1 on the
CYBER 205, and $l = 0$ on the ICL DAP. The lower of the two values
for l applies to problems with $n < 100$. Temperton (1979b) has timed a
SERIFACR(l) program on the CRAY-1 and measured the optimum value
of $l = 1$ for $n = 32, 64$ and 128. This agrees with our figure except for $l = 128$,
where figure 5.26 predicts $l = 2$ as optimal. This discrepancy is probably
because Temperton uses the Buneman form of cyclic reduction (see Hockney
1970) which increases the computational cost of cyclic reduction and tends
to move the optimum value of l to smaller values. For a given problem size
(value of n), figure 5.26 shows more serial computers (smaller $n_{1/2}$) to the

---

† Hockney (1970, 1980) quoted $4.5l$ for the leading term of the first bracket. This was
because scalar cyclic reduction (6 operations per point) was assumed for the
solution of the tridiagonal systems, instead of Gaussian elimination as assumed here
(5 operations per point). Other assumptions can make minor and unimportant
differences to this equation.

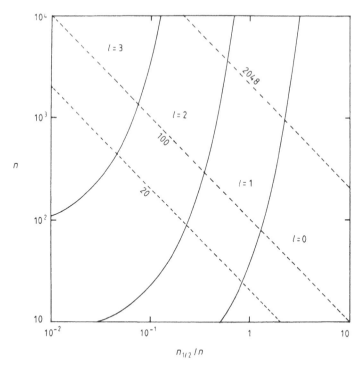

FIGURE 5.26   The $(n_{1/2}/n, n)$ parameter plane for the SERIFACR(l) algorithm. The full curves delineate regions where the stated values of l lead to the minimum execution time. The broken lines are lines of constant $n_{1/2}$ corresponding to the CRAY-1 ($n_{1/2} = 20$), CYBER 205 ($n_{1/2} = 100$), and the average performance of the ICL DAP ($n_{1/2} = 2048$). (From Hockney (1983), courtesy of IEEE.)

left and more parallel computers (larger $n_{1/2}$) to the right. We see, therefore, that the more parallel the computer, the smaller is the optimum value of l.

In the SERIFACR algorithm the vectors are laid out along one or other side of the mesh and never exceed a vector length of n. It is an algorithm suited to computers that perform well on such vectors, i.e. those that have $n_{1/2} < n$, and/or which have a natural parallelism (or vector length) which matches n. The latter statement refers to the fact that some computers (e.g. CRAY X-MP) have vector registers capable of holding vectors of a certain length (64 on the CRAY X-MP). There is then an advantage in using an algorithm that has vectors of this length and therefore fits the hardware design of the computer. For example, the SERIFACR algorithm would be particularly well suited for solving a 64×64 Poisson problem on the CRAY X-MP using vectors of maximum length 64, particularly as this machine is working at better than 80% of its maximum performance for vectors of this length. On other computers, such as the CYBER 205, there are no vector registers and $n_{1/2} \approx 100$. For these machines it is desirable to increase the vector length as much as possible, preferably to thousands of elements. This means implementing the FACR algorithm in such a way that the parallelism is proportional to n^2 rather than n. That is to say, the vectors are matched to the size of the whole two-dimensional mesh, rather than to one of its sides. The PARAFACR algorithm that we now describe is designed to do this.

(ii) The PARAFACR algorithm
Each of the stages of the FACR algorithm can be implemented with vector lengths proportional to n^2.

(a) *Modify* RHS—at each level, r, of cyclic reduction, the modification of the right-hand side can be done in parallel on all the $n^2 2^{-r}$ mesh points that are involved. Hence, the timing formula becomes

$$t_a = \sum_{r=1}^{l} (n_{1/2} + n^2 2^{-r})(3 \times 2^{r-1} + 2). \tag{5.148a}$$

(b) *Fourier analysis*—the $n2^{-l}$ transforms of length n are performed in parallel as in SERIFACR, but now we use a parallel algorithm, PARAFT, for performing the FFT with a vector length of n. The vector length for all lines becomes $n^2 2^{-l}$ and the timing equation is

$$t_b = (n_{1/2} + n^2 2^{-l})4 \log_2 n. \tag{5.148b}$$

The factor 4 replaces the 2.5 in (5.145c) because extra operations are introduced in order to keep the vector length as high as possible in the PARAFT algorithm (see §5.5.4). We also note that the factor n has moved inside the parentheses in comparing equation (5.145c) with (5.148b), because the vector length has increased from $n2^{-l}$ to $n^2 2^{-l}$.

(c) *Solve harmonic equations*—the harmonic equations are solved in parallel as in SERIFACR, but we use a parallel form of cyclic reduction, PARACR, instead of Gaussian elimination, for the solution of the tridiagonal systems (see §5.4.3). For the special case of the coefficients previously noted, there are three parallel operations at each of $\log_2 n$ levels of cyclic reduction. The vector length is $n^2 2^{-l}$ giving

$$t_c = (n_{1/2} + n^2 2^{-l})3 \log_2 n. \tag{5.148c}$$

(d) *Fourier synthesis*—as stage (b):

$$t_d = (n_{1/2} + n^2 2^{-l})4 \log_2 n. \tag{5.148d}$$

(e) *Filling-in*—at each level, $r, n2^{-r}$ tridiagonal systems of length n are to be solved. Using PARACR, as in stage (c), the vector length is $n^2 2^{-r}$. Afterwards a further two operations are required per point, which may also be done in parallel, giving

$$t_e = \sum_{r=1}^{l} (n_{1/2} + n^2 2^{-r})(3 \times 2^{r-1} \log_2 n + 2). \tag{5.148e}$$

Summing the above, we find that the time per mesh point for the PARAFACR algorithm is proportional to

$$n^{-2} t_{\text{PARAFACR}} = s + (n_{1/2}/n^2) q'' \tag{5.149a}$$

where

$$\begin{aligned} s &= \tfrac{1}{2}(3 \log_2 n + 1)l + 4 + (11 \log_2 n - 4)2^{-l} \\ q'' &= 4l + (3 \log_2 n + 1)(2^l - 1) + 11 \log_2 n. \end{aligned} \tag{5.149b}$$

The equal performance line between the level l and $l+1$ algorithms is given by

$$n_{1/2}/n^2 = [(11 \log_2 n - 4)2^{-(l+1)} - \tfrac{1}{2}(3 \log_2 n + 1)]/[4 + (3 \log_2 n + 1)2^l]. \tag{5.149c}$$

The form of equation (5.149c) leads us to choose to plot the results for the PARAFACR algorithm on the $(n_{1/2}/n^2, n)$ parameter plane, and this is done in figure 5.27. We find that the equal performance lines are approximately vertical in this plane, and conclude that $l = 2$ is optimal for $n_{1/2} < 0.1 n^2$, $l = 1$ for $0.1 n^2 < n_{1/2} < n^2$, and $l = 0$ for $n_{1/2} > n^2$. There are no circumstances when more than two levels of reduction are worthwhile, thus justifying our use of the unstabilised FACR algorithm. In particular, for a processor array with as many or more processors than mesh points ($N \geqslant n^2$), we take $n_{1/2} = \infty$ and find $l = 0$. This case corresponds to the solution of a 64×64 problem on the ICL DAP which is an array of 64×64 processors. The broken line for $n_{1/2} = 100$ is shown in figure 5.27, corresponding to the CYBER 205. For all but the smallest meshes (i.e. for $n \geqslant 30$) we find $l = 2$ optimal. The line for $n_{1/2} = 20$ is also given, from which we conclude $l = 2$ is optimal in all circumstances if this algorithm is used on the CRAY-1.

(iii) *SERIFACR/PARAFACR comparison*

So far we have considered the choice of the best value of l for each algorithm. Having optimised each algorithm, we now consider which is the best algorithm to use. This is done by plotting t_{SERIFACR} and t_{PARAFACR} against

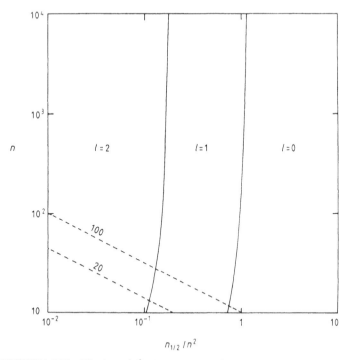

FIGURE 5.27   The $(n_{1/2}/n^2, n)$ parameter plane for the PARAFACR(l) algorithm. Notation as in figure 5.26. (From Hockney (1983), courtesy of IEEE.)

$(n_{1/2}/n)$ for a series of values of n, in order to determine approximately which algorithms abut each other in different parts of the parameter plane. One can then calculate the equal performance line between PARAFACR(l) and SERIFACR(l') from

$$n_{1/2}/n = (a - b)/(c - d) \qquad (5.150a)$$

where

$$a = \tfrac{1}{2}(3 \log_2 n + 1)l + 4 + (11 \log_2 n - 4)2^{-l}$$
$$b = 4l' + 4 + (1 + 5 \log_2 n)2^{-l'}$$
$$c = 4l' - 8 + 8 \times 2^{l'} + 5 \times 2^{l'} + 5 \log_2 n \qquad (5.150b)$$
$$d = [4l + (3 \log_2 n + 1)(2^l - 1) + 11 \log_2 n]/n.$$

The interaction of the two algorithms is shown in figure 5.28 on the $(n_{1/2}/n, n)$ parameter plane. This division between the two algorithms is about vertical in this plane showing that SERIFACR is the best algorithm for smaller $n_{1/2}$ (less than $0.4n$, the more serial computers), and that PARAFACR

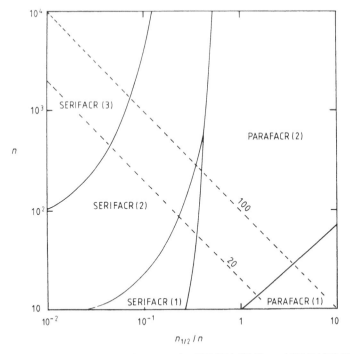

FIGURE 5.28   Comparison between the SERIFACR(l) and PARAFACR(l), showing the regions of the $(n_{1/2}/n, n)$ parameter plane where each has the minimum execution time. (From Hockney (1983), courtesy of IEEE.)

is the best for large $n_{1/2}$ (greater than $0.4n$, the more parallel computers). Lines of constant $n_{1/2}$ are shown for the CRAY-1 and CYBER 205. We conclude that SERIFACR should be used on the CRAY-1, except for small meshes with $n < 64$ when PARAFACR(2) is likely to be better. On the CYBER 205, PARAFACR is preferred except for very large meshes when SERIFACR(2) $(300 < n < 1500)$ or SERIFACR(1) $(n > 1500)$ is better.

5.6.3   Three-dimensional methods

The number of different strategies for solving the three-dimensional equivalent of the difference equation (5.119b) is large. We indicate here only the most obvious possibilities and leave it to the reader to estimate the best procedure for his purpose using the results of §5.6.1 and §5.6.2 as appropriate.

Iterative methods should only be considered for the case of general coefficients. The point SOR method obviously generalises to three dimensions. The SLOR method of line relaxation may also be used in three dimensions by assuming that all surrounding lines are correct and making a line correction. The adjustment of a whole plane of values, assuming the adjacent planes are correct (i.e. SPOR), may be performed if a program is available for solving a plane of values. This, of course, is the two-dimensional problem of §5.6.1 and §5.6.2, and an appropriate method should be selected. It will be iterative if the coefficients are general, or a direct method if the two-dimensional problem is the two-dimensional Poisson equation. Clearly the best choice will depend strongly on the coefficients of the difference equation. There is an advantage in solving as many of the dimensions of the problem as possible with direct methods. This will be possible for any dimension which has only even derivatives differenced on a uniform mesh.

In discussing wholly direct methods in three dimensions we limit ourselves to the solution of the discrete Poisson equation [generalisation to three dimensions of equation (5.131)]. The following possibilities suggest themselves:

(a) MFT—the multiple Fourier transform method of (5.133) or (5.139b) can be applied if a three-dimensional transform replaces the two-dimensional transform previously discussed. This method is used in the P^3M algorithm of Hockney and Eastwood (1988). The method does not have the minimum execution time but would be used if other circumstances demand a knowledge of the three-dimensional wave spectrum of the source that is calculated by the algorithm. This occurs frequently in plasma physics.

(b) $FACR(l)$—this method generalises to three dimensions if we replace the one-dimensional Fourier transform in the x direction with a two-dimensional transform in the (x, z) plane. Cyclic reduction is then performed plane-wise in the y direction. Since Fourier transformation is now twice as expensive per mesh point (data must be transformed in both x and z directions), the optimum value of l will be larger than for the two-dimensional case. A FACR(0) procedure was adopted by Kascic (1983) to solve a 64^3 problem on the CYBER 205, although he used a form of LU decomposition for the solution of the tridiagonal systems.

(c) *1D transform*—if a one-dimensional Fourier transform is performed, say in the z direction, then each harmonic obeys a discrete Helmholtz equation $(\nabla^2 \phi - k\phi = f)$ in the other two dimensions. This may be solved by any Helmholtz-solver. If this happens to be a FACR(l) procedure with Fourier transformation in the x direction and cyclic reduction in the y direction, one has in fact re-invented option (b) above.

In considering the implementation of the above options on particular computers, one must consider the use of n, n^2 and n^3 parallelism. In the simplest case of MFT we may transform n lines of length n in parallel using the best serial algorithm and repeat this n times in all three directions. Alternatively we may transform n^2 lines of length n in parallel using the best serial algorithm and repeat this in all directions. Finally, we may transform n^2 lines in parallel using the parallel algorithm PARAFT and obtain a parallelism of n^3. Similar considerations can be applied to the levels of parallelism used in the implementation of the iterative methods. There is clearly not space here to develop and contrast these alternatives; however all the necessary data and principles have been developed earlier in the chapter for the reader to do it himself for his particular problem.

6 Technology and the Future

Over the last five years, since the publication of the first edition of this book, we have seen major advances in the use of parallel processing. Up until that time the dominant architectures in the supercomputer market were the pipelined vector machines. such as the CRAY-1. The 1980s, however, have seen many manufacturers turning to replication in order to meet computational demands. All of the vector computer manufacturers of the late 1970s have looked for more performance by including multiple processors, usually coupled by shared memory. Another major development has been the INMOS transputer and other similar single-chip solutions to the use of multiple processors in a single system. The T800 transputer, for example, can perform at a continuous rate of 1 Mflop/s, and can be connected into four-connected networks of other transputers. The N-cube (Emmen 1986) processor can perform at about 0.5 Mflop/s and can be connected in ten-connected networks (hypercubes containing up to 1024 processors). In both cases 1000 chips is not an excessive number to be included into a single system, and therefore these CMOS VLSI devices will compete in performance with the more expensive ECL supercomputers.

Indeed, it is instructive to look at the technological aspects of the improvements in supercomputer performance. In the CRAY-1 range of machines, for example, although performance is up by a factor of six during this period, only a factor of 1.5 of this is due to an improvement in clock rates; the other factor of 4 comes from the use of multiple processors. This amply illustrates the diminishing returns in very high-speed logic implementations of large monolithic machines. It is quite likely that gate delays in this range of machines have improved by considerably more than the 1.5 times that the clock rate would imply; however, the length of wire in this range of machines has not been significantly reduced. It is this that is limiting the greater improvements in clock rate. The CRAY-2, by clever refrigeration engineering, has been able to reduce its clock because of a more compact physical size. This engineering, however, does not tend to produce cheap machines.

It can be said, therefore, that replication has become accepted as a necessary vehicle (desirable or not) towards greater computational power. Where we should go now that this barrier has been overcome is the purpose of this chapter. We will restrict ourselves mainly to architectural considerations, although the impact on algorithms and applications will not be ignored, as the latter is the main driving force behind the continuing quest for more computational power.

From the introduction to this book we have seen that increases in computational power have averaged a factor of ten every five years or so (see figure 1.1). This has been due to demand and there is no reason to suspect that this demand should suddenly abate. In fact the signs indicate that this demand for more computational power is increasing. The US Department of Defense launched a programme to develop VLSI processors, with capacities of 3×10^9 operations per second (Sumney 1980). Other more important applications are also begging for more computational power (Sugarman 1980), many of which are important to our life style (e.g. modelling energy resources, weather and climate, and even people in computer-assisted tomography).

Like all new developments, the use of parallelism in computer systems started at the top end of the computer market. The scientific main-frame or supercomputer, as it now seems to be called, is costly but provides a state-of-the-art performance (currently around 500 Mflop/s). However, as with all developments that prove cost effective, they soon find niches in a more general market, and become accepted in wider ranges of applications.

Applications which will provide large markets in the future have moved away from the scientific and simulation areas and are likely to be concerned with the more immediate aspects of many people's lives. For example, expert systems and database applications will become more frequently used. The Japanese fifth-generation program, and following this the UK Alvey and EEC ESPRIT programs, has provided a large amount of activity in this area. Other examples are: human interfaces, such as speech input and natural language understanding; image processing applications, for office systems or factor automation; and of course other signal processing applications for systems as diverse as mobile cellular radio and radar guidance systems.

There is no doubt that the provision of processing power for these applications will come from parallel processing. It is also obvious that this demand is being driven by the cheap processing power provided by the high-volume VLSI end of the semiconductor market. We therefore direct most of this chapter towards the technological trends, their effectiveness and limitation in the parallel and distributed systems that are described in Chapters 2 and 3 of this book.

6.1 CHARACTERISATION

Recent advances in integrated circuit design, in particular the levels of integration now possible, will allow many new architectural ideas to be put into practice. Technology also becomes a key component in some of the parameters we have defined. In particular the specific performance of a parallel computer, defined in §1.3.4 as the performance per unit parallelism, will be very technology-dependent. The most obvious contribution to this will be made in terms of circuit speed. This may, in well designed systems, be measured by the propagation delay of a single logic gate, τ_d. More indirect influences will be made in terms of power dissipation, as this is a major factor in determining packing densities and levels of integration.

The power required per gate, P_D, is made up from a static and transient power requirement. In the case of a 'demand' logic gate the static requirement is very small and the gate only draws power when switched. The power requirements here will therefore be a function of clocking frequency. Other logic is termed 'loss load' and here the transients are small compared with the static power requirements. In general the power requirement varies with logic output, thus average power dissipation is usually given.

In any technology there will always be some trade-off between power and gate delay, as devices will switch faster if driven harder. The operation point on this power delay curve is sometimes fixed in production, alternatively it may be varied using resistors external to the chip. The product $\tau_d P_D$ is another parameter associated with a given technology and gives a measure of the switching energy of a gate, e.g.

$$E_{sw} = \tau_d P_D. \tag{6.1}$$

To illustrate the importance of switching energy or power dissipation parameters, we will consider the power requirements for a single integrated circuit. In general the power limit for a single integrated circuit is around 2 W, although this may be more than doubled using bonded heat sink packaging and possibly fluid cooling. However, assuming 2 W per package, how many gates can be assembled onto a single chip? For the loss load gate this is given by equation (6.2) where $\bar{P}_D$ is the average power dissipation per gate:

$$N_g \leqslant 2/\bar{P}_D. \tag{6.2}$$

In the case of demand gates, power dissipation will depend on the average clocking frequency per gate, $\bar{f}_c$. Thus assuming two logic transitions per cycle N_g will be limited by equation (6.3).

$$N_g \leqslant (\bar{f}_c E_{sw})^{-1}. \tag{6.3}$$

The average clocking frequency will generally be less than the system clock, as not all gates will make transitions every clock period. This is particularly important in memory technology, where only a few memory cells are accessed in each memory cycle. However, because of circuit size, static memory cells usually use loss load logic. Tables 6.1 and 6.2 summarise the above relationships.

There are other considerations which limit the number of gates on a single chip, the most obvious being gate area. Gate area is a deciding factor for the cost of the chip, it measures the 'silicon real estate' required. However as minimum dimensions on devices approach the 1 μm barrier, area usage is more dominated by interconnections, as will be discussed later.

Thus we now have three parameters which characterise a given technology: gate delay τ_d, power dissipation per gate P_D (alternatively switching energy $E_{sw} = \tau_d P_D$) and finally gate area. In the following discussion we will use these as figures of merit in the comparison of technologies. We will also show how advances in processing techniques will affect these parameters for a given technology.

The technologies we consider fall broadly into two categories: logic made from bipolar transistors and logic made from field effect transistors, in particular MOSFETS. In the latter, current flow is in the plane of the chip, whereas in bipolar technologies, currents also flow perpendicular to the surface between regions which have been doped by different impurities, as

TABLE 6.1 Levels of integration possible for different switching energies and clocking frequencies for a demand gate technology.

f_c (MHz)	E_{sw} (pJ)					
	10^{-1}	1	10	10^2	10^3	10^4
10^{-1}	10^8	10^7	10^6	10^5	10^4	10^3
1	10^7	10^6	10^5	10^4	10^3	10^2
10	10^6	10^5	10^4	10^3	10^2	10
10^2	10^5	10^4	10^3	10^2	10	1
10^3	10^4	10^3	10^2	10	1	—

VLSI   LSI   MSI   SSI

TABLE 6.2   Levels of integration possible for a loss load gate technology as a function of average gate power dissipation.

$\bar{P}_D$ (mW)	200	20	2	0.2	0.02
N_g	10	10^2	10^3	10^4	10^5

	SSI	MSI	LSI	VLSI	

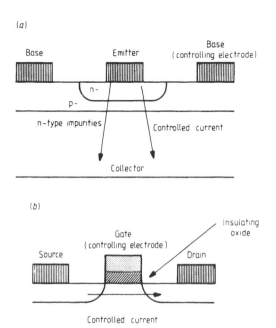

FIGURE 6.1   Cross section of the two basic switching transistors: (a) bipolar transistor; (b) MOSFET.

shown in figure 6.1. 'Free electrons' are introduced into the silicon crystal lattice by n-type impurities and 'free holes' are created in the lattice by p-type impurities. Holes are the absence of an electron, which move and create currents exactly as free electrons but of course in the opposite direction. They are, however, less mobile. It is because the boundaries between these regions can be controlled with great precision (~ 0.1 μm) when compared with planar dimensions (~ 1 μm) that bipolar technologies tend to be faster. Their logic circuits, however, are more complex.

There are many different technologies within these two broad categories, depending on circuits, materials and processing features. For example, many of the experimental technologies rely on novel processing steps. These reduce device dimensions and consequently reduce power and increase speed.

6.2   BIPOLAR TECHNOLOGIES (TTL, ECL, I²L)

The three major bipolar technologies are transistor–transistor logic (TTL), emitter coupled logic (ECL) and integrated injection logic (I²L). ECL is sometimes referred to as current mode logic (CML) and I²L as merged transistor logic (MTL).

TTL was the first integrated circuit logic to be produced. In 1964 Texas Instruments announced the first TTL family of circuits. The basic gate is the NAND gate, which operates with logic voltage levels of 0 and 2 V for logic 0 and 1 respectively. TTL is slow and has a high power dissipation; because of this it is now virtually obsolete. However other TTL series, namely Schottky TTL and low-power Schottky TTL, and FAST, are still used in small- and medium-scale applications.

The TTL market, however, is rapidly declining; the replacement technologies are provided by customisable chips, which provide a higher packing density for small random logic (glue logic). One low-cost solution is the PAL, which is a field programmable logic array, blown in a similar manner to an EPROM. If more complex circuits are required, the extremely bouyant market in gate arrays is providing the solution. These are user-customisable circuits (usually CMOS but also made in bipolar technologies), which require the last fabrication stage to be generated from the users' own requirement specification. These chips contain from 500–20 000 logic gates, arranged in regular arrays, and the users' circuits are wired in metal to connect the gates as required. This process is usually automatic, and can be generated from a computer-captured circuit diagram. Turn-around times can be very rapid, as the metallisation pattern can be implemented by exposing the silicon wafers using electron beams. The entry level costs for a run can also be very cheap, and at least one company offers a service for a few hundreds of pounds.

ECL is another bipolar logic family, but one which prevents the transistors from going into the saturation regime. Current is switched in the basic ECL circuit, which is the logic inverter. The inverter can be augmented to give the dual OR/NOR gate. Logical 0 and 1 are represented by -1.7 and -0.8 V respectively. ECL is often customised by the manufacturer for particular requirements. However, this does not mean that the entire chip will be built to a customer's specification, as gate array technology using OR/NOR gates is now a standard technology.

The principal advantage of ECL is its very small gate delay time, which can be as small as 100 ps for off-the-shelf components. Its other advantage is a large fan-out ratio, although a speed penalty will be paid for this. The disadvantages of ECL are its large power dissipation, around 1 mW per gate for sub-nanosecond delays, and its relatively large gate area ($> 300 \ \mu m^2$).

Currently therefore, ECL is not a good candidate for LSI circuits. Typical packages contain around 1000 gates.

6.3 MOS TECHNOLOGIES (NMOS AND CMOS)

The field effect transistor (FET) is a relatively simple device (figure 6.1(b)), in which a field generated by a gate electrode controls the flow of current in a channel between source and drain electrodes. The most common FET is called the MOSFET as the gate electrode is formed by a sandwich of metal–oxide–semiconductor. More recently, polysilicon has been substituted for metal in the fabrication process, leaving the metal for a second level of interconnections. The channel is formed from either n- or p-type charge induced in doped silicon of the opposite polarity, giving the NMOS and PMOS technologies. Although NMOS is still used, the third and major MOS technology uses both NMOS and PMOS transistors. This is called complementary MOS or CMOS. This is a good VLSI technology as it has very favourable power dissipation properties.

The major advantages of all MOS technologies are their relatively simple fabrication processes and high packing densities. Thus MOS has been used almost exclusively for LSI applications, giving extremely low-cost products (e.g. microprocessors and large memory chips). The disadvantage of MOS devices is their slow operating speed, although MOS technologies are rapidly gaining ground in this respect. Typical CMOS circuits will operate at around 20–40 MHz clock rates.

The basic NMOS gate is the inverter, which uses only two transistors, one of which is always on, either by biasing as in enhancement mode logic or by process control as in depletion mode logic. Thus this transistor acts as a fixed high-value resistor, while the second acts as a switch (figure 6.2). The operation of the gate is simple: the two transistors act as a resistance network, giving a voltage division between supply and ground. Thus when T_1 is on ($V_{in} \approx 5$ V), V_{out} is near ground potential ($V_{out} \approx 0$ V) and when T_1 is off ($V_{in} \approx 0$ V), V_{out} is near supply potential ($V_{out} \approx 5$ V). NAND and NOR gates may be constructed using pairs of switching transistors (T_{1A} and T_{1B}), connected either in parallel or series to ground (figure 6.3).

Because of the large resistances involved, MOS technologies consume very little power. For example when T_1 is on ($V_{in} \approx 5$ V) the total resistance between supply and ground is typically 100 kΩ. Thus with a supply of 5 V,

$$P_D = 0.25 \text{ mW},$$

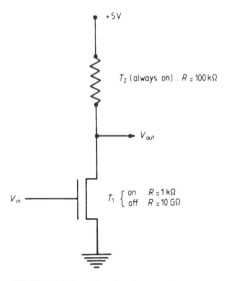

FIGURE 6.2   Circuit of a MOS inverter.

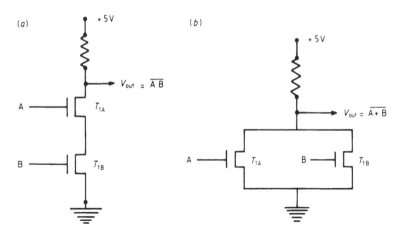

FIGURE 6.3   Combining MOS transistors to give dual input logic functions: (a) NAND gate; (b) NOR gate.

and similarly when T_1 is off

$$P_D = 2.5 \text{ nW}.$$

Thus it can be seen that NMOS gates only draw significant power when their output is low; also transient power spikes are not significant.

The packing density of the basic NMOS inverter gate using 2 μm rules is about 25 000 gates per mm^2 based on the geometries of an inverter, although this density is rarely found in practice.

In NMOS circuits, because transistor T_2 does not switch but only acts as a passive resistor, there is an asymmetry in the rise and fall times associated with this technology. The pull-down is active and typically requires 1–5 ns, whereas the pull-up is passive and requires 4–20 ns. This is shown in figure 6.5(a), which shows transient voltage and current characteristics. (Note: gate delay in MOS technologies is very sensitive to gate loading; these figures represent well-loaded gates.)

CMOS technology avoids the long pull-up delay by using active pull-up and pull-down. As PMOS has opposite polarity to NMOS, they can be used in a complementary pair to form the basic inverter gate (see figure 6.4(a)). This forms an almost symmetrical inverter (there are differences in carrier mobility) in which only one gate is ever turned on. The NAND and NOR gates are constructed using two complementary pairs of transistors. The NAND gate, shown in figure 6.4(b), uses the p-type transistors in parallel and n-type

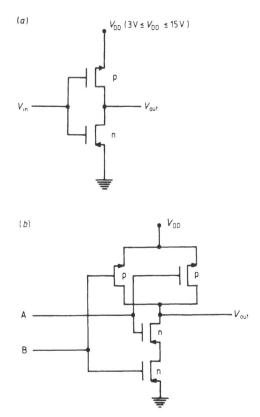

FIGURE 6.4 Complementary MOS (CMOS) logic circuits: (a) basic inverter; (b) NAND gate.

transistors in series. The NOR gate is constructed with series p-type and parallel n-type transistors.

Another attractive feature of CMOS technology is that it only draws current when switching (figure 6.5(b)). This is simply verified by noting that in any complementary transistor pair, one device is always off. This means that the power dissipation will be a function of average clocking rate. This is a great advantage in memory technology, where currently NMOS is approaching the thermal barrier of about 1–2 W per package (Wollesen 1980). CMOS has slightly poorer packing densities than NMOS and is a more expensive process. However, CMOS is now more competitive and has become the major process of the 1980s.

One disadvantage of all MOS processes, which is not shared by bipolar devices, is that they cannot be run hot. MOS devices suffer a decrease in speed of a factor of two over a 100°C temperature rise. This means that MOS technologies are more firmly bound to the 1–2 W per package thermal dissipation, whereas bipolar technologies are not.

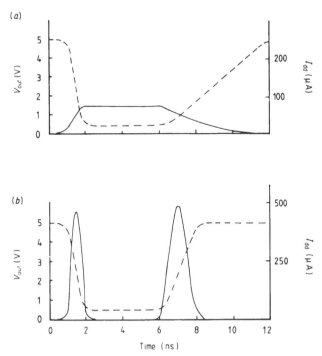

FIGURE 6.5 Dynamic characteristics of MOSFET switching gates: (a) NMOS; (b) CMOS. Broken curves V_{out}, full curves I_{DD}.

6.4 SCALING TECHNOLOGIES

Parallel computers, especially large replicated designs, are very well suited to the continuous revolution that is taking place in the micro-electronics industry. Very large-scale integration (VLSI), with 10^5 or even 10^6 gates on a single integrated circuit, is now commonplace and the technology marches on with the continuing advances in processing facilities, in particular the resolution of printing or 'writing' of the circuits on the silicon slice.

The effects of scaling down device sizes have been a topic of much interest and, for MOS transistors, rules which maintain well behaved devices have been known for some time (Dennard *et al* 1974, Hoeneisen and Mead 1972). For a scaling factor of K (> 1), if all horizontal and vertical distances are scaled by $1/K$ and substrate doping levels are scaled by K, then using voltage levels also scaled by $1/K$ the characteristics of devices should scale as given in table 6.3 (Hayes 1980).

As we have already seen, it is desirable to decrease the propagation delay, but not at the expense of the power dissipated. With these scaling rules, the delay is scaled by $1/K$ and the power dissipated per device is scaled by a factor of $1/K^2$, both very favourable. However it should be noted that the packing density is increased by $1/K^2$ and thus the power density remains the same. Thus it is possible to increase packing densities by device scaling, without the problems of approaching the thermal dissipation barrier.

There are problems in this scaling, as the current density increases by a factor of K, which may cause reliability problems. If current densities become

TABLE 6.3 Characteristic scaling factors for a MOS technology, scaled by K^{-1} in both horizontal and vertical dimensions and voltages, and with impurity densities scaled by K.

Characteristic	Scaling
Device current I	K^{-1}
Propagation delay τ_d	K^{-1}
Power dissipation P_D	K^{-2}
Power delay product $\tau_d P_D$	K^{-3}
Resistance R	K
Interconnect current density J	K
Packing density	K^{-2}
Power density	1
IR drop	1
RC constant	1

too high, metal connections will migrate with the current flow. Other problems occur because voltages must be scaled with devices, reducing the difference between logic levels. Noise levels remain constant due to the thermal energy of the discrete particles, and this can be a very critical problem. Scaling devices without scaling voltages lead to a K^3 increase in power density.

A final problem in device scaling occurs in the increased relative size and delay of the devices required to drive the external environment. Thus the full benefit of scaling may be found inside the chip but as soon as signals must leave the chip, only diminishing returns for scaling will be observed.

Although scaling is very attractive and may seem to be unlimited, there are certain fundamental limitations due to the quantum nature of physics. These are discussed in some detail in Mead and Conway (1980). In practice these limits should not be approached until devices shrink to 0.25 μm size, a further ten-fold reduction over current mainstream processes.

In contrast, the scaling for bipolar transistors is neither so rigorous nor so well defined. Bipolar transistors may be scaled without scaling vertical dimensions; indeed with present technologies it would be very difficult to scale this dimension to a very great extent. As transit times in bipolar transistors are dependent on the vertical dimension, ECL propagation delay times will not scale linearly with planar dimensions. However there will be some reduction in propagation delay due to capacitative effects and power scaling will be similar to MOS devices.

A paper by Hart et al (1979) has looked at the simulation of both bipolar and MOS devices as they are scaled down. Their results are summarised in figures 6.6 and 6.7. Experimental sub-micron devices which have already been fabricated confirm these simulated trends (e.g. Fang and Rupprecht 1975, Sakai et al 1979). How soon such devices will be in production is a difficult question to answer. The seemingly simple scaling rules presuppose many improvements in processing technologies and there is a great deal of development between the yields suitable for experimental devices and the yields suitable for VLSI production.

6.5   THE PROBLEMS WITH SCALING

In Chapters 2, 3 and 5 we have seen that communication is fundamental to both computer and algorithm design. Also from the above section we have hinted at communication problems found at the chip level. Therefore the communication problem is not something which magically disappears when we wave the VLSI wand; its emphasis is merely shifted from the system level to the microelectronic level. Thus communication is the key to any successful computer design.

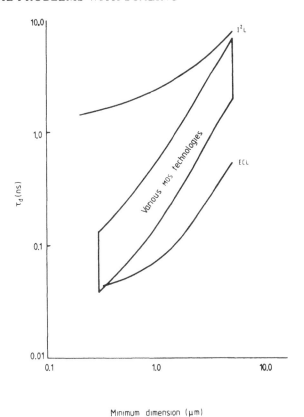

Minimum dimension (μm)

FIGURE 6.6 Simulated scaling of gate delay for ECL, I²L and various MOS technologies (data taken from Hart *et al* 1979).

Communication in an electronic digital computer is the propagation of 'square' and hence high-frequency electronic signals, along wires or printed circuits, or in impurity patterns in silicon. The fact that these signals take a finite time to propagate between the various components in a computer and that this time could be significant was discovered the hard way in one of the early MU designs. Manchester University had severe timing problems in constructing the first ATLAS computer. Since then the physical layout of a computer has become more and more important and is now a major design consideration.

The CRAY-1 epitomises this design limitation and all features of the machine are designed to minimise and equalise propagation delay. The most obvious feature is the strikingly unusual cabinet shape (see figure 2.1). This can be considered as a topological transformation of a rectangular cabinet which reduces backwiring lengths. Figure 6.8 gives an equal area transformation

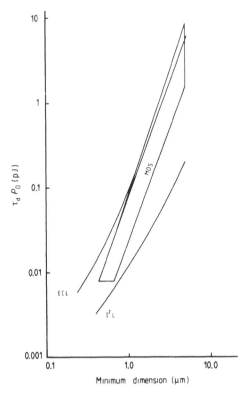

FIGURE 6.7   Simulated scaling of power gate delay product for ECL,
I²L and various MOS technologies (data taken from Hart *et al* 1979).

similar to that used in the CRAY cabinet. It can be seen that one dimension
of the backwiring plane is reduced by a half. The ideal shape is one with
complete spherical symmetry as this minimises wire lengths. Perhaps the
CRAY-2 will come in this shape!

Other features in the CRAY-1 which minimise propagation delay are
matched transmission lines which are resistively terminated and the extremely
high chip packing densities. These features cause considerable cooling
problems, and in excess of 100 kW of heat power must be extracted from
the 100 or so cubic feet of cabinet space. To give some feeling for this power
density, imagine putting a 1 kW electric element into a biscuit tin and trying
to keep it just above room temperature. Even with such a feat of refrigeration
50% of the 12.5 ns clock period on the CRAY-1 is an allowance for
propagation delay.

To given an idea of the scales involved in this problem consider the
following: in a perfect transmission line signals travel at the speed of light

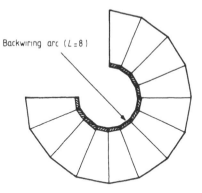

FIGURE 6.8   A constant area mapping used to reduce backplane wire lengths.

which is 952×10^6 ft/s, i.e. a little under 1 ft/ns. In practice, transmission lines have capacitance and resistance and the speed of propagation of a signal will be attenuated. For an RC network the diffusion equation describes the propagation of signals:

$$RC\frac{dV}{dt} = \frac{d^2V}{dx^2},$$
(6.4)

where R and C are the resistances and capacitances per unit length. The diffusion delay varies quadratically with length for constant R and C, and is the critical delay for on-chip signal propagation.

Returning to table 6.3, we see that although the RC delay remains constant during scaling, this figure is based on a scaled-down wire. It is reasonable to assume that the chip itself will remain the same size, so as to reap the benefits of the increased packing density. If this is so, distances will be relatively larger, with the net effect that tracks spanning the chip will be slower to respond by a factor of K^2, using the same drive capability. Indeed the situation is worse than this, for this absolute delay must really be compared to the now faster gate delay, which has scaled by K^{-1}, giving an overall perceived degradation of K^3 in the responsiveness of a global wire. Although the effects of this unfavourable scaling are only just being felt, this factor is likely to have a major effect on the design of systems exploiting the new technologies. They will be blindingly fast at a local level, but increasingly sluggish at the global or off-chip levels. The implications of this are now considered.

As processing techniques improve and more devices are integrated onto single circuits, the scaling factors described above will have a major effect on the architectures implemented, and indeed on the design methodologies. The problems that must be overcome include:

(a) increase in design effort;
(b) greater wire delays;
(c) greater wire density;
(d) discrepancy between on- and off-chip performance.

The latter problem arises because of the constant view of the world that is presented by external components, such as pads, pins, and circuit tracking. It introduces a discrepancy between on- and off-chip performance. On-chip circuits may cycle at 25–50 MHz, but it is difficult to cycle pad driver outputs at this rate, without using excessive power.

The problems associated with wire density can be illustrated by considering the effects of scaling on the interconnection of abstract modules on a chip. If we assume the same scaling as for the electrical parameters, then we could obtain K^2 more modules of a given complexity on a scaled chip. If we assume that each module is connected to every other module on the chip, then for n modules n^2 wires are required. The scaled circuit requires $(nK^2)^2$ wires giving a K^4 increase in wire density for a corresponding K^2 increase in module density. This is obviously a worse case, as in general the modules will not all be fully connected. However, the best case, which maintains a balance between wire and gate density, would require each module to be connected to one and only one other module. As mentioned earlier, the problem of communications is thus traded between system and implementation levels; the linear network that would result is only suitable for a few applications (see §3.3). However, this is known to be a major problem, and mandates design styles and architectures to be adapted to meet or ameliorate it. The solutions, as in most engineering situations, are found in squeezing the problem on all fronts, which may (for example) include the introduction of new interconnect techniques employing optical connections (Goodman *et al* 1984).

Looking at these problems in terms of design, we can find styles which can be used to advantage at both the circuit and systems levels. At the circuit level we can employ large and regular gate structures, which overcome the problems of wire density. Effectively, they increase module size and ameliorate the wire density problem by containing a regular network of local or bused connections. Examples are: RAM; ROM; programmable logic array (PLA); switch array; gate matrix.

To illustrate this consider figure 6.9, which shows an early and a modern microprocessor. In the former, the module is effectively the gate, and logic is built up randomly, as one would do using TTL chips. In the latter, there is much more structure evident, including large blocks of RAM and ROM, and regular datapaths in the ALUS.

Although ROM and RAM are ideal structures for VLSI, due to their regular structure, there is little processing that can be accomplished using RAM alone; processing architectures that also exploit this fine and regular structure are also prime contenders for good VLSI chips. At the system level, therefore, we can use the techniques which create this regular and preferably local structure. Pipelining and replication have been extensively discussed as architectural techniques for increasing machine performance without requiring increased component speed. It is fortuitous, therefore, that these same techniques provide for regular structure and can be designed with regular and local communications. Figure 6.10 illustrates a pipelined data path, where it can

(a)

FIGURE 6.9   A comparison of different generations of microprocessors.
(a) An Intel 8080 8-bit microprocessor.

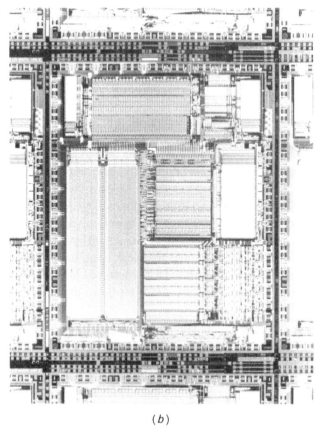

(b)

FIGURE 6.9 cont.   (b) A T800 transputer 32/64-bit microprocessor.

be seen that the introduction of registers for storing partial results within the data path creates local temporal regions instead of a global temporal region. The clocks required to synchronise the flow of data through the pipeline are still global signals; however, their latency does not prejudice the operation of the system, providing any delays to different modules can be equalised. Indeed, if required, the system may be made asynchronous or self-timed, by providing a local agreement between stages as to when data should be passed. This requires a handshake between adjacent modules. Asynchronous pipelines are a programming technique which is described in §4.5.2.

Figure 6.11 illustrates the chip floor-plan of a typical processor array, which uses replication as a means of improving performance. Again it can be seen that connections between the replicated modules have been chosen to reflect the planar nature of the medium and are therefore local. The same arguments concerning control apply equally to the processor array structure.

(a)

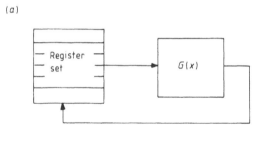

(b)

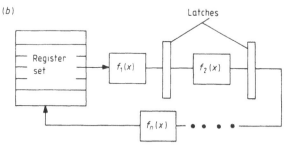

FIGURE 6.10 A diagram comparing the conventional ALU (a) with a pipelined ALU (b). It can be seen that the latter has locally distributed communication.

Control bus

FIGURE 6.11 Diagram showing the floor-plan and regular local communications in the RPA PE chip.

Either a synchronous system is used, in which case a clock and even the control word will be globally distributed from a single source. Otherwise the system may be made self-timed, in which case each processor would have its own clock and local control store. The former is typical of a SIMD computer and the latter would correspond to a multi-transputer chip. Such designs have been considered by INMOS, and (moreover) it has been proposed that they could be automatically generated from OCCAM programs by a silicon compiler (Martin 1986).

6.6   SYSTEM PARTITIONING

It seems ironic that the problems found in VLSI systems' design seem to reflect those in society at large. Chips are rapidly evolving into a two-class society—as in systems comprising many VLSI circuits there are the privileged, who can communicate locally or on-chip, and the unfavoured, who must resort to slow off-chip communications. With these inequalities the partitioning of a system becomes very important. There are three main considerations when partitioning a system into integrated circuits. These are yield, pin-out and power dissipation, which are considered below. However, the problems above concerning the discrepancy between on- and off-chip performance will also constrain the partitioning. Points of maximum bandwidth must now be maintained on-chip.

It is interesting to note that components from most semiconductor companies maintain the point of maximum bandwidth in a system off-chip, and then add complexity to the system in order to minimise the bandwidth through this bottleneck. The interface is, of course, the memory–processor interface, which carries code and data between memory parts and processor parts in all microprocessor systems. The complexity introduced to combat the excessive bandwidth at this point includes on-chip cache memories, instruction prefetch and pipelining, and more complex instruction sets. There are of course exceptions to this approach, which involve an integration of memory and processor function, and a deliberate attempt to reduce unnecessary complexity. A good example of this style can be found in the INMOS transputer, described in §3.5.5.

6.6.1   Yield
Ideally one would like to implement the whole of a system onto a single integrated circuit, as driving signals off-chip is slow, consumes a great deal of power, is inherently unreliable and requires a large volume to implement.

However, defects are introduced into the silicon during processing which can cause circuits not to work as expected. These defects include:

(a) crystal defects in the materials used;
(b) defects in the masks used to pattern the silicon;
(c) defects introduced during processing (e.g. foreign particles);
(d) defects introduced by handling (e.g. scratching, photoresist damage);
(e) defects introduced by uneven processing (e.g. metal thinning);
(f) pinholes between layers;
(g) crystal defects introduced during processing.

Because of the presence of these defects, only a fraction of the chips on a processed wafer will be completely functional (assuming a correct design). For example, consider a 3 inch wafer with 100 chip sites fabricated in a MOS technology. A typical yield for such wafers would be around 30%, implying an average of 30 working chips per wafer. The number of defects occurring on a wafer is usually assumed to be distributed randomly and expressed as a number per cm^2. During the last 20 years, major advances have been made in clean room technology, equipment used, and materials and masks. As a consequence it is now possible to produce reasonable yields on chips up to and even over 1 cm^2.

The simplest yield model assumes a random distribution of point defects and, furthermore, assumes that a single defect anywhere on the chip will cause that chip to fail. In this case the probability of finding any defects on a given chip can be calculated using the Poisson distribution and the parameters D (the defect density) and A (the area of the chip). The probability of a chip being good using Poisson statistics is given by

$$P(D,A) = \exp(-DA).$$

This model, however, does not accurately reflect the real behaviour of a fabrication process, although it gives a fair approximation to the region of very low yield. Chips of an area of many times $1/D$ will have a vanishingly small probability of not containing a defect. Areas must be kept to a few times $1/D$ to give reasonable yields. The defect densities found in a good process will vary from $1-5\,cm^{-2}$. Thus chips of around 1 cm^2 will give modest yields.

In reality the defects are not randomly distributed and many more defects will be found in the periphery of the wafer when compared with the centre. Also the assumption that all defects can be modelled as points is not valid and many defects are large compared with the feature sizes found on the chip. These are called area defects. Mathematically, points are very much easier to deal with than areas. Area defects may be modelled with modifications to Poisson statistics by considering the area to contain a number of point defects. This effect is therefore modelled as a defect clustering, and the distribution can no longer be considered as being random. The system can be modelled by the superposition of a number of random distributions. The final flaw in the simple Poisson model is that the defects are distributed over a number of process steps, some of which are more critical than others. For example, a masking defect on the metal 1–metal 2 interface via the mask will not be fatal unless it is included in the union of metal 1 and metal 2. It is clear from this that not all defects will cause a chip fatality.

Although it may not be possible to avoid defects entirely, it is possible to design circuits which can operate in the presence of some defects. These techniques, which are considered in §6.7 below, enable larger chips to be economically fabricated. The techniques have been successfully exploited in memory design for some time. The regular structure of memory gives particularly efficient schemes for avoiding faults introduced during processing. The ultimate goal, of course, is to achieve the situation where the complete silicon wafer can be used for a single system. At current levels of complexity, this would imply systems containing hundreds of millions of gates. The need for replication in such systems is clear.

6.6.2 Pin-out

Another limitation on the partitioning of systems is the number of pins that may be provided between the integrated circuit and the outside world. This is constrained by packaging and power dissipation requirements, and also contributes to the bandwidth limitation at the chip boundary.

A good design for a VLSI chip is therefore a portion of the system that is to some extent self-contained, and has as few wires as possible to the outside world. However, this often conflicts with other design requirements for VLSI, as found in partitioned, regularly connected structures such as grid-connected processor arrays. The problem is that as more and more processors are included onto a single silicon substrate, then the bandwidth required between chips also becomes larger. This increase can either vary with the area of processors enclosed, as in the case of external memory connections found in many SIMD chips, or as the perimeter of the array, which is a partitioning of the grid network.

External connections to a chip are expensive; on-chip pads consume a relatively large area, and the size of a pad (100–$150 \ \mu m^2$) must remain constant, despite any scaling of the circuit. The driver circuits also remain constant in size, to maintain drive capabilities. Pad drivers also consume a large amount of power, which can also cause considerable noise on the supply rails if many pads change state at the same time (as in a bus for example).

Externally, large packages are expensive and consume valuable board area. Although packaging technologies are now providing dense, low-area chip carriers, such as pin grid arrays, and surface mount packages, the density of pins makes circuit boards more expensive due to the additional layers required to handle the track density.

Single-chip packaging is not the only boundary that can be drawn in system partitioning, as many chips may be mounted on a substrate before packaging. Manufacturers are experimenting with packages containing hybrids for commercial use. They have been used in military applications for some time. Examples are ceramic thick film and even metallised silicon. Although these techniques are relatively new and still expensive when compared to conventional packaging technologies; however, their obvious advantages will force their development.

To find a good partitioning in a system, graph theory may be used. At a given level of description, any system may be described by a connected graph, where the system components are the nodes (these may be gates or functional blocks, for example) and the wires connecting them are the edges of the graph. A good system partitioning divides the graph into subgraphs, where each subgraph contains a high degree of connectivity and the graph formed by the partitioning has a low degree of connectivity.

The situation is not quite so simple as this, however, as no system has only a single implementation and there are always a number of trade-offs that may be made to reduce the pin-out of a given partitioning. For example, at the expense of additional delay, signals may be encoded and then decoded on-chip prior to use. Also a single pin may be used for many signals by time-multiplexing the data.

The INMOS link implemented on the transputer provides a good example of such a trade-off. The transputer has four links to connect it to other transputers, each implemented as a pair of wires transmitting a byte of data in an 11-bit data packet. The link is bidirectional and can transmit data and acknowledge packets in any order. A large proportion of chip area has been used to optimise the speed of this link (20 MHz). This is a good compromise, however, as the alternative parallel implementation would have required at least ten wires per link. Scaling this figure by a factor of eight to provide the four bidirectional links gives a large difference in pin count. Thus, some speed has been lost in return for a massive pin reduction and a modest use of chip area, the non-critical resource.

6.6.3 Power dissipation

Power dissipation is the final constraint on system partitioning. There is a limit of about 2 W per package for conventional packaging technologies, i.e. printed circuit boards and forced air or convected cooling. Beyond this, more sophisticated packaging technologies are required, such as hybrid ceramic substrates, cooling fins, heat clamps, and liquid immersion technology. All of these techniques increase system costs.

As an aside, low-voltage, low-temperature CMOS provides a very fast technology, which is now comparable to ECL. ETA are building supercomputers from VLSI CMOS gate arrays, using liquid nitrogen temperatures (77 K) and low voltages. They are obtaining 100–200 MHz system speeds.

In any system there is a dynamic power dissipation (P_d), which is due to capacitative loading and is proportional to the number of gates (N_g), their average frequency of operation $\langle f \rangle$, and the voltage through which they are switched (V_{DD}):

$$P_d = N_g \langle f \rangle C V_{DD}^2$$

where C is the load capacitance.

This is the major dissipation in a CMOS circuit. However, in NMOS, in addition to this, there is a constant power dissipation on all gates whose output is low. This is given by

$$P_d = N_g \langle V_{DD} - V_{out} \rangle / R_{dep}.$$

A typical value for R_{dep} in an NMOS circuit (the effective gate load) is about 50 kΩ, giving a power dissipation for a single gate of about 0.1 mW.

6.6.4 Techniques for reducing power dissipation

There are a number of techniques that can be used in order to reduce the power consumption of an integrated circuit. In general these are more appropriate to the loss load technology, such as NMOS.

(1) The first technique is to reduce pin-out. Pad drivers may consume many orders of magnitude more power than logic gates. If there are many drivers, then this can represent a large proportion of the power dissipated on-chip.

(2) Pad drivers can be designed so that the power spikes on transitions are minimised. This can be achieved using non-overlapping phases on the push–pull output stage.

(3) Another technique is to use dynamic logic, which has no static power dissipation. This technique gives a space reduction in CMOS systems, with little power reduction. In NMOS, however, it gives considerable improvements in power dissipation.

(4) There are techniques for reducing the static loss in NMOS systems, such as transistor switch arrays, which again use no static power. These can be

used in CMOS also, although an n-well process is preferred, as the 'good' devices are n-channel, which have the better mobility.

6.7 WAFER-SCALE INTEGRATION

Wafer-scale integration (WSI) is the effective use of an entire processed wafer of silicon on a single system. Many of the disadvantages encountered in the scaling of VLSI technologies will disappear (some will remain). However, in order to utilise a circuit with certain faults, special techniques need to be adopted.

There has always been a trend in IC manufacture to get as much circuitry as possible onto a single chip. The two to three orders of magnitude increase in complexity that has been achieved over the last decade in the semiconductor industry (Augarten 1983) has been brought about by decreases in circuit feature sizes. Chip sizes have remained more or less constant, at a little under 1 cm^2. However, because of the statistics of yield, the harvest of good chips from a wafer will decrease rapidly with larger chip sizes. Therefore, the leading edge of the industry, typified by memory manufacturers, has used fault-tolerant techniques and redundancy to increase yields of the larger chips. By using extensive redundancy, and relaxing design rules where critical, it is feasible to increase the chip area further, until the circuit covers the complete wafer. The earliest reported attempt at a whole wafer circuit was over 20 years ago. In the UK, Sinclair Research has started a project to investigate the feasibility of wafer-scale products. This project is still continuing under the company Anamartic (without fault). The first products produced will be solid state mass memory devices (Pountain 1986b). Later products will have processing power distributed throughout this memory; they will be parallel computers.

Potential advantages of wafer-scale circuits are:

(a) higher speed through shorter interconnect and smaller loads;
(b) lower power through fewer pad drivers;
(c) greater density of interconnections possible;
(d) smaller volume;
(e) higher reliability through fewer mechanical connections;
(f) lower system costs.

Against these can be set the known disadvantages:

(a) yield statistics work against very large circuits;
(b) high power densities;

(c) testing difficulties;
(d) configuration costs (may need special process);
(e) slower and more expensive prototyping;
(f) problems in mixing technologies;
(g) not appropriate for all systems.

Many of the potential difficulties in wafer-scale integration can be ameliorated by the choice of the correct architecture. It is clear, however, that these architectures must be regular and replicated.

6.7.1   Architecture for WSI

Architectures suitable for WSI are regular, contain few highly replicated modules and ideally contain only regular local communications. In this way both design effort, a major consideration, and the adverse effects of wire scaling are minimised. Some good architectures for WSI are *memory, systolic arrays* and *processor arrays*. But how good are these for processing data?

(i) Memory
Memory is one of the most ideal structures for WSI implementation. It is very regular, consists of a very small replicated module (a transistor and capacitor in dynamic RAM) and has regular interconnections, i.e. word and bit lines. It is not surprising, therefore, that it has been one of the most successful products to exploit redundancy. Although many experiments have been undertaken to construct wafer-scale circuits, it is in the area of yield enhancement of marginal state-of-the-art chips that the successes in memory design have been made commercially. However, memory does not perform processing, and the availability of cheap memory has perpetuated the von Neumann architecture, with its separate processor, control and memory unit.

(ii) Systolic arrays
Systolic arrays are regular arrays of simple finite state machines (FSMs), where each finite state machine in the array is identical. The name is derived from 'systole' a medical term describing the heart. A systolic algorithm relies on data from different directions arriving at cells in the array at regular intervals and being combined. Some state other than for the control of the local FSM may be involved. The algorithm therefore relies on the position of data and its propagation through the array. Although these machines are regular and contain simple cells, they do not perform very general purpose computations, or not until the FSM becomes a stored program machine, when they can be considered as processor arrays. See Moore *et al* (1987) for more information.

(iii) Processor arrays

When the replicated element becomes a computer, rather than a simple finite state machine, the structure can be called a processor array. In replicated computer systems, the most important issues are a matter of scale (see Chapter 3). Whereas a processor containing small numbers may communicate efficiently using bused systems or by shared memory, in systems using larger numbers the inherent sequentiality of these methods produces bottlenecks. The designer is then forced to consider systems in which data is switched between processors. This switching takes place either in a fixed topology network, in which distances grow with the number of processors involved, or in programmable connection networks, in which the costs of the switch follow a square law. In silicon, the planar nature of the connection medium will tend to favour a planar network.

6.7.2 Review of techniques

The techniques for increasing the area of chips of adequate yield fall into two categories—fault-tolerant circuit techniques and redundancy.

(i) Fault-tolerant circuits

Circuits can be designed to tolerate faults, usually by including redundant coding into the circuit operation. In this way certain faults or fault combinations can be detected and masked during the circuit's normal operation. These techniques are usually expensive in area and are not appropriate by themselves for building whole wafer circuits. They can, however, be used to ensure the integrity of critical circuits using other schemes. One area where this technique is particularly effective is in the implementation of memory circuits, where redundancy codes can be used to detect and correct a single bit in error in any word (unit of protection) in the memory. The overhead for this is $(\log_2 n)/n$, 50% for 8-bit data.

(ii) Modular redundancy

Modular redundancy requires a circuit to be duplicated, triplicated or otherwise replicated, and combined with circuits to detect differences in outputs. With two circuits, if the circuits agree over a range of tests, then we can be confident that they are functioning. However, the yield of the functioning circuit has been decreased. With triple circuits, any two of three agreeing can be confidently accepted as a correct result. A voter circuit is therefore required for this. A single voter will not, however, detect a wide range of common failure modes. This can be overcome using triple redundancy and triple voter circuits. However, this technique imposes a massive overhead, which may be unacceptable. It is typically used in situations where continuous operation is required from a system, where provision has been made to off-line and repair the faulty module when detected.

(iii) Other redunacncy schemes

Other redundancy schemes are only appropriate where there is a high degree of regularity within a circuit. These are, however, the most suitable schemes and can be exploited by all regular architectures, such as memory or arrays of processor of some description. Because of the regular structure, it is possible to implement a set of spares, which may be included in the system in the event of failure of one of the system's modules. An example would be the use of spare rows and/or columns of processors in a rectangular array of processors. This is now done routinely by most memory manufacturers and involves little overhead. The spares can be included in the system in a permanent (fabrication step after test) or volatile manner. In the latter case the system can be 'repaired' after in-field faults.

An example of such a replacement scheme is shown in figures 6.12 and 6.13, taken from experimental work performed at Southampton University (Bentley and Jesshope 1986). Figure 6.12 gives a schematic diagram of one quadrant of the wafer shown in figure 6.13. Each of the blocks is serial memory, and the overall system provides a parallel-access, solid state disc memory. In each quadrant one of eight virtual columns can be addressed and enabled onto the output bus. However, each cell receives four adjacent

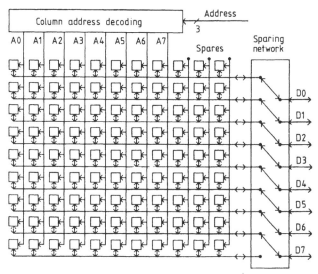

FIGURE 6.12   A schematic diagram of the operation of a wafer-scale memory circuit containing two-dimensional redundancy.

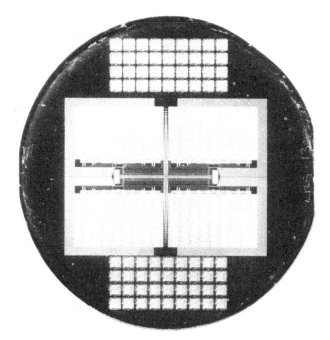

FIGURE 6.13   A plate of the fabricated wafer-scale circuit shown in figure 6.11. (Approximate diameter 7.5 cm.)

word lines, and coded configuration control information indicating the number of bad cells to the left (0, 1, 2, or 3 bad). Using this control information, the appropriate decoded word line is used to select the cell. Because each row has its own control bus, each row may discard any three bad cells. Should a row contain more than three bad cells, an error control signal will tell the sparing network to ignore that row. Only one complete bad row may be dropped in each quadrant. Thus each quadrant of the memory may be configured by any eight good cells from eleven in each row and eight good rows from the nine implemented.

The wafer can be self-configuring, as when in test mode, a coded signal can be written to all blocks in parallel. This signal is recognised after having exercised the bus and storage logic, and a status latch stores a good/bad status on leaving the test mode. It is this status latch that is then used to generate the control signals in each row. Obviously the status latch and control logic have not been tested by this process, but there will always be a core of logic in any self-tested system, which must be verified by externally verified test equipment.

This experimental wafer does not have the correct communication structure for a grid-connected array. However, modifications to this scheme have been proposed (Jesshope and Bentley 1986, 1987) and are being implemented in silicon. These schemes would allow arrays of up to 16×16 RPA PES to be fabricated on a single 3 inch wafer, using 3 μm CMOS technology. More aggressive design rules (1 μm, say) and a larger wafer (5 inches, say), could be expected to give a 25-fold increase in density over the above estimate. This would give up to 200 Mflop/s for floating-point operations and up to 2 Gflop/s for 16-bit integer operations—quite an impressive performance for a computer that you could slip into your pocket!

We do not have the space for a complete treatment of this subject here, but for further reading those interested are referred to the recent book edited by Jesshope and Moore (1986). This is the proceedings of a conference held at Southampton University, at which no fewer than three prototype wafer designs were displayed; one from Sinclair Research was demonstrated while configuring itself. Like the Southampton design, this demonstration was a memory circuit, but configured from a linear chain of cells, constructed from a spiral of good sites on the wafer.

6.8   THE LAST WORD

With technology making advances that seem to have no prospective limit, there is no doubt that, with all their problems, parallel computers are here to stay. It is perhaps unfortunate that software technology has not kept pace with the rapid advances made in hardware developments. The reason for this seems to be the driving force behind these developments; the almost insatiable thirst for more processing power. This has led to the rejection of the more abstract developments in software technologies, in favour of highly optimised, but mature (long in the tooth one might say), systems. However, with applications for the superspeed computers opening out into areas where systems are more complex, there has been a large impetus in the development of more abstract programming methodologies (Harland 1984, 1986). What is required before these developments mature is a raising of the level of abstraction of the hardware of the machine, while at the same time exploiting the nature of the technology, its powers and limitations. This is not a trivial task, however, as both sides of the equation are rapidly moving targets. The recent trend in higher education, in which these disciplines are converging (information engineering), is a good omen for the future in this respect.

Appendix

SYNTAX OF ASN—AN ALGEBRAIC-STYLE STRUCTURAL
NOTATION

The syntax of the structural notation for computer architecture is defined
below in the Backus Naur Form (BNF) as used in the ALGOL60 report
(Backus *et al* 1960). Angular brackets ($\langle \rangle$) delimit metalinguistic terms and
the vertical bars (|) separate alternatives (read as 'or'). The double-colon
equals (:: =) is to be read as 'may be'.

A.1   MISCELLANEOUS

$\langle$ empty $\rangle$:: =
$\langle$ digit $\rangle$:: = 0|1|2|3|4|5|6|7|8|9
$\langle$ lower case letter $\rangle$:: = a|b|c|d|e|f|g|h|i|j|k|l|m|n|o|p|q|r|s|t|u|v|w|x|y|z
$\langle$ pipelined $\rangle$:: = p|$\langle$ empty $\rangle$
$\langle$ SI prefix $\rangle$:: = K|M|G|T|$\langle$ empty $\rangle$
$\langle$ unsigned integer $\rangle$:: = $\langle$ digit $\rangle$|$\langle$ unsigned integer $\rangle\langle$ digit $\rangle$
$\langle$ power $\rangle$:: = $\langle$ unsigned integer $\rangle$|$\langle$ empty $\rangle$|$\langle$ lower case letter $\rangle$
$\langle$ multiplier $\rangle$:: = $\langle$ lower case letter $\rangle$|$\langle$ unsigned integer $\rangle\langle$ SI prefix $\rangle$
         |$\langle$ unsigned integer $\rangle\cdot\langle$ unsigned integer $\rangle$|$\langle$ comment $\rangle$
$\langle$ factor $\rangle$:: = $\langle$ multiplier $\rangle^{\langle \text{power} \rangle}$|$\langle$ factor $\rangle$ * $\langle$ factor $\rangle$
$\langle$ comment $\rangle$:: = ($\langle$ any sequence of symbols $\rangle$)|$\langle$ empty $\rangle$
$\langle$ statement separator $\rangle$:: =   ;
$\langle$ statement $\rangle$:: = $\langle$ definition $\rangle$|$\langle$ highway definition $\rangle$|$\langle$ structure $\rangle$

Examples
$\langle$ empty $\rangle$:: =
$\langle$ digit $\rangle$:: = 3; 9
$\langle$ lower case letter $\rangle$:: = c; z
$\langle$ pipelined $\rangle$:: =   ; p
$\langle$ unsigned integer $\rangle$:: = 34; 128

$\langle\,\text{power}\,\rangle ::= \;; 2; s$
$\langle\,\text{multiplier}\,\rangle ::= n; 16; 8G; 9.5$
$\langle\,\text{factor}\,\rangle ::= n*m; 64^2; n*128*32^2$
$\langle\,\text{comment}\,\rangle ::= \quad ; (\text{bipolar ECL})$

A.2   E UNITS

$\langle\,\text{E symbol}\,\rangle ::= B|Ch|D|E|F|IO|P|U|S$
$\langle\,\text{E identifier}\,\rangle ::= \langle\,\text{E symbol}\,\rangle\langle\,\text{pipelined}\,\rangle | \langle\,\text{E identifier}\,\rangle\langle\,\text{digit}\,\rangle$
$\langle\,\text{operation time in ns}\,\rangle ::= \langle\,\text{multiplier}\,\rangle | \langle\,\text{comment}\,\rangle$
$\langle\,\text{bit width of operation}\,\rangle ::= \langle\,\text{multiplier}\,\rangle$
$\qquad | \langle\,\text{bit width of operation}\,\rangle, \langle\,\text{multiplier}\,\rangle | \langle\,\text{comment}\,\rangle$
$\langle\,\text{E unit}\,\rangle ::= \langle\,\text{E identifier}\,\rangle {}^{\langle\,\text{operation time in ns}\,\rangle}_{\langle\,\text{bit width of operation}\,\rangle} \langle\,\text{comment}\,\rangle$

Examples
$\langle\,\text{E identifier}\,\rangle ::= E; F12; Bp46$
$\langle\,\text{operation time in ns}\,\rangle ::= t; (4 \text{ milliseconds})$
$\langle\,\text{bit width of operation}\,\rangle ::= b; (4 \text{ Bytes}); 16, 32$
$\langle\,\text{E unit}\,\rangle ::= E; F12^{100}_{32}(*); S(\text{omega})$

A.3   M UNITS

$\langle\,\text{M symbol}\,\rangle ::= M|O$
$\langle\,\text{M identifier}\,\rangle ::= \langle\,\text{M symbol}\,\rangle\langle\,\text{pipelined}\,\rangle | \langle\,\text{M identifier}\,\rangle\langle\,\text{digit}\,\rangle$
$\langle\,\text{access time in ns}\,\rangle ::= \langle\,\text{multiplier}\,\rangle | \langle\,\text{comment}\,\rangle$
$\langle\,\text{number of words}\,\rangle ::= \langle\,\text{multiplier}\,\rangle *$
$\qquad | \langle\,\text{number of words}\,\rangle\langle\,\text{number of words}\,\rangle | \langle\,\text{comment}\,\rangle *$
$\langle\,\text{bits in word accessed}\,\rangle ::= \langle\,\text{multiplier}\,\rangle | \langle\,\text{comment}\,\rangle$
$\langle\,\text{size of memory}\,\rangle ::= \langle\,\text{number of words}\,\rangle\langle\,\text{bits in word accessed}\,\rangle$
$\qquad | \langle\,\text{comment}\,\rangle$
$\langle\,\text{M unit}\,\rangle ::= \langle\,\text{M identifier}\,\rangle {}^{\langle\,\text{access time in ns}\,\rangle}_{\langle\,\text{size of memory}\,\rangle} \langle\,\text{comment}\,\rangle$

Examples
$\langle\,\text{M identifier}\,\rangle ::= Mp; M1; M2; M3; O16$
$\langle\,\text{access time in ns}\,\rangle ::= 100; (4 \text{ milliseconds})$
$\langle\,\text{number of words}\,\rangle ::= n*; (1, 2 \text{ or } 4 \text{ MBytes})*;$
$\langle\,\text{bits in word accessed}\,\rangle ::= b; 32$
$\langle\,\text{size of memory}\,\rangle ::= n*b; n*32; \;; n*; *32; 2K*8*64$
$\langle\,\text{M unit}\,\rangle ::= M; O16^{100}_{n*32}; M_{2K*8}(2716 \text{ EPROM}); M_{8*64*64}$

A.4   COMPUTERS

$\langle$ control symbol $\rangle ::= I | Iv | C | Cv$
$\langle$ control identifier $\rangle ::= \langle$ control symbol $\rangle \langle$ pipelined $\rangle$
          $| \langle$ control identifier $\rangle \langle$ digit $\rangle$
$\langle$ clock time in ns $\rangle ::= \langle$ multiplier $\rangle | \langle$ comment $\rangle$
$\langle$ bits in instruction $\rangle ::= \langle$ multiplier $\rangle$
          $| \langle$ bits in instruction $\rangle , \langle$ multiplier $\rangle | \langle$ comment $\rangle$
$\langle$ number of streams $\rangle ::= \langle$ multiplier $\rangle * | \langle$ comment $\rangle * | \langle$ empty $\rangle$
$\langle$ instruction streams $\rangle ::= \langle$ number of streams $\rangle \langle$ bits in instruction $\rangle$
$\langle$ connectivity $\rangle ::= \langle$ multiplier $\rangle$-nn$| \langle$ comment $\rangle$
$\langle$ control type $\rangle ::= a | h | l | r | \langle$ comment $\rangle$
$\langle$ controlled elements $\rangle ::= [\langle$ structure $\rangle]_{\langle \text{control type} \rangle}^{\langle \text{connectivity} \rangle} | \langle$ empty $\rangle$
$\langle$ computer $\rangle ::= \langle$ control identifier $\rangle_{\langle \text{instruction streams} \rangle}^{\langle \text{clock time in ns} \rangle} \langle$ comment $\rangle$
          $\langle$ controlled elements $\rangle$

Examples
$\langle$ control identifier $\rangle ::= I; Ivp3; Cvl2$
$\langle$ controlled elements $\rangle ::= \quad ; [64\bar{P}]; [C1[64\bar{P}], C2]; [10Fp - 4M]$
$\langle$ connectivity $\rangle ::= (2D \text{ hexagonal mesh}); c\text{-nn}; 2\text{-nn}$
$\langle$ control type $\rangle ::= a; h; l; r; \quad ;$
$\langle$ computer $\rangle ::= C1[64^2\bar{P}]_i^{2\text{-nn}}; C1[16\bar{F} \times 17\bar{M}]; C4_8^{250};$
          $Ivp[10Fp - 4M]_h; Ivp_{16,32}^{12}; Ip_{50*64}(HEP)$

A.5   DATA PATHS

$\langle$ highway identifier $\rangle ::= H \langle$ unsigned integer $\rangle \langle$ comment $\rangle$
          $| \langle$ comment $\rangle$
$\langle$ time per word in ns $\rangle ::= \langle$ multiplier $\rangle | \langle$ comment $\rangle$
$\langle$ data bits $\rangle ::= \langle$ multiplier $\rangle | \langle$ comment $\rangle$
$\langle$ address bits $\rangle ::= \langle$ multiplier $\rangle | \langle$ comment $\rangle$
$\langle$ number of paths $\rangle ::= \langle$ multiplier $\rangle | \langle$ comment $\rangle$
$\langle$ size of path $\rangle ::= \langle$ number of paths $\rangle * \{ \langle$ data bits $\rangle + \langle$ address bits $\rangle \}$
          $| \langle$ number of paths $\rangle * \langle$ data bits $\rangle | \langle$ data bits $\rangle$
          $| \langle$ data bits $\rangle + \langle$ address bits $\rangle$
$\langle$ data bus $\rangle ::= \dfrac{\langle \text{time per word in ns} \rangle}{\langle \text{size of path} \rangle} | \langle$ data bus $\rangle - | - \langle$ data bus $\rangle$
$\langle$ cross connection $\rangle ::= \dfrac{\langle \text{time per word in ns} \rangle}{\langle \text{size of path} \rangle} \times \quad | \langle$ cross connection $\rangle \times$
          $| \times \langle$ cross connection $\rangle$
$\langle$ connection $\rangle ::= \langle$ data bus $\rangle | \langle$ data bus $\rangle \langle$ highway identifier $\rangle \langle$ data bus $\rangle$

$|\langle$cross connection$\rangle|\langle$cross connection$\rangle\langle$highway identifier$\rangle$
$\langle$cross connection$\rangle$
$\langle$no connection$\rangle::=$ $|$
$\langle$simplex left$\rangle::=\langle\langle$connection$\rangle$
$\langle$simplex right$\rangle::=\langle$connection$\rangle\rangle$
$\langle$duplex$\rangle::=\langle$simplex left$\rangle\rangle|\langle$simplex left$\rangle,\langle$simplex right$\rangle$
$\langle$half duplex$\rangle::=\langle$simplex left$\rangle/\langle$simplex right$\rangle$
$\langle$data path$\rangle::=\langle$connection$\rangle|\langle$simplex left$\rangle|\langle$simplex right$\rangle$
$|\langle$duplex$\rangle|\langle$half duplex$\rangle|\langle$no connection$\rangle$
$\langle$highway definition$\rangle::=\langle$highway identifier$\rangle=\langle$structure$\rangle$

Examples
$\langle$highway identifier$\rangle::=$ H; H3 (twisted pair)
$\langle$time per word in ns$\rangle::=$ t; 12.5; (1ms)
$\langle$data bits$\rangle::=$ 64; n
$\langle$address bits$\rangle::=$ 16; a
$\langle$number of paths$\rangle::=$ 4; n
$\langle$size of path$\rangle::=4*\{64+16\}; 4*64; 64; 64+16$
$\langle$data bus$\rangle::=-;\dfrac{100}{4*\{64+16\}}$
$\langle$cross connection$\rangle::=\times;\times\times\times$
$\langle$connection$\rangle::=-$H3 (twisted pair)$---;\times\times$ (Banyan Network)$\times\times$;
$-$H3$-$
$\langle$no connection$\rangle::=$ $|$
$\langle$simplex left$\rangle::=\langle---;\langle\times\times;\langle-;\langle\times$
$\langle$simple right$\rangle::=---\rangle;\times\times\rangle;-\rangle;\times\rangle$
$\langle$duplex$\rangle::=\langle-\rangle;\langle\frac{-}{16},\frac{-}{32}\rangle;\langle-\frac{150}{64}-\rangle$
$\langle$half duplex$\rangle::=\langle-/-\rangle;\langle-----$H2$--/--$H3$----\rangle$
$\langle$data path$\rangle::=|;-\rangle;--;\times$

$\langle$highway definition$\rangle::=$ H3 $=\{\{\frac{-}{64}\rangle,\langle\frac{-}{8}\}/\langle\frac{-}{24}\}$

A.6   STRUCTURES

$\langle$unit$\rangle::=\langle$E unit$\rangle|\langle$M unit$\rangle|\langle$computer$\rangle|\langle$lower case letter$\rangle$
$\langle$primary$\rangle::=\langle$unit$\rangle|\{$structure$\}^{\langle\text{connectivity}\rangle}{}_{\langle\text{pipelined}\rangle}$
$|\langle$parallel structure$\rangle\langle$pipelined$\rangle$
$\langle$secondary$\rangle::=\langle$primary$\rangle|\langle$factor$\rangle\langle$primary$\rangle|\overline{\langle$factor$\rangle\langle$primary$\rangle}$
$|\langle$empty$\rangle$

⟨ structure ⟩:: = ⟨ secondary ⟩|⟨ structure ⟩⟨ data path ⟩⟨ secondary ⟩
⟨ concurrent list ⟩:: = ⟨ structure ⟩|⟨ concurrent list ⟩,⟨ structure ⟩
⟨ sequential list ⟩:: = ⟨ structure ⟩|⟨ sequential list ⟩/⟨ structure ⟩
⟨ list ⟩:: = ⟨ concurrent list ⟩|⟨ sequential list ⟩
⟨ parallel structure ⟩:: = { ⟨ list ⟩ }$^{⟨connectivity⟩}$
⟨ definition ⟩:: = ⟨ secondary ⟩ = ⟨ structure ⟩

Examples

⟨ unit ⟩:: = E; M_{8*64}; C[64$\bar{P}$]; a

⟨ primary ⟩:: = { − E − M − a }p; { −E1−, E2 }; { E1/E2/E3 }; IO

⟨ secondary ⟩:: = 16∗32$^2\bar{P}$; 4{ 3F, 2P }

⟨ structure ⟩:: = −E−M−a; { E−{ −M1−M2−,− ⟩ }−−−M3 }−a;
   { 32$^2\bar{P}$ }$^{1\text{-}nn}$

⟨ concurrent list ⟩:: = I, E, M

⟨ sequential list ⟩:: = 3E1/E2/E3

⟨ parallel structure ⟩:: = { I, E, M }; { 3E1/E2/E3 }; { −E−, |M−,− }

⟨ definition ⟩:: = 3E1 = { E(+), E(∗), E(÷) }; E2 = { F(∗)/F(+)/B };
   C = I[64$\bar{P}$]$_i^{2\text{-}nn}$; M1 = { −M$_{1K*}^{100}$−,−− ⟩ }−M$_{|M*64}^{\{1ms\}}$

References

The following is an alphabetical list of papers and books referred to in the text. The abbreviations follow the British Standards Institution (1975) rules which are also an American ANSI standard. They are used by INSPEC in Physics Abstracts (published by the Institute of Electrical Engineers, London).

Abel N E, Budnik P P, Kuck D J, Muraoka Y, Northcote R S and Wilhelmson R B 1969 TRANQUIL: A language for an array processing computer *AFIPS Conf. Proc.* **34** 57–75

ACM 1978 Special issue on computer architecture *Commun. Assoc. Comput. Mach.* **21** (1)

Adams L and Ortega J 1982 A multi-color SOR method for parallel computation *Proc. 1982 Int. Conf. Parallel Processing* (Silver Spring, MD: IEEE Comput. Soc.) 53–6

Agarwal R C and Burrus C S 1974 Fast convolution using Fermat number transforms with application to digital filtering *IEEE Trans. Acoustics, Speech and Signal Processing* **ASSP-22** 87–97

—— 1975 Number theoretic transforms to implement fast digital convolution *Proc. IEEE* **63** 550–60

Alexander T 1985 Cray's way of staying super-duper *Fortune* March 18 66–76

Amdahl G 1967 The validity of the single processor approach to achieving large scale computing capabilities *AFIPS Conf. Proc. Spring Joint Comput. Conf.* **30** 483–5

Amdahl G M, Blaauw G A and Brooks F P Jr 1964 Architecture of the IBM System/360 *IBM J. Res. Dev.* **8** 87–101

American National Standards Institute (ANSI) 1985 Fortran 8X *Document X3J3/S8* version 95

Amini M and Hockney R W 1979 Computer simulation of melting and glass formation in a potassium chloride microcrystal *J. Non-Cryst. Solids* **31** 447–52

Anderson D W, Sparacio F J and Tomasulo R M 1967 The IBM System/360 Model 91: machine philosophy and instruction handling *IBM J. Res. Dev.* **11** 8–24

Arnold C N 1982 Performance evaluation of three automatic vectoriser packages *Proc. 1982 Int. Conf. Parallel Processing* (Silver Spring, MD: IEEE Comput. Soc.) 235–42

—— 1983 Vector optimisation on the CYBER 205 *Proc. 1983 Int. Conf. Parallel Processing* (Silver Spring, MD: IEEE Comput. Soc) 530–6

Arvind D K, Robinson I N and Parker I N 1983 A VLSI chip for real time image processing *Proc. IEEE Symp. on Circuits and Systems* 405–8

Aspinall D 1977 Multi-micro systems *Infotech State of the Art Conference: Future Systems* (Maidenhead: Infotech) vol. 2 45–62

—— 1984 Cyba-M *Distributed Computing* ed F B Chambers, D A Duce and G P Jones (London: Academic) 267–76

Auerbach 1976a *Auerbach Corporate EDP Library of Computer Technology Reports* (18 volumes) (Pennsauken, NJ: Auerbach Publishers Inc.)

—— 1976b Cray Research Inc. Cray-1 *Auerbach Corporate EDP Library of Computer Technology Reports* (Pennsauken, NJ: Auerbach Publishers Inc.)

Augarten S 1983 *State of the art* (US: Tickner and Fields)

Auslander L and Cooley J W 1986 On the development of fast parallel algorithms for Fourier transforms and convolution *Proc. Symp. Vector and Parallel Processors for Scientific Calculation (Rome) 1985* (Rome: Accademia Nazionale dei Lincei)

Auslander L, Cooley J W and Silberger A J 1984 Numerical stability of fast convolution algorithms for digital filtering *IEEE Workshop on VLSI Signal Processing* 172–213 (London: IEEE)

Austin J H Jr 1979 The Burroughs Scientific Processor *Infotech State of the Art Report: Supercomputers* vol. 2 ed C R Jesshope and R W Hockney (Maidenhead: Infotech Int. Ltd) 1–31

Baba T 1987 Microprogrammable Parallel Computer (Cambridge, MA: MIT)

Babbage C 1822 A note respecting the application of machinery to the calculation of astronomical tables *Mem. Astron. Soc.* 1 309

—— 1864 *Passages from the Life of a Philosopher* (Longman, Green, Longman, Roberts and Green) Reprinted 1979 (New York: Augustus M Kelly)

Babbage H P 1910 Babbage's Analytical Engine *Mon. Not. Roy. Astron. Soc.* 70 517–26, 645

Backus J W, Bauer F L, Green J, Katz C, McCarthy J, Naur P, Perlis A J, Rutishauser H, Samelson K, Vauquois B, Wegstein J H, van Wijngaarden A and Woodger M 1960 Report on the algorithmic language ALGOL60 *Numer. Math.* 2 106–36

Barlow R H, Evans D J, Newman I A and Woodward M C 1981 The NEPTUNE parallel processing system *Internal Report* Department of Computer Studies, Loughborough University, UK

Barlow R H, Evans D J and Shanehchi J 1982 Performance analysis of algorithms on asynchronous parallel processors *Comput. Phys. Commun.* 26 233–6

Barnes G H, Brown R M, Kato M, Kuck D J, Slotnick D L and Stokes R A 1968 The ILLIAC IV computer *IEEE Trans. Comput.* C-17 746–57

Batcher K E 1968 Sorting networks and their applications *AFIPS Conf. Proc.* 32 307–14

—— 1979 The STARAN Computer *Infotech State of the Art Report: Supercomputers* vol. 2 ed C R Jesshope and R W Hockney (Maidenhead: Infotech Int. Ltd) 33–49

—— 1980 Design of a massively parallel processor *IEEE Trans. Comput.* C-29 1–9

Bell C G and Newell A 1971 *Computer Structures: Readings and Examples* (New York: McGraw-Hill)

Ben-Ari M 1982 *Principles of Concurrent Programming* (London: Prentice-Hall)

Benes V 1965 *Mathematical Theory of Connecting Networks and Telephone Traffic* (New York: Academic)

Bentley L and Jesshope C R 1986 The implementation of a two-dimensional redundancy scheme in a wafer scale, high speed disc memory *Wafer Scale Integration* (Bristol: Adam Hilger)

Berg R O, Schmitz H G and Nuspl S J 1972 PEPE—an overview of architecture, operation and implementation *Proc. IEEE Natl. Electron. Conf.* **27** 312–7

Berger H H and Wiedman S K 1972 Merged-transistor-logic MTL—a low cost bipolar logic concept *IEEE J. Solid St. Circuits* **SC-7** 340–6

Bergland G D 1968 A fast Fourier transform algorithm for real-valued series *Commun. Assoc. Comput. Mach.* **11** 703–10

Berney K 1984 IBM eyes niche in burgeoning supercomputer market *Electronics* July 12 45–6

Blackman R B and Tukey J W 1959 *The Measurement of Power Spectra* (New York: Dover Publications Inc.)

Bloch E 1959 The engineering design of the STRETCH computer *Proc. East. Joint Comp. Conf.* (New York: Spartan Books) 48–58

Booch G 1986 Object oriented development *IEEE Trans. Software Eng.* **SE-12** 211–21

Bossavit A 1984 The 'Vector Machine': an approach to simple programming on CRAY-1 and other vector computers *PDE Software: Modules, Interfaces and Systems* ed B Engquist and T Smedsaas (Amsterdam: Elsevier Science BV, North-Holland) 103–21

Bourne S R 1982 *The Unix System* (International Computer Science Series) (London: Addison-Wesley)

Bracewell R 1965 *The Fourier Transform and Its Applications* (New York: McGraw-Hill)

Brigham E O 1974 *The Fast Fourier Transform* (Englewood Cliffs, NJ: Prentice-Hall)

British Standards Institution 1975 The abbreviation of titles of periodicals: part 2. Word-abbreviation list *Br. Stand. Specif.* BS4148 Part 2

Brownrigg D R K 1975 Computer modelling of spiral structure in galaxies *PhD Thesis*, University of Reading

Bruijnes H 1985 Anticipated performance of the CRAY-2 *NMFECC Buffer* **9** (6) 1–3

Bucher I Y 1984 The computational speed of supercomputers *Supercomputers: Design and Applications* ed K Hwang (Silver Spring, MD: IEEE Comput. Soc.) 74–88

Bucher I Y and Simmons M L 1986 Performance assessment of supercomputers *Vector and Parallel Processors: Architecture, Applications and Performance Evaluation* ed M Ginsberg (Amsterdam: North-Holland) (Also preprint LA-UR-85-1505, LANL, USA)

Budnik P and Kuck D J 1971 The organisation and use of parallel memories *IEEE Trans. Comput.* **C-20** 1566–9

Buneman O 1969 A compact non-iterative Poisson-solver *Stanford University Institute for Plasma Research Report No* 294

Burks A W 1981 Programming and structural changes in parallel computers *Conpar 81* ed W Händler (Berlin: Springer) 1–24

Burks A W and Burks A R 1981 The ENIAC: first general-purpose electronic computer *Ann. Hist. Comput.* **3** (4) 310–99

Burks A W, Goldstine H H and von Neumann J 1946 Preliminary discussion of the logical design of an electronic computing instrument in *John von Neumann, Collected Works* vol. 5 ed A H Taub (Oxford: Pergamon) 35–79

Burns A 1985 *Concurrent programming in Ada* (Ada Companion Series) (Cambridge: Cambridge University Press)

Burroughs 1977a Burroughs scientific processor—file memory *Burroughs Document* 61391B

—— 1977b Burroughs scientific processor—overview, perspective, architecture *Burroughs Document* 61391A

—— 1977c Burroughs scientific processor—fault tolerant features *Burroughs Document* 61391C

—— 1977d Burroughs scientific processor—control program *Burroughs Document* 61391D

—— 1977e Burroughs scientific processor—implementation of FORTRAN *Burroughs Document* 61391E

—— 1979 Final report NASF feasibility study *NASA Contractor Report* NAS2-9897

Burrus C S 1977 Index mappings for multidimensional formulation of the DFT and convolution *IEEE Trans. Acoust. Speech Signal Process.* **25** 239–42

Burrus C S and Eschenbacher P W 1981 An in-place in-order prime factor FFT algorithm *IEEE Trans. Acoust. Speech Signal Process.* **29** 806–17

Bush V 1931 The differential analyser: a new machine for solving a differential equation *J. Franklin Inst.* **212** 447–88

—— 1936 Instrumental analysis *Bull. Amer. Math. Soc.* **42** 649–69

Bustos E, Lavers J D and Smith K C 1979 A parallel array of microprocessors—an alternative solution to diffusion problems *COMPCON'79 (Fall) Digest* 380–9

Buzbee B L, Golub G H and Neilson C W 1970 On direct methods for solving Poisson's equations *SIAM J. Numer. Anal.* **7** 627–56

Calahan D A 1977 Algorithmic and architectural issues related to vector processors *Proc. Int. Symp. Large Eng. Syst.* (New York: Pergamon) 327–39

—— 1984 Influence of task granularity on vector multiprocessor performance *Proc. 1984 Int. Conf. Parallel Processing* (Silver Spring, MD: IEEE Comput. Soc.) 278–84

Calahan D A and Ames W G 1979 Vector processors: models and applications *IEEE Trans. Circuits and Systems* **CAS-26** 715–26

Campbell L W (ed) 1987 *Draft Proposal of Revised American National Standard X3.9–198X* (obtainable from Global Engineering Documents, PO Box 2584, Santa Ana, CA 92787)

Cannon S F 1983 VLSI chips increase processing power in array processors *Comput. Technol. Rev.* Winter 294–5

CDC 1978 STAR FORTRAN language, version 2, reference manual *CDC Publication* 60386200

—— 1979 Final report feasibility study for NASF *NASA Contractor Report* NAS2-9896

—— 1983 CDC CYBER 200 Model 205 computer system hardware reference manual *Publication* 60256020 (St Paul, Minnesota: Control Data Corporation)

Chambers F B, Duce D A and Jones G P (eds) 1984 *Distributed Computing APIC Studies in Data Processing* vol. 20 (London: Academic)

Charlesworth A E 1981 An approach to scientific array processing: the architectural design of the AP-120B/FPS-164 family *IEEE Comput.* **14** (9) 18–27

Charlesworth A E and Gustafson J L 1986 Introducing replicated VLSI to supercomputing: the FPS-164/MAX scientific computer *IEEE Comput.* **19** (3) 10–23

Chen S S 1984 Large-scale and high-speed multiprocessor system for scientific applications: CRAY X-MP Series *High-Speed Computation* ed J S Kowalik (NATO ASI Series vol. 7) (Berlin: Springer) 59–67

Chen S S, Dongarra J J and Hsiung C C 1984 Multiprocessing linear algebra algorithms on the CRAY X-MP-2: experiences with small granularity *J. Parallel Distrib. Comput.* **1** 22–31

Cheung T and Smith J E 1984 An analysis of the Cray X-MP memory system *IEEE Proc. 1984 Int. Conf. Parallel Process.* 499–505

Clapp R M, Duchesnau L, Volz R A, Mudge T N and Schultze T 1986 Towards real-time performance benchmarks for Ada *Commun. ACM* **29** 760–78

Clark K and Gregory S 1984 Parlog parallel programming in logic *Imperial College Research Report* **84/4**

Clementi E, Corongin G, Detrich L, Chin S and Domingo L 1984 Parallelism in quantum chemistry: hydrogen bond study in DNA base pairs as an example *Int. J. Quant. Chem.: Quant. Chem. Symp.* **18** 601–18

Clos C 1953 A study of non-blocking switching networks *Bell. Syst. Tech. J.* **32** 406–24

Cooley J W 1982 Rectangular transforms for digital convolution on the research signal processor *IBM J. Res. Dev.* **26** 424–30

Cooley J W, Lewis P A W and Welch P D 1967 The fast Fourier transform algorithm and its applications *IBM Research Paper* RC-1743 (Yorktown Heights, NY: IBM Watson Research Center)

—— 1979 The fast Fourier transform algorithm: programming considerations in the calculation of sine, cosine and Laplace transforms *J. Sound and Vibrations* **12** 315–37

Cooley J W and Tukey J W 1965 An algorithm for the machine calculation of complex Fourier series *Math. Comput.* **19** 297–301

Cornell J A 1972 Parallel processing of ballistic missile defense radar data with PEPE *COMPCON '72 Digest* 69–72

Cox B 1986 *Object oriented programming* (London: Addison-Wesley)

Crane B A, Gilmartin M J, Huttenhoff J H, Rux P T and Shively R R 1972 PEPE computer architecture *COMPCON '72 Digest* 57–60

Crane B A and Githens J A 1965 Bulk processing in distributed logic memory *IEEE Trans. Electron. Comput.* **EC-14** 186–96

Cray 1976 *CRAY-1 computer system, reference manual* Publication 2240004 (Mendota Heights, Minnesota: Cray Research Inc.)

—— 1980 *CRAY-1 computer systems library reference manual* Publication SR-0014 (Mendota Heights, Minnesota: Cray Research Inc.)

—— 1981 Cray announces breakthrough in computer development *Cray Channels* **3** (2) 12–13

—— 1982 *CRAY X-MP computer systems: mainframe reference manual* Publication HR-0032 (Mendota Heights, Minnesota: Cray Research Inc.)

—— 1984a *CRAY computer systems, model 48 mainframe reference manual* Publication HR-0097 (Mendota Heights, Minnesota: Cray Research Inc.)

—— 1984b *The CRAY X-MP series of computer systems* Publication MP-2101 (Mendota Heights, Minnesota: Cray Research Inc.)

—— 1985 Introducing the CRAY-2 computer system *Cray Channels* Summer 2–5

Curington I J and Hockney R W 1986 Synchronization and pipeline overhead measurements on the FPS-5000 MIMD computer *Parallel Computing 85* ed M Feilmeier, G R Joubert and U Schendel (Amsterdam: North-Holland) 469–76

Dagless E L 1977 A multi-microprocessor—CYBA-M *IFIP Information Processing 77* ed B Gilchrist (Amsterdam: North-Holland) 843–8

Dahl O J, Dijkstra E W and Hoare C A R 1972 *Structured Programming* (New York: Academic)

Danielson G C and Lanczos C 1942 Some improvements in practical Fourier analysis and their application to x-ray scattering from liquids *J. Franklin Inst.* **233** 365–80 and 435–52

Darlington J and Reeve M 1981 ALICE—a multiprocessor reduction machine for the parallel evaluation of functional programming languages *Proc. ACM Conf. Functional Programming Languages and Computer Architectures (New Hampshire)* 65–75

Deb A 1980 Conflict free access of arrays: a counter example *Information Processing Lett.* **10** 20

Dennard R H, Gaesslen F H, Kuhn L and Yu H N 1974 Design of ion implanted MOSFETS with very small physical dimensions *IEEE J. Solid State Circuits* **SC-9** 256–68

Dennis J B 1980 Data Flow Supercomputers *IEEE Computer* **13**(11) 48–56

Dijkstra E W 1975 Guarded commands nondeterminancy and formal derivation of programs *Commun. Assoc. Comput. Mach.* **18** 453–7

Dongarra J J 1985 Performance of various computers using standard linear equations software in a Fortran environment *Argonne National Laboratory Report* MCSD-TM-23 (August) (An early version appeared in *ACM SIGNUM Newsletter* 1984 **19** 23–6)

—— 1986 How do the 'minisupers' stack up? *IEEE Computer* **19** (3) 93–100

Dongarra J J, Bunch J R, Moler C B and Stewart G W 1979 *LINPACK Users' Guide* (Philadelphia, PA: SIAM Publications)

Dongarra J J and Eisenstat S C 1984 Squeezing the most out of an algorithm in CRAY Fortran *ACM Trans. Math. Software* **10** 221–30

Dongarra J J and Hewitt T 1985 Implementing dense linear algebra algorithms using multitasking on the CRAY X-MP-4 (or approaching the gigaflop) *Argonne National Laboratory Report* MCSD-TM-25

Dongarra J J and Sorensen D C 1985 Parallel linear algebra algorithms for the Denelcor HEP *MIMD Computation: HEP Supercomputer and its Applications* ed J S Kowalik (Cambridge, MA: MIT) 275–94

—— 1987 Algorithm design for high performance computers *IBM Europe Inst. 1986*

Duce D A 1984 *Distributed Computing Systems Programme* (IEE Digital Electronics and Computing Series) vol. 5 ed S L Hurst and M W Sage (London: Peter Peregrinus for IEE)

Duff M J B 1978 The review of the CLIP image processing system *Proc. Natl. Comput. Conf. (USA)* 1055–60

—— 1980a CLIP-4 *Image Processing Computers* ed T S Huang (Berlin: Springer)

—— 1980b CLIP-4 *Special Computer Architecture for Pattern Processing* ed K S Fu (Boca Raton, Florida: CRC Press)

Dungworth M 1979 The CRAY-1 computer system *Infotech State of the Art Report: Supercomputers* vol. 2 ed C R Jesshope and R W Hockney (Maidenhead: Infotech Int. Ltd) 51–76

Dunwell S W 1956 Design objectives for the IBM STRETCH computer *Proc. East. Joint Comp. Conf.* (New York: Spartan) 20–2

Eastwood J W 1976 Optimum P^3M algorithms for molecular dynamic simulations *Computational Methods in Classical and Quantum Physics* ed M B Hooper (London: Advance) 206–28

Eastwood J W and Brownrigg D R K 1979 Remarks on the solution of Poisson's equation for isolated systems *J. Comput. Phys.* **32** 24-38

Eastwood J W and Jesshope C R 1977 The solution of elliptic partial differential equations using number theoretic transforms with application to narrow or limited computer hardware *Comput. Phys. Commun.* **13** 233-9

Eckert J P Jr, Weiner J R, Welsh H F and Mitchell H F 1951 The UNIVAC system, in Bell and Newell 1971 157-69

Eden R C, Welch B M, Zucca R and Long S I 1979 The prospects for ultra-highspeed vlsi GaAs digital logic *IEEE Trans. Electron. Devices* **ED-26** 299-317

Emmen A 1986 Boost your PC to 8 Mflops with the N-Cube 4 *Supercomputer* **16** (Nov) 7-9

Enslow P H Jr (ed) 1974 *Multiprocessors and Parallel Processing* (New York: Wiley)

—— 1977 Multiprocessor organization—a survey *Comput. Surv.* **9** 103-29

Ericsson T and Danielson P E 1983 LIPP—A SIMD architecture for image processing *Proc. ACM 10th Int. Symp. Comput. Archit.* 395-400

Evans D J 1980 On the solution of certain Toeplitz tridiagonal linear systems *SIAM J. Numer. Anal.* **17** 675-80

—— (ed) 1982 *Parallel Processing Systems* (Cambridge: Cambridge University Press)

—— 1984a Parallel SOR iterative methods *Parallel Comput.* **1** 3-18

—— 1984b New algorithms for partial differential equations *Parallel Comput. 83* ed M Feilmeier, G Joubert and U Schendel (Amsterdam: North-Holland) 3-56

Falk H 1976 Reaching for the gigaflop *IEEE Spectrum* **13** (10) 65-70

Fang F F and Rupprecht H S 1975 High performance mos integrated circuits using ion implanted techniques *IEEE J. Solid St. Circuits* **SC-10** 205-11

Farmwald P M 1984 The S-1 Mark IIA supercomputer *High-Speed Computation* ed J S Kowalik (Berlin: Springer) 145-55

Feierbach G and Stevenson D 1977 A feasibility study of programmable switching networks for data routing *Institute for Advanced Computation, California, USA, Phoenix Report* 3

—— 1979a The ILLIAC IV *Infotech State of the Art Report: Supercomputers* vol. 2 ed C R Jesshope and R W Hockney (Maidenhead: Infotech Int. Ltd 77-92)

—— 1979b The Phoenix project *Infotech State of the Art Report: Supercomputers* vol. 2 ed C R Jesshope and R W Hockney (Maidenhead: Infotech Int. Ltd) 91-115

Feigenbaum E A and McCorduck P 1983 *The Fifth Generation—artificial intelligence and Japan's challenge to the world* (London: Michael Joseph)

Feilmeier M, Joubert G and Shendel U (eds) 1984 *Parallel Computing 83* (Amsterdam: North-Holland)

—— 1986 *Parallel Computing 85* (Amsterdam: North-Holland)

Feng T Y 1981 A summary of interconnection networks *IEEE Computer* **14**(12) 12-27

Flanders P M 1980 Musical bits—a generalised method for a class of data movements on the DAP *ICL RADC Document* CM-70

—— 1982a Languages and techniques for parallel array processors *PhD Thesis* Queen Mary College, London

—— 1982b A unified approach to a class of data movements on an array processor *IEEE Trans. Comput.* **C-31** 405-8

Flanders P M, Hunt D J, Reddaway S F and Parkinson D 1977 Efficient high speed computing with the distributed array processor *High Speed Computer and Algorithm Organisation* (London: Academic) 113–28

Flynn M J 1972 Some computer organizations and their effectiveness *IEEE Trans. Comput.* **C-21** 948–60

Forsythe G E and Wasow W R 1960 *Finite-difference Methods for Partial Differential Equations* (London: Wiley)

Foster C C 1976 *Content-Addressable Parallel Processors* (New York: Van Nostrand Reinhold)

Foster M J and Kung H T 1980 The design of special-purpose vlsi chips *IEEE Computer* **13** (1) 26–40

Fotheringham J 1961 Dynamic storage allocation in the ATLAS computer, including use of backing store *Commun. Assoc. Comput. Mach.* **3** 435–6

Fountain T J 1983 The development of the CLIP 7 image processing system *Pattern Recog. Lett.* **5** 331–9

Fountain T J and Goetcherian V 1980 CLIP 4 Parallel processing system *IEEE Proc. E* **129** 219–24

Fox G C and Otto S W 1984 Algorithms for concurrent processors *Physics Today* **37** (5) 50–9

FPS 1976a *Processor Handbook* Publication 7259-02 (Beaverton, Oregon: Floating Point Systems Inc.)

—— 1976b *AP-120B Math Library* Publication 7288-02 (Beaverton, Oregon: Floating Point Systems Inc.)

—— 1979a *AP Math Library Vol. 2* Publication 860-7288-005 (Beaverton, Oregon: Floating Point Systems Inc.)

—— 1979b *AP Math Library Vol. 3* Publication 860-7288-006 (Beaverton, Oregon: Floating Point Systems Inc.)

—— 1984a *FPS-5000 System Introduction Manual Publication* 860-7437-041A (Portland, Oregon: Floating Point Systems Inc.) March

—— 1984b *FPS Arithmetic Coprocessor MATHLIB Publication* 860-7400-001A (Portland, Oregon: Floating Point Systems Inc.) March

—— 1984c *FPS Arithmetic Coprocessor Hardware Reference Manual Publication* 860-7400-005A (Portland, Oregon: Floating Point Systems Inc.) October

—— 1985a *CPFORTRAN Users' Guide Publication* 800-1018-002A (Portland, Oregon: Floating Point Systems Inc.) April

—— 1985b *A Strategic Approach to Scientific and Engineering Supercomputing Document* FPS SEC 5227 (Portland, Oregon: Floating Point Systems Inc.)

Fuller S H, Ousterhout, J, Raskin L, Rubinfeld P, Sindhu P and Swan R 1978 Multi-microprocessors: an overview and working example *Proc. IEEE* **66** (2) 216–28

Fuss D and Hollenberg J 1984 A comparison between Japanese and American supercomputers *Supercomputer* **3** 8–9

Gajski D, Kuck D, Lawrie D and Sameh A 1983 Cedar—a large scale multiprocessor *IEEE Proc. 1983 Int. Conf. on Parallel Processing* 524–9 (London: IEEE)

Gannon D B and Van Rosendale J 1984 On the impact of communication complexity on the design of parallel numerical algorithms *IEEE Trans. Comput.* **C-33** 1180–94

Gentleman W M and Sande G 1966 Fast Fourier transforms—for fun and profit, 1966 Fall Joint Computer Conference *AFIPS Proc.* **29** 563–78

Gentzsch W 1984 Vectorization of computer programs with application to computational fluid dynamics *Notes on Numerical Fluid Mechanics* vol. 8 (Braunschweig-Weisbaden: Vieweg)

Gill S 1951 A process for the step-by-step integration of differential equations in an automatic digital computing machine *Proc. Cambridge Philos. Soc.* **47** 96–108

Goke L R and Lipovski G J 1973 Banyan networks for partitioning multiprocessor systems *1st Annual Symp. on Computer Architecture (IEEE/ACM)*, *Florida* **1** 21–8

Goldberg A and Robson D 1984 *Smalltalk-80* (London: Addison-Wesley)

Goldstine H H and Goldstine A 1946 The Electronic Numerical Integrator and Computer (ENIAC) *Mathematical Tables and Aids to Computation* **2** (15) 97–110 (Reprinted in 1975 in *The Origins of Digital Computers* ed B Randell (Berlin: Springer) 333–47)

Good I J 1958 The interaction algorithm and practical Fourier analysis *J. R. Stat. Soc. Ser. B* **20** 361–72

—— 1971 The relationship between two fast Fourier transforms *IEEE Trans. Comput.* **20** 310–7

Goodman J W, Leonberger F I, Kung S Y and Athale R 1984 Optical interconnections for VLSI systems *Proc. IEEE* **72** 850–66

Gorsline G W 1980 *Computer Organization* (Englewood Cliffs, NJ: Prentice-Hall)

Gostick R W 1979 Software and algorithms for the distributed array processor *ICL Tech. J.* **1** 116–35

Gottlieb A, Grishman R, Kruskal C P, McAuliffe K P, Rudolph L and Snir M 1983 The NYU Ultracomputer—designing an MIMD shared memory computer *IEEE Trans. Comput.* **C-32** 175–89

Graham W R 1970 The parallel and pipeline computers *Datamation* **16** (4) 68–71

Gray G M D 1984 *Logic algebra and databases* (Chichester: Ellis Horwood)

Gregory J and McReynolds R 1963 The SOLOMON computer *IEEE Trans. Electron. Comput.* **EC-12** 774–81

Gries D 1976 An illustration of current ideas on the derivation of correctness proofs and correct programs *IEEE Trans. Software Eng.* **SE-2**

Grosch G E 1979 Performance analysis of Poisson solvers on array computers *Infotech State of the Art Report: Supercomputers* vol. 2 ed C R Jesshope and R W Hockney (Maidenhead: Infotech Int. Ltd) 149–81

Gurd J R 1987 Data flow architectures *Major Advances in Parallel Processing* ed C R Jesshope 113–132 (preprint) (Aldershot: Technical)

Gurd J R and Watson I 1980 Dataflow system for high-speed parallel computation *Comput. Des.* **19** (6) 91 and (7) 97

Gustafson J 1985 The measurement of floating-point operations per second *Checkpoint: FPS Tech. J.* **3** (7) 9–13

Haley A C D 1956 DEUCE: a high speed general purpose computer *Proc. IEE* **103** Part B, Suppl. 2, 165–73

Händler W (ed) 1981 *Conpar 81: Lecture Notes in Computer Science* **111** (Berlin: Springer)

—— 1984 Simplicity and flexibility in concurrent computer architecture *High-Speed Computation* ed J S Kowalik (NATO ASI Series vol. 7) (Berlin: Springer) 68–88

Händler W, Bode A, Fritsch G, Henning W and Volkert J 1985 A tightly coupled and hierarchical multiprocessor architecture *Comput. Phys. Commun.* **37** 87–93

Händler W, Hofmann F and Schneider H J 1975 A general-purpose array with a broad spectrum of applications *Comput. Archit.* vol. 4 ed W Händler (Berlin: Springer) 311–35

Harland D M 1984 *Polymorphic programming languages design and implementation* (Chichester: Ellis Horwood)

—— 1985 Towards a language for concurrent processes *Software Practice and Experience* **15** 839–88

—— 1986 *Concurrency and programming languages* (Chichester: Ellis Horwood)

Hart K and Slob A 1972 Integrated injection logic: a new approach to LSI *IEEE J. Solid State Circuits* **SC-7** 101–9

Hart P A H, Van't Hof T and Klaassen F M 1979 Device down scaling and expected circuit performance *IEEE Trans. Electron. Devices* **ED-26** 421–9

Harte J 1979 The FPS AP-120B array processor *Infotech State of the Art Report: Supercomputers* vol. 2 ed C R Jesshope and R W Hockney (Maidenhead: Infotech. Int. Ltd) 183–203

Hartree D R 1946 The Eniac, an electronic computing machine *Nature* **158** 500–6

—— 1950 *Calculating Instruments and Machines* (Cambridge: Cambridge University Press)

Hayes J 1980 MOS scaling *IEEE Computer* **13** (1) 8–13

Heller D 1978 A survey of parallel algorithms in numerical linear algebra *SIAM Rev.* **20** 740–77

Henderson P 1980 *Functional Programming* (Englewood Cliffs, NJ: Prentice-Hall)

—— 1986 Functional Programming formal specification and rapid prototyping *IEEE Trans. Software Eng.* **SE-12** 241–50

Herzog U 1984 Performance modelling and evaluation for concurrent computer architectures *High-Speed Computation* ed J S Kowalik (NATO ASI Series vol. 7) (Berlin: Springer) 177–89

Hey A J G, Jesshope C R and Nicole D A 1986 High performance simulation of lattice physics using enhanced transputer arrays *Computing in High Energy Physics* ed L O Hertzberger and W Hoogland (Amsterdam: North-Holland)

Higbie L C 1972 The OMEN computers: associative array processors *COMPCON '72 Digest* 287–90

—— 1979 Vectorisation and conversion of FORTRAN programs for the CRAY-1 (CFT) compiler *Cray Research Inc., Document* 2240207

Hillis W D 1985 *The connection machine* (Cambridge, MA: MIT)

Hillis W D and Steele G L Jr (1985) *The connection machine Lisp manual* (Cambridge, MA: Thinking Machine Corp)

Hintz R G and Tate D P 1972 Control Data STAR-100 processor design *COMPCON '72 Digest* 1–4

Hoare C A R 1978 Communicating sequential processes *Commun. ACM* **21** (8) 666–77

—— 1986 *Communicating Sequential Processes* (Englewood Cliffs, NJ: Prentice-Hall)

Hockney R W 1965 A fast direct solution of Poisson's equation using Fourier analysis *J. Assoc. Comput. Mach.* **12** 95–113

—— 1970 The potential calculation and some applications *Methods Comput. Phys.* **9** 135–211

—— 1977 Supercomputer architecture *Infotech State of the Art Conference: Future Systems* Chairman F Summer (Maidenhead: Infotech) 277–305

—— 1978 Computers, compilers and Poisson solvers *Computers, Fast Elliptic Solvers, and Applications* ed U Schumann (London: Advance) 75–97

—— 1980 Rapid elliptic solvers *Numerical Methods in Applied Fluid Dynamics* ed B

Hunt (London: Academic) 1–48

—— 1982 Characterization of parallel computers and algorithms *Comput. Phys. Commun.* **26** 285–91

—— 1983 Characterizing computers and optimising the FACR (l) Poisson-solver on parallel unicomputers *IEEE Trans. Comput.* **C-32** 933–41

—— 1984a Performance of parallel computers *High-Speed Computation* ed J S Kowalik (Berlin: Springer) 159–75

—— 1984b The $n_{1/2}$ method of algorithm analysis *PDE Software: Modules, Interfaces and Systems* ed B Engquist and T Smedsaas (Amsterdam: North-Holland) 429–44 (also in *Proc. 18th Annu. Hawaii Int. Conf. Syst. Sciences 1985* vol. 2 ed B D Shriver (North Hollywood: Western Periodicals) 8–20)

—— 1985a (r_∞, $n_{1/2}$, $s_{1/2}$) Measurements on the 2-CPU Cray X-MP *Parallel Computing* **2** 1–14

—— 1985b MIMD computing in the USA—1984 *Parallel Computing* **2** 119–36

—— 1985c Performance characterization of the HEP *MIMD Computation: HEP Supercomputer and its Applications* ed J S Kowalik (Cambridge, MA: MIT) 59–90

—— 1985d *Novel Computer Architectures Report* RCS 188 (Reading University Computer Science Department, Reading, Berks, UK) (also in *Proc. Symp. Vector and Parallel Processors for Scientific Calculations (Rome) 1985* (Rome: Accademia Nazionale dei Lincei)

—— 1986 Parallel Computers: architecture and performance *Parallel Computing 85* ed M Feilmeier, G R Joubert and U Schendel (Amsterdam: North-Holland) 33–69

—— 1987a Algorithmic phase diagrams *IEEE Trans. Comput.* **C-36** 231–3

—— 1987b Parametrization of computer performance *Parallel Computing* **5** 97–103

—— 1987c Characterizing overheads on VM-EPEX and multiple FPS-164 processors *Proc. IBM Europe Inst. 1986* (Oberlech, Austria) (Published 1988 as *Parallel Systems and Computation* ed G Paul and G S Almasi (Amsterdam: Elsevier Science) 255–72)

—— 1987d Synchronization and Communication overheads on the /CAP multiple FPS-164 computer system *Parallel Computing* (submitted for publication)

Hockney R W and Curington I J 1986 $f_{1/2}$: a parameter to characterize memory and communication bottlenecks *Parallel Computing* (submitted for publication)

Hockney R W and Eastwood J W 1988 *Computer Simulation Using Particles* (Bristol: Adam Hilger)

Hockney R W and Jesshope C R 1981 *Parallel Computers: architecture, programming and algorithms* 1st edn (Bristol: Adam Hilger)

Hockney R W and Snelling D F 1984 Characterizing MIMD computers: e.g. Denelcor HEP *Parallel Computing 83* ed M Feilmeier, G R Joubert and U Schendel (Amsterdam: North-Holland) 521–6

Hoeneisen B and Mead C A 1972 Fundamental limitations in microelectronics *Solid State Electron.* **15** 819–29

Holland J H 1959 A universal computer capable of executing an arbitrary number of subprograms simultaneously *Proc. East. Joint Comput. Conf.* **16** 108–13

Howarth D J, Jones P D and Wyld M T 1962 The ATLAS scheduling system *Comput. J.* **5** 238–44

Howarth D J, Payne R B and Sumner F H 1961 The Manchester University ATLAS operating system, Part II: User's description *Comput. J.* **4** 226–9

Hue S (Chairman) 1979 FORTRAN language requirements *Fourth Report of the Language Working Group of the Advanced Computing Committee, Department of*

Energy, USA

Hunt D J 1978 *UK Patent Application No.* 45858/78

—— 1979 Application techniques for parallel hardware *Infotech State of the Art Report: Supercomputers* vol. 2 ed C R Jesshope and R W Hockney (Maidenhead: Infotech Int. Ltd) 205–19

Hwang K and Briggs F A 1984 *Computer Architecture and Parallel Processing* (New York: McGraw-Hill)

Ibbett R N 1982 *The Architecture of High Performance Computers* (London: Macmillan)

IBM 1980 Josephson computer technology *IBM J. Res. Dev.* **24** (2)

ICL 1979a DAP: FORTRAN language reference manual *ICL Tech. Pub.* TP 6918

—— 1979b DAP: Introduction to FORTRAN programming *ICL Tech. Pub.* TP 6755

—— 1979c DAP: APAL language *ICL Tech. Pub.* TP 6919

IEEE 1978 Special issue on fine line devices *IEEE Trans. Electron. Devices* **ED-25** (4)

—— 1979 Special issue on VLSI *IEEE Trans. Electron. Devices* **ED-26** (4)

—— 1983 A proposed standard for binary floating-point arithmetic (draft 10.0) *IEEE Computer Society Microprocessor standards Committee Task P754 Publication* P754 (January)

Infotech 1976 Multiprocessor systems *Infotech State of the Art Report: Multiprocessor Systems* ed C H White (Maidenhead: Infotech Int. Ltd)

INMOS 1984 *Occam Programming Manual* (Englewood Cliffs, NJ: Prentice-Hall)

—— 1985 IMSt414 Transputer Reference Manual (Bristol: INMOS Ltd)

—— 1986 *Occam 2 Product Definition* (Bristol: INMOS Ltd) (preliminary)

Iverson K E 1962 *A Programming Language* (London: Wiley)

—— 1979 Operators *ACM Trans. Program. Lang. Syst.* **1** 161–76

Jacobi C G J 1845 Über eine Neue Auflösungsart der bei der Methode der Kleinsten Quadrate Vorkommenden Lineären Gleichungen *Astr. Nachr.* **22** (523) 297–306

James R A and Parkinson D 1980 Simulation of galactic evolution on the ICL distributed array processor *IUCC Bull. Cambridge University Library* **2** 111–4

Jensen C 1978 Taking another approach to supercomputing *Datamation* **24** (2) 159–75

Jesshope C R 1980a The implementation of fast radix-2 transforms on array processors *IEEE Trans. Comput.* **C-29** 20–7

—— 1980b Some results concerning data routing in array processors *IEEE Trans. Comput.* **C-29** 659–62

—— 1980c Data routing and transpositions in processor arrays *ICL Tech. J.* **2** 191–206

—— 1982 Programming with a high degree of parallelism in Fortran *Comput. Phys. Commun.* **26** 237–46

—— 1984 A reconfigurable processor array *Supercomputers and Parallel Computation* ed D J Paddon (Oxford: Clarendon) 35–40

—— 1985 The RPA—optimising a processor array architecture for implementation using VLSI *Comput. Phys. Commun.* **37** 95–100

—— 1986a Computational physics and the need for parallelism *Comput. Phys. Commun.* **41** 363–75

—— 1986b Communications in wafer scale systems *Wafer Scale Integration* ed C R Jesshope and W R Moore (Bristol: Adam Hilger) 65–71

—— 1986c Building and binding systems with transputers *Proc. IBM Europe Inst. 1986 Parallel Computing (Austria)* to be published

—— 1987a The RPA as an intelligent transputer memory system *Systolic Arrays* ed

W Moore, A McCabe and R Urquhart (Bristol: Adam Hilger)

—— (ed) 1987b *Major Advances in Parallel Processing* (London: Technical)

Jesshope C R and Bentley L 1986 Low-cost restructuring techniques for WSI *Electron. Lett.* **22** 439–41

—— 1987 Techniques for a wafer scale RPA *IEE Proc. E* part E **134**(2) 87–92

Jesshope C R and Craigie J A I 1980 Small is O.K. too (another matrix algorithm for the DAP) *DAP Newsletter* **4** 7–12

Jesshope C R and Hockney R W 1979 Analysis and Bibliography *Infotech State of the Art Report: Supercomputers* vol. 1 (Maidenhead: Infotech Int. Ltd)

Jesshope C R and Moore W R (eds) 1986 *Wafer Scale Integration* (Bristol: Adam Hilger)

Jesshope C R, Rushton A J, Cruz A and Stewart J M 1986 The structure and application of RPA—a highly parallel adaptive architecture *Highly Parallel Computers* (Amsterdam: Elsevier) 81–95

Jesshope C R and Stewart J M 1986 MIPSE—a microcode development system for the RPA computer system *Software Eng. 86* ed D Barnes and P Brown (London: Peter Peregrinus)

Johnson H W and Burrus C S 1983 The design of optimal DFT algorithms using dynamic programming *IEEE Trans. Acoust. Speech Signal Process.* **31** 378–87

Johnson P M 1978 An introduction to vector processing *Comput. Design* **17** (2) 89–97

Jones A K and Schwarz P 1980 Experience using multiprocessor systems—a status report *Comput. Surv.* **12** 121–65

Jones G 1985 *Programming in Occam* Oxford University Computing Laboratory Technical Monograph PRG-43

Jordan H F 1978 A special purpose architecture for finite element analysis *IEEE Proc. 1978 Int. Conf. on Parallel Processing* 263–6 (London: IEEE)

Jordan T L and Fong K 1977 Some linear algebraic algorithms and their performance on the CRAY-1 *High-Speed Computer and Algorithm Organization* ed D J Kuck *et al* (New York: Academic) 313–6

Jump J R and Ahuja S R 1978 Effective pipelining of digital systems *IEEE Trans. Comput.* **C-27** 855–65

Kamiya S, Isobe F, Takashima H and Takiuchi M 1983 Practical vectorization techniques for the FACOM VP *Information Processing 83* ed R E A Mason (Amsterdam: Elsevier, North-Holland) 389–94

Kascic M J Jr 1979 Vector processing on the CYBER 200 *Infotech State of the Art Report: Supercomputers* vol. 2 ed C R Jesshope and R W Hockney (Maidenhead: Infotech Int. Ltd) 237–70

—— 1984a Anatomy of a Poisson solver *Parallel Computing 83* ed M Feilmeier, G Joubert and U Schendel (Amsterdam: North-Holland) 173–9

—— 1984b A performance survey of the CYBER 205 *High-Speed Computation* ed J S Kowalik (NATO ASI Series vol. 7) (Berlin: Springer) 191–209

Kashiwagi H 1984 Japanese super-speed computer project *High-Speed Computation* ed J S Kowalik (Berlin: Springer) 117–25

Kautz W H 1971 An augmented content-addressed memory array for implementation with large-scale integration *J. Assoc. Comput. Mach.* **18** 19–31

Keller R M 1976 Look-ahead processors *Comput. Surv.* **7** 177–95

Kershaw D 1982 Solution of a single tridiagonal linear system and vectorization of the ICCG algorithm on the Cray-1 *Parallel Computations* ed G Rodrigue (London:

Academic) 85–99

Kilburn T, Edwards D B G and Aspinall D 1960 A parallel arithmetic unit using a saturated transistor fast-carry circuit *Proc. IEE* Part B **107** 573–84

Kilburn T, Edwards D B G, Lanigan M J and Sumner F H 1962 One-level storage system *Inst. Radio Eng. Trans.* **EC-11** (2) 223–35 Reprinted in Bell and Newell 1971 Ch 23, 276–90

Kilburn T, Howarth D J, Payne R B and Sumner F H 1961 The Manchester University ATLAS operating system Part I: Internal Organization *Comput. J.* **4** 222–5

Kimura T 1979 Gauss–Jordan elimination by vlsi mesh-connected processors *Infotech State of the Art Report: Supercomputers* vol. 2 ed C R Jesshope and R W Hockney (Maidenhead: Infotech Int. Ltd) 273–90

Kogge P M 1981 *The Architecture of Pipelined Computers* (New York: McGraw-Hill)

Kogge P M and Stone H S 1973 A parallel algorithm for the efficient solution of a general class of recurrence equations *IEEE Trans. Comput.* **C-22** 786–93

Kolba D P and Parks T W 1977 A prime factor FFT algorithm using high-speed convolution *IEEE Trans. Acoust. Speech Signal Process.* **25** 281–94

Kondo T, Nakashima T, Aoki M and Sudo T 1983 An lsi adaptive array processor *IEEE J-SSC* **18** 147–56

Korn D G and Lambiotte J J Jr 1979 Computing the fast Fourier transform on a vector computer *Math. Comput.* **33** 977–92

Kowalik J S (ed) 1984 *High-Speed Computation* (NATO ASI Series F: Computer and System Sciences vol. 7) (Berlin: Springer)

—— (ed) 1985 *Parallel MIMD Computation: HEP Supercomputer and its Applications* (Cambridge, MA: MIT)

Kowalik J S and Kumar S P 1985 Parallel algorithms for recurrence and tridiagonal systems *Parallel MIMD Computation: HEP Supercomputer and Applications* (Cambridge, MA: MIT) 295–307

Kuck D J 1968 ILLIAC IV software and application programming *IEEE Trans. Comput.* **C-17** 758–70

—— 1977 A survey of parallel machine organisation and programming *Comput. Surv.* **9** 29–59

—— 1978 *The Structure of Computers and Computations* vol. 1 (New York: Wiley) p 33

—— 1981 Automatic program restructuring for high-speed computation *Lecture Notes in Computer Science 111: Conpar 81* ed W Händler (Berlin: Springer) 66–77

Kuck D J, Lawrie D H and Sameh A (eds) 1977 *High Speed Computers and Algorithm Organization* (New York: Academic)

Kulisch U W and Miranker W L (eds) 1983 *A New Approach to Scientific Computation* (New York: Academic)

Kung H T 1980 The structure of parallel algorithms *Advances in Computers* ed Yovits (New York: Academic) 65–112

Ladner R E and Fisher M J 1980 Parallel prefix computation *J. Assoc. Comput. Mach.* **27** 831–8

Lambiotte J R Jr and Voight R G 1975 The solution of tridiagonal linear systems on the CDC STAR-100 computer *ACM Trans. Math. Software* **1** 308–29

Lang T 1976 Interconnections between processors and memory modules using the shuffle exchange network *IEEE Trans. Comput.* **C-25** 496–503

Lang T and Stone H S 1976 A shuffle-exchange network with simplified control *IEEE Trans. Comput.* **C-25** 55–65

Larson J L 1984 An introduction to multitasking on the CRAY X-MP-2 multiprocessor *IEEE Comput.* **17** (7) 62–9

—— 1985 Practical concerns in multitasking on the CRAY X-MP *ECMWF Workshop on Using Multiprocessors in Meteorological Models Report* November 92–110 (Reading, UK: European Centre for Medium Range Weather Forecasts)

Lavington S H 1978 The Manchester Mark I and ATLAS: a historical perspective *Commun. Assoc. Comput. Mach.* **21** 4–12

Lawrie D H 1975 Access and alignment of data in an array processor *IEEE Trans. Comput.* **C-24** 1145–55

Lawrie D H, Layman T, Baer D and Randal J M 1975 GLYPNIR—a programming language for ILLIAC IV *Commun. Assoc. Comput. Mach.* **17** 157–64

Lawson C, Hanson R, Kincaid D and Krogh F 1979 Basic linear algebra sub-programs for Fortran usage *Assoc. Comput. Math. Trans. Math. Software* **5** (3) 308–71

Lazou C 1986 *Supercomputers and their Use* (Oxford: Oxford University Press)

Lea R M 1986 WASP A WSI associative string processor for structured data processing *Wafer Scale Integration* ed C R Jesshope and W R Moore (Bristol: Adam Hilger)

Lee C Y and Paull M C 1963 A content addressable distributed logic memory with applications to information retrieval *Proc. IEEE* **51** 924–32

Lee K Y, Abu-Sufah W and Kuck D J 1984 On modelling performance degradation due to data movement in vector machines *Proc. 1984 Int. Conf. Parallel Processing* (Silver Spring, MD: IEEE Comput. Soc.) 269–77

Lenfant J 1978 Parallel permutations of data: a Benes network control algorithm for frequently used bijections *IEEE Trans. Comput.* **C-27** 637–47

Leondes C and Rubinoff M 1952 DINA—a digital analyser for Laplace, Poisson, diffusion and wave equations *AIEE Trans. Commun. and Electron.* **71** 303–9

Linback J R 1984 CMOS gates key to 'affordable' supercomputer *Electronics Week* October 29 17–19

Lincoln N R 1982 Technology and design trade-offs in the creation of a modern supercomputer *IEEE Trans. Comput.* **C-31** 349–62

Lint B and Agerwala T 1981 Communication issues in the design and analysis of parallel algorithms *IEEE Trans. Software Eng.* **SE-7** 174–88

Lorin H 1972 *Parallelism in Hardware and Software: Real and Apparent Concurrency* (Englewood Cliffs, NJ: Prentice-Hall)

Lubeck O, Moore J and Mendez P 1985 A benchmark comparison of three supercomputers: Fujitsu VP-200, Hitachi S810/20 and Cray X-MP/2 *IEEE Computer* **18** (12) 10–24

McClellan J H and Rader C M 1979 *Number Theory in Digital Signal Processing* (Englewood Cliffs, NJ: Prentice-Hall)

McCrone J 1985 The dawning of a silent revolution *Computing: The Magazine* July 18 12–13

McIntyre D 1970 An introduction to the ILLIAC IV computer *Datamation* **16** (4) 60–7

McLaughlin R A 1975 The IBM 704: 36-bit floating-point money maker *Datamation* **21** (8) 45–50

McMahon F 1972 (Code and information available from L-35, LLNL, PO Box 808, CA 94550, USA. See also Riganah and Schneck 1984.)

Madsen N and Rodrigue G 1976 A comparison of direct methods for tridiagonal systems on the CDC STAR-100 *Reprint* UCRL-76993 *Lawrence Livermore Laboratory Rev.* **1**

Maples C, Rathbun W, Weaver D and Meng J 1981 The design of MIDAS—a Modular Interactive Data Analysis System *IEEE Trans. Nucl. Sci.* **NS-28** 3746–53

Martin A J 1986 Compiling communicating processes into delay insensitive VLSI circuits *J. Distrib. Computer*

Mashburn H H 1982 The C.mmp/Hydra project: an architectural overview *Computer Structures: principles and examples* ed D P Siewiorek, G C Bell and A Newell (New York: McGraw-Hill) 350–70

Masson G M, Gingher G C and Nakamura S 1979 A sampler of circuit switching networks *IEEE Computer* **12** (6) 32–48

Matisso J 1980 Overview of Josephson technology logic and memory *IBM J. Res. Dev.* **24** 113–29

Matsuura T, Miura K and Makino M 1985 Supervector performance without toil—Fortran implemented vector algorithms on the VP-100/200 *Comput. Phys. Commun.* **37** 101–7

May D and Taylor R 1984 Occam—an overview *Microprocessors and microsystems* **8** (2) 73–9

Mead C A and Conway L A 1980 *Introduction to VLSI Systems* (Reading, MA: Addison-Wesley)

Metcalf M and Reid J K 1987 *Fortran 8X Explained* (Oxford and New York: Oxford University Press)

Metropolis N, Howlett J and Rota G-C (eds) 1980 *A History of Computing in the Twentieth Century—a collection of essays* (New York: Academic)

Meyrowitz N (ed) 1986 OOPSLA Object-oriented programming systems languages and applications *Conf. Proc. Sigplan Not.* **21** (21)

Miklosko J and Kotov V E (eds) 1984 *Algorithms, Software and Hardware of Parallel Computers* (Berlin: Springer; Bratislava: Veda)

Millard W 1975 Hyperdimensional μP collection seen functioning as mainframe *Digital Design* **5** (11) 20

Millstein R E 1973 Control structures in ILLIAC IV FORTRAN *Commun. Assoc. Comput. Mach.* **16** 622–7

Miranker W L 1971 A survey of parallelism in numerical analysis *SIAM Rev.* **13** 524–47

Miura K and Uchida K 1984 FACOM vector processor VP-100/VP-200 *High-Speed Computation* ed J S Kowalik (NATO ASI Series vol. F7) (Berlin: Springer) 127–38

Moore W R, McCabe A and Urquhart R 1987 *Systolic Arrays* (Bristol: Adam Hilger)

Morrison P and Morrison E 1961 *Charles Babbage and his Calculating Engines* (New York: Dover)

Motegi M, Uchida K and Tsuchimoto T 1984 The architecture of the FACOM vector processor *Parallel Computing 83* ed M Feilmeier, J Joubert and U Schendel (Amsterdam: Elsevier, North-Holland) 541–6

Moto-oka T (ed) 1982 *Fifth Generation Computer Systems* (New York: Elsevier, North-Holland)

—— 1984 Japanese project on fifth generation computer systems *High-Speed Computation* ed J S Kowalik (Berlin: Springer) 99–116

Murphey J O and Wade R M 1970 The IBM 360/195 *Datamation* **16** (4) 72–9

Nagashima S, Inagami Y, Odaka T and Kawabe S 1984 Design considerations for a high speed vector processor: the HITACHI S-810 *Proc. Int. Conf. CD* (MD: IEEE)

Nassimi D and Sahni S 1980 An optimal routing algorithm for mesh-connected parallel computers *J. Assoc. Comput. Mach.* **27** 6–29

NCR 1984 *Geometric arithmetic parallel processor* NCR45CG72 Product Specification (Dayton, Ohio: NCR)

von Neumann J 1945 First draft of a report on EDVAC *The Origins of Digital Computers* ed B Randell (Berlin: Springer) 355–64

—— 1966 A system of 29 states with a general transition rule *Theory of Self-Reproducing Automata* ed A W Burks (Urbana, Ill: University of Illinois Press) 305–17. Paper first published 1952

Neves K W 1984 Vectorization of scientific software *High-Speed Computation* (NATO ASI Series vol. F7) ed J S Kowalik (Berlin: Springer) 277–91

Nicole D A and Lloyd K 1985 *Switches for reconfigurable transputer networks* (Southampton University internal report)

Nussbaumer H J 1982 *Fast Fourier Transform and Convolution Algorithms* (Berlin: Springer)

Organick E I 1973 *Computer Systems Organisations—85700/6700 series* (New York: Academic)

Orszag S A 1971 Numerical simulation of incompressible flows within simple boundaries, I: Galerkin (Spectral) Representations *Stud. Appl. Math.* **50** 293–327

Parker D S Jr 1980 Notes on shuffle/exchange-type switching networks *IEEE Trans. Comput.* **C-29** 213–22

Parnas D 1972 On the criteria to be used in decomposing systems into modules *Commun. Assoc. Comput. Mach.* **15** 1053–8

Paul G and Wilson M W 1975 The VECTRAN language: an experimental language for vector/matrix array processing *IBM Research Rep.* 320–34

—— 1978 An introduction to VECTRAN and its use in scientific applications programming *Proc. 1978 LASL Workshop on Vector and Parallel Processors* LA-7491-C

Pease M C 1967 Matrix inversion using parallel processing *J. Assoc. Comput. Mach.* **14** 757–64

—— 1968 An adaption of the fast Fourier transform for parallel processing *J. Assoc. Comput. Mach.* **15** 252–64

—— 1969 Organisation of large scale Fourier processors *J. Assoc. Comput. Mach.* **16** 474–82

—— 1977 The indirect binary *n*-cube microprocessor array *IEEE Trans. Comput.* **C-26** 458–73

Perrott R H 1978 Scientific computing in the 1980s: programming languages *Nat. Bureau Standards 17th Ann. Tech. Symp. Tools for Improved Computing* 65–69

—— 1979 A language for array and vector processors *ACM Trans. Prog. Lang. Syst.* **1** 117–95

Perrott R H and Stevenson D K 1979 *A survey of ILLIAC IV programming languages and their uses* (Sunnyvale, CA: Institute for Advanced Computation)

Petersen W P 1978 Complex fast Fourier transform binary Radix subroutine (CFFT2) *CRAY Research Inc. Publication* 2240203

Peyton-Jones S L 1987a Using Futurebus in a fifth generation computer *Major Advances in Parallel Processing* ed J R Jesshope (London: Technical)

—— 1987b The implementation of functional languages (Englewood Cliffs, NJ: Prentice-Hall)

Pollard J M 1971 The fast Fourier transform in a finite field *Math. Comput.* **25** 365–74

Poole W G Jr and Voight R G 1974 Numerical algorithms for parallel and vector computers: an annotated bibliography *Comput. Rev.* **15** 379–88

Pountain D 1986a *A tutorial Introduction to Occam Programming* Preliminary (INMOS Ltd: Bristol)

—— 1986b Integration on a new scale *Byte* November 351–6

Queyssac D 1979 Projecting vLsI's impact on microcomputers *IEEE Spectrum* **16** (5) 38–41

Rader C M 1972 Discrete convolution via Mersenne transforms *IEEE Trans. Comput.* **C-21** 1269–73

Ramamoorthy C V and Li H F 1977 Pipeline architecture *Comput. Surv.* **9** 61–102

Randell B (ed) 1975 *The Origins of Digital Computers—Selected Papers* 2nd edn (Berlin: Springer)

Reddaway S F 1973 DAP—a distributed array processor *1st Annual Symp. on Comput. Architecture (IEEE/ACM), Florida*

—— 1979 The DAP approach *Infotech State of the Art Report: Supercomputers* vol. 2 ed C R Jesshope and R W Hockney (Maidenhead: Infotech Int. Ltd) 311–29

—— 1984 Distributed array processor, architecture and performance *High-Speed Computation* ed J S Kowalik (NATO ASI Series vol. 7) (Berlin: Springer) 89–98

Reid J K and Wilson A 1985 The array features in FORTRAN 8X with examples of their use *Comput. Phys. Commun.* **37** 125–32

Rem M and Mead C A 1979 Highly concurrent structures with global communication *IEEE J. Solid State Circuits* **SC-14** 455–62

Rhodes G D 1986 Task times *Ada User* **7** (3)

Riganati J P and Schneck P 1984 Supercomputing *IEEE Computer* **17** (10) 97–113

Ritchie D M and Thompson K 1974 The UNIX time sharing system *Commun. Assoc. Comput. Mach.* **17** 365–75

Roberts J D 1977 A fast discrete Fourier transform algorithm suitable for a pipeline vector process *University of Reading Computer Science Report* RCS 82

—— 1980 How to design super DAPs? *University of Reading Computer Science Report* RCS 140

Robinson I N and Moore W R 1982 A parallel processor architecture and its implementation in silicon *IEEE Proc. CICC* 405–8

Rodrigue G (ed) 1982 Parallel Computations vol. 1 in *Computational Techniques* ed B J Alder and S Fernbach (New York: Academic)

Roscoe A W and Hoare C A R 1986 *The Laws of Occam Programming* (Oxford University Computing Laboratory Programming Research Group)

Rosen S 1969 Electronic computers: an historical survey *Comput. Surv.* **1** 7–36

Runge C 1903 Über die Zerlegung einer Empirische Gegebener Periodischer Funktionen in Sinuswellen *Z. Math. Phys. (Ger)* **48** 443–56

—— 1905 Über die Zerlegung einer Empirischen Funktion in Sinuswellen *Z. Math. Phys. (Ger)* **52** 117–23

Runge C and König H 1924 Die Grundlehren der Mathematischen Wissenschaften Band XI *Vorlesungen uber Numerisches Rechnen* (Berlin: Springer)

Rushton A and Jesshope C R 1986 RPA an architecture in need of wafer scale integration *Wafer Scale Integration* ed C R Jesshope and W R Moore (Bristol: Adam Hilger) 148–58

Russell R M 1978 The CRAY-1 computer system *Commun. Assoc. Comput. Mach.* **21** 63–72

Sakai T, Yamamoto Y, Kobayashi H Y, Ishitani T and Sudo T 1979 Elevated electrode integrated circuits *IEEE Trans. Electron. Dev.* **ED-26** 379–84

Sameh A H 1977 Numerical parallel algorithms—a survey *High Speed Computer and Algorithm Organization* ed D J Kuck *et al* (New York: Academic) 207–28

Satyanarayanan M 1980 *Multi-processors: a comparative study* (Englewood Cliffs, NJ: Prentice-Hall)

Schendel U 1984 *Introduction to Numerical Methods for Parallel Computers* (Chichester: Ellis Horwood/John Wiley)

Schönauer W 1987 *Scientific Computing on Vector Computers* (Amsterdam: Elsevier/ North Holland)

Schumann U (ed) 1978 *Computers, Fast Elliptic Solvers and Applications* (London: Advance)

Schwartz J T 1980 Ultracomputers *ACM Trans. Prog. Lang. Syst.* **2** 484–521

Seitz C L 1985 The cosmic cube *Commun. Assoc. Comput. Mach.* **28** 22–33

Sguazzero P (ed) 1986 *Vector and Parallel Processors for Scientific Computation* (Rome: Accademia Nazionale dei Lincei)

Shapiro E Y 1984 A subset of concurrent Prolog and its interpretor *ICOT Institute for New Generation Technology Technical Report* TR003

Shapiro H D 1978 Theoretical limitation on the efficient use of parallel memories *IEEE Trans. Comput.* **C-27** 421–8

Shibata H, Iwasaki H, Yamada K, Oku T and Tarui Y 1979 A new fabrication method of small-dimension devices—multiple-wall self-aligned devices *IEEE Trans. Electron. Dev.* **ED-26** 604–10

Shooman W 1960 Parallel computing with vertical data *AFIPS Conf. Proc.* 110–5

—— 1970 Orthogonal processing *Parallel Processor System Techniques and Applications* (New York: Spartan Books)

Shore J E 1973 Second thoughts on parallel processing *Comput. Elect. Eng.* **1** 95–109

Siegel H J 1979 Interconnection networks for SIMD machines *IEEE Computer* **12** (6) 57–65

—— H J 1985 *Interconnection networks for large scale parallel processing* (Lexington, MA: Lexington Books)

Siewiorek D P, Bell C G and Newell A 1982 *Computer Structures: Principles and Examples* (New York: McGraw-Hill)

Singleton R C 1967 On computing the fast Fourier transform *Commun. Assoc. Comput. Mach.* **10** 647–54

—— 1969 An algorithm for computing the mixed radix fast Fourier transform *IEEE Trans. Audio Electroacoustics* **17** 93–103

Slotnick D L 1967 Unconventional systems *AFIPS Conf. Proc.* **30** 477–81

—— 1971 The fastest computer *Sci. Am.* **224** (2) 76–87

Slotnick D L, Borck W C and McReynolds R C 1962 The SOLOMON computer *AFIPS Conf. Proc.* **22** 97–107

Smith B J 1978 A pipelined shared resource MIMD computer *IEEE Proc. 1978 Int. Conf. on Parallel Processing* 6–8

Sorensen D C 1984 Buffering for vector performance on a pipelined MIMD machine *Parallel Computing* **1** 143–64

van der Steen A 1986 Results on the Livermore loops on some new supercomputers *Supercomputer* **12** 13–14

Stevens K G Jr 1974 CFD—a FORTRAN based language for ILLIAC IV *Reference Manual* (Moffet Field, CA: NASA Ames Research Center)

—— 1975 CFD—a FORTRAN-like language for the ILLIAC IV *SIGPLAN Not.*

10 (3) 72–80

—— 1979 Numerical aerodynamic simulation facility project *Infotech State of the Art Report: Supercomputers* vol. 2 ed C R Jesshope and R W Hockney (Maidenhead: Infotech Int. Ltd) 331–42

Stone H S 1970 A logic-in-memory computer *IEEE Trans. Comput.* **C-19** 73–8

—— 1971 Parallel processing with the perfect shuffle *IEEE Trans. Computers* **C-20** 153–61

—— 1973 An efficient parallel algorithm for the solution of a tridiagonal system of equations *J. Assoc. Comput. Mach.* **20** 27–38

—— 1975 Parallel tridiagonal solvers *Assoc. Comput. Mach. Trans. Math. Software* **1** 289–307

Strakos Z 1985 Performance of the EC2345 array processor *Computers and Artificial Intelligence* **4** 273–84

—— 1987 Effectivity and optimizing of algorithms and programs on the host-computer/array-processor system *Parallel Computing* **4** 189–207

Stumpff K 1939 *Tafeln und Aufgaben zur Harmonischen Analyse und Periodogramm-rechnung* (Berlin: Springer)

Sugarman R 1980 Superpower computers *IEEE Spectrum* **17** (4) 28–34

Sumner F H (ed) 1982 *State of the Art Report: Supercomputer Systems Technology* Series 10 No 6 (Maidenhead: Pergamon Infotech Ltd)

Sumner F H, Haley G and Chen E C Y 1962 The central control unit of the ATLAS computer *Proc. IFIP Congr.* 657–62

Sumney L W 1980 VLSI with a vengeance *IEEE Spectrum* **17** (4) 24–7

Swan R J, Fuller S H and Siewiorek D P 1977 Cm*: a modular multi-microprocessor *Proc. National Computer Conference* **46** 637–44

Swarztrauber P N 1974 A direct method for the discrete solution of separable elliptic equations *SIAM J. Numer. Anal.* **11** 1136–50

—— 1977 The methods of cyclic reduction, Fourier analysis and the FACR algorithm for the discrete solution of Poisson's equation on a rectangle *SIAM Rev.* **19** 490–501

—— 1979a The solution of tridiagonal systems on the CRAY-1 *Infotech State of the Art Report: Supercomputers* vol. 2 ed C R Jesshope and R W Hockney (Maidenhead: Infotech Int. Ltd) 343–59

—— 1979b A parallel algorithm for solving general tridiagonal equations *Math. Comput.* **33** 185–99

—— 1982 Vectorising FFTs *Parallel Computations* ed G Rodrigue (London: Academic) 51–83

—— 1984 FFT algorithms for vector computers *Parallel Computing* **1** 45–63

Sweet R A 1974 A generalised cyclic-reduction algorithm *SIAM J. Numer. Anal.* **11** 506–20

—— 1977 A cyclic-reduction algorithm for solving block tridiagonal systems of arbitrary dimension *SIAM J. Numer. Anal.* **14** 706–19

Tamura H, Kamiya S and Ishigai T 1985 FACOM VP-100/200: Supercomputers with ease of use *Parallel Computing* **2** 87–107

Taylor G S 1983 Arithmetic on the ELXSI 6400 *IEEE Proc. 6th Ann. Symp. on Computer Architecture* 110–5 (London: IEEE)

Tedd M, Crespi-Reghizzi S and Natali A 1984 Ada for Multiprocessors *Ada Companion Series* (Cambridge: Cambridge University Press)

Temperton C 1977 Mixed radix fast Fourier transforms without reordering *ECMWF Report No. 3* European Centre for Medium-range Weather Forecasting, Shinfield

Park, Reading, Berks, UK

—— 1979a Direct methods for the solution of the discrete Poisson equation: some comparisons *J. Comput. Phys.* **31** 1–20

—— 1979b Fast Fourier transforms and Poisson solvers on the CRAY-1 *Infotech State of the Art Report: Supercomputers* vol. 2 ed C R Jesshope and R W Hockney (Maidenhead: Infotech Int. Ltd) 359–79

—— 1980 On the FACR (*l*) algorithm for the discrete Poisson equation *J. Comput. Phys.* **34** 314–29

—— 1983a Self-sorting mixed-radix fast Fourier transforms *J. Comput. Phys.* **52** 1–23

—— 1983b A note on prime factor FFT algorithm *J. Comput. Phys.* **52** 198–204

—— 1983c Fast mixed-radix real Fourier transforms *J. Comput. Phys.* **52** 340–50

—— 1984 Fast Fourier transforms on the CYBER 205 *High Speed Computation* ed J S Kowalik (NATO ASI Series vol. 7) (Berlin: Springer) 403–16

—— 1985 Implementation of a self-sorting in-place prime factor FFT algorithm *J. Comput. Phys.* **58** 283–99

—— 1988 Implementation of a prime factor FFT algorithm on the CRAY-1 *Parallel Computing* **6** 99–108

Theis D J 1974 Vector supercomputers *IEEE Comput.* **7** (4) 52–61

Thomas L H 1963 Using a computer to solve problems in physics *Application of Digital Computers* (New York: Ginn)

Thompson C D 1977 Generalised connection networks for parallel processor intercommunication *Department of Computer Science Report, Carnegie-Mellon University, USA*

Thornton J E 1964 Parallel operation in the control data 6600 *AFIPS Conf. Proc.* **26** (II) 33–40

—— 1970 *Design of a Computer—The Control Data 6600* (Glenview, Illinois: Scott, Foresman and Company)

Thurber K J 1976 *Large-scale Computer Architectures: Parallel and Associative Processors* (Rochelle Park, NJ: Hayden)

Thurber K J and Wald L D 1975 Associative and parallel processors *Comput. Surv.* **7** 215–55

Touretsky D S 1974 *LISP A Gentle Introduction to Symbolic Computation* (London: Harper and Row)

Treleaven P C 1979 Exploiting program concurrency in computing systems *IEEE Computer* **12** (1) 42–50

Treleaven P C, Brownbridge D R and Hopkins R P 1982 Data-driven and demand-driven computer architecture *Comput. Surv.* **14** 93–143

Turing S 1959 *Alan M Turing* (Cambridge: W Heffer and Sons)

Turn R 1974 *Computers in the 1980s* (New York: Columbia University Press)

Turner D A 1982 in *Functional Programming and its Applications: An Advanced Course* (Cambridge: Cambridge University Press)

Uchida S 1983 Towards a new generation of computer architecture *LSI Architecture* ed B Randell and PC Treleaven (Englewood Cliffs, NJ: Prentice-Hall) 95–422

Uhrich M L 1969 Fast Fourier transforms without sorting *IEEE Trans. Audio Electroacoustics* **17** 170–2

Unger S H 1958 A computer oriented towards spatial problems *Proc. Inst. Radio Eng. (USA)* **46** 1744–50

US Department of Defense 1978 *STEELMAN Requirements for high order computer programming languages*

USAS 1966 USA standard FORTRAN *USAS X3.9 1966* (New York: USA Standards

Institute)

—— 1978 USA standard FORTRAN *USA X3.9 1978* (New York: USA Standards Institute)

Vajtersic M 1982 Parallel Poisson and biharmonic solvers implemented on the EGPA multi-processor *Proc. 1982 Int. Conf. on Parallel Processing* (Silver Spring, MD: IEEE Comput. Soc.) 72–81

—— 1984 Parallel marching Poisson solvers *Parallel Computing* 1 325–30

VAPP 1982 Proceedings Vector and Parallel Processors in Computational Science (Chester 1981) *Comput. Phys. Commun.* 26 217–479

—— 1985 Proceedings Vector and Parallel Processors in Computational Science (Oxford 1984) *Comput. Phys. Commun.* 37 1–386

Varga R S 1962 *Matrix Iterative Analysis* (Englwood Cliffs, NJ: Prentice-Hall)

Vick C R and Cornell J A 1978 PEPE architecture—present and future *AFIPS Conf. Proc.* 47 981–1002

Waksman A 1968 A permutation network *J. Assoc. Comput. Mach.* 15 159–63

Wang H H 1980 On vectorizing the fast Fourier transform *BIT* 20 233–43

—— 1985 A parallel method for tridiagnal equations *ACM Trans. Math. Software* 7 170–83

Ware F, Lin L, Wong R, Woo B and Hanson C 1984 Fast 64-bit chipset gangs up for double precision floating-point work *Electronics* 57 (14) July 12 99–103

Waser S 1978 High speed monolithic multipliers for real-time digital signal processing *IEEE Comput.* 11 (10) 19–29

Watanabe T 1984 Architecture of supercomputers—NEC supercomputer SX system *NEC Res. Dev.* 73 1–6

Watson I and Gurd J R 1982 A practical dataflow computer *IEEE Computer* 15 (2) 51–7

Watson W J 1972 The Texas Instruments advanced scientific computer *COMPCON '72 Digest* 291–4

Wetherell C 1980a Array processing for FORTRAN *Lawrence Livermore Laboratory Computer Documentation* UCID-30175

—— 1980b Design considerations for array processing languages *Software Practice and Experience* 10 265–71

Whittaker E T and Robinson G 1944 *Calculus of Observations* (Glasgow: Blackie) 260–83

Widdoes L C Jr and Correll S 1979 The S-1 project: developing high performance digital computers *Energy Technol. Rev.* (Lawrence Livermore Laboratory) September 1–15

Wilkes M V and Renwick W 1949 The EDSAC, an electronic calculating machine *J. Sci. Instrum.* 26 385–91

Wilkes M V, Wheeler D J and Gill S 1951 *The Preparation of Programs for an Electronic Digital Computer* (Cambridge, MA: Addison-Wesley)

Wilkinson J H 1953 The pilot ACE *Computer Structures: Readings and Examples* Ch. 11 ed C G Bell and A Newell (New York: McGraw-Hill) 193–9

Williams F C and Kilburn T 1949 A storage system for use with binary digital computing machines *Proc. IEE* 96 (3) 81–100

Williams S A 1979 The portability of programs and languages for vector and array processors *Infotech State of the Art Report: Supercomputers* vol. 2 ed C R Jesshope and R W Hockney (Maidenhead: Infotech Int. Ltd) 382–93

Winograd S 1967 On the time required to perform multiplication *J. Assoc. Comput. Mach.* **14** 793–802

—— 1978 On computing the discrete Fourier transform *Math. Comput.* **32** 175–99

—— 1980 *Arithmetic complexity of computations* Monograph No 33 (CBMS-NSF Conf. Series Applied Math.) (Philadelphia: SIAM)

Winston P H and Horn B K P 1981 *LISP* (London: Addison-Wesley)

Wirth N 1981 *Programming in Modula 2* 2nd edn (Berlin: Springer)

Wittmayer W R 1978 Array processor provides high throughput rates *Comput. Design* **17** (3) 93–100

Wolfe M 1986 Software optimisation for supercomputers *Supercomputers* ed S Fernback (Amsterdam: North-Holland) 221–38

Wollesen D L 1980 cmos lsi—the computer component process of the 80s *IEEE Computer* **13** (2) 59–67

Wulf W A and Bell C G 1972 C.mmp: a multi-mini-processor *Proc. AFIPS Fall Joint Computer Conference* **41** (2) 765–77

Yasamura M, Tanaka Y, Kanada Y and Aoyama A 1984 Compiling algorithms and techniques for the S-810 vector processor *IEEE Proc. 1984 Int. Conf. on Parallel Processing* 285–90 (London: IEEE)

Yau S S and Fung H S 1977 Associative processor architecture—a survey *Comput. Surv.* **9** 3–27

Zakharov V 1984 Parallelism and array processing *IEEE Trans. Comput.* **C-33** 45–78

Zuse K 1958 Die Feldrechenmaschine *Math. Tech. Wirt-Mitteilungen (Wien)* **5** (4) 213–20

Index